Introduction to Python Network Automation Volume II

Stepping up: Beyond the Essentials for Success

Second Edition

Brendan Choi

Apress®

Introduction to Python Network Automation Volume II: Stepping up: Beyond the Essentials for Success, Second Edition

Brendan Choi
Sydney, NSW, Australia

ISBN-13 (pbk): 979-8-8688-0390-1 ISBN-13 (electronic): 979-8-8688-0391-8
https://doi.org/10.1007/979-8-8688-0391-8

Copyright © 2024 by Brendan Choi

This work is subject to copyright. All rights are reserved by the Publisher, whether the whole or part of the material is concerned, specifically the rights of translation, reprinting, reuse of illustrations, recitation, broadcasting, reproduction on microfilms or in any other physical way, and transmission or information storage and retrieval, electronic adaptation, computer software, or by similar or dissimilar methodology now known or hereafter developed.

Trademarked names, logos, and images may appear in this book. Rather than use a trademark symbol with every occurrence of a trademarked name, logo, or image we use the names, logos, and images only in an editorial fashion and to the benefit of the trademark owner, with no intention of infringement of the trademark.

The use in this publication of trade names, trademarks, service marks, and similar terms, even if they are not identified as such, is not to be taken as an expression of opinion as to whether or not they are subject to proprietary rights.

While the advice and information in this book are believed to be true and accurate at the date of publication, neither the authors nor the editors nor the publisher can accept any legal responsibility for any errors or omissions that may be made. The publisher makes no warranty, express or implied, with respect to the material contained herein.

    Managing Director, Apress Media LLC: Welmoed Spahr
    Acquisitions Editor: Celestin Suresh John
    Development Editor: James Markham
    Copy Editor: Kezia Endsley
    Editorial Assistant: Gryffin Winkler

Cover designed by eStudioCalamar
Cover image designed by Freepik (www.freepik.com)

Distributed to the book trade worldwide by Springer Science+Business Media New York, 1 New York Plaza, Suite 4600, New York, NY 10004-1562, USA. Phone 1-800-SPRINGER, fax (201) 348-4505, e-mail orders-ny@springer-sbm.com, or visit www.springeronline.com. Apress Media, LLC is a California LLC and the sole member (owner) is Springer Science + Business Media Finance Inc (SSBM Finance Inc). SSBM Finance Inc is a **Delaware** corporation.

For information on translations, please e-mail booktranslations@springernature.com; for reprint, paperback, or audio rights, please e-mail bookpermissions@springernature.com.

Apress titles may be purchased in bulk for academic, corporate, or promotional use. eBook versions and licenses are also available for most titles. For more information, reference our Print and eBook Bulk Sales web page at http://www.apress.com/bulk-sales.

Any source code or other supplementary material referenced by the author in this book is available to readers on GitHub. For more detailed information, please visit https://www.apress.com/gp/services/source-code.

If disposing of this product, please recycle the paper

*To Mom and in remembrance of Dad,
your boundless love and unwavering patience have been the true guiding lights, illuminating every step of this profound journey. Within these pages, this book stands as a testament to the steadfast faith and unwavering encouragement you both selflessly offered. I am deeply thankful for the profound impact you've had on this endeavor.*

Table of Contents

About the Author xi

About the Technical Reviewer xiii

Acknowledgments xv

Introduction xvii

Chapter 1: Building a Python Network Automation Lab Environment 1
    Embarking on the Next Phase of Your Python Network Automation Journey 2
    Empowering Your Network Automation Journey with GNS3 and Cisco CML Integration 3
    Before Integrating Cisco CML Images in GNS3 4
    Integrating Cisco CML L2 and CML L3 Images on Your GNS3 6
    Installing Cisco CML L2 Switch on GNS3 8
    Quick Ping Test on CML L2 Switch 18
    Installing the Cisco CML L3 Router on GNS3 Using the Import Appliance Option 24
        Quick Ping Test on the CML L3 Router 38
    Building a CML Lab Topology 40
    Summary 50
    Storytime 1: Still Enjoying the Most Expensive Sandwich in the World 51

Chapter 2: Python Network Automation Labs - Basic Telnet 53
    Python Network Automation: Main Logical Topology 54
    Python Network Automation: Network Engineer's interface 56
    Telnet Lab 1: Interactive Telnet Session to Cisco Devices on a Python Interpreter 60
    Telnet Lab 2: Configure a Single Switch with a Python Telnet Template 72
    Telnet Lab 3: Configure Random VLANs Using a for Loop 79
    Telnet Lab 4: Configure Random VLANs Using a while Loop 86

TABLE OF CONTENTS

Telnet Lab 5: Configure 100 VLANs Using the for ~ in range Loop... 96

Telnet Lab 6: Add a Privilege 3 User on Multiple Devices Using IP Addresses
from an External File... 102

Telnet Lab 7: Taking Backups of running-config (or startup-config) to Local Server Storage... 112

Summary... 119

Chapter 3: Python Network Automation Labs: SSH in Action, paramiko and netmiko Labs.. 121

The Enduring Role of SSH in Network Infrastructure: Navigating the
Evolution of Remote Management Tools.. 122

Python Network Automation Labs Using the paramiko and netmiko Libraries...................... 123

Checking the SSH in Action on a Linux Server.. 124

Checking the SSH in Action on Cisco Devices... 127

Python SSH Labs: paramiko... 129

paramiko Lab 1: Configure the Clock and Time Zone of All Devices
Interactively in the Python Interpreter 129

paramiko Lab 2: Configuring an NTP Server on Cisco Devices Without
User Interaction (NTP Lab) 162

paramiko Lab 3: Create an Interactive paramiko SSH Script to Save Running
Configurations to the TFTP Server 175

Python SSH Labs: netmiko... 186

    Installing the netmiko Library... 187

netmiko Lab 1: netmiko Uses a Dictionary for Device Information, Not a JSON Object........... 188

netmiko Lab 2: Develop a Simple Port Scanner Using a Socket Module and
Then Develop a netmiko Disable Telnet Script... 203

netmiko Lab 3: config compare 213

Summary... 226

Chapter 4: Python Network Automation Labs cron 229

Unveiling the Power of cron in Python Networking... 229

Cloning a GNS3 Project for the Next Lab... 230

Router Failure Simulation, P1 Network Recovery 238

Lab Pre-Task, Deleting VPCS and Adding More L2 Switches 244

Quick-Start to the Linux Scheduler, cron 250

Ubuntu LTS Task Scheduler: crontab 251

  Fedora Task Scheduler: crond 258

  Learn cron Job Definitions with Examples 267

 Summary 270

Chapter 5: Python Network Automation Labs: SNMP Discovery with Python 273

 Bridging Theory and Practice: Learning About SNMP with Python 274

 Quick-Start Guide to SNMP 275

  Quick SNMP History 276

  Types of System Monitoring 277

  Comparison Table: SNMP vs. Syslog vs. APIs 279

  SNMP Standard Operation 281

  SNMP Polling vs. the Trap (Event Reporting) Method 283

  SNMP Versions 283

  Understanding the SNMP Protocol in the TCP/IP Stack 284

  About SNMP Message Types 285

  Understanding SNMP's SMI, MIB, and OID 286

  SNMP Access Policy 286

  Synchronous vs. Asynchronous Communication While Using SNMP Traps 287

  SNMP Tools and Python Integrations 288

 Using Python to Run an SNMPv3 Query 290

  SNMP-Related Python Libraries 290

  Understanding MIBs and OIDs and Browsing OIDs Directly 291

  Learning to Use SNMPwalk on a Linux Server 292

  Installing and Configuring SNMP Server on Fedora 293

 Configuring Cisco Routers and Switches to Support SNMPv3 Using a Python Script 296

  Troubleshooting SSH Connections Between Fedora and Cisco Devices 298

  Configuring SNMP Agents (engineID, group, and user) on Cisco Devices 312

  Doing the SNMPwalk on the Linux Server 325

 Borrowing Some Example Python SNMP Code 331

  SNMP Lab Source Code 332

vii

TABLE OF CONTENTS

    SNMPv3 Python Code to Query Cisco Device Information 335

    Using Python SNMP Code to Run a Query for an Interface Description 342

  Summary 347

Chapter 6: Python Network Automation Virtual Lab 1: Ansible and pyATS in Python virtualenv 349

  Python Virtual Lab 1: Ansible and pyATS in Python virtualenv 350

    Navigating the Ansible and pyATS Networking Tools 351

    Lab 1: Quick-Start Guide to Ansible in virtualenv 352

  Installing virtualenv and Getting Started with Ansible 356

    Lab 2: Quick-Start Guide to pyATS (Genie) in virtualenv 375

  Summary 401

  Storytime 2: Doubling Down! 401

Chapter 7: Python Network Automation Virtual Lab 2: Sendmail Email Notification and Twilio SMS Notification on Docker 405

  Cost-Efficient IT Operations: Leveraging Docker for Python Networks 406

  Expense Optimization in Python Networking: Docker, Kubernetes Impact 408

  Python Network Automation Virtual Lab 2: Docker and Twilio SMS 409

  Lab 1: Sendmail Lab Using an Imported Docker Image 410

    Docker Components, Account Registration, and Installation 412

    Docker Installation Procedure 412

    Test-Driving Docker 415

  Lab 2: Sendmail Python Lab Using Docker 424

    Lab Prerequisites 424

  Lab 3: Sendmail Email Notification Script Development in Docker 436

  Lab 4: CPU Utilization Monitoring Lab: Send an SMS Message Using Twilio 447

  Lab 5: TWILIO Account Creation: Install Twilio Python Module and Set Up an SMS Message 448

  Lab 6: CPU Utilization Monitoring Lab with an SMS Message 456

  Summary 471

Chapter 8: Upgrading Multiple Cisco IOS Routers 473

Empowering Python Network Automation: Unveiling OOP and Real-world Scenarios 473

    Applying OOP Concepts to Your Network 474

Managing Flow Control and User Input: UID, PWD, and Information Collection 482

    Proactive Patch Management and Zero Trust: Shaping Network Security 491

Router IOS Upgrade Lab Preparation 492

    Cisco Catalyst 8000v IOS XE Software and Download 493

    Cisco Catalyst 8000v Installation on VMware Workstation 495

How an IOS (IOS XE/XR) Is Upgraded on Cisco Devices 509

    Tasks Involved in a Cisco IOS Upgrade 510

    Summary 514

Chapter 9: Cisco IOS-XE Upgrade Tools Development Part 1 515

Mastering Cisco IOS Upgrades using Python: A Comprehensive Guide for Network Engineers 515

    Tools Development 1: Network Connectivity and Socket Validation Tool 518

Tools Development 2: Login Credentials and User Input Collector 536

Tools Development 3: Collect a New IOS File Name and MD5 Value from a CSV File 543

Tools Development 4: Check the MD5 Value of the New IOS on the Server 556

Tools Development 5: Check the Flash Size on Cisco Routers 564

Tools Development 6: Make Backups of running-config, Interface Status, and Routing Table 570

Summary 575

Chapter 10: Cisco IOS-XE Upgrade Tools Development: Part 2 577

Cisco IOS Upgrade Application Development (Continued) 578

    Tools Development 7: IOS Uploading and More Pre-check Tools Development 578

    Tools Development 8: Check the New IOS MD5 Value on the Cisco Device's Flash 597

    Tools Development 9: Options to Stop or Reload the Routers 611

    Tools Development 10: Check the Reloading Device and Perform a Post-Reload Configuration Verification 630

    Summary 645

    Storytime 3: IT Evolution: Unraveling the Cloud's Impact with Chicken Littles 646

TABLE OF CONTENTS

Chapter 11: Upgrading the Application and Routers 649

Python Mastery for Network Automation: Building the Final IOS Upgrade Tool 650

Summary 714

Chapter 12: Installing NetBox (IPAM/DCIM) with Python 715

What Is NetBox, and Why Do You Need It? 715

    Some NetBox Features 718

    What NetBox Is Not 718

    Server Specifications, Dependencies, and Functions 718

NetBox Concept: The "Site" Is the Core of NetBox 720

NetBox IP Address Management (IPAM): High-Level Overview 720

Changing NetBox into Multi-tenancy IPAM 721

NetBox 3 Manual Installation on Ubuntu 22.04 LTS 722

NetBox 3 Automated Python Script Installation on Ubuntu 22.04 LTS 742

Summary 765

Storytime 4: Embarking on the Python Programming Odyssey:
A Journey of Network Automation and Career Advancement 766

Index 769

About the Author

Brendan (Byong Chol) Choi, working as a Principal ICT Infrastructure Architect at NSW Telco Autority, has over two decades of hands-on experience in the dynamic world of the ICT industry. He holds certifications in Cisco, VMware, Fortinet, and ITIL. Brendan has lent his expertise to prominent enterprises such as Cisco Systems, Dimension Data (NTT), Fujitsu, and more recently, leading Australian IT integrators like Telstra and NTT. His focus on optimizing enterprise IT infrastructure management and refining business processes sees him utilizing a diverse range of both open-source and proprietary tools. Beginning his journey in the trenches of Cisco TAC Frontline, Brendan transitioned to pivotal IT engineering roles, navigating a spectrum of emerging and legacy technologies. Stemming from a traditional infrastructure background, he ardently explores emerging IT domains like the cloud, IoT, DevOps, data center and the transformative technologies linked to the fourth industrial revolution.

Brendan's literary contributions include *Python Network Automation: Building an Integrated Virtual Lab, Introduction to Python Network Automation: The First Journey*, and his primary work, *Introduction to Ansible Network Automation: The Practical Primer*. These works are crafted to resonate with and address the current enterprise IT landscape, sharing invaluable industry insights with the wider IT community. His legacy of training over 200 network and systems engineers in Python and Ansible network automation is complemented by his passion for disseminating industry-acquired knowledge through social media, blogging, and his YouTube channel. His curiosity spans across diverse domains such as private and public cloud, enterprise networking, security, virtualization, Linux, automation, and the transformative technologies driving the fourth industrial revolution. Amid a cacophony of disingenuous voices in the IT industry, "live by your words" remains Brendan's guiding principle. His unwavering dedication to enterprise infrastructure management echoes through his commitment to continuous learning, extensive knowledge sharing, and his profound contributions to the rich ICT world.

About the Technical Reviewer

Radhika Shitlani is a network professional with over eight years of experience in network design, implementation, and automation. She has been using Python and contributing to it since its inception in the network domain. She has extensive experience in automating network feature testing and building tools to automate daily tasks in network operations. Currently, she designs and automates networks in data centers of global cloud leaders.

An enthusiast, she is always striving to learn and upgrade her skills. Apart from automating things in Python, she enjoys adventure sports and has completed the first (beginner) level of flying training. She is currently undergoing scuba diving training and dreams of becoming a professional sea diver.

Acknowledgments

I am deeply grateful to my IT industry mentors—Lesek Geba and Truyen Nguyen—and my former managers—Kai Schweisfurth and Ty Starr. Their extensive industry expertise and wisdom, amassed over 40 years in ICT industry, have been invaluable in shaping the first and second editions of this book. I am also immensely grateful to Justin (Cheol Yoon) Cheong for his guidance in navigating the world of authorship. I extend my deepest appreciation to all of them.

My heartfelt thanks also go to my wife, Sue, and our children, Hugh, Leah, and Caitlin. Their unwavering support has been my bedrock throughout the creation of this book. Their love and understanding have been my unwavering pillars. I extend sincere gratitude to my extended family, friends, and colleagues for their steadfast support. Lastly, I want to express my gratitude to my readers for embarking on this incredible network automation journey with me.

(Brendan Choi, 2024)

Introduction

Python's integration into enterprise network administration has surged in popularity among leading IT organizations. It is increasingly recognized as a pivotal skillset for network engineers in a growing number of organizations, gradually revolutionizing the landscape of network administration and ICT (Information and Communications Technology) infrastructure management.

Written by an ICT expert for real ICT professionals, this book aims to distill core practical knowledge devoid of false assumptions. It's founded on the belief that the book's target audience enjoys hands-on learning approaches over pure theory. This book is not for talkers, idealists, or theorists who merely preach, but for those ready to act. The book is committed to providing an authentic experience, guiding readers through the actual journey of acquiring network automation skills using Python and its encompassing technologies from scratch. Understanding that individuals in IT come from diverse backgrounds and varying skill levels, this book operates under the notion that its targeted readers are beginner to intermediate users in IT concerning Python Network Automation. It encourages readers to build everything themselves for a first-hand, immersive experience, emphasizing the importance of holistic comprehension rather than a specialized approach within the enterprise IT and network automation ecosystem. In essence, this book operates on the principle that seeing is believing!

In its inaugural edition, *Introduction to Python Network Automation: The First Journey*, this book offered readers a structured path to establish a robust foundation in Python Network Automation. This second edition revisits the original, bolstering the structured learning path with additional new content. The original version was an invaluable resource for IT professionals and students aiming to enhance their automation skills. This revised edition represents a refined iteration, combining five foundational learnings—Python, Linux, basic networking, essential virtualization, and enterprise networking lab building techniques—across two comprehensive guides. Tailored specifically for networking students and IT engineers, it provides hands-on experience in constructing a Python Network Automation lab from the ground up. With practical examples derived from actual enterprise infrastructure, this edition offers valuable insights into leveraging Python effectively within real enterprise network

INTRODUCTION

management scenarios. With the aim of improving readability and ease of use, the book has been split into two volumes: Volume I *Laying the Foundation: Essential Skills for Growth* and Volume II *Advancing Further: Beyond the Basics for Success* You're currently reading Volume II.

While many resources delve into Python-based network automation, few adequately equip students and engineers for Python Network Automation within a comprehensive IT infrastructure management framework. The success of the first edition among genuine ICT professionals stemmed from this book's direct, relatable, and holistic approach to learning network automation using Python and related technologies. Unlike resources that promise a zero-to-hero journey but falter due to their narrow focus, this book stresses the interconnectedness of technologies in the enterprise IT ecosystem. Solely comprehending Python network applications focused on software programmability offers limited mileage; a broader perspective becomes essential.

Numerous network automation books tantalize with the allure of Python Network Automation without emphasizing the no pain, no gain reality. Conversely, this book adopts a genuine approach, immersing readers in an authentic experience that exposes them to the challenges of acquiring network automation skills using Python. Readers engage in installing and configuring nearly everything themselves, learning from mistakes, overcoming obstacles, and ultimately mastering essential IT skills. This journey empowers them to overcome initial hurdles in Python Network Automation, fostering the resilience and proficiency necessary to develop functional Python network applications.

By engaging with this book, readers will gain valuable knowledge and well-rounded skills for Python-based network automation. The book offers a structured learning path, gradually building proficiency in Python and its encompassing technologies. It begins with essential Linux administration skills, progresses through Python basics pertinent to enterprise network automation, explores foundational enterprise network labs, delves into basic networking concepts, and culminates in integrating various technologies, developing real Python networking applications, and optimizing code in a production-safe environment—all achievable from a single laptop. Readers can learn, write, and test their Python applications without causing major outages to network services. This journey provides insights into streamlining enterprise network management processes, transitioning from manual tasks to semi- or full automation, thus enhancing efficiency and productivity. Throughout this book, I share industry insights gained over the last two decades in the ICT industry, equipping readers with real working knowledge and essential skills to apply Python in network automation, empowering them to navigate the ever-evolving landscape of enterprise networking administration confidently.

INTRODUCTION

Designed for diligent readers seeking to enhance their network automation skills with Python, this book caters to IT students, network engineers, developers managing IP services, networking devices, servers, cloud, and data centers. Technical leaders implementing network automation, mentors training team members, instructors teaching network automation, and Cisco Network Academy students pursuing network administration certifications will also find value in its pages. It's tailored for those interested in integrating network automation into their development process, offering practical knowledge across their enterprise network. Leveraging Python, it effectively teaches network automation techniques and encompassing technologies.

Whether you've finished the Part I book and are returning, or you've skipped ahead to this book for fresh insights, be prepared to elevate your Enterprise Network Automation expertise. Join me on this journey to hone your network automation skills and discover new possibilities in enterprise network management.

CHAPTER 1

Building a Python Network Automation Lab Environment

Welcome to *Introduction to Python Network Automation II: Stepping Up: Beyond the Essentials for Success,* 2nd edition. Here, you'll seamlessly pick up where you left off in the Part I book and continue your Python Network Automation journey. This chapter takes a decisive step into the world of Cisco CML image integration within GNS3, focusing on the vital process of installing and testing both CML L2 and L3 images. It aims to enable access to the functionalities of CML L2 and L3 images. The primary goal is to enhance the GNS3 `pynetauto-lab` project's topology with the latest CML images in order to support comprehensive network features and allow communication to Python scripting servers. This process involves downloading, integrating, and configuring these images for future development labs. The chapter highlights the challenges involved in the CML image-integration process, offering effective solutions to building a strong foundation for subsequent network automation labs. It's important to note that this chapter primarily focuses on setting up the `pynetauto-lab` topology, where all network devices are initialized with basic configurations, and their connectivity is tested with the Linux servers, which function as the Python servers. The chapter is mostly scripting-free, except for basic interactive `tclsh` commands used for multiple IP ping tests. The exclusion of scripting tasks ensures that all efforts are geared toward perfecting the GNS3 project, serving as the springboard for your exploration into Python Network Automation scripting.

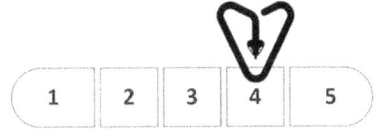

CHAPTER 1   BUILDING A PYTHON NETWORK AUTOMATION LAB ENVIRONMENT

Embarking on the Next Phase of Your Python Network Automation Journey

As you embark on this phase of your journey through this Python Network Automation book, dedication and hard work are essential. To maximize your experience in this book, it's crucial to have a solid understanding of the material covered in the Part I tome (*Laying the Foundation: Essential Skills for Growth*). It is worthwhile to take a moment to reflect on your progress in Part I and set your course for this book.

In the initial stage of your Python Network Automation journey, you gained hands-on skills that were fundamental for installing, configuring, and operating an integrated virtual lab using a variety of open-source and proprietary software. You also took your first steps into basic Python scripting based on your newfound knowledge. While there were numerous exciting topics that book could have explored, such as DevOps concepts, Ansible, ACI, REST API, HTML, Postman, YANG, NETCONF, Nornir, NAPALM, and more, the focus is on helping you build a solid foundation in technologies that are the pillars of Python Network Automation. **It's best to aim for long-term wins, not quick and short-lived victories.**

You've likely heard the saying, "Give a man a fish, and you feed him for a day; teach a man to fish, and you feed him for a lifetime." In the context of Python network application development, this book's philosophy is clear: teaching network engineers to use Ansible empowers them to become semi-professional soccer players, while teaching them in Python programming equips them as full-time professional soccer players. Even more powerful is conquering both Python coding and Ansible YAML playbook writing, which could make an average network engineer into a potential superstar of the ICT industry. It is your responsibility to expand upon the skills you gained through various learning materials, charting your path, and applying these skills to your work. While this book delves into the teachings of Python Network Automation, I've also written a sequel to cover the beginner topics in Ansible network automation. When you have the time and more passion for study, consider reading the sequel, *Introduction to Ansible Network Automation: A Practical Primer* (2023, 1st Edition) by Brendan Choi & Erwin Medina.

Throughout your journey so far, you've honed diverse IT skills that have laid a solid foundation for the chapters ahead, which will predominantly revolve around Python Network Automation tasks—central to both this book's focus and your primary interest. Regrettably, I've had to guide you through what some might call "the hard way" or "the way not many want to go" but rest assured, the Part I book will leave a lasting impact on your IT career. **Unlike the snake-oil promises found in many study resources, the knowledge you've gained here is genuine and enduring.** If you've smoothly navigated through the Part I without encountering significant technical challenges, you should find the following chapters equally manageable and considerably more enjoyable. It's as if you've emerged from the dense undergrowth, and your second half of the journey promises to be much smoother. However**, if you haven't yet fully set up your lab environment or if you've only partially completed the exercises, I strongly encourage you to revisit and complete these essential tasks, as they form the foundational stepping stones for what awaits you ahead.**

Empowering Your Network Automation Journey with GNS3 and Cisco CML Integration

Within this chapter, you pick up where you left off at the end of Chapter 11 in the Part I book, continuing your journey of further developing the pynetauto-lab project topology. In this chapter, I momentarily divert your attention from scripting tasks to carefully preparing and optimizing your network topology. This preparatory step is pivotal to your survival and success in the remaining chapters, getting ready for the scripting challenges and further skills conditioning that await you in the remainder of this book. This groundwork in this chapter serves as a solid foundation for writing and testing Python code specifically tailored to a selective network automation applications. Whether you are an experienced network engineer with a wealth of hands-on experience across diverse network platforms or a newcomer embarking on your career journey, this chapter will help you prepare a well-thought-out, yet simple lab environment for Python Network Automation labs.

For those eager to explore Cisco's offerings, a yearly subscription to Cisco Modeling Labs–Personal (CML) is an option. As of October 2023, the subscription-based CML was available for $199, granting access to a rich set of resources. While this may appear as an investment, especially for aspiring network students, it is significantly more cost-effective compared to the traditional hardware purchases that network engineers had to make in the past. If you have a way to access the CML image files, there are more cost-effective ways to set up your lab, a topic this chapter delves into. As mentioned in Part I, you'll use CML L2 and L3 images in your labs for learning. Cisco CML, like many Cisco products, is a vendor-specific Virtual Network Emulator designed for studying cutting-edge Cisco technologies. The networking concepts you learn using this Cisco tool apply universally to all other vendor technologies, given that Cisco's certifications have been around since the '90s and have long been regarded as industry-standard networking certification for the ICT industry, like how AWS and Azure certifications hold prominence in the newer cloud services certifications.

However, this book is going against the grain and aims for a hybrid lab setup, utilizing Cisco CML's router (L3) and switch (L2) images within GNS3 and VMware Workstation. Note that a similar lab deployment approach can be applied in other enterprise network emulators, such as EVE-NG. The installation and integration procedures for CML images in GNS3 are not rocket science, but some parts of the integration can be a little tricky, as you are relying on software to make your lab work seamlessly. As announced by VMare by Broadcom on May 14, 2024, VMware Workstation 17 Pro for personal use is now free. Note, this development occurred after the completion of this book.

It's important to understand that CML is Cisco's proprietary software, subject to legal distribution restrictions. To access all the latest CML images, subscribing to CML through the Cisco Learning Network Store is the recommended method. It's important to put forward a disclaimer that any Cisco IOS, CML, or other proprietary software mentioned in this book must be sourced by the reader. This book does not include the supply of software, and readers are encouraged to explore their channels for obtaining all software. For more information on CML, visit https://developer.cisco.com/modeling-labs/.

Before Integrating Cisco CML Images in GNS3

Before Cisco CML L2 and L3 image integrations can begin on GNS3, you must download the required files and save them to the Downloads folder on your Windows host, as shown in Figure 1-1. GNS3 will initially search for the required files in two specific locations: in your Downloads and in the IOS and QEMU folders under C:\Users\your_name\GNS3\images\.

CHAPTER 1 BUILDING A PYTHON NETWORK AUTOMATION LAB ENVIRONMENT

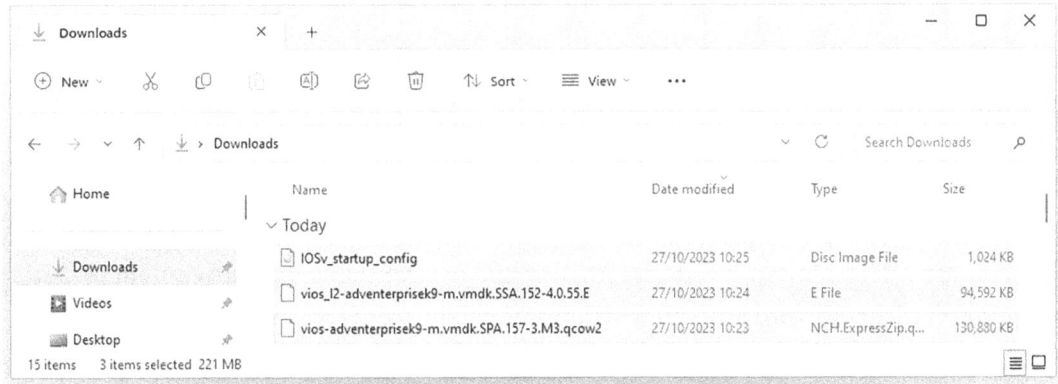

Figure 1-1. *CML L2 switch image, saved to the Downloads folder*

For the L2 image integration, a single file is needed. In this book, `vios_l2-adventerprisek9-m.vmdk.SSA.152-4.0.55.E` is used to serve as the layer 2 switch. In contrast, the L3 image integration requires two files: one image file and one startup file. The specifics of the files are tabulated in Table 1-1 to give you more clarity about the file requirements. For this book, the `vios-adventerprisek9-m.vmdk.SPA.157-3.M3.qcow2` and `IOSv_startup_config` files are required to enable you to use the Cisco L3 routers in the lab environment. However, if you are using custom CML images, you can choose to use a GNS3 appliance template method using a `cisco-iosv` file to import it through the device template. You will soon see how to do this for L3 image integration. As you are dealing with multiple files to make the Cisco router to boot up, there are file version conflicts and you will have to find the working sets and ensure that they are fully compatible. You do not need the latest and greatest software for your learning, but it's recommended to use the latest images whenever possible for your GNS3 setup.

Table 1-1. *CML Switch and Router Files Used for GNS3 Integration*

Description	File Name	File Size
Cisco CML L2 Switch:	`vios_l2-adventerprisek9-m.vmdk.SSA.152-4.0.55.E`	92.3MB
Cisco CML L3 Router:	`vios-adventerprisek9-m.vmdk.SPA.157-3.M3.qcow2`	127MB
	`IOSv_startup_config.img`	1MB
	`cisco-iosv` (for importing appliances)	3KB

While this book utilizes older CML image versions, for the specific use cases to learn Python Network Automation and network application development, the feature requirements are the bare minimum. This is not a joke, for the most part, you could still replace the older IOS 3745 image for the L3 router function during your studies, so, the latest and greatest is always on the cutting edge. However, this is not necessary. The primary focus in this book is on enterprise network automation. While routing and switching concepts are extremely important to learn, they are less central to the core content of this book. The CML images provide you with fundamental L2 and L3 Cisco platforms without breaking the bank and they facilitate lab-generated traffic across any network topologies you desire to run on a single laptop. Any recent versions of CML L2 and L3 images can be used in your lab, as long as they can serve as the endpoint to be managed, saving you the time and resources typically associated with running full-scale hardware-based labs. Once again, for seamless CML integrations, ensure that these files are in your Downloads folder, so you can continue the topology preparation tasks in this chapter (see Figure 1-1).

Integrating Cisco CML L2 and CML L3 Images on Your GNS3

At the end of Chapter 11 in the Part 1 book, you copied and created a new project called pynetauto-lab. Reopen that project now, as this is the starting point of your tasks in this chapter. If you are continuing your work from Part I, continue working from your GNS3, but if you have just powered on your PC, then first launch GNS3 using the GNS3 shortcut icon on your desktop. Ensure that you are running the same topology in your desktop environment. If the lab environment is set up correctly, GNS3 should been running, and the GNS3 VM installed on VMware Workstation should also start up automatically after a few seconds.

Tip

Securing the Right Appliances and Conquering CML Integration Challenges

The Part I book covers how to update GNS3 appliances. However, if you prefer to download individual appliances, you have that option too. Simply visit GNS3.com's Marketplace and locate your appliance icons (see Figure 1-2), then download the required appliance files. Keep in mind that while some appliances are open-source, some are proprietary, and the Download links will only connect you to the vendor websites. Many appliances requiring proprietary images will redirect you to the vendor sites (like in CML). In CML's case, the GNS3 site provides some supplementary files but not the actual software images. You will need a valid support contract or active subscription account to download the actual working images.

To download appliances, choose Marketplace ➤ Appliances.

`https://www.gns3.com/marketplace/appliances`

Figure 1-2. *GNS3 Marketplace – the Cisco IOSv2 and IOSv appliance icons*

In older GNS3 versions, appliance templates were installed and ready for immediate use. However, in more recent versions, you have the option to import and update the appliance templates through GNS3's Update from Online Registry option within the New Template feature or import a specific appliance template for each device you want to add. To do each appliance import, simply select Import Appliance to point it to a downloaded appliance template file. Obtaining the correct templates and image files can sometimes be the primary hurdle that prevents you from progressing through this book. This is particularly relevant to networking students striving to secure positions in ICT jobs.

Additionally, another significant hurdle that may impede readers' progress in this book is encountering challenges during software integration, both major and minor. Issues with software integration, such as file corruption, software failures, or system incompatibilities, can be quite discouraging and can disrupt your study plans. For instance, integrating CML images with GNS3 might fail multiple times, requiring numerous attempts to achieve success. This process involves multiple installation and uninstallation cycles, which might lead you to consider giving up after just one or two attempts. It's important to note that CML images were not initially designed for GNS3, leading to occasional integration challenges during the initial lab setup. If your initial attempts at integrating CML images with GNS3 encounter difficulties, do not be disheartened. Persistence is crucial, and you can continue trying multiple times until the integration succeeds. Should you find it challenging to resolve these issues, consider verifying the file integrity, investigating potential bugs in your GNS3 installation, and seeking assistance from the GNS3 community via their discussion board. (`https://www.gns3.com/community/discussions`).

Installing Cisco CML L2 Switch on GNS3

You will install the Cisco CML L2 switch image on GNS3 first. When you complete the installation tasks described here, you will be able to run the virtual switches on GNS3, which was technically not possible on the older IOS image integration. To begin the integration process, your first step is to install the Cisco CML L2 switch image within GNS3. This procedure is subject to your host PC being connected to the Internet, as it requires access to certain resources and updates from online repositories.

Upon successful installation, a world of possibilities unfolds. With the CML L2 switch in place, you will have the capability to create and operate virtual switching labs. These labs provide a dynamic environment for testing and experimenting with various L2 networking technology scenarios, making it an invaluable resource for network administrators, engineers, and students alike. Unlike the older IOS router's NM-16ESW module-based L2 switching, the CML L2 switch offers enhanced performance and expanded capabilities, almost identical to production switches. It empowers you to configure, manage, and troubleshoot complex networking concepts, providing a platform for in-depth exploration of enterprise switching technologies.

Now, let's import a CML L2 image in preparation for the topology build at the end of this chapter. You are expected to continue the tasks in your last cloned lab project, `pynetauto-lab`.

#	Task
①	In the GNS3 main window, choose File and click + New template, as shown in Figure 1-3. 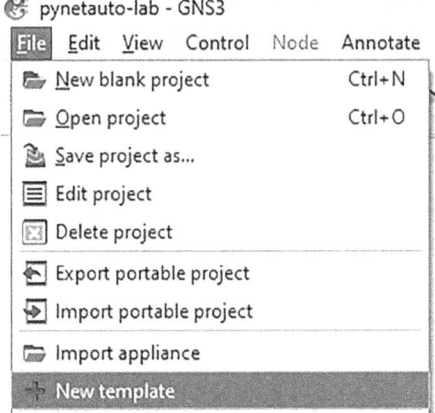 ***Figure 1-3.*** *CML L2 installation, adding a new template*
②	When the New Template wizard opens, leave the default selection of Install an Appliance from the GNS3 Server (Recommended) and click the Next button (see Figure 1-4). 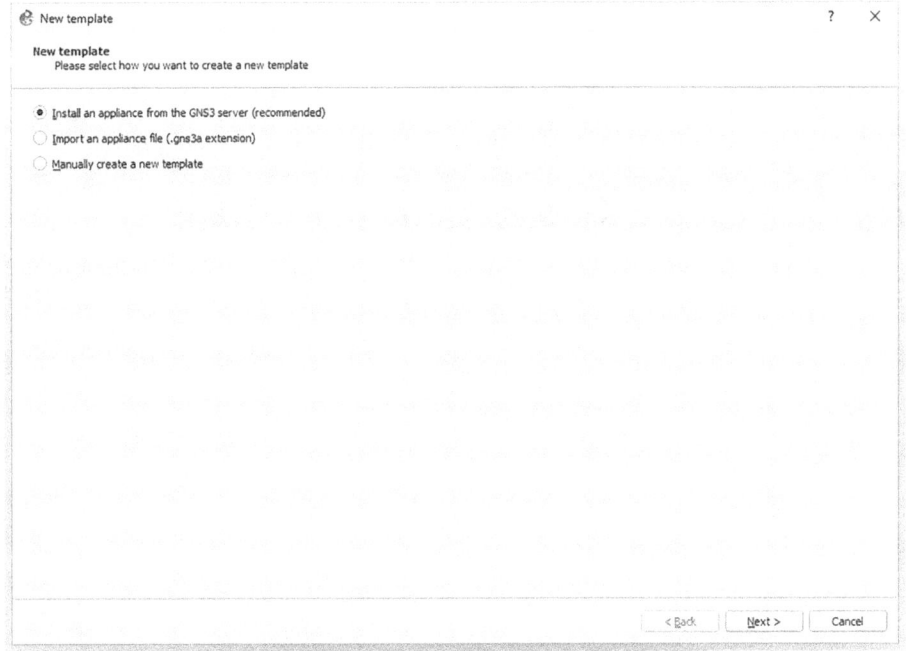 ***Figure 1-4.*** *CML L2 installation, new template selection*

(continued)

CHAPTER 1   BUILDING A PYTHON NETWORK AUTOMATION LAB ENVIRONMENT

#	Task
③	In the Appliances from Server window, you should have updated all device templates online in the last chapter of the Part I book, so open the Switch drop-down menu and select the Cisco IOSv2 option, as depicted in Figure 1-5. Now click the Install button.

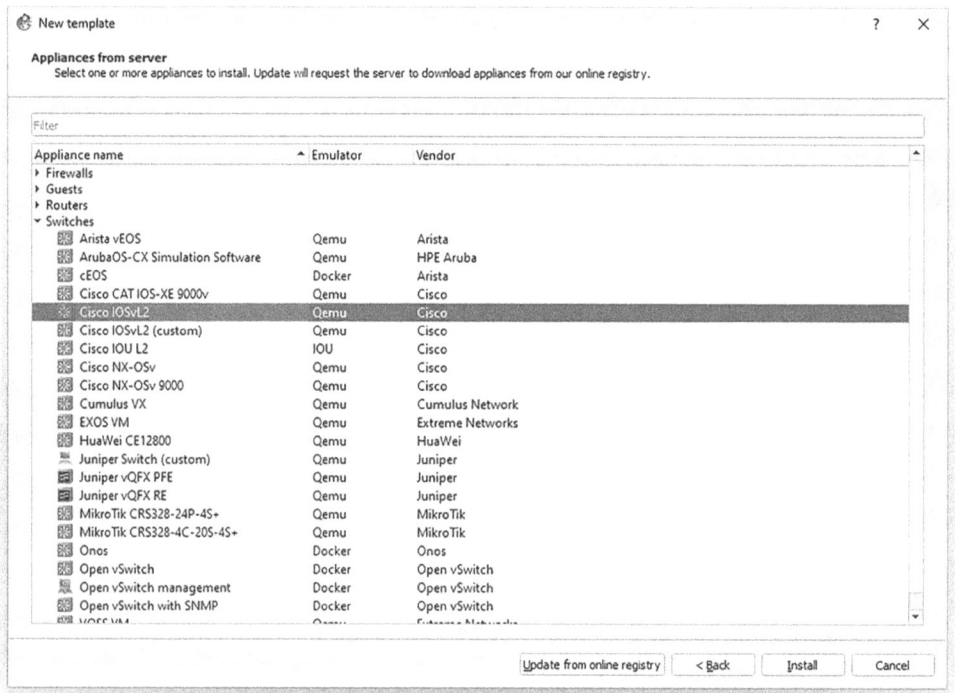

Figure 1-5. *CML L2 installation, appliances from server*

(continued)

CHAPTER 1   BUILDING A PYTHON NETWORK AUTOMATION LAB ENVIRONMENT

#	Task
④	In the Server window, Install the Appliance on the GNS3 VM (Recommended) should be selected already. Click the Next button (see Figure 1-6).

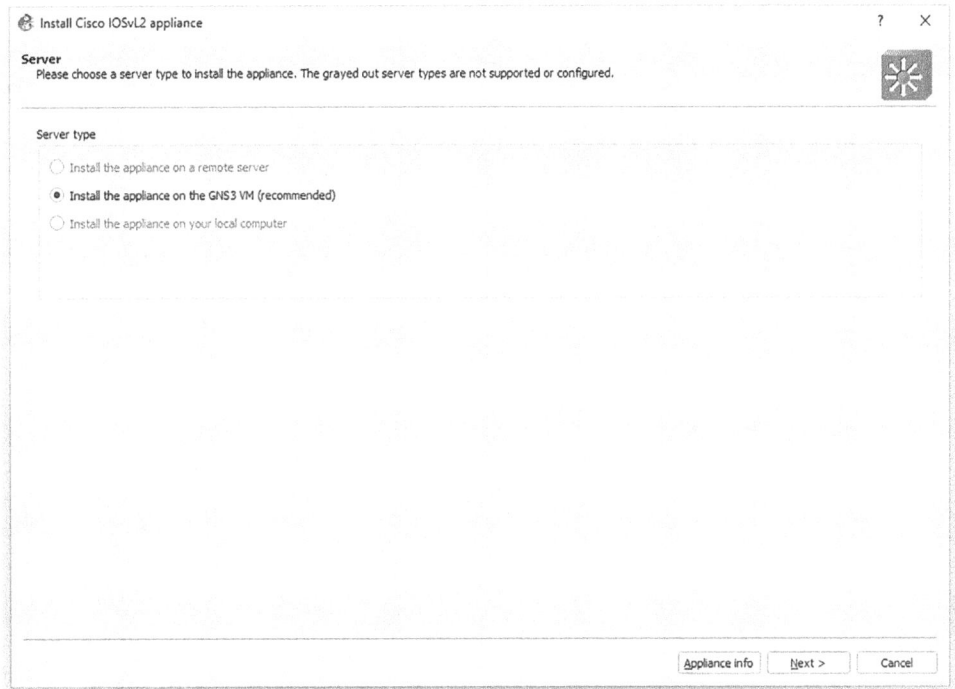

Figure 1-6. *CML L2 installation, server*

(*continued*)

CHAPTER 1   BUILDING A PYTHON NETWORK AUTOMATION LAB ENVIRONMENT

#	Task
⑤	In the Qemu Settings window, leave the default choice and click the Next button (see Figure 1-7).

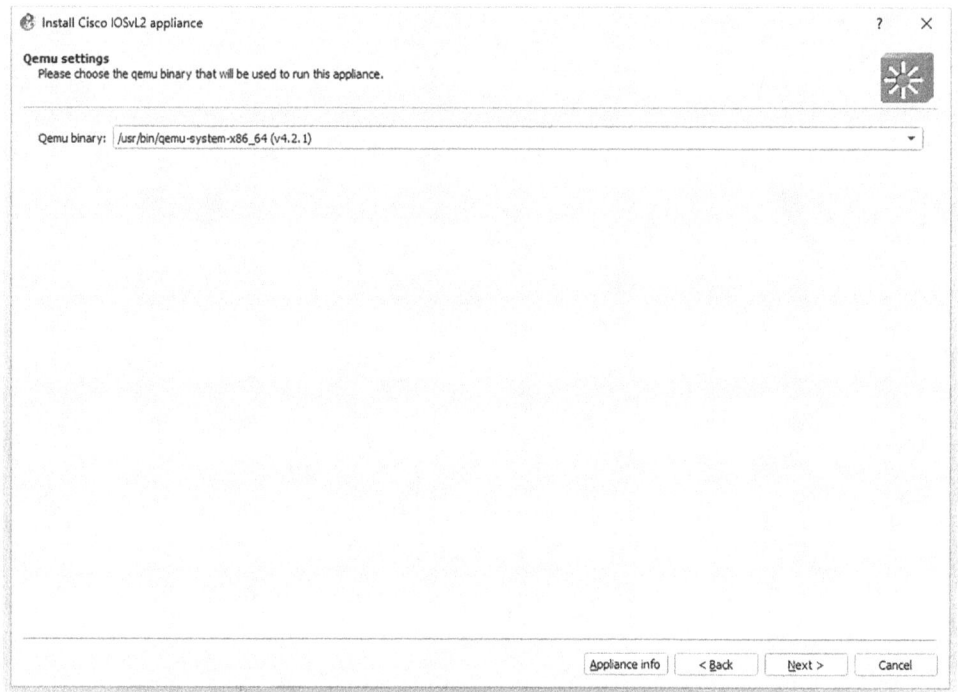

Figure 1-7. *CML L2 installation, Qemu settings*

(*continued*)

CHAPTER 1 BUILDING A PYTHON NETWORK AUTOMATION LAB ENVIRONMENT

#	Task
6	Under Required Files, if you have placed the CML L2 .qcow2 file in the Downloads folder as instructed earlier, the installation wizard will automatically detect the L2 image file and display it in green. The missing files are marked in red. Select your image by highlighting it and click the Next button again (see Figure 1-8).

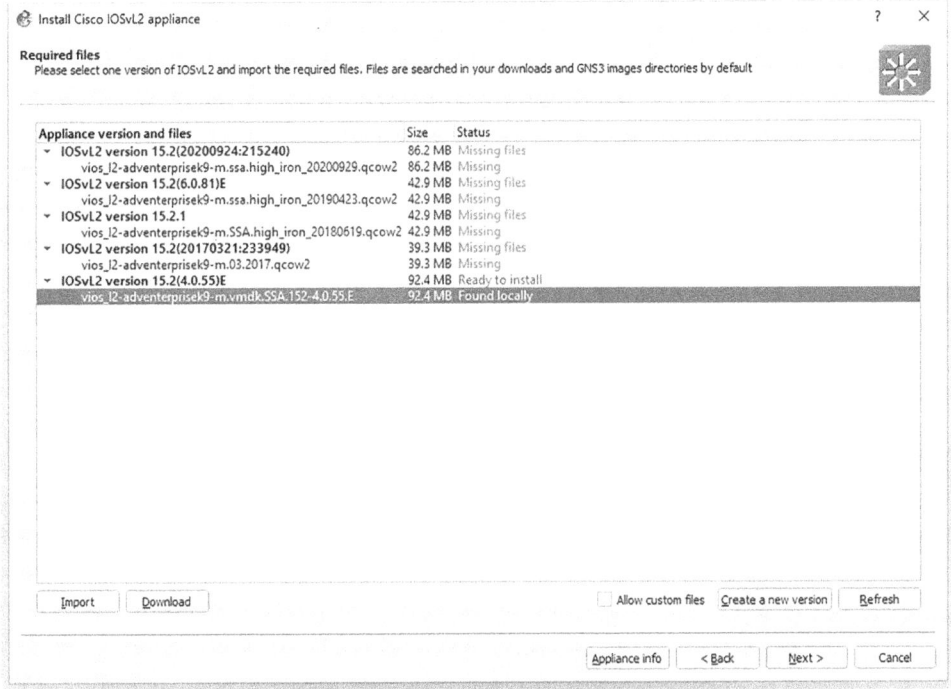

Figure 1-8. *CML L2 installation, required files*

(*continued*)

13

CHAPTER 1   BUILDING A PYTHON NETWORK AUTOMATION LAB ENVIRONMENT

#	Task
⑦	When the Appliance window pops up, click the Yes button (see Figure 1-9). 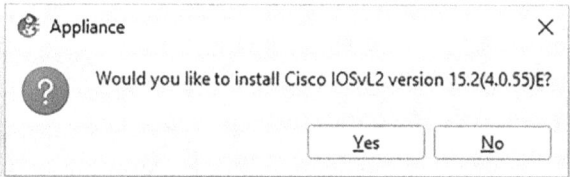 *Figure 1-9. CML L2 installation, appliance install pop-up*
⑧	To complete the installation, click the Finish button on the wizard, as shown in Figure 1-10. 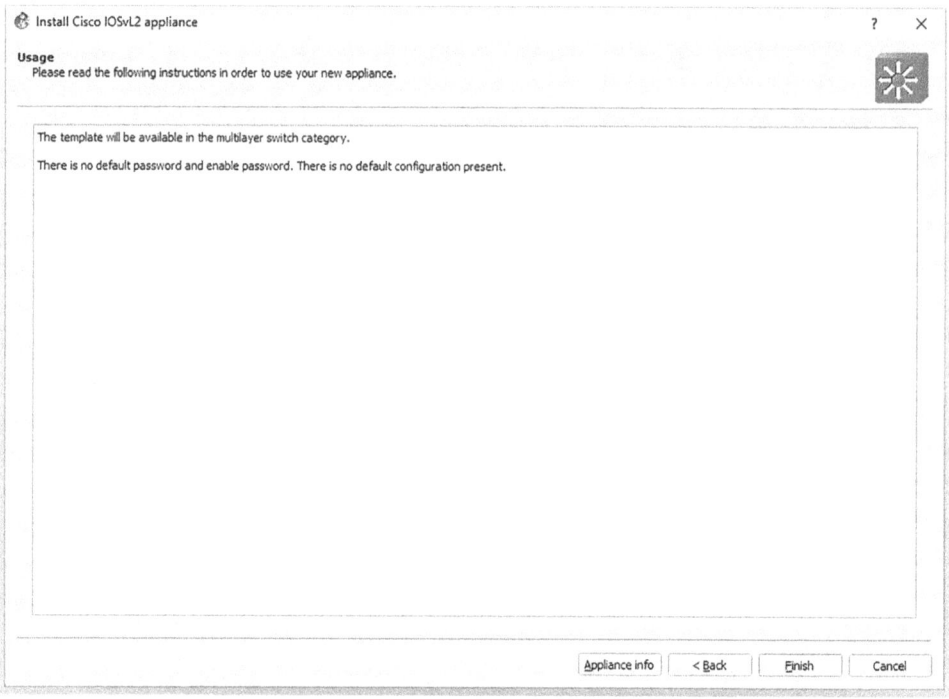 *Figure 1-10. CM L2 installation, usage*

(continued)

CHAPTER 1 BUILDING A PYTHON NETWORK AUTOMATION LAB ENVIRONMENT

#	Task
9	The Add Template window pops up. Click the OK button to finish the installation (see Figure 1-11).

Figure 1-11. *CML L2 installation, Add Template pop-up*

10	Now you are back on the GNS3 main window. Click the Switch icon button on the left, which looks like two opposite-pointing arrows, to open the switches list (see Figure 1-12).

Figure 1-12. *CML L2 installation, select the switch icon on GNS3*

11	If Cisco CML IOSvL2 was installed correctly, you should see your new L2 switch installed under Switches, as highlighted in Figure 1-13.

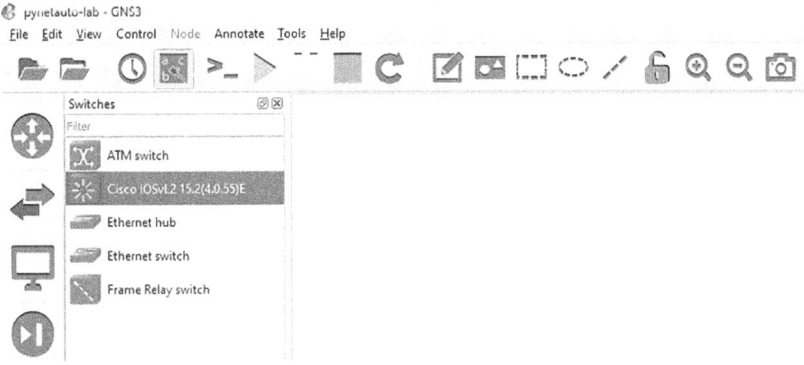

Figure 1-13. *CML L2 installation, selecting the Cisco IOSvL2 switch*

(*continued*)

15

CHAPTER 1   BUILDING A PYTHON NETWORK AUTOMATION LAB ENVIRONMENT

#	Task
⑫	Now, drag and drop the selected IOSvL2 switch to the `pynetauto-lab` topology canvas, as shown in Figure 1-14.

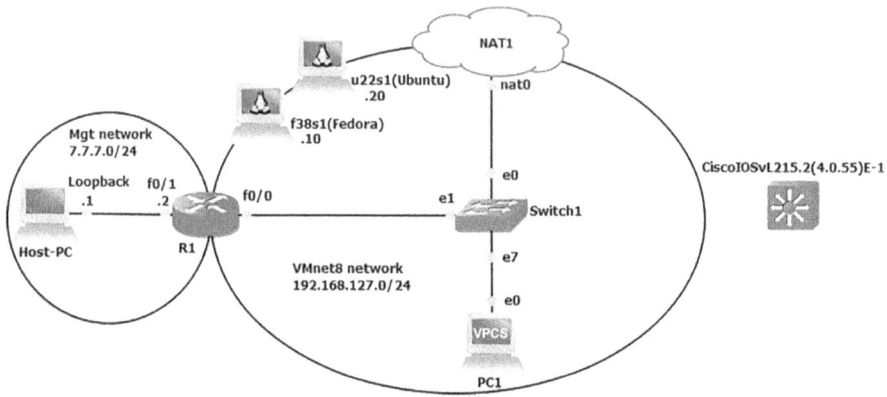

Figure 1-14. *GNS3 pynetauto-lab topology, dragging and dropping a CML switch*

⑬	Use the Add Link tool to establish a connection between Gi0/0 of the new CML switch and Switch1's e2 interface. Next, right-click the CML Switch icon and modify its hostname to SW1. To power on SW1, right-click again and select the Start option, as depicted in Figure 1-15. **Ensure that R1 and PC1 in GNS3 are powered on and validate that the Linux servers, f38s1, and the u22s1 VMs are running in the VMware Workstation.**

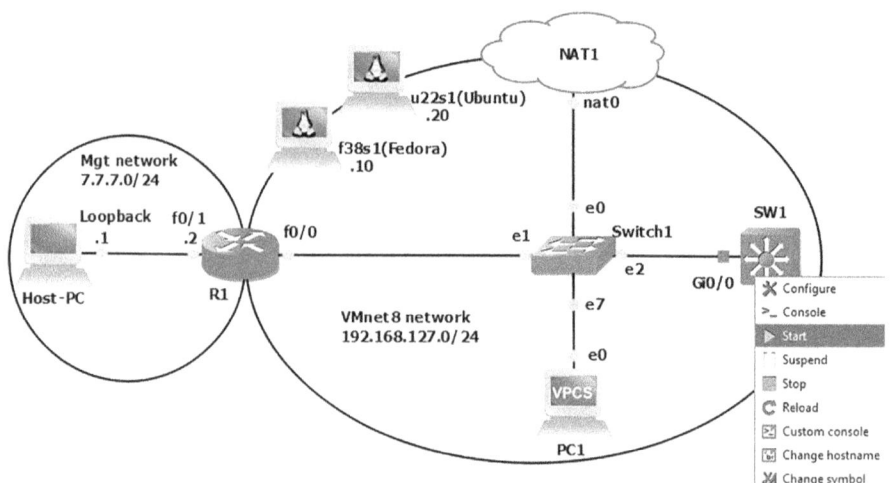

Figure 1-15. *GNS3 pynetauto-lab topology, powering on SW1*

(continued)

CHAPTER 1 BUILDING A PYTHON NETWORK AUTOMATION LAB ENVIRONMENT

#	Task
⑭	After powering on SW1, access its console to observe the booting process like the one depicted in Figure 1-16. Allow SW1 to complete its POST (Power on Self-Test) and give the operating system a couple of minutes to stabilize. 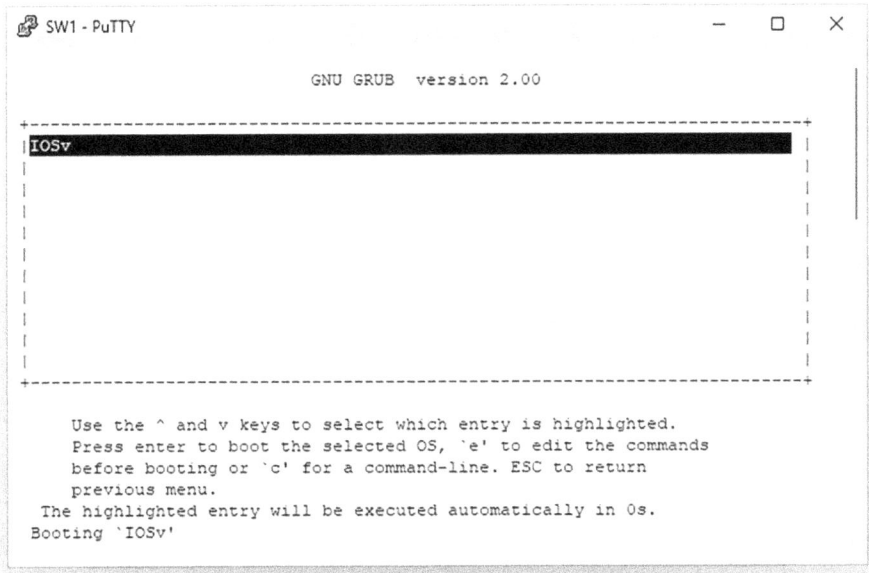 ***Figure 1-16.*** *SW1, powered-on CML switch screen*

(*continued*)

CHAPTER 1 BUILDING A PYTHON NETWORK AUTOMATION LAB ENVIRONMENT

#	Task
⑮	After booting up, run the show version command on the console. Your new switch's console screen should look similar to Figure 1-17.

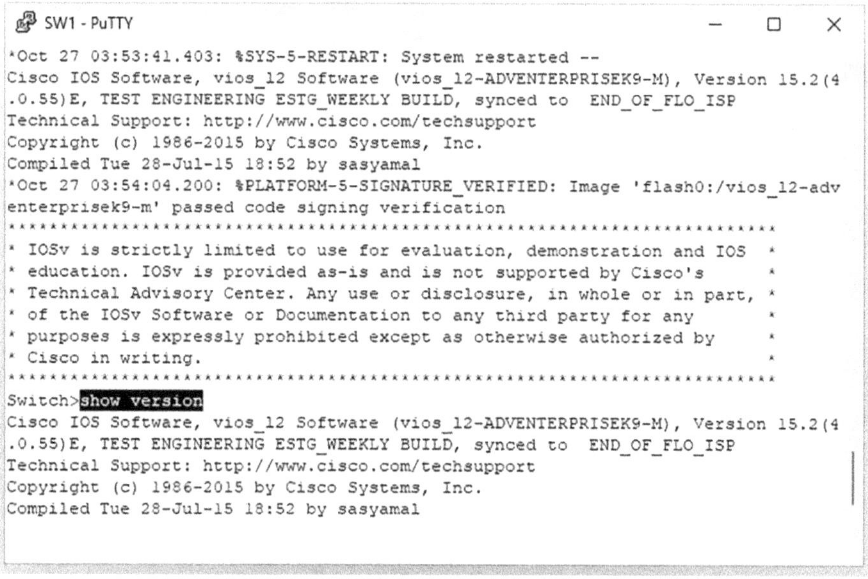

Figure 1-17. SW1, run the show version command

You have successfully integrated the CML L2 switch in GNS3. Prepare for hands-on switching labs. This virtual switch serves as a crucial tool for addressing a wide array of Python Network Automation topics. Additionally, within these settings, the CML L2 switch offers the advantage of being used to prepare for your Cisco certifications.

Quick Ping Test on CML L2 Switch

At the end of the CML L2 switch installation, it is crucial to conduct a ping (ICMP) test to ensure that the switch is functioning correctly within the lab environment. This step is integral to confirming that the switch can effectively communicate with other devices on the network. Once again, use a series of ping tests. *Ping* is a fundamental network tool that allows devices to send and receive ICMP messages, as well as query for information about network connectivity. These tests provide valuable insights into the operational status of the switch and its ability to establish connections with other devices.

#	Task
①	On the SW1 console, quickly configure the switch using the provided Cisco commands Each command line includes an explanation to assist new network students. ``` Switch>enable # Enter Privileged EXEC mode Switch#configure terminal # Enter Configuration Mode SW1(config)#hostname SW1 # Rename hostname SW1(config)#interface GigabitEthernet 0/0 # Enter interface Gi0/0 config mode SW1(config-if)#no negotiation auto # Disable duplex auto negotiation SW1(config-if)#duplex full # Change duplex to full SW1(config)#spanning-tree vlan 1 root primary # Designate this switch as the root primary SW1(config)#interface vlan 1 # Enter interface vlan 1 config mode SW1(config-if)#ip address 192.168.127.101 255.255.255.0 # Configure an IP address SW1(config-if)#no shutdown # Bring up this interface SW1(config-if)#end # Go back to Privileged EXEC mode ```

②	When activating Gi0/0 on SW1, you might encounter duplex mismatch errors from R1's f0/0 interface. To prevent this recurring CDP error message on SW1, it's necessary to manually set R1's f0/0 interface speed to 100 and duplex to full. If you do not encounter this error, you can proceed to the next step.
	
	``
	SW1(config-if)#
	*Oct 27 04:17:04.201: %CDP-4-DUPLEX_MISMATCH: duplex mismatch discovered on GigabitEthernet0/0 (not full duplex), with R1 FastEthernet0/0 (half duplex).
	``
	
	Open the console for R1 and confirm or change the interface speed and duplex settings:
	
	``
	R1#conf t # Enter Global Configuration mode
	R1(config)#interface f0/0 # Enter interface configuration mode
	R1(config-if)#duplex full # Change interface duplex to full (default auto)
	R1(config-if)#speed 100 # Change interface speed to 100
	R1(config-if)# do write memory # save the configuration in current mode
	``

(continued)

CHAPTER 1 BUILDING A PYTHON NETWORK AUTOMATION LAB ENVIRONMENT

#	Task
③	Go to PC1 and execute the `ip dhcp` command to acquire an IP address. Alternatively, if you are continuing from the previous book, run the `show ip` command to verify PC1's IP address (see Figure 1-18). PC1> **ip dhcp** # Request an IP address from the DHCP server PC1> **show ip** # Display IP Settings of PC1 (VPCS) 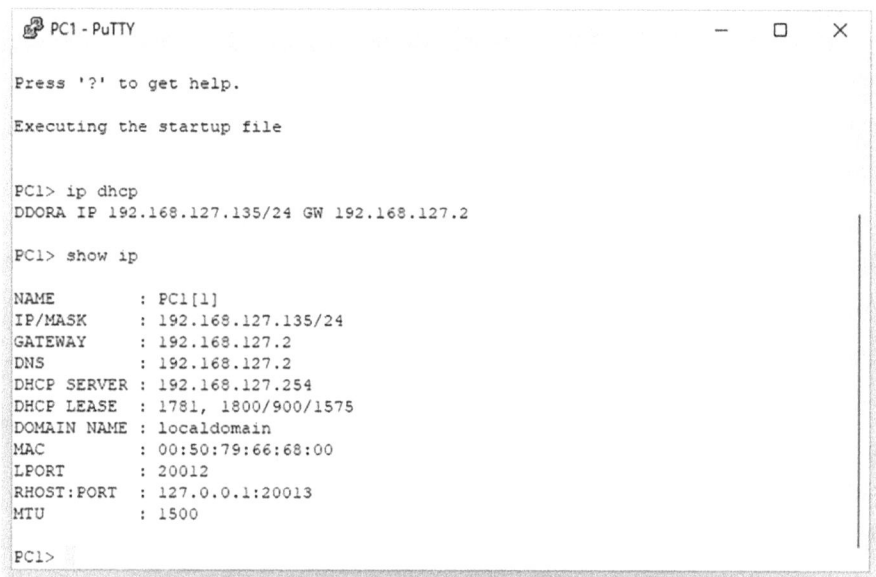 *Figure 1-18. PC1, running ip dhcp and show ip commands* Note that if PC1 encounters difficulty obtaining a new IP address from the DHCP server, you need to troubleshoot the issue before proceeding. Typically, restarting the application and rebooting the host PC can help resolve this issue 90 percent of the time.

(continued)

#	Task

④	From the SW1 console (192.168.127.101), you'll use an interactive Tcl shell (`tclsh`) to check the communication between all local devices, using their respective IP addresses. For this `tclsh` ping test, the target IP addresses are GW (VMnet8) 192.168.127.2; F38s1 (Fedora 38 Server) 192.168.127.10; U22s1 (Ubuntu 22 Server) 192.168.127.20; R1 192.168.127.134; and PC1 192.168.127.135.
	
	You can enter each line into the console, or simply cut and paste the whole command. The same commands can be found in `ch1_tclsh_ping_mip.txt`.
	
	tclsh
	
	``tcl
	set ip_list {
	  192.168.127.2
	  192.168.127.10
	  192.168.127.20
	  192.168.127.134
	  192.168.127.135
	}
	
	set timeout 2
	foreach ip $ip_list {
	  if {[catch {ping $ip timeout $timeout} result]} {
	    puts "Ping to $ip failed: $result"
	  } else {
	    puts "Ping to $ip succeeded"
	  }
	}
	``
	
	Refer to Figure 1-19 for the `tclsh` command run example. You can use Ctrl+Z or Ctrl+C to exit the Tcl shell.

(continued)

CHAPTER 1 BUILDING A PYTHON NETWORK AUTOMATION LAB ENVIRONMENT

#	Task

```
SW1 - PuTTY                                              —   □   ×
SW1(tcl)#
SW1(tcl)#tclsh
SW1(tcl)#
SW1(tcl)## List of IP addresses to ping
SW1(tcl)#set ip_list {
+>     192.168.127.2
+>     192.168.127.10
+>     192.168.127.20
+>     192.168.127.134
+>     192.168.127.135
+>}
    192.168.127.2
    192.168.127.10
    192.168.127.20
    192.168.127.134
    192.168.127.135

SW1(tcl)#
SW1(tcl)## Set the timeout value in seconds
SW1(tcl)#set timeout 2
2
SW1(tcl)#
SW1(tcl)#foreach ip $ip_list {
+>     if {[catch {ping $ip timeout $timeout} result]} {
+>         puts "Ping to $ip failed: $result"
+>     } else {
+>         puts "Ping to $ip succeeded"
+>     }
+>}
```

Figure 1-19. *SW1, tclsh cut and paste command*

After pressing the Enter key, you should expect all ping tests to succeed. Note that Figure 1-20 displays the perfect ping result from the second ping test, not the first.

(continued)

CHAPTER 1 BUILDING A PYTHON NETWORK AUTOMATION LAB ENVIRONMENT

#	Task

```
SW1 - PuTTY                                              —  □  ×
+>}
Type escape sequence to abort.
Sending 5, 100-byte ICMP Echos to 192.168.127.2, timeout is 2 seconds:
!!!!!
Success rate is 100 percent (5/5), round-trip min/avg/max = 3/3/7 msPing to 192.
168.127.2 succeeded

Type escape sequence to abort.
Sending 5, 100-byte ICMP Echos to 192.168.127.10, timeout is 2 seconds:
!!!!!
Success rate is 100 percent (5/5), round-trip min/avg/max = 3/82/396 msPing to 1
92.168.127.10 succeeded

Type escape sequence to abort.
Sending 5, 100-byte ICMP Echos to 192.168.127.20, timeout is 2 seconds:
!!!!!
Success rate is 100 percent (5/5), round-trip min/avg/max = 3/4/8 msPing to 192.
168.127.20 succeeded

Type escape sequence to abort.
Sending 5, 100-byte ICMP Echos to 192.168.127.134, timeout is 2 seconds:
!!!!!
Success rate is 100 percent (5/5), round-trip min/avg/max = 3/4/6 msPing to 192.
168.127.134 succeeded

Type escape sequence to abort.
Sending 5, 100-byte ICMP Echos to 192.168.127.135, timeout is 2 seconds:
!!!!!
Success rate is 100 percent (5/5), round-trip min/avg/max = 3/4/7 msPing to 192.
168.127.135 succeeded

SW1(tcl)#
```

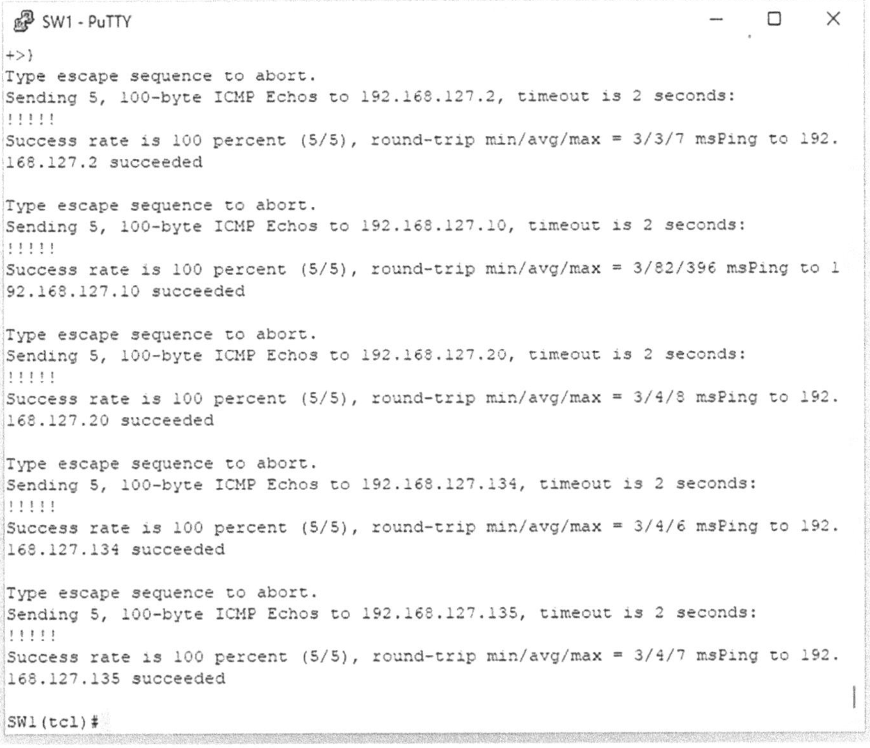

Figure 1-20. *SW1, tclsh ping command output*

⑤ Finally, when you are happy with your ping result, save SW's running configuration by running the following command, or alternatively use `write memory` or `wri`. This ensures that your current configuration remains in place even after a reboot or power cycle.

SW1# **copy running-config startup-config** # save configuration

Congratulations, you've completed the integration and network verification of the CML L2 switch image in GNS3. This GNS3 appliance equips you with the ability to deploy multiple Cisco switches, offering a robust environment for *labbing* Python Network Automation scenarios. **Labbing is a jargon commonly used in the IT field, particularly among IT technicians, network engineers, and professionals involved in network infrastructure or system administration.** It refers to the act of creating, configuring, and experimenting with network topologies and systems, or performing

CHAPTER 1   BUILDING A PYTHON NETWORK AUTOMATION LAB ENVIRONMENT

Proof of Concept (PoC) labs for testing or certification preparation. While labbing is not in the dictionary, within the context of IT, it's widely understood and accepted as the act of engaging in practical, hands-on lab work to learn, test, or certify IT systems.

Now, let's move forward and proceed with the integration of the L3 router image. This next step will equip you with a Cisco router for network emulation labs and to experiment with Python Network Automation techniques.

Installing the Cisco CML L3 Router on GNS3 Using the Import Appliance Option

Let's follow the installation process for the Cisco CML L3 image on GNS3. These tasks will enable you to run the CML L3 routing device or Cisco router within your GNS3 environment. As with the previous CML L2 integration, a reliable Internet connection is a must to complete the required tasks successfully. It's recommended that you use a connection to your home network during the installation rather than a public or work network to ensure a trouble-free experience.

CHAPTER 1 BUILDING A PYTHON NETWORK AUTOMATION LAB ENVIRONMENT

Warning

For custom CML L3 image users.

When integrating a custom Cisco CML L3 image by selecting the + New Template method, you may come across a problem when Cisco IOSv is selected. During the installation process, GNS3 won't be able to recognize your L3 image file, as seen in Figure 1-21.

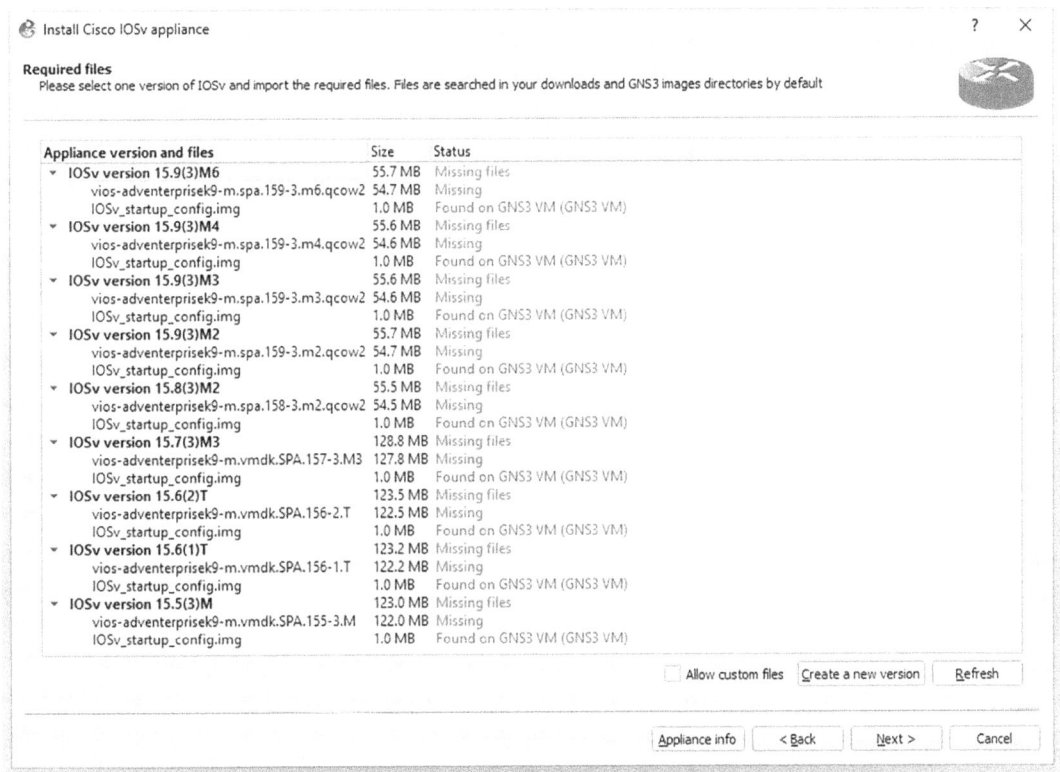

Figure 1-21. GNS3, wrong appliance template selected, no image match

To work around this problem, you may instead have to select the Cisco IOSv (Custom) option while selecting the appliances to install. Refer to Figure 1-22.

25

CHAPTER 1 BUILDING A PYTHON NETWORK AUTOMATION LAB ENVIRONMENT

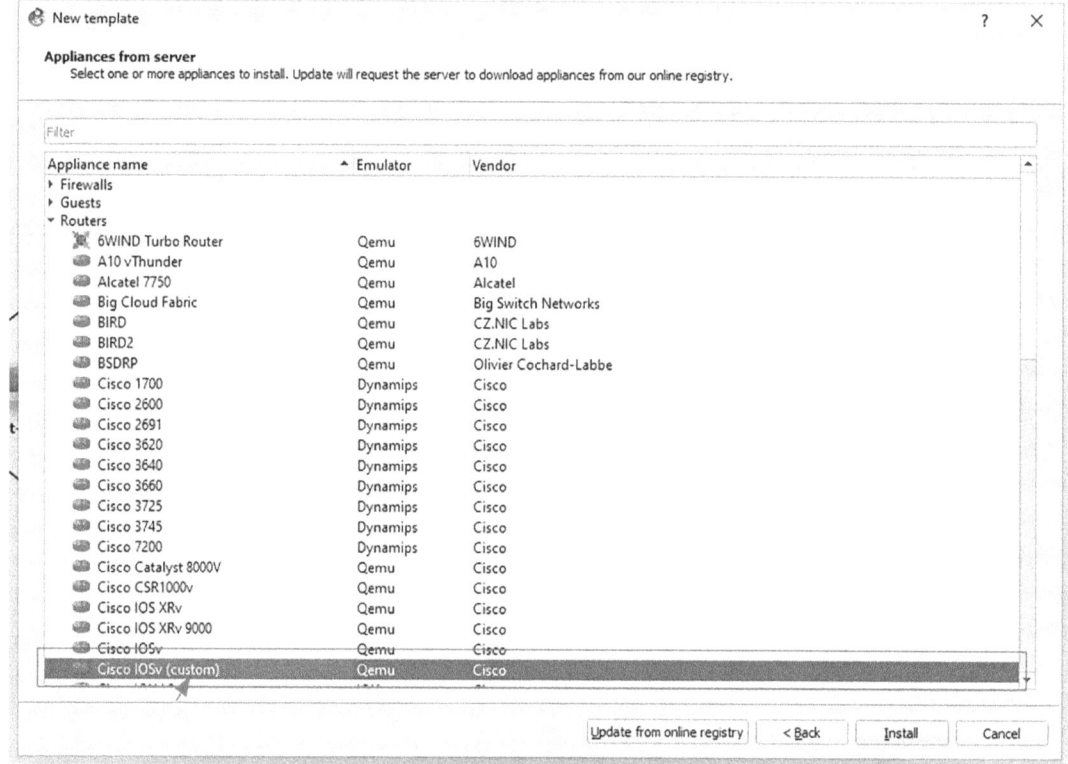

Figure 1-22. *GNS3, when using a custom CML L3 image, try Cisco IOSv (Custom)*

You should see a window similar to the one in Figure 1-23 if you have a valid custom image file.

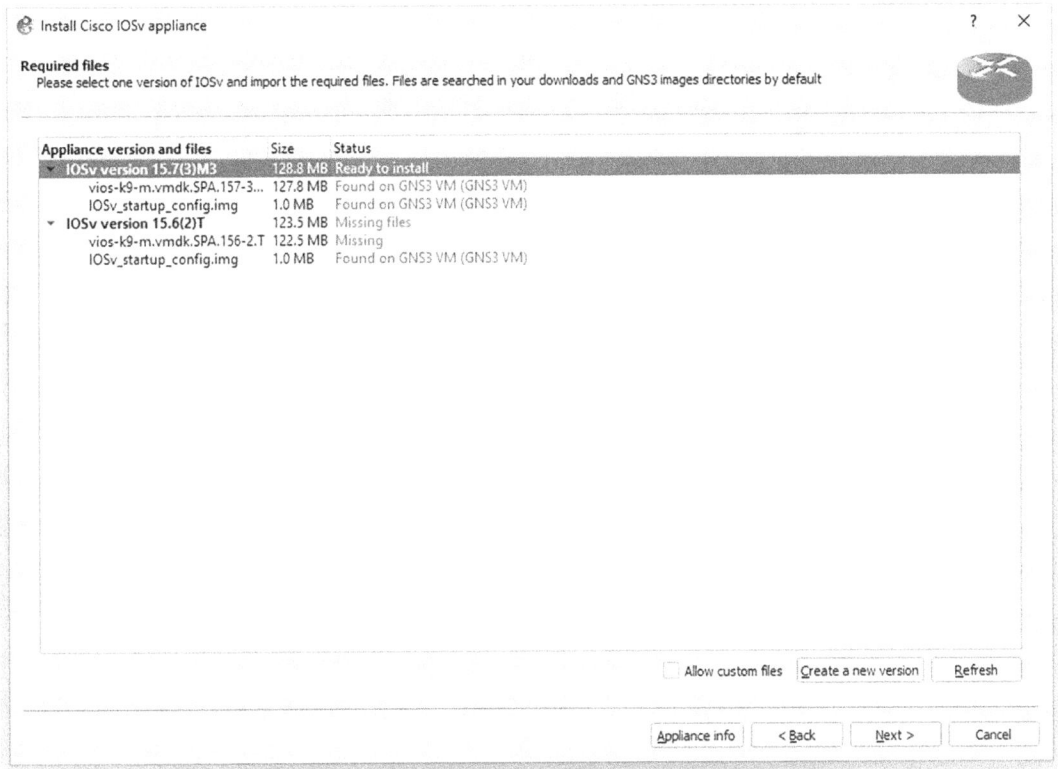

Figure 1-23. *GNS3, custom image and startup files matched*

When you integrated the CML L2 image, you used the + New template option to integrate it with GNS3. This time, I show you how to integrate the L3 image using the import appliance method, so you know how to use both methods. In this example, I am using a `cisco-iosv` GNS3 appliance file, which is compatible with my Cisco IOSv3 file. You must locate and download a compatible file from GNS3 Marketplace or the developer's GitHub. If you are using the same image file as demonstrated here and are unable to find the correct `cisco-iosv` GNS3 appliance file, this file is found in the `ch1` folder if you need it. So, for your CML L3 integration to go without a hiccup, you need all three required files under the `Downloads` folder—`cisco-iosv`, `vios-adventerprisek9-m.vmdk.SPA.157-3.M3.qcow2`, and `IOSv_startup_config.img`. The version and file compatibility are extremely important here. So take special care when locating your files.

Next, let's integrate the CML L3 image with GNS3 and finish the integration process.

CHAPTER 1   BUILDING A PYTHON NETWORK AUTOMATION LAB ENVIRONMENT

#	Task
①	In the GNS3 main window, choose File and select Import Appliance, as depicted in Figure 1-24. 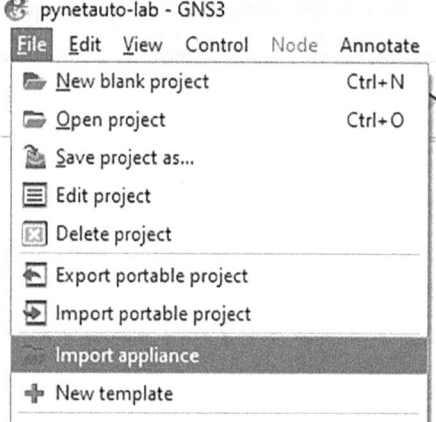 *Figure 1-24. CML L3 installation, select Import Appliance*
②	From the Import Appliance window, select the `cisco-iosv` you downloaded. Select the file and click the Open button to continue (see Figure 1-25). 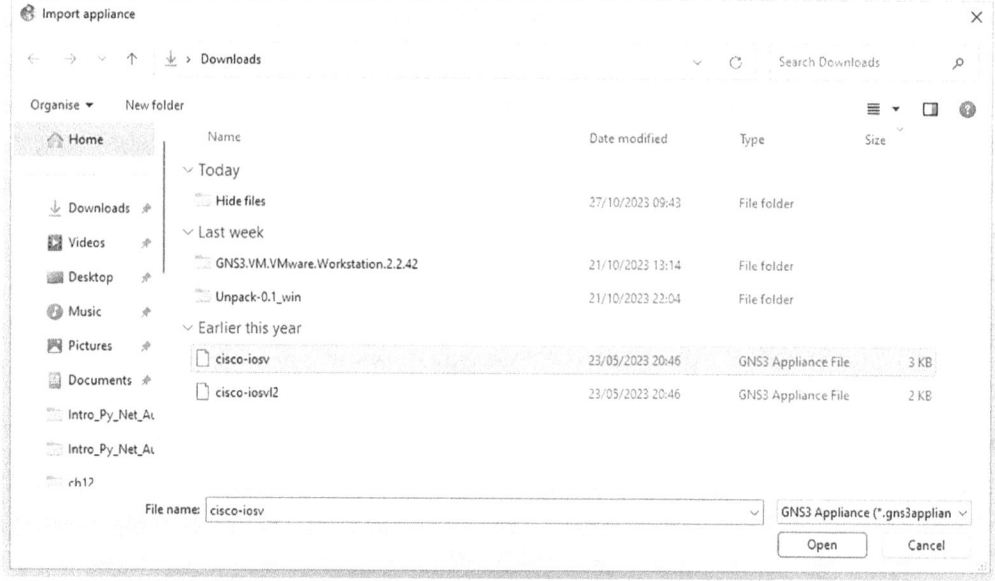 *Figure 1-25. CML L3 installation, select the cisco-iosv file*

(*continued*)

CHAPTER 1 BUILDING A PYTHON NETWORK AUTOMATION LAB ENVIRONMENT

#	Task
③	When the Server window opens, you'll see that Install the Appliance on the GNS3 VM (Recommended) is the only selection, so click the Next button again (see Figure 1-26).

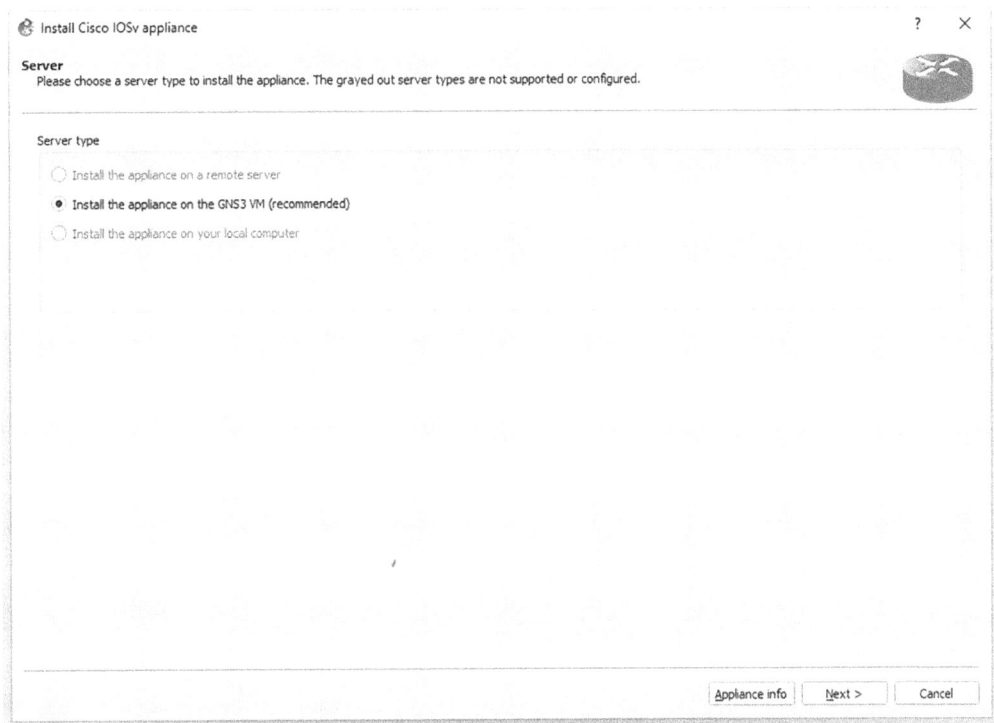

Figure 1-26. CML L3 installation, appliance server selection

(*continued*)

CHAPTER 1   BUILDING A PYTHON NETWORK AUTOMATION LAB ENVIRONMENT

#	Task
④	Next, you will be prompted with the Qemu settings selection. Keep the default selection and click the Next button (see Figure 1-27).

Figure 1-27. *CML L3 installation, Qemu settings selection*

(continued)

CHAPTER 1 BUILDING A PYTHON NETWORK AUTOMATION LAB ENVIRONMENT

#	Task
⑤	When you reach the required files, you will notice that your image files for both the `startup_config` and CML L3 are correctly detected and matched. You should see Found Locally on both and Ready to Install on the image version line, as shown in Figure 1-28. Highlight your IOSv version and then click the Next button.

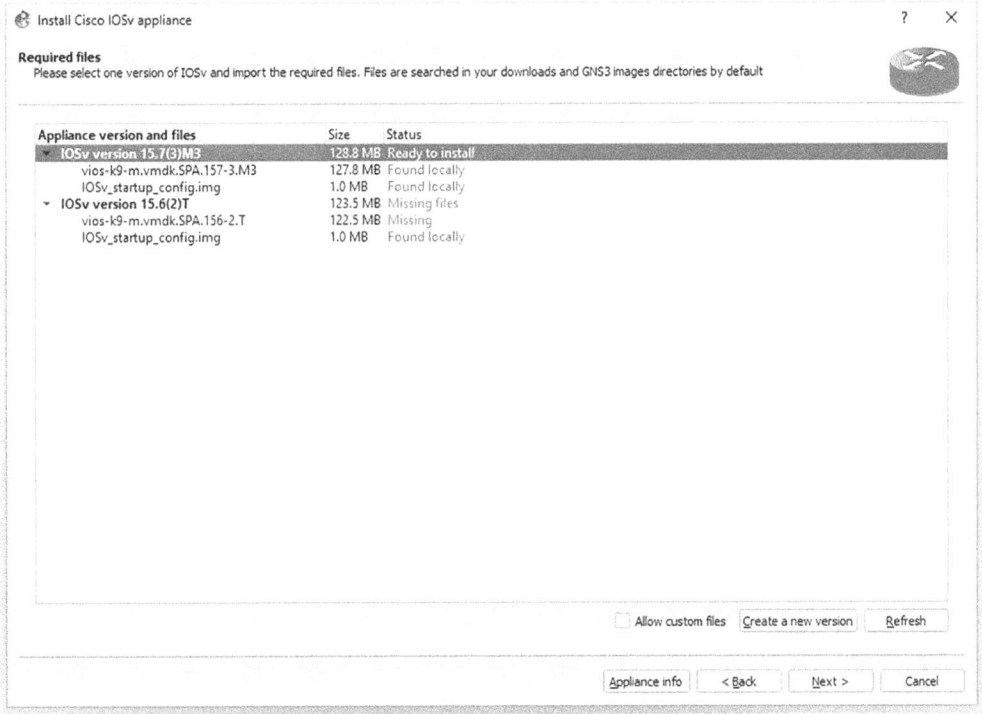

Figure 1-28. CML L3 installation, required files found

⑥	You want to install this image on GNS3, so click the Yes button (see Figure 1-29).

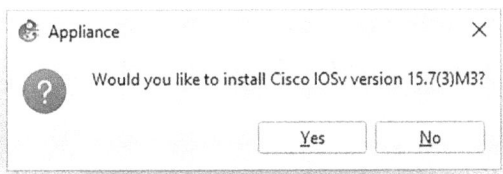

Figure 1-29. CML L3 installation, Install Cisco IOSv

(*continued*)

CHAPTER 1 BUILDING A PYTHON NETWORK AUTOMATION LAB ENVIRONMENT

#	Task
(7)	On the Usage screen, click the Finish button to complete your installation (see Figure 1-30).

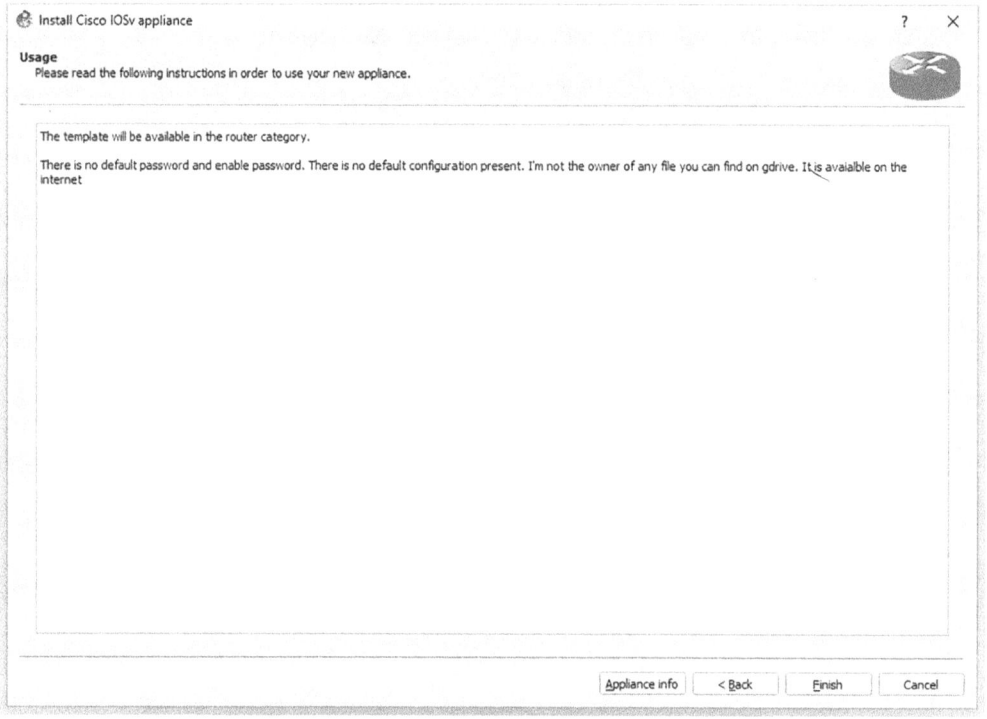

Figure 1-30. CML L3 installation, usage

(8)	On the Add Template window, click the OK button to complete the installation (see Figure 1-31).

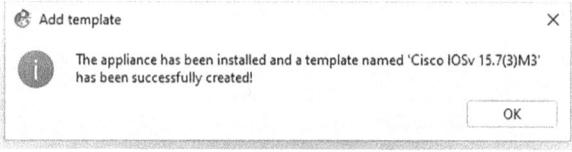

Figure 1-31. CML L3 installation, complete message

(continued)

#	Task
⑨	Go back to the main GNS3 window, and this time click the Router icon that looks like a circle with four arrows (see Figure 1-32).  ***Figure 1-32.*** *CML L3 installation, selecting the router icon on GNS3*
⑩	The Routers list appears, as shown in Figure 1-33, and you will notice your new IOSv router in the list. Select it by highlighting the Cisco IOSv icon. 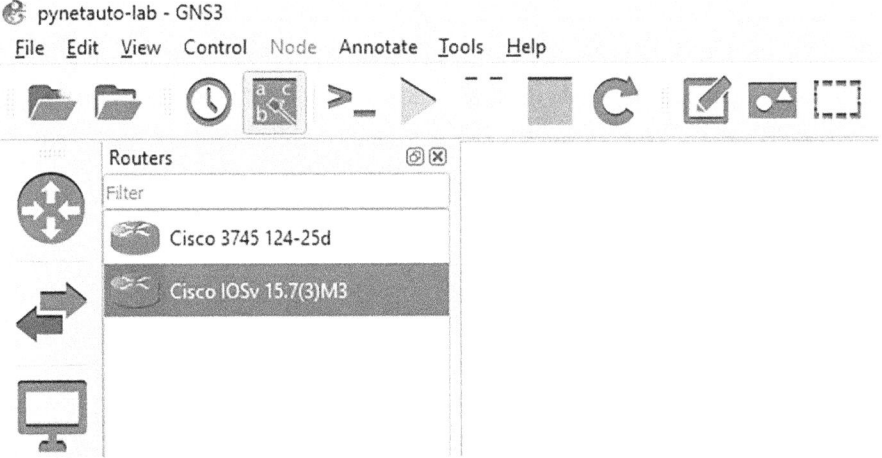 ***Figure 1-33.*** *CML L3 installation, selecting the CML router*

(*continued*)

CHAPTER 1 BUILDING A PYTHON NETWORK AUTOMATION LAB ENVIRONMENT

#	Task
⑪	Now drag and drop a new router onto the `pynetauto-lab` project topology (see Figure 1-34).

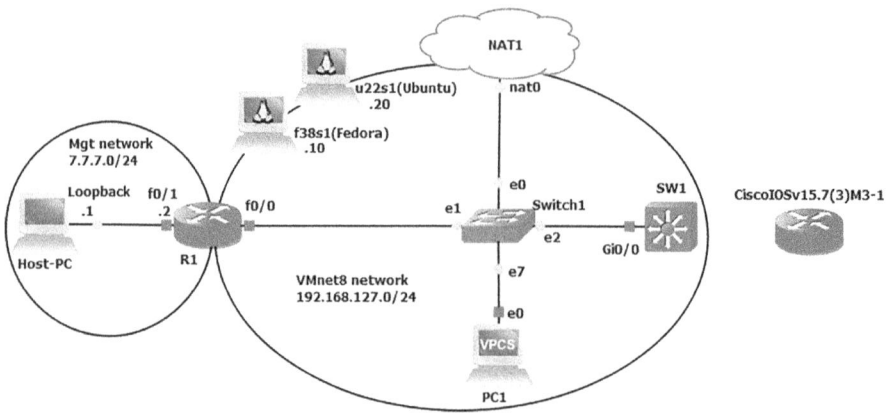

Figure 1-34. *GNS3, pynetauto-lab topology, adding a CML router*

⑫	Rename the router R2. Use the Add Link (connector) tool to connect R2's Gi0/0 interface to SW1's Gi0/1 interface. You can make the connection while the other device is powered on. Now power on R2 and all other devices if they are not powered on already. Your latest GNS3 topology with all devices powered on should look like Figure 1-35.

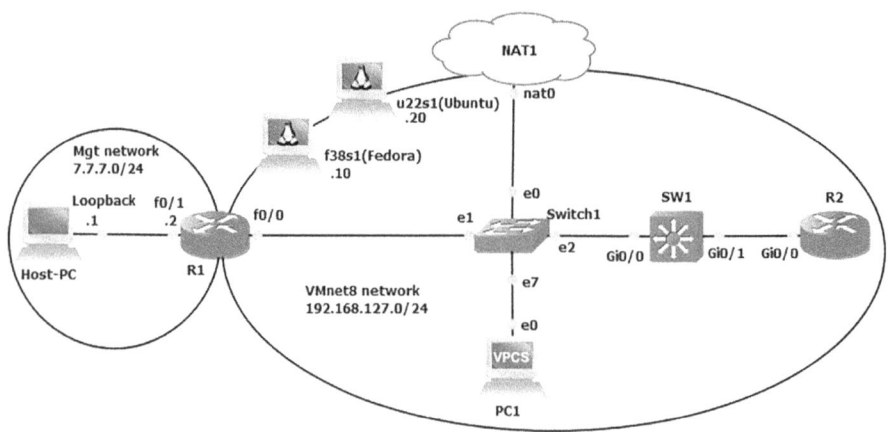

Figure 1-35. *GNS3, pynetauto-lab topology after adding and powering on R2*

(continued)

#	Task
⑬	The Topology Summary window shows the virtual connection between the devices (see Figure 1-36).

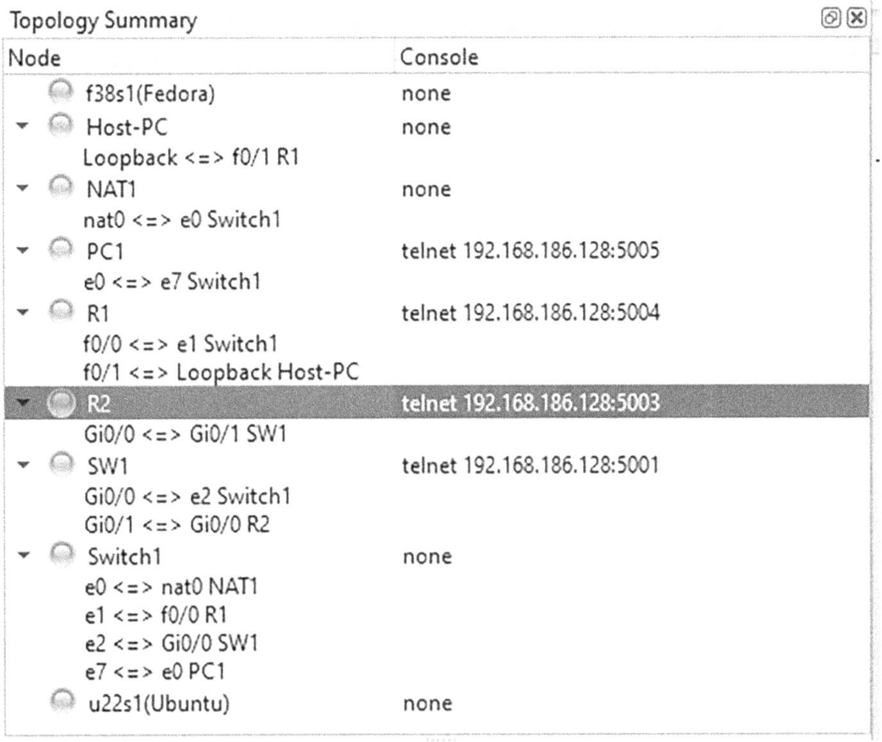

```
Topology Summary
Node                              Console
    f38s1(Fedora)                 none
  ▾ Host-PC                       none
    Loopback <=> f0/1 R1
  ▾ NAT1                          none
    nat0 <=> e0 Switch1
  ▾ PC1                           telnet 192.168.186.128:5005
    e0 <=> e7 Switch1
  ▾ R1                            telnet 192.168.186.128:5004
    f0/0 <=> e1 Switch1
    f0/1 <=> Loopback Host-PC
  ▾ R2                            telnet 192.168.186.128:5003
    Gi0/0 <=> Gi0/1 SW1
  ▾ SW1                           telnet 192.168.186.128:5001
    Gi0/0 <=> e2 Switch1
    Gi0/1 <=> Gi0/0 R2
  ▾ Switch1                       none
    e0 <=> nat0 NAT1
    e1 <=> f0/0 R1
    e2 <=> Gi0/0 SW1
    e7 <=> e0 PC1
    u22s1(Ubuntu)                 none
```

Figure 1-36. pynetauto-lab project, Topology Summary window

(*continued*)

CHAPTER 1   BUILDING A PYTHON NETWORK AUTOMATION LAB ENVIRONMENT

#	Task

The Servers Summary window shows where each device is hosted (see Figure 1-37).

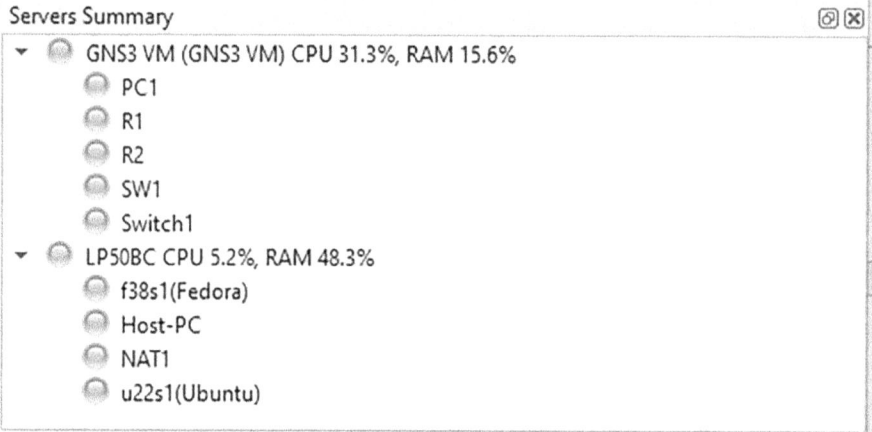

Figure 1-37.   pynetauto-lab project, Servers Summary window

⑭   Finally, when the router boots up, you will be prompted with the GNU GRUB version 2.00 screen, as shown in Figure 1-38. You have completed all Cisco CML L3 installation tasks.

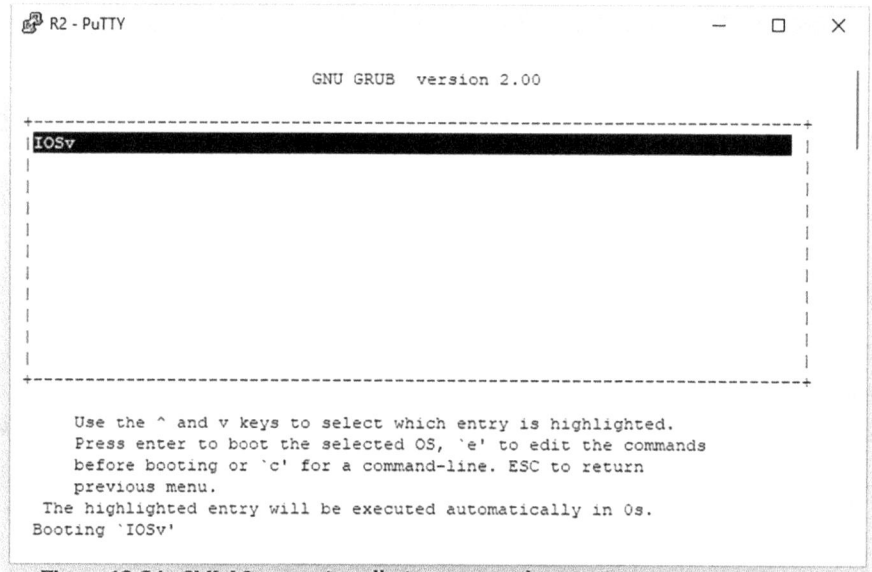

Figure 1-38.   CML L3 router console, GNU GRUB version 2.00 message

Hopefully, you didn't encounter too much trouble following this processes. Unlike the CML L2 image installation, the CML L3 image can be tricky for new GNS3 users due to file and software compatibility mismatch, as well as potential network connection problems. Before you proceed to jump to the ICMP (ping) test, here's one of the many warnings users may encounter during CML image integrations.

Warning

Troubleshooting CML image integration in GNS3

Be aware that if your host PC is connected to the Internet through a company proxy, you might encounter certain challenges during your installation. One error message I usually encounter on the work network is captured and displayed for your reference (see Figure 1-39). If you see this message, that means you have an issue with your Internet. This is just one of many GNS3 errors you may encounter during your integration tasks.

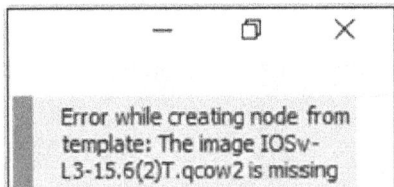

Figure 1-39. *GNS3, CML L3 image error*

The process of integrating CML images may seem straightforward until you encounter errors. When trying to drag and drop the image icon onto the GNS3 Topology canvas, many errors will appear as it tries to pull files from Internet repositories. Even after successfully importing the templates, if you still cannot complete the integration, use the Delete Template option, as shown in Figure 1-40, and try the integration processes again. Remember that the initial integration can be the most challenging, so persist until you know your CML image integrations are complete and reliable.

CHAPTER 1   BUILDING A PYTHON NETWORK AUTOMATION LAB ENVIRONMENT

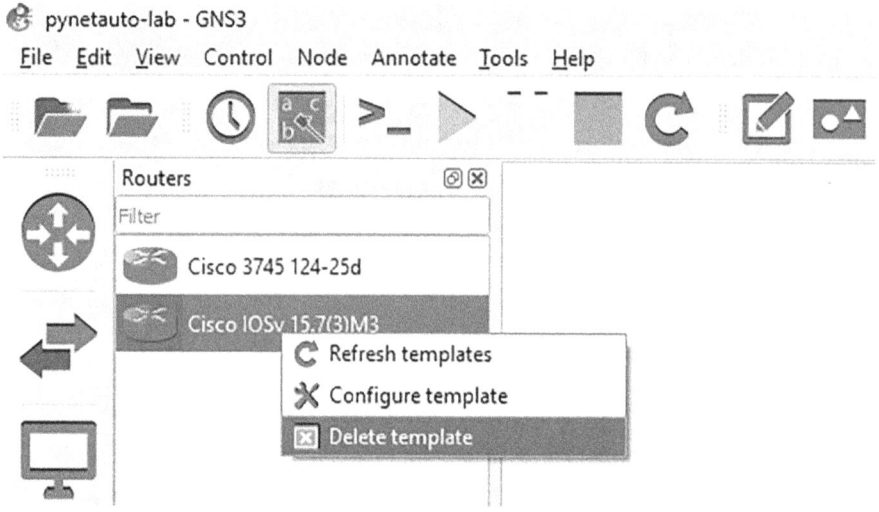

Figure 1-40. GNS3, deleting a CML template

Quick Ping Test on the CML L3 Router

Now, let's quickly configure your first CML router, R2, and perform some ping tests to all reachable local IP addresses on the local network.

#	Task
①	From the R2 console, use the following Cisco commands to configure the router. The router name is R2 and its GigabitEthernet 0/0 interface should be assigned with 192.168.127.3/24. Make sure you unshut the port to make it live on the local network. Router#**conf t** # Enter Global Configuration mode Router(config)#**hostname R2** # Rename hostname R2(config)#**int g0/0** # Enter interface Gi0/0 config mode R2(config-if)#**ip add 192.168.127.3 255.255.255.0** # Configure an IP address R2(config-if)#**no shut** # Bring up this interface R2(config-if)#**end** # Go back to Privileged EXEC mode R2#**wri** # Save the configuration

(continued)

#	Task

②	Next, perform the ping test to check the network connectivity to all the local devices. You can borrow the same tcl commands you used while testing SW1 (192.168.127.101) connectivity. You should expect to see the responses from all the destinations, and notice that the first `arp` packet drops as expected (see Figure 1-41).
	
	tclsh
	
	set ip_list {
	  192.168.127.2
	  192.168.127.10
	  192.168.127.20
	  192.168.127.101 # added SW1 IP
	  192.168.127.134
	  192.168.127.135
	}
	
	set timeout 2
	
	foreach ip $ip_list {
	  if {[catch {ping $ip timeout $timeout} result]} {
	    puts "Ping to $ip failed: $result"
	  } else {
	    puts "Ping to $ip succeeded"
	  }
	}

(continued)

CHAPTER 1　BUILDING A PYTHON NETWORK AUTOMATION LAB ENVIRONMENT

#	Task
	After the `tclsh` interactive ping test, you should see ping results like Figure 1-41. Use Ctrl+Z or Ctrl+C to exit the Tcl shell session.

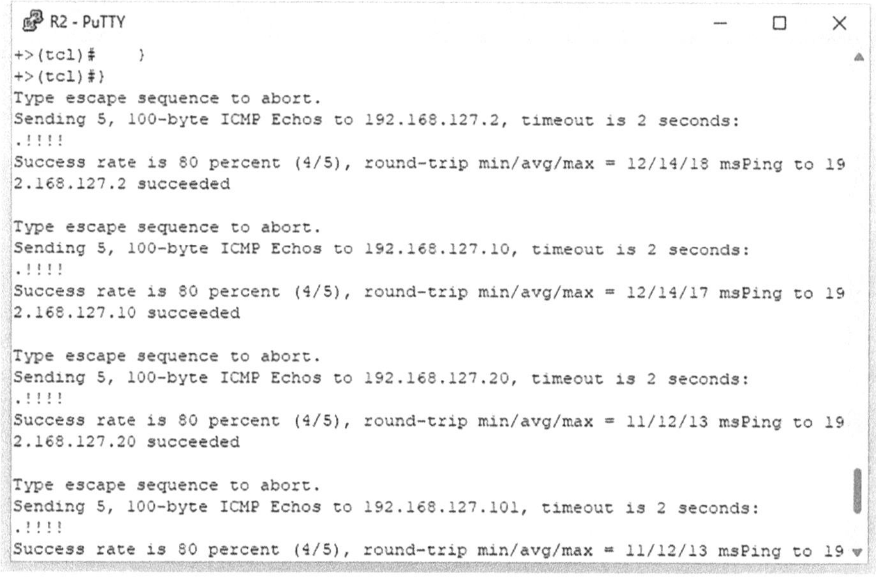

```
R2 - PuTTY                                                    —    □    ×
+>(tcl)#    }
+>(tcl)#}
Type escape sequence to abort.
Sending 5, 100-byte ICMP Echos to 192.168.127.2, timeout is 2 seconds:
.!!!!
Success rate is 80 percent (4/5), round-trip min/avg/max = 12/14/18 msPing to 19
2.168.127.2 succeeded

Type escape sequence to abort.
Sending 5, 100-byte ICMP Echos to 192.168.127.10, timeout is 2 seconds:
.!!!!
Success rate is 80 percent (4/5), round-trip min/avg/max = 12/14/17 msPing to 19
2.168.127.10 succeeded

Type escape sequence to abort.
Sending 5, 100-byte ICMP Echos to 192.168.127.20, timeout is 2 seconds:
.!!!!
Success rate is 80 percent (4/5), round-trip min/avg/max = 11/12/13 msPing to 19
2.168.127.20 succeeded

Type escape sequence to abort.
Sending 5, 100-byte ICMP Echos to 192.168.127.101, timeout is 2 seconds:
.!!!!
Success rate is 80 percent (4/5), round-trip min/avg/max = 11/12/13 msPing to 19
```

Figure 1-41. R2, ping test output for local network

You have now successfully installed and integrated both CML L2 and L3 images on GNS3. You have also added one CML L2 switch, SW1, and one CML L3 router, R2, to the pynetauto-lab project topology and completed the local connectivity tests. Ready to improve the topology and wrap up this chapter? Whenever you're set, let's dive in!

Building a CML Lab Topology

Before moving on to the hands-on lab tasks in the next chapter, it's crucial to further develop the pynetauto-lab project topology by adding two more network devices to complete the topology. This step serves as the foundation for your exploration of various networking concepts while learning about Python network application development. Creating a personalized Proof of Concept (POC) lab on a single PC like this offers numerous benefits. Getting the lab topology right at this stage ensures a robust environment for testing, experimentation, and learning. It's the canvas for scenarios ranging from basic configurations to real-world challenges, allowing you to refine your

CHAPTER 1   BUILDING A PYTHON NETWORK AUTOMATION LAB ENVIRONMENT

understanding of networking principles and sharpen your Python programming skills in practical network automation. **This careful topology development sets the stage for a dynamic testing ground, enabling you to test, learn, and innovate in the world of networking and Python automation.** So, let's continue building the foundation for a more enjoyable learning experience by adding more devices to the pynetauto-lab GNS3 project. This process assumes that you're continuing from the previous steps.

#	Task
①	Continuing from the previous section, add one more IOSvL2 switch and one more IOSv router to the topology, as shown in Figure 1-42. Use lowercase letters to rename the switch to sw2 and the router to r3. The device names intentionally use lowercase here, although it is a best practice to follow a strict naming convention for your network and systems devices. In production, this convention is not followed by all engineers and, yes, I am introducing an annoyance into your lab with a purpose. **You do not want a perfect Python Network Automation lab; you want a perfect production like a Python Network Automation lab.** Hence, to emulate the imperfections of a production environment, you purposely do not follow a strict device naming convention. Wear your traditional Cisco Network Engineer hat, and let's dive into configuring Cisco routers and switches right away. 1-1. Connect Gi0/0 of sw2 to Gi0/2 of SW1 and Gi0/1 of sw2 to Gi0/1 of r3. Also, connect Gi0/0 of r3 to Gi0/1 of R2, as depicted in Figure 1-42. 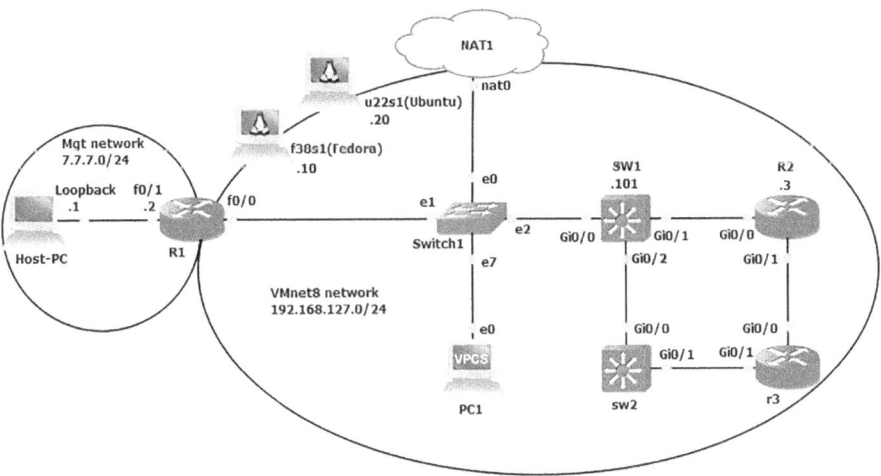 *Figure 1-42. Improving pynetauto-lab project topology, add sw2 and r3*

(continued)

CHAPTER 1 BUILDING A PYTHON NETWORK AUTOMATION LAB ENVIRONMENT

#	Task
	1-2. For the physical connections, refer to the Topology Summary window (see Figure 1-43).

Topology Summary	
Node	Console
◯ f38s1(Fedora)	none
▾ ◯ Host-PC	none
Loopback <=> f0/1 R1	
▾ ◯ NAT1	none
nat0 <=> e0 Switch1	
▾ ◯ PC1	telnet 192.168.186.128:5005
e0 <=> e7 Switch1	
▾ ◯ R1	telnet 192.168.186.128:5004
f0/0 <=> e1 Switch1	
f0/1 <=> Loopback Host-PC	
▾ ◯ R2	telnet 192.168.186.128:5003
Gi0/0 <=> Gi0/1 SW1	
Gi0/1 <=> Gi0/0 r3	
▸ ◯ r3	telnet 192.168.186.128:5008
▾ ◯ SW1	telnet 192.168.186.128:5001
Gi0/0 <=> e2 Switch1	
Gi0/1 <=> Gi0/0 R2	
Gi0/2 <=> Gi0/0 sw2	
▸ ◯ sw2	telnet 192.168.186.128:5010
▾ ◯ Switch1	none
e0 <=> nat0 NAT1	
e1 <=> f0/0 R1	
e2 <=> Gi0/0 SW1	
e7 <=> e0 PC1	
◯ u22s1(Ubuntu)	none

Figure 1-43. pynetauto-lab, Topology Summary

(continued)

#	Task

	1-3. For device-to-server allocation, refer to the Servers Summary window (see Figure 1-44).
	
	Servers Summary
	▼ ◯ GNS3 VM (GNS3 VM) CPU 88.2%, RAM 17.7%
	    ◯ PC1
	    ◯ R1
	    ◯ R2
	    ◯ r3
	    ◯ SW1
	    ◯ sw2
	    ◯ Switch1
	▼ ◯ LP50BC CPU 12.1%, RAM 51.4%
	    ◯ f38s1(Fedora)
	    ◯ Host-PC
	    ◯ NAT1
	    ◯ u22s1(Ubuntu)
	
	Figure 1-44. *pynetauto-lab, Servers Summary*
	
	Note that the running configuration files of all devices—R1, R2, r3, SW1, and sw2—are saved in the ch1 folder of this book. You can refer to the initial running configuration files for your reference.
	
	(https://github.com/pynetauto/apress_pynetauto_ed2.0/tree/main/source_codes/Part2/ch1)

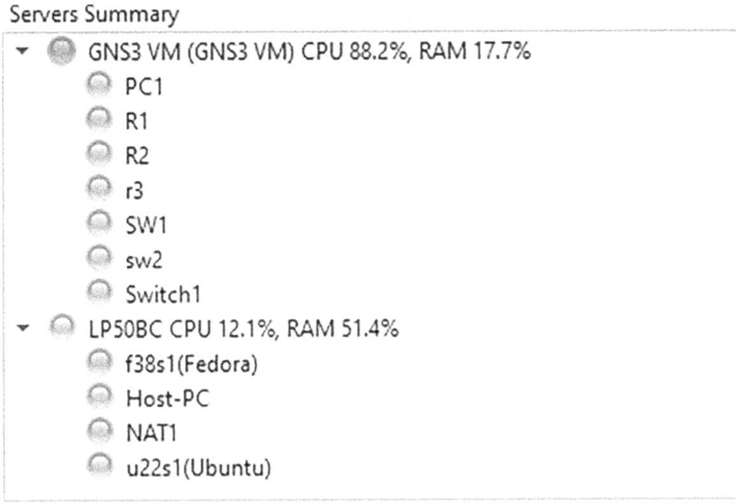

(continued)

CHAPTER 1   BUILDING A PYTHON NETWORK AUTOMATION LAB ENVIRONMENT

#	Task

②	In preparation for Chapter 2 and beyond, you must reconfigure the existing devices (R1, R2, and SW1) and add new configurations to the newly added devices (r3 and sw2). Let's go through each configuration and ensure that the routers and switches are correctly configured and fully working.
	
	 First, open the R1 console from GNS3 and assign a static IP address to its interface f0/0, so its IP address does not change every time the DHCP lease time elapses. It is currently assigned 192.168.127.134, but change this to a static IP address of **192.168.127.133**. Leave the f0/1 configuration as is.
	
	 R1, IOS router configuration changes:
	 R1#**conf t**
	 R1(config)#**int f0/0** # Enter interface configuration mode
	 R1(config-if)#**no ip add dhcp** # Disable DHCP IP allocation from DHCP server
	 R1(config-if)#**ip add 192.168.127.133 255.255.255.0** # Manually assign IP Address
	 R1(config-if)#**speed 100** # Fix speed if you have not done so yet above
	 R1(config-if)#**duplex full** # Fix duplex if you have not done so yet above
	 R1(config-if)# **line con 0** # the console line (console port 0)
	 R1(config-if)# **logging synchronous** # Change line automatically during log message
	 R1(config-if)# **line vty 1 15** # virtual terminal (telnet or SSH) lines
	 R1(config-if)# **logging synchronous** # improve the readability of log messages
	 R1(config-if)#**exit** # Move back to Global configuration mode
	 R1(config)#**ip domain-lookup** # Enable DNS lookup
	 R1(config)#**ip name-server 192.168.127.2** # Assign Default Gateway as the name-server
	 R1(config)#**ip route 0.0.0.0 0.0.0.0 192.168.127.2** # Add a default route
	 R1(config)#**router ospf 1** # Configure basic ospf

(continued)

#	Task
	R1(config-router)#**network 0.0.0.0 255.255.255.255 area 0** # add all network to ospf area 0 R1(config-router)#**end** # Move back to Exec Privilege mode R1#**wri** # Save current running configuration to startup configuration After saving the configuration, perform some basic ping tests to the default gateway (192.168.127.2) and any other devices on the network and Google DNS server (8.8.8.8).

③	Now open the SW1 console from GNS3, then add a more basic configuration to the switch. This switch is on the network and its vlan 1 has been assigned an IP of 192.168.127.101, so there's no need for another IP address. Because this switch will be used as an L2 device, there is no need to configure DNS, static routes, or IP routing.
	
	SW1, L2 CML switch configuration changes:
	SW1#**conf t** # Enter Global configuration mode
	SW1(config)#**enable password cisco123** # Configure enable password
	SW1(config)#**username jdoe pri 15 password cisco123** #configure privilege 15 user with pwd
	SW1(config)#**line vty 0 15** # virtual terminal (telnet or SSH) lines
	SW1(config-line)#**login local** # Allow locally configured account
	SW1(config-line)#**transport input all** # use both telnet and SSH connection
	SW1(config-line)#**logging synchronous** # Change line automatically during log message
	SW1(config-line)#**no exec-timeout** # No connection time-out, for Lab use only
	SW1(config-line)#**line console 0** # the console line (console port 0)
	SW1(config-line)#**logging synchronous** # improve the readability of log messages
	SW1(config-line)#**no exec-timeout** # No connection time-out, for Lab use only
	SW1(config-line)#**end** # Move back to Exec privilege mode
	SW1#**wri** # Save running-config to startup-config
	
	It is recommended to conduct essential ping tests as needed.

(continued)

CHAPTER 1 BUILDING A PYTHON NETWORK AUTOMATION LAB ENVIRONMENT

#	Task
④	Let's apply a similar configuration to the newly added switch, sw2. This switch will also be configured as a layer 2 switch, so there's no need to enable IP routing or configure DNS. No explanation accompanies this configuration, as it is almost identical to sw1's configuration.

sw2 L2 CML switch configuration (new):
```
Switch>en
Switch#conf t
Switch(config)#hostname sw2
sw2(config)#interface vlan 1
sw2(config-if)#ip add 192.168.127.102 255.255.255.0
sw2(config-if)#no shut
sw2(config-if)#enable password cisco123
sw2(config)#username jdoe pri 15 pass cisco123
sw2(config)#line vty 0 15
sw2(config-line)#login local
sw2(config-line)#transport input all
sw2(config-line)#logging synchronous
sw2(config-line)#no exec-timeout
sw2(config-line)#line console 0
sw2(config-line)#logging synchronous
sw2(config-line)#no exec-timeout
sw2(config-line)#end
sw2#wri
```

A best practice involves performing necessary ping tests.

(continued)

#	Task
⑤	Next, R2's console from GNS3. Assign a DNS server and then add a static route to allow smooth communication. R2's Gi0/0 has been manually configured with an IP address of 192.168.127.3/24, so it does not rely on the DHCP services. Configure the rest of R2's essential configuration as shown here: **R2, L3 CML router configuration changes:** R2#**conf t** # Enter Global configuration mode R2(config)#**interface g0/1** # Enter interface configuration mode R2(config-if)#**ip add 172.168.1.1 255.255.255.0** # Assign IP address to Gi0/1 interfacing r3 R2(config-if)#**no shut** # Bring up the interface R2(config-if)#**exit** # Exit interface configuration mode R2(config)#**ip domain-lookup** # Enable DNS lookup R2(config)#**ip name-server 192.168.127.2** # Assign Default Gateway as DNS server R2(config)#**ip route 0.0.0.0 0.0.0.0 192.168.127.2** # Add a static Gateway of last resort R2(config)#**enable password cisco123** # Configure enable password R2(config)#**username jdoe privilege 15 password cisco123** # configure level 15 user with pwd R2(config)#**line vty 0 15** # Enter Virtual Terminal Line configuration mode R2(config-line)#**login local** # Allow locally configured account R2(config-line)#**transport input all** # use both telnet and SSH connection R2(config-line)#**logging synchronous** # Change line automatically during log message R2(config-line)#**no exec-timeout** # No connection time-out, for Lab use only R2(config-line)#**line console 0** # Enter Line Console 0 config mode R2(config-line)#**logging synchronous** # Change line automatically during console message

(continued)

CHAPTER 1   BUILDING A PYTHON NETWORK AUTOMATION LAB ENVIRONMENT

#	Task
	R2(config-line)#**no exec-timeout** # No connection time-out, for Lab use only R2(config-line)#**end** # Move back to Exec privilege mode R2#**wri** # Save running-config to startup-config It's advisable to carry out fundamental ping tests as necessary.
⑥	Now, refer to the following configuration and configure the last device in this topology, r3. **r3, L3 CML router configuration (new):** Router#**conf t** Router(config)#**hostname r3** r3(config-if)#**interface GigabitEthernet 0/0** r3(config-if)#**ip address 172.168.1.2 255.255.255.0** r3(config)#**interface GigabitEthernet 0/1** r3(config-if)#**ip add 192.168.127.4 255.255.255.0** r3(config-if)#**no shut** r3(config-if)#**exit** r3(config)#**ip domain-lookup** r3(config)#**ip name-server 192.168.127.2** r3(config)#**ip route 0.0.0.0 0.0.0.0 192.168.127.2** r3(config)#**enable password cisco123** r3(config)#**username jdoe pri 15 password cisco123** r3(config)#**line vty 0 15** r3(config-line)#**login local** r3(config-line)#**transport input all** r3(config-line)#**logging synchronous** r3(config-line)#**no exec-timeout** r3(config-line)#**line console 0** r3(config-line)#**logging synchronous** r3(config-line)#**no exec-timeout** r3(config-line)#**end** r3#**wri** Engaging in required basic ping tests is the recommended best practice.

(continued)

CHAPTER 1　BUILDING A PYTHON NETWORK AUTOMATION LAB ENVIRONMENT

#	Task
⑦	The lab topology has changed, transitioning one of the background shapes from an oval to a rectangle, thus offering a refreshed appearance. Moreover, IP addresses have been added to increase the usability during your lab sessions (see Figure 1-45). While I'll provide specific lab topologies with the correct IP addresses, consider this topology as a fundamental reference throughout the book, as it will remain largely unchanged for most of your Python network application development. Also, because this topology in printing format will be tiny and less useful, I have placed the Figure 1-45.jpg file in the ch1/pynetauto-lab device initial config folder for your reference. Refer to Figures 1-43 and 1-44 for guidance on device connections, and feel free to further annotate your topology to assist with IP addressing in upcoming labs. 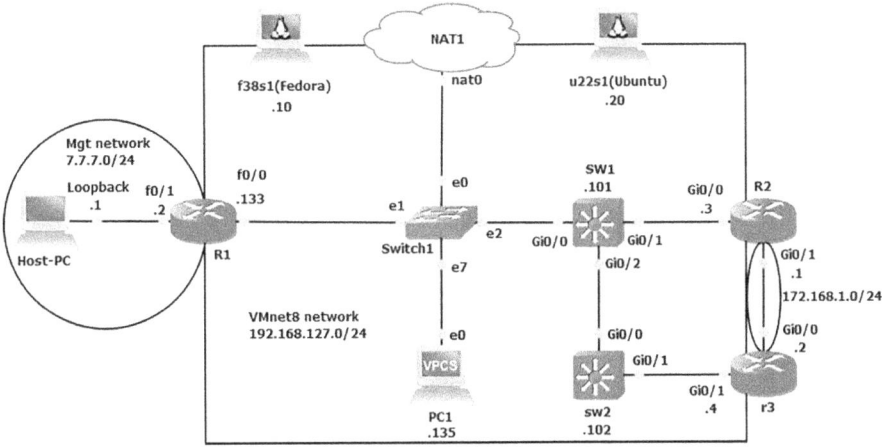 ***Figure 1-45.*** *Configuring the CML lab topology, final topology*

(continued)

#	Task
⑧	The IP addresses used to fully prepare the `pynetauto-lab` project topology are listed here. Your objective is to become acquainted with these IP addresses, their associated devices, and their interfaces. Instead of tabulating the information, I've presented it in plain text, enabling seamless copying and pasting into Python scripts for swift testing. This approach allows you to swiftly retrieve these addresses whenever required for your experiments. **192.168.127.3** #R2 g0/0 **192.168.127.4** #r3 g0/1 **192.168.127.10** #f38s1 **192.168.127.20** #u22s1 **192.168.127.101** #SW1 vlan1 **192.168.127.102** #sw vlan1 **192.168.127.133** #R1 f0/0 **192.168.127.135** #PC1 **172.168.1.1** #R2 g0/1 **172.168.1.2** #r3 g0/0 **7.7.7.1** #Host PC MS lo0 **7.7.7.2** #R1 f0/1

While this chapter didn't involve any Python scripting, it could have been indeed a challenging lab for some, especially for readers with little Cisco networking background. If you've made it through the integration of the CML L2 and L3 images and have invested several hours in prepping this lab, you're now well-prepared to embark on an exciting journey of developing and writing Python code to automate your networking tasks. If you've managed to follow along and have reached this point successfully, congratulations! You can look forward to the remainder of the book, which is filled with Python coding adventures.

Summary

Chapter 1 marked a crucial phase in the exploration of Cisco CML image integration within GNS3. The chapter focused on the installation and validation of both CML L2 and L3 images, aiming to unlock the potential of these images. The primary goal was to enhance the GNS3 `pynetauto-lab` topology, allowing for robust network functionality and direct communication with Python scripting servers and the Internet. The chapter entailed a vital integration process that included downloading, integrating, and configuring these images for subsequent development labs. You will perhaps encounter and address challenges inherent in the CML image integration process, so the chapter offered effective solutions to build a strong foundation for upcoming network automation labs. The focus was on setting up the `pynetauto-lab` topology, configuring all network devices with basic settings, and testing their connectivity using interactive Tcl shell commands. While the chapter primarily remained scripting-free, basic interactive `tclsh` commands were used for multiple IP ping tests. This approach directed all efforts toward refining the core GNS3 project, `pynetauto-lab`, which will serve as the springboard to guide you into the realm of Python Network Automation scripting.

Storytime 1: Still Enjoying the Most Expensive Sandwich in the World

Figure 1-46. *Certified $1600 Sandwich*

Ah, the quest for a Cisco Badge number! Flashback to 2008 when Cisco Certifications had more market value and credibility. It was a journey filled with expensive hardware and, well, some of the world's priciest sandwiches—twice, no less! You see, the Cisco CCIE Lab exam day is no casual walk in the park; it's more of a full-day extravaganza. When you step up to take the Lab exam, be prepared to fork over a cool US$1600 (as of October 2023, previously US$1500) for each attempt. Cisco's exam proctor does try to make up for it by treating you to a gourmet chicken or vegetarian sandwich for lunch. But if, by some bizarre twist of fate, you happen to not pass the exam, that sandwich becomes one of the most expensive meals you've ever had.

CHAPTER 1   BUILDING A PYTHON NETWORK AUTOMATION LAB ENVIRONMENT

Now, what about the over $10,000 I once invested in Cisco hardware to build my very own CCIE dream lab? Let's just say those gadgets are now happily resting in the dusty corner of my garage. The longer they sit there, unpowered, the more I save on my electricity bill. After all the passion, time, and money poured into it, I began to resent those sky-high electricity bills just to keep the lab running. That's when I put on my thinking cap and got creative, finding ways to build cost-effective labs that could rival your wildest dreams, all while keeping those bills in check. Over the years, I've dabbled with older equipment, experimented with virtualization technologies, played around with network device simulators and emulators, and more recently, delved into the enchanting realms of public and private cloud technologies. Through this journey, I've realized there's simply no more cost-effective way to set up a study environment than the methods I've shared in this book.

After weathering my fair share of emerging technologies in large organizations, I've now found myself working for a small but capable MSP. Here, we host our own private/hybrid cloud and venture into the vast landscapes of Azure and AWS Cloud. And let me tell you, the cloud isn't exactly a budget-friendly playground. It's not for the faint of heart, especially for students and budding network engineers trying to make ends meet. Yes, AWS offers free-tier accounts for the first year, and Azure throws monthly credits at partners, but when it comes to running a decently beefy lab, those free tiers and credits can seem about as expensive as those sandwiches!

So, in my quest for the most cost-effective labs, I find myself still here, sticking to my trusty laptop, and I must say, I rather enjoy my lab adventures on a single laptop. Sometimes, you just can't beat the classics, can you? As for my pursuit of the number, well, that ended over ten years ago. You can say that my "Cisco Fan Boy" days are well over the expiry date, but I've found an extended passion, "network automation." The Cisco CCIE labbing legacy lives on, as I've written and published five books (this Part II is my fifth book) based on my learning experiences and recouped all the investments in my training and old Cisco equipment. The sandwiches, though? I've had my fill of those! (Jan 2024)

CHAPTER 2

Python Network Automation Labs - Basic Telnet

This chapter introduces a series of Telnet labs that utilize Python's `telnetlib` library, providing an exploration of the network engineer's thought process behind selected networking tasks and translating them into actionable lines of Python code, all from a single point of control. Although Telnet, as a protocol, has fallen out of favor due to its insecure features, this provides you with an opportunity to use this outdated protocol to master converting network engineer's tasks into actionable lines of the Python programming language. Even though Telnet is not as enticing as SSH and APIs, this chapter contains some exciting network automation topics, including establishing network topologies, discussing network engineer's tools to interface with network devices, configuring VLANs using loops, adding role-based user accounts to restrict access, performing backups of multiple devices' running configurations to a local server directory, and carrying out post-change verifications all through a single pane of glass—Python server.

The main goal of this chapter is to steer readers away from direct console or Telnet (or SSH) access to target devices, encouraging access to managed devices via Python applications. This imparts valuable insights into transitioning manual tasks into Python scripting. The labs in this chapter rely on a single Telnet interface, exploring various networking ideas and the conversion of physical tasks into scripts, thereby progressing from manual administrative interactions by engineers to programmed, actionable Python code. This foundational knowledge lays the groundwork for future chapters, which will expand into the Python SSH library, bridging more networking concepts with Python in subsequent chapters.

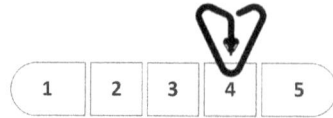

Python Network Automation: Main Logical Topology

Because this marks the first lab after the completion of the CML images integration, I am providing you with the comprehensive logical topology encompassing all the pertinent information you've been working with. This will expedite your navigation, minimizing the time spent needed to search for connectivity-related details. Presented here is the main logical topology of the `pynetauto-lab` project. While the network has been flattened (simplified) to streamline and focus on the Python labs, the topology itself remains quite sophisticated. Nonetheless, not all resources will be utilized simultaneously; the lab approach is modular, focusing on specific devices at a time, allowing for a focused study of relevant topics, concepts, and skillsets.

For each lab scenario, you will be provided with specific topologies based on this main topology. Therefore, there's no need to power on all the virtual machines at once. Expect to invest most of your time scripting and testing on the Linux Python servers using Python 3, which forms the core content of this book. The primary objective is to steer you away from the Cisco command-line interface and instead, help your journey into Python coding for managing deployed routers and switches.

If you've closely followed this book and the Part I book, you should already be familiar with the IP addressing and the corresponding device interface assignments. If you've followed the author's recommendation for a hands-on approach with this book, you must be astounded by what you have built and accomplished up to this point! However, if you still feel less confident about the topology information, I encourage you to spend some time getting familiar with the details presented in this topology (refer to Figure 2-1 and Table 2-1).

CHAPTER 2 PYTHON NETWORK AUTOMATION LABS - BASIC TELNET

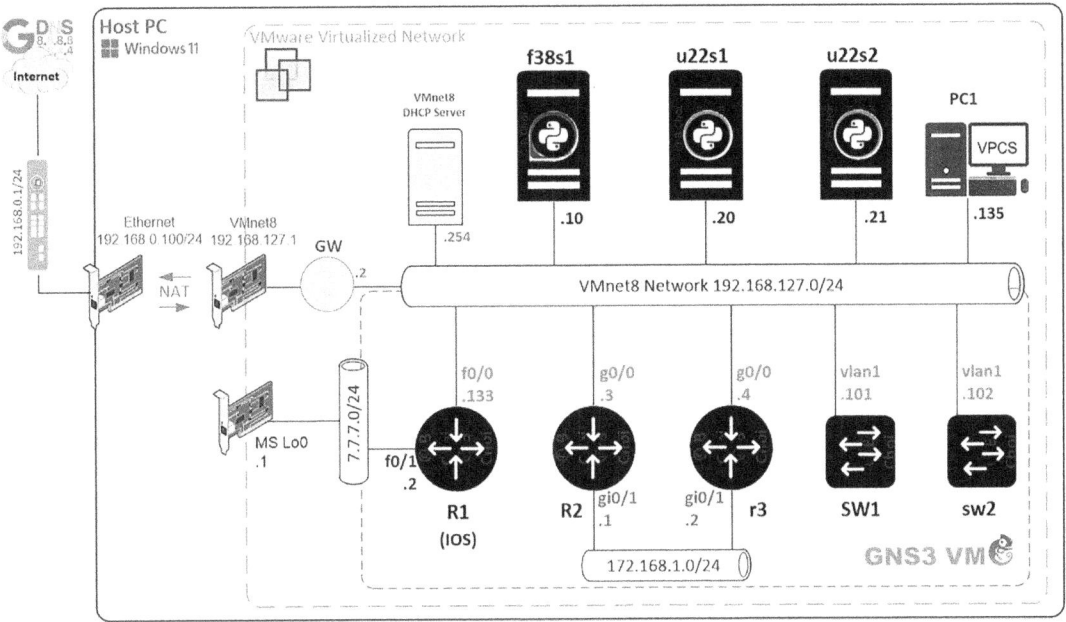

Figure 2-1. pynetauto-lab, main logical topology

Table 2-1. pynetauto-lab, IP Addresses in Use

Hostname	Role	Interface	IP Address (all /24 Subnet)
R1	Router	F0/0	192.168.127.133
		F0/1	7.7.7.2
R2	Router	G0/0	192.168.127.3
		G0/1	172.168.1.1
r3	Router	G0/0	192.168.127.4
		G0/1	172.168.1.1
SW1	Switch	VLAN1	192.168.127.101
sw2	Switch	VLAN2	192.168.127.102
f38s1	Server	ens160	192.168.127.10
u22s1	Server	ens33	192.168.127.20

(*continued*)

Table 2-1. (*continued*)

Hostname	Role	Interface	IP Address (all /24 Subnet)
u22s2	Server	ens33	192.168.127.21
PC1	VPCS	PC1	192.168.127.135
GW	Gateway	VMnet8	192.168.127.2
DHCP	DHCP server	n/a	192.168.127.254
Host PC - Lo0	Management	Ms Loopback0	7.7.7.1
Host PC – VMnet8	VMnet8	VMnet8	192.168.127.1
Host PC – Ethernet	Host PC physical connection	Ethernet	192.168.0.1 (Update to your network)

Python Network Automation: Network Engineer's interface

This book has been carefully crafted to replicate a physical equipment lab, offering a realistic experience without the need for actual hardware, concerns about various connector types, or concerns about expensive electricity utility bills. The current lab setup on your Windows PC supports coding on three distinct operating systems: Windows 11 for the host PC, Fedora 38, and Ubuntu 22.04 LTS Linux servers. Additionally, you have the option to import any operating system supported by GNS3 appliances as demonstrated with the `NetworkAutomation-1` server, which ran the Ubuntu 20 Server. With the current settings, you have the flexibility to add various types of virtual machines on the VMware Workstation and appliances on GNS3. In Chapter 2 of the Part 1 book, Python 3 was introduced on the Windows OS. Subsequent chapters (Chapter 4, 5, and 6 of Part I) primarily focused on exercises conducted through the Linux command-line interface (CLI). Typically, companies rarely install Python on a Windows server for use in a production environment; instead, the vast majority of Python applications run on secure and robust Linux servers. This is despite Microsoft's introduction of the Windows Subsystem for Linux (WSL), which is a highly practical tool for running Linux systems on Windows for learning, but perhaps not for the enterprise-level production environment.

The modern-day surge in network automation aims to streamline network engineering tasks by enabling a select group of engineers to handle more workload with less headcount, using Application Programming Interfaces (APIs), and reducing reliance on opening CLI console, Telnet, or SSH terminal windows. While Graphical User Interfaces (GUI) have been dominant in the last decade, there's a notable transition from GUI to API. In the coming decade, it's plausible that APIs might supersede GUIs as the favored interface among network engineers. The ongoing shift to API usage suggests an eventual transition where all network engineers will need to adapt to using APIs, moving away from GUIs.

Engineers currently find themselves in a transitional phase. One significant challenge for engineers is that not all enterprise-level vendor equipment offers robust working APIs. At times, APIs might not even be provided or enabled to ensure device security. However, the landscape is rapidly changing, and it's anticipated that soon, all enterprise networking devices will need to support comprehensive API features due to market demand. However, industry insiders anticipate that the traditional methods of accessing and managing network devices will coexist for many years to come. Despite your primary use of keyboard input for writing Python code and accessing API libraries, unless you are specifically an API developer, you may not fully grasp the inner workings of each API. **Network automation essentially mirrors the problem-solving approach of seasoned network engineers**. For example, tools like Red Hat's Ansible depend on agentless SSH connections rather than APIs. In this context, the API functions to reinterpret this model for machine-to-machine interaction, operating without the reliance on a keyboard and screen. Additionally, even if the APIs are used, the code still needs to be validated by developers or smart network engineers like you, and, mark this word, all applications must run on some type of computing system. It could be Windows, Linux, or serverless applications such as Amazon Lambda, Azure Functions, Google Cloud Functions, and everything in between—they all have to run on a computer. Furthermore, even when APIs are utilized, the code still requires validation by developers or astute network engineers like yourself. It's crucial to highlight that all applications must operate on a computing system, whether it's Windows, Linux, or an application running on a serverless platforms such as Amazon Lambda, Azure Functions, Google Cloud Functions, or a myriad of other platforms in between. Ultimately, these applications need a computing environment to function.

As interfaces and methodologies in computing continue to evolve, the core of running code on a computer remains an essential aspect of computing operations. While the potential replacement of engineers and developers by code-writing AI and systems like ChatGPT is a possibility, this transition is far from flawless. Entrusting an enterprise network entirely to an AI bot is still a long way from being a foolproof solution, especially considering the blame game often played out during Post Incident Reporting (PIR) meetings following major outages.

Should AI generate, test, and implement code that affects network changes, accountability for disruptions becomes complex. Legal considerations indicate a continued need for human network engineers to ensure system stability and assume responsibility for mishaps. In the ICT industry, causing significant outages resulting in the loss of application services can lead to severe consequences. Imagine making a change that triggers a major network outage in a crucial setting like a major airport, credit card company, or bank. Severe repercussions have been faced by some in such scenarios. Personal responsibility is crucial, and those held accountable might face significant consequences for these incidents. I even know a couple of people who had to walk the plank.

However, holding AI accountable in the same way as a human is not practical yet. The ethical and accountability considerations for AI in complex enterprise network outage scenarios, especially those concerning critical infrastructure, remain subjects of ongoing discussion and development. Therefore, the jobs of network engineers are likely to continue, as it's improbable that all roles will be replaced by AI any time soon. Job specifics in network engineering may change, titles might evolve as in the case of Site Reliability Engineer (SRE), Network Automation Engineer, and Network DevOps Engineer, and the types of engineers might diversify, but there will be a need for more hybrid engineers proficient in both traditional and modern methods of interfacing with networking technologies.

This chapter is dedicated to writing Python Telnet scripts for communication to network devices in your lab. Despite the increased utilization of APIs in the management of enterprise networking devices, many old and newer devices still lack API support. This absence requires remote connections via Telnet or SSH. While GUI represents an alternative, many networking vendors still have substantial room for improvement in integrating all platform features into visually represented and clickable menus. The prevalence of API tools has led to a gradual decline in the utilization of GUI over the past few years. Consequently, many competent network engineers persist in working with the command-line interface (CLI). Some individuals are devoted advocates of the CLI and passionately defend its use. However, I propose a shift from exclusive reliance on the command line, advocating for the gradual development of scripts to interface with routers and switches. Python continues to stand out as a pivotal tool in network automation, relying on Telnet and SSH libraries. The foundational skills in Python code endure, while the modes of access and libraries continue to evolve.

On a highly secured client environment, management of core network and system devices is confined to designated servers known as *jump hosts*. These jump hosts can be based on either Windows or Linux operating systems. The GNS3 topology within the pynetauto-lab emulates this specific setup. To manage devices in the network, you'll need to use PuTTY on your host PC to SSH into the u22s1 server, which functions as both the Python server and jump host.

The f38s1 server in the network provides essential IP services that will be relied upon for various lab scenarios. Throughout this chapter, Python scripts are written using Windows Notepad++ (or Microsoft VS Code), or you can enter them directly into the text editor on the u22s1 server via an SSH session. For those proficient in typing, Python scripts can also be entered directly into the Ubuntu Server. However, if preferred, writing the scripts in Windows Notepad, Notepad++, or any other IDE text editor and then later pasting them into the u22s1 server is also an acceptable method.

Tip

Accessing the source code

While I encourage you to type each line of code for a genuine coding experience, if you need to save time and expedite your learning, you can use the cut-and-paste method to run the code in your environment. All the Python code (scripts) used in this book are accessible through the author's or publisher's GitHub repositories.

Author's GitHub: https://github.com/pynetauto/apress_pynetauto_ed2.0/tree/main/source_codes/Part2

Apress GitHub: https://github.com/Apress

The next section powers up all devices depicted in the lab-specific topology, shown in Figure 2-2, and delves into writing Python code. It explores how Telnet Python scripts interact with Cisco routers and switches.

CHAPTER 2   PYTHON NETWORK AUTOMATION LABS - BASIC TELNET

Telnet Lab 1: Interactive Telnet Session to Cisco Devices on a Python Interpreter

For this lab, you need to power on the u22s1 Linux server (192.168.127.20), SW1 (192.168.127.101), and R2 (192.168.127.3). Also, power on the IOS router, R1 (192.168.127.133), and later check the OSPF neighborship and perform ICMP testing to R2's two new loopback interfaces. As mentioned, you only need to power on the required devices for each lab. Figure 2-2 highlights the devices powered on and in use in Telnet Lab 1.

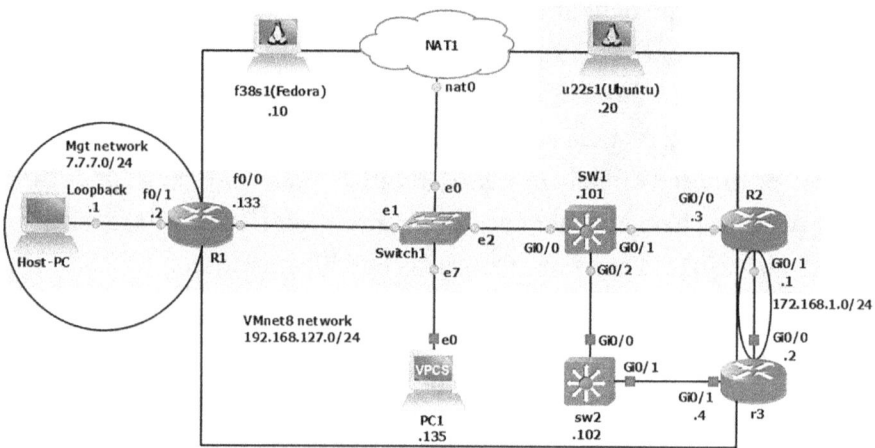

Figure 2-2. *Telnet Lab 1, powered-on devices*

Figure 2-2 is only for a first-time reference and in subsequent labs, you will reference the Devices in Use topologies (see Figure 2-3). If you are using another subnet to connect to your NAT (VMnetwork8), your IP addresses will be different from this book, and you will have to replace the IP addresses accordingly.

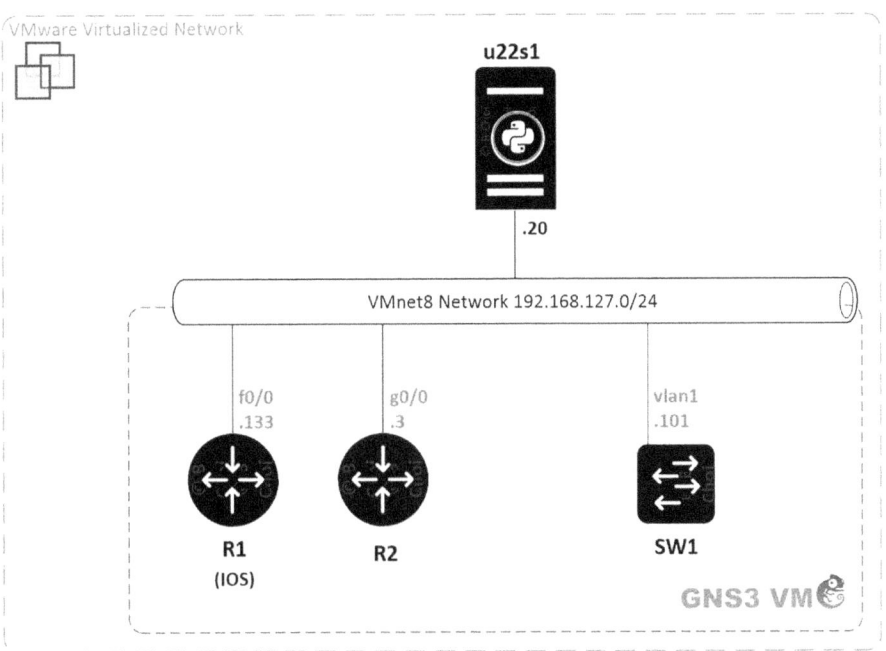

Figure 2-3. *Telnet Lab 1, devices in use*

You will be creating a Telnet script to execute from the Python server, u22s1, connecting to the R2 router to implement the following configuration changes:

Telnet Lab 1 objectives:

1. Configure Loopback 0 with an IP address of 2.2.2.2/32.

2. Set up Loopback 1 with an IP address of 4.4.4.4/32.

3. Implement OSPF 1 with the network 0.0.0.0 and a subnet mask of 255.255.255.255 in area 0.

By the end of this lab, R2 must contain these configurations in the running-config. Additionally, you must copy and modify your script to conduct configuration validation tests, such as the ping, show ip interface brief, and show ip ospf neighbor commands, without directly logging into R1.

CHAPTER 2   PYTHON NETWORK AUTOMATION LABS - BASIC TELNET

#	Task
①	Open PuTTY on your host PC and SSH into the u22s1 (192.168.127.20) server. If you have been assigned a different IP address, use your server IP address. If you forgot which IP address was assigned to the u22s1 server, open the Linux console from the VMware Workstation and log in; then run the `ip address show` command. Now log in with your username and password. `login as:` **jdoe** `jdoe@192.168.127.20's password:` **********

②	Currently, to run Python from u22s1, you have to issue the `python3` command. Because there is no Python 2.7 installed and Python 3 is the only version you are going to be using here, you can use an alias command to shorten `python3` to `python`.
	
	 2-1. Follow the instructions here to use the alias in Linux:
	
	 ```jdoe@u22s1:~$ python --version```
	 `Command 'python' not found, did you mean:`
	 `  command 'python3' from deb python3`
	 `  command 'python' from deb python-is-python3`
	 `jdoe@u22s1:~$` **python3 --version**
	 `Python 3.10.12`
	 `jdoe@u22s1:~$` **users**
	 `jdoe`
	 `jdoe@u22s1:~$` **pwd**
	 `/home/jdoe`
	 `jdoe@u22s1:~$` **nano /home/jdoe/.bashrc**
	
	 Once you open the `.bashrc` file in the `/home/user/` directory, add a new line, `alias python=python3`, to the end of the `.bashrc` file. Once you finish adding a line, press Ctrl+X to save and exit the file.
	
	 # alias ADDED by jdoe
	 alias python=python3

(continued)

#	Task

	2-2. After adding the alias line, run the source ~/.bashrc command to restart the user's bashrc. You can now use the python or python3 command to run Python 3 on your u22s1 server.
	
	``
	jdoe@u22s1:~$ source ~/.bashrc
	jdoe@u22s1:~$ python -V # no need to append '3' after 'python'
	Python 3.10.12
	jdoe@u22s1:~$ python # This should work now
	Python 3.10.12 (main, Jun 11 2023, 05:26:28) [GCC 11.4.0] on linux
	Type "help", "copyright", "credits" or "license" for more information.
	>>> # Use 'Ctrl+Z' to exit or type in 'exit ()' and Enter
	``

③	As you did in the IOS Telnet lab, go to https://docs.python.org/3/library/telnetlib.html and scroll down to the bottom of the page. Locate the sample Telnet code. You can alternatively reuse the old code from the R1 IOS lab from Chapter 10 of the Part I book. Copy this code to make your first Telnet script. This code will be used throughout the Telnet labs, so short explanations of each line appear in the given Python script. Because you are using Cisco devices, more explanation is provided related to Cisco Telnet sessions.
	
	``python
	# Telnet Example: A simple example illustrates the typical use
	import getpass # Import the getpass module for password input
	import telnetlib # Import the telnetlib module for telnet operations
	
	HOST = "localhost" # IP address or hostname of the device to connect to
	user = input("Enter your remote account: ") # Prompt for user input to enter the UserID
	password = getpass.getpass() # Use the getpass.getpass() function to securely input the password
	
	tn = telnetlib.Telnet(HOST) # Establish a Telnet connection
	tn.read_until(b"login: ") # Wait until the login prompt is received
	``

(continued)

#	Task
	`tn.write(user.encode('ascii') + b"\n")` # Enter the user ID and encode it in ASCII format to send it to the telnet virtual terminal `if password:` `tn.read_until(b"Password: ")` # If a password is requested, wait for the password prompt `tn.write(password.encode('ascii') + b"\n")` # Encode the entered password in ASCII and send it to the telnet virtual terminal # Perform operations (e.g., sending commands) within the telnet session `tn.write(b"ls\n")` # 'ls' is a Linux command; it can be removed or modified as it's not used in Cisco devices `tn.write(b"exit\n")` # Send the 'exit' command to terminate the telnet virtual terminal `print(tn.read_all().decode('ascii'))` # Read and display the output of the commands run in this session

④	Make a new directory called ch2_telnet, change the working directory, and then create the first CML Telnet script using the touch command. My file is called add_lo_ospf1.py.
	
	`jdoe@u22s1:~$ `**`mkdir ch2_telnet && cd ch2_telnet`**
	`jdoe@u22s1:~/ch2_telnet$ `**`pwd`**
	`/home/jdoe/ch2_telnet`
	`jdoe@u22s1:~/ch2_telnet$ `**`touch add_lo_ospf1.py`**
	`jdoe@u22s1:~/ch2_telnet$ `**`ls`**
	`add_lo_ospf1.py`

(continued)

#	Task

⑤	To speed up the lab, use the nano text editor on the Linux server. First, use nano add_lo_ospf1.py to open the blank Python script. Then, copy and paste the Telnet example by using the right-click menu. Finally, modify the sections highlighted in the following source code:
	
	``python
	import getpass
	import telnetlib
	
	HOST = "192.168.127.3"
	user = input("Enter your username: ")
	password = getpass.getpass()
	
	tn = telnetlib.Telnet(HOST)
	
	tn.read_until(b"Username: ")
	tn.write(user.encode('ascii') + b"\n")
	if password:
	    tn.read_until(b"Password: ")
	    tn.write(password.encode('ascii') + b"\n")
	
	tn.write(b"enable\n")
	tn.write(b"cisco123\n")
	tn.write(b"conf t\n")
	tn.write(b"int loopback 0\n")
	tn.write(b"ip add 2.2.2.2 255.255.255.255\n")
	tn.write(b"int loopback 1\n")
	tn.write(b"ip add 4.4.4.4 255.255.255.255\n")
	tn.write(b"router ospf 1\n")
	tn.write(b"network 0.0.0.0 255.255.255.255 area 0\n")
	
	tn.write(b"end\n")
	tn.write(b"exit\n")
	
	print(tn.read_all().decode('ascii'))
	``

(continued)

CHAPTER 2   PYTHON NETWORK AUTOMATION LABS - BASIC TELNET

#	Task
⑥	Before running the first script from the u22s1 server, open the R2 console on GNS3 and run the debug `telnet` command to monitor the Telnet activities on the R2 console. If you are connecting via SSH, you will also have to run the `terminal monitor` command to display the debug outputs. R2#**debug telnet** Incoming Telnet debugging is on
⑦	On u22s1, run your script using the python add_lo_ospf1.py command. When prompted for the username and password, enter your router username and password. Once the script runs successfully, your screen should look like Figure 2-4. jdoe@u22s1:~/ch2_telnet$ **python add_lo_ospf1.py** Enter your username: **jdoe** Password: ******** 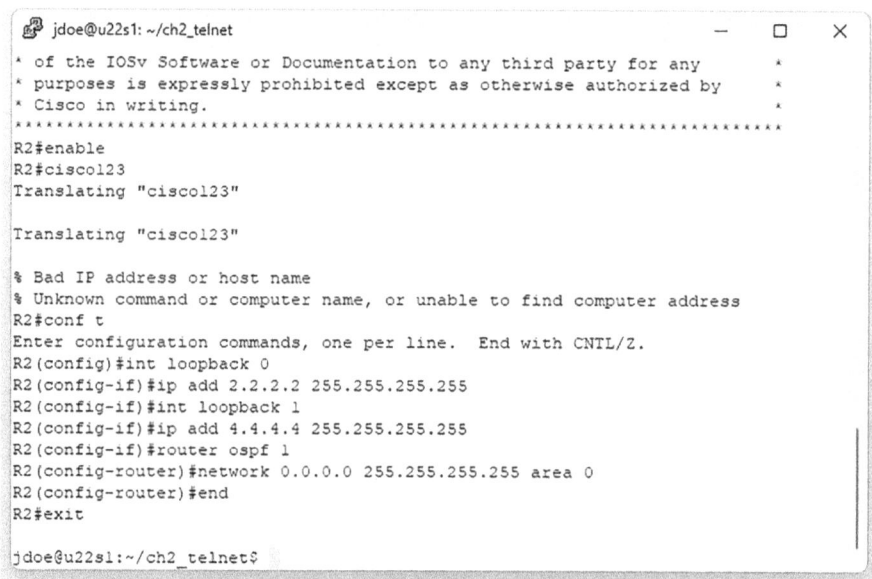 ***Figure 2-4.*** *Telnet Lab 1 - u22s1, execute add_lo_ospf1.py script*

(*continued*)

#	Task
⑧	Go to R2 and check the console. You can see the Telnet activities that have just taken place. Your debugging message should look like Figure 2-5.

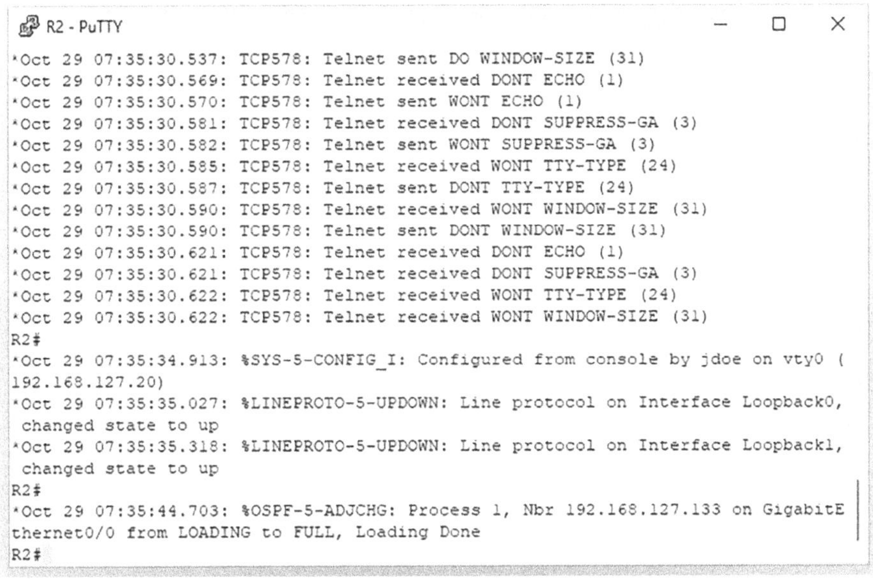

Figure 2-5. Telnet Lab 1: R2, telnet debugging message

⑨	On the u22s1 server, copy and modify the first script to run the show ip interface brief command and confirm the changes from u22s1. I aim to help you slowly move away from direct logins. When you run the script, you should be able to confirm the interface status from your Linux Python server (see Figure 2-6). jdoe@u22s1:~/ch2_telnet$ **cp add_lo_ospf1.py sh_ip_int_bri1.py** jdoe@u22s1:~/ch2_telnet$ **nano sh_ip_int_bri1.py** jdoe@u22s1:~/ch2_telnet$ **cat sh_ip_int_bri1.py** import getpass import telnetlib HOST = "192.168.127.3" user = input("Enter your username: ") password = getpass.getpass() tn = telnetlib.Telnet(HOST)

(continued)

#	Task
	``
tn.read_until(b"Username: ")
tn.write(user.encode('ascii') + b"\n")
if password:
  tn.read_until(b"Password: ")
  tn.write(password.encode('ascii') + b"\n")
```
**`tn.write(b"show ip interface brief\n")`**
```
tn.write(b"exit\n")

print(tn.read_all().decode('ascii'))
```
jdoe@u22s1:~/ch2_telnet$ **python sh_ip_int_bri1.py**
Enter your username: **jdoe**
Password: **\*\*\*\*\*\*\*\***

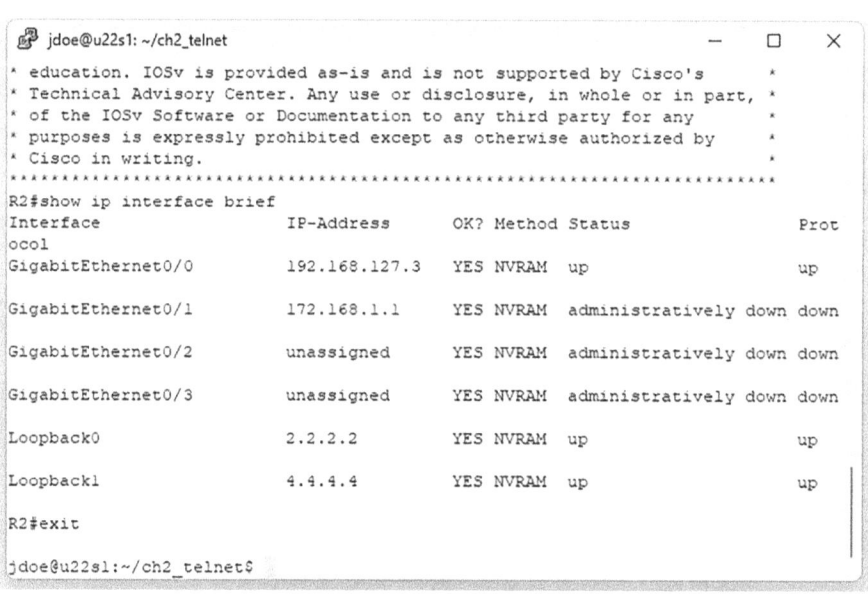

*Figure 2-6. Telnet Lab 1: u22s1, R2's show ip interface brief output* |

*(continued)*
```

#	Task
| ⑩ | As you saw in Step 8, the Figure 2-5 output displayed the *Oct 29 07:35:44.703: %OSPF-5-ADJCHG: Process 1, Nbr 192.168.127.133 on GigabitEthernet0/0 from LOADING to FULL, Loading Done message. This message contained a familiar IP address, 192.168.127.133, which is R1 IOS router's f0/0 interface IP address. This means the OSPF neighborship has formed between R1 and R2 and their routing information is shared.<br><br>10-1. Let's quickly run the show ip ospf neighbor command and make R1 test the reachability to the two loopback interfaces just created on R2—2.2.2.2 and 4.4.4.4. This should prove my theory without opening the console or Telnet/SSH into R1. In other words, write another Python Telnet script to run the commands from your Python server, u22s1.<br><br>```
jdoe@u22s1:~/ch2_telnet$ ls
add_lo_ospf1.py   sh_ip_int_bri1.py
jdoe@u22s1:~/ch2_telnet$ cp sh_ip_int_bri1.py sh_ip_ospf_ping1.py
jdoe@u22s1:~/ch2_telnet$ nano sh_ip_ospf_ping1.py
jdoe@u22s1:~/ch2_telnet$ cat sh_ip_ospf_ping1.py
import getpass
import telnetlib

HOST = "192.168.127.133"
user = input("Enter your username: ")
password = getpass.getpass()

tn = telnetlib.Telnet(HOST)

tn.read_until(b"Username: ")
tn.write(user.encode('ascii') + b"\n")
if password:
    tn.read_until(b"Password: ")
    tn.write(password.encode('ascii') + b"\n")

tn.write(b"show ip ospf neighbor\n")
tn.write(b"ping 2.2.2.2\n")
output1 = tn.read_until(b"ms", timeout=10).decode('utf-8')
print(output1)
``` |

(continued)
```

CHAPTER 2　PYTHON NETWORK AUTOMATION LABS - BASIC TELNET

#	Task

```
tn.write(b"ping 4.4.4.4\n")
output2 = tn.read_until(b"ms", timeout=10).decode('utf-8')
print(output2)

tn.write(b"exit\n")

tn.close()
```

10-2. Once you have modified your Python script and executed the Python script, you should receive output like Figure 2-7.

jdoe@u22s1:~/ch2_telnet$ **python sh_ip_ospf_ping1.py**

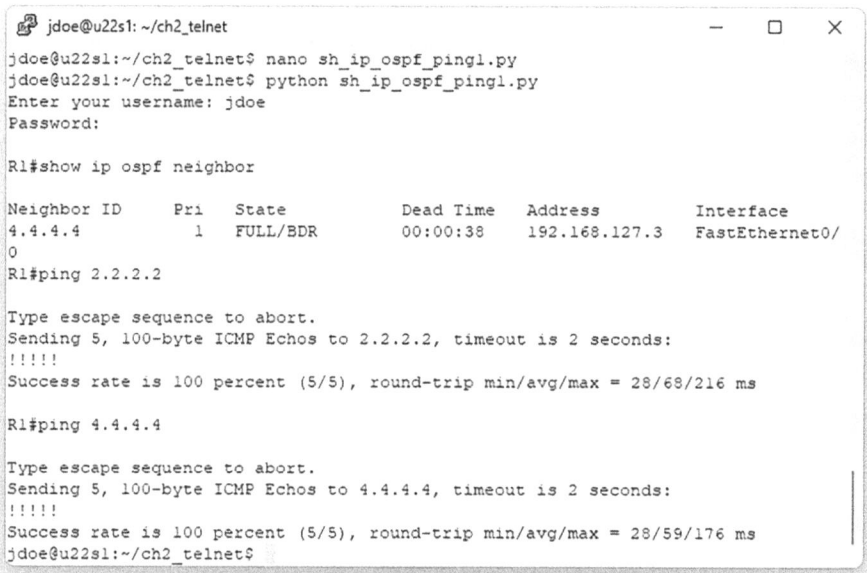

Figure 2-7. *Telnet Lab 1 - u22s1, R1's show ip ospf neighbor and ping output*

(continued)

#	Task
⑪	Do not forget to disable the debugging you have left on R2. Once again, use the copy and modify method or modify the previous Python script to run the `undebug all` command on R2 from the Python server, u22s1 (see Figure 2-8). jdoe@u22s1:~/ch2_telnet$ **cp sh_ip_int_bri1.py undebug_all.py** jdoe@u22s1:~/ch2_telnet$ **nano undebug_all.py** jdoe@u22s1:~/ch2_telnet$ **cat undebug_all.py** [...*omitted for brevity*] **tn.write(b"undebug all\n")** tn.write(b"exit\n") [...*omitted for brevity*] jdoe@u22s1:~/ch2_telnet$ **python undebug_all.py** Enter your username: **jdoe** Password: ******** 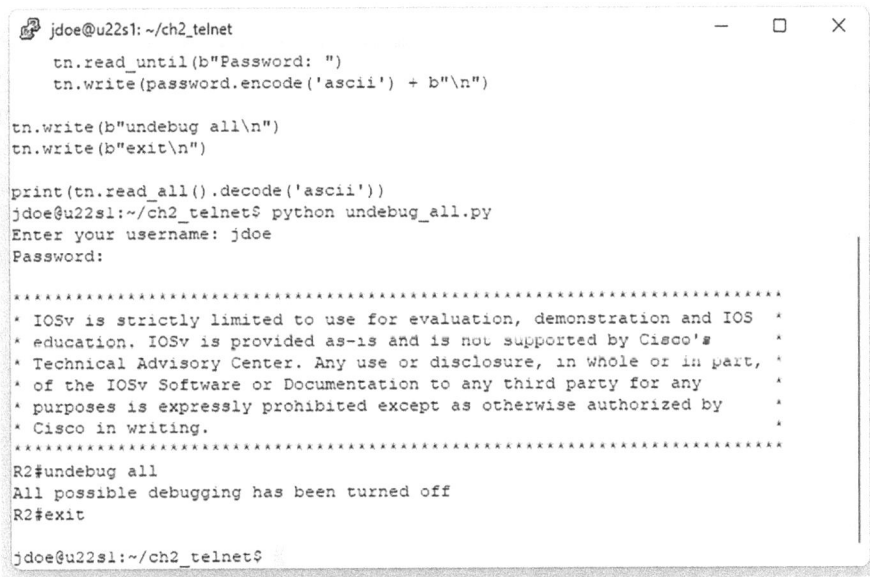 *Figure 2-8. Telnet Lab 1: u22s1, disable debugging on R2*

(*continued*)

CHAPTER 2 PYTHON NETWORK AUTOMATION LABS - BASIC TELNET

Although all the source code used in this book is provided on the author's GitHub and on the publisher's official code repository, it is recommended that you follow these tasks yourself in front of your computer. If you enjoyed your first Telnet lab, you will enjoy the second lab even more.

Telnet Lab 2: Configure a Single Switch with a Python Telnet Template

In this lab, the task is to configure new VLANs on SW1 via a Python Telnet script. To begin, power on the u22s1 Linux server (192.168.127.20) and SW1 (192.168.127.101). For efficiency, the initial script called add_lo_ospf1.py will be duplicated and renamed add_vlans_single.py (see Figure 2-9).

Upon completion of this lab, SW1 should be configured with configurations specified in the Telnet Lab 2 objectives, with the correct VLANs assigned to designated GigabitEthernet ports.

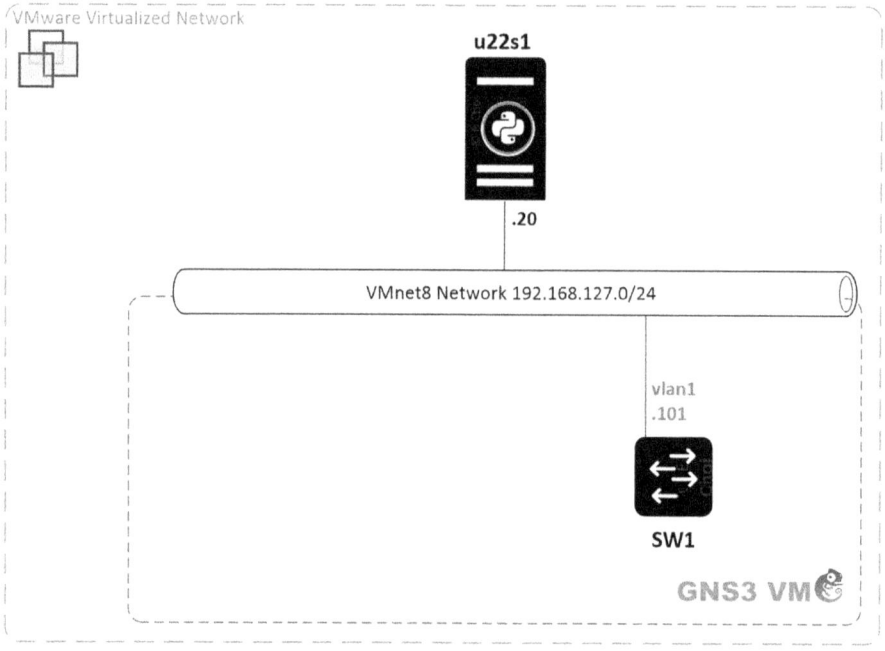

Figure 2-9. Telnet Lab 2, devices in use

CHAPTER 2   PYTHON NETWORK AUTOMATION LABS - BASIC TELNET

Telnet Lab 2 objectives:

1. Create VLANs 2 to 5 (in sequential order, 2-3-4-5) with the associated VLAN descriptions:

    - VLAN 2, Data_vlan_2
    - VLAN 3, Data_vlan_3
    - VLAN 4, Voice_vlan_4
    - VLAN 5, Wireless_vlan_5

2. Set GigabitEthernet ports 1/0 to 1/3 as access switch ports, configured with Wireless_vlan_5.

3. Configure GigabitEthernet ports 2/0 to 2/3 as switch ports, assigned as Data_vlan_2 (Data VLAN) and voice_vlan_4 (Auxiliary VLAN).

4. Activate all configured interfaces using the no shut command.

Let's dive back into coding—time to begin writing more code.

#	Task
①	SSH into the u22s1 server (192.168.127.20) and continue working in the same directory as the first lab. Follow these instructions to copy the first Python script and rename it add_vlans_single.py. jdoe@u22s1:~/ch2_telnet$ **cp add_lo_ospf1.py add_vlans_single.py** jdoe@u22s1:~/ch2_telnet$ **nano add_vlans_single.py** jdoe@u22s1:~/ch2_telnet$ **cat add_vlans_single.py**
②	Open the newly created .py file with a nano text editor. Press Ctrl+K to copy/cut lines of code and press Ctrl+U to paste information. Note that the IP address of SW1 is 192.168.127.101. Make sure you update the IP address next to HOST with this IP address. ```\nimport getpass\nimport telnetlib\nHOST = "192.168.127.101"\nuser = input("Enter your username: ")\npassword = getpass.getpass()\ntn = telnetlib.Telnet(HOST)\n```

(*continued*)

#	Task
| | ```python
tn.read_until(b"Username: ")
tn.write(user.encode('ascii') + b"\n")
if password:
  tn.read_until(b"Password: ")
  tn.write(password.encode('ascii') + b"\n")
# Get into config mode
tn.write(b"conf t\n")
# configure 4 VLANs with VLAN names
tn.write(b"vlan 2\n")
tn.write(b"name Data_vlan_2\n")
tn.write(b"vlan 3\n")
tn.write(b"name Data_vlan_3\n")
tn.write(b"vlan 4\n")
tn.write(b"name Voice_vlan_4\n")
tn.write(b"vlan 5\n")
tn.write(b"name Wireless_vlan_5\n")
tn.write(b"exit\n")
# configure Gi1/0 - Gi1/3 as access switch ports and assign vlan 5 for wireless APs
tn.write(b"interface range gi1/0 - 3\n")
tn.write(b"switchport mode access\n")
tn.write(b"switchport access vlan 5\n")
tn.write(b"no shut\n")
#configure gi2/0 - gi2/3 as access switch ports and assign vlan 2 for data and vlan 4 for voice
tn.write(b"interface range gi2/0 - 3\n")
tn.write(b"switchport mode access \n")
tn.write(b"switchport access vlan 2\n")
tn.write(b"switchport voice vlan 4\n")
tn.write(b"no shut\n")
tn.write(b"end\n")
tn.write(b"exit\n")
print(tn.read_all().decode('ascii'))
``` |

(*continued*)
```

CHAPTER 2　PYTHON NETWORK AUTOMATION LABS - BASIC TELNET

#	Task
③	Let's begin by checking the network connectivity from the u22s1 server to SW1 at 192.168.127.101. While numerous Python modules and methods are available for this task, let's keep it simple and use just two lines of code utilizing the Python os module. If the ping is successful, proceed to the next step. jdoe@u22s1:~/ch2_telnet$ **nano ping_sw1.py** jdoe@u22s1:~/ch2_telnet$ **cat ping_sw1.py** **import os** **os.system("ping -c 4 192.168.127.101")** jdoe@u22s1:~/ch2_telnet$ **python ping_sw1.py** PING 192.168.127.101 (192.168.127.101) 56(84) bytes of data. 64 bytes from 192.168.127.101: icmp_seq=1 ttl=255 time=28.4 ms *[...omitted for brevity]*
④	On u22s1, enable the debug telnet command on SW1 to capture Telnet activities during the script execution. After debugging has been enabled from u22s1, open SW1's console window. First, copy the undebug_all.py file change to the HOST IP address and change undebug all to debug telnet. Then run the script to enable Telnet debugging. After executing your command, your u22s1's screen should look like Figure 2-10. jdoe@u22s1:~/ch2_telnet$ **cp undebug_all.py debug_telnet.py** jdoe@u22s1:~/ch2_telnet$ **nano debug_telnet.py** jdoe@u22s1:~/ch2_telnet$ **cat debug_telnet.py** *[...omitted for brevity]* HOST = "**192.168.127.101**" *[...omitted for brevity]* **tn.write(b"debug telnet\n")** **tn.write(b"exit\n")** *[...omitted for brevity]* jdoe@u22s1:~/ch2_telnet$ **python debug_telnet.py** Enter your username: **jdoe** Password: *******

(*continued*)

CHAPTER 2   PYTHON NETWORK AUTOMATION LABS - BASIC TELNET

#	Task
	![screenshot of terminal showing telnet debug session on SW1]

```
tn.read_until(b"Password: ")
tn.write(password.encode('ascii') + b"\n")

tn.write(b"debug telnet\n")
tn.write(b"exit\n")

print(tn.read_all().decode('ascii'))
jdoe@u22s1:~/ch2_telnet$ python debug_telnet.py
Enter your username: jdoe
Password:

*****************************************************
* IOSv is strictly limited to use for evaluation, demonstration and IOS *
* education. IOSv is provided as-is and is not supported by Cisco's    *
* Technical Advisory Center. Any use or disclosure, in whole or in part,*
* of the IOSv Software or Documentation to any third party for any     *
* purposes is expressly prohibited except as otherwise authorized by   *
* Cisco in writing.                                                    *
*****************************************************
SW1#debug telnet
Incoming Telnet debugging is on
SW1#exit

jdoe@u22s1:~/ch2_telnet$
```

Figure 2-10.   Telnet Lab 2, u22s1 - enable debug telnet on SW1

| ⑤ | You have checked the connectivity and enabled debug telnet. Place the u22s1 SSH PuTTY screen and the SW1 GNS3 console screen side-by-side and run the python add_vlans_single.py command to run the script (see Figure 2-11). This script will add new VLANs and configure the designated switch ports with their VLANs.

jdoe@u22s1:~/ch2_telnet$ **python add_vlans_single.py**
Enter your username: **jdoe**
Password:********

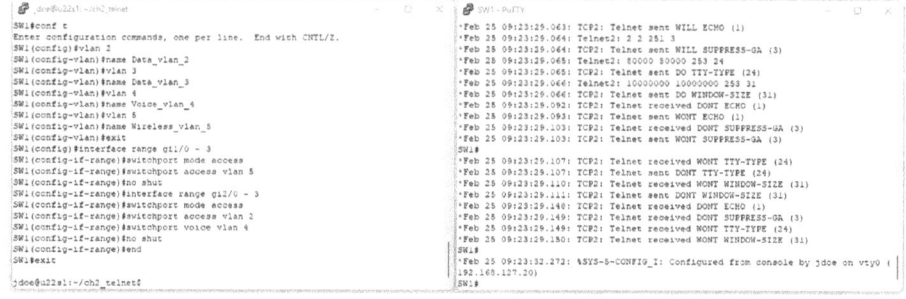

Figure 2-11.   Telnet Lab 2, u22s1 - run add_vlans_single.py script

As soon as you enter the username and password, you should see the Telnet message scrolling up on SW1's console. Whereas on u22s1's screen, it will output the commands that have been executed on SW1.

(continued)

#	Task
6	Upon checking the SW1's console window for debugging information, you should see a similar Telnet log message as shown here: ``` SW1# *Oct 29 12:45:46.180: Telnet2: 1 1 251 1 *Oct 29 12:45:46.181: TCP2: Telnet sent WILL ECHO (1) *Oct 29 12:45:46.181: Telnet2: 2 2 251 3 *Oct 29 12:45:46.182: TCP2: Telnet sent WILL SUPPRESS-GA (3) [...omitted for brevity] *Oct 29 12:45:51.917: %SYS-5-CONFIG_I: Configured from console by jdoe on vty0 (192.168.127.20) SW1# ```
7	From the u22s1 server, now check the VLANs configured on SW1's switch ports. Make use of the show vlan, show ip interface brief, and show run commands to check the switch VLAN configuration changes and the port's new VLAN assignments. You must perform the checks from u22s1, not through the SW1 console window. 7-1. Can you recall the show flash Telnet script that you worked with in Chapter 11 of the Part I book? There is a file named ch11_r1_cf1.py in the ch11 directory on this server. Copy it and modify the file, so you can easily run all commands at once and display the output for your review. ``` jdoe@u22s1:~/ch2_telnet$ cp ../ch11/ch11_r1_cf1.py telnet_lab2_check.py jdoe@u22s1:~/ch2_telnet$ cat telnet_lab2_check.py jdoe@u22s1:~/ch2_telnet$ nano telnet_lab2_check.py import telnetlib host, port, username, password = "192.168.127.101", 23, "jdoe", "cisco123" tn = telnetlib.Telnet(host, port) ```

(*continued*)

#	Task
	``
tn.read_until(b"Username: ")
tn.write(username.encode('utf-8') + b"\n")
if password:
  tn.read_until(b"Password: ")
  tn.write(password.encode('utf-8') + b"\n")

commands = [
  "show vlan",
  "show ip interface brief",
  'terminal length 0', # required to capture all 'show run' output
  "show run"
]
output = ""
  for command in commands:
  tn.write(command.encode('utf-8') + b"\n")
  tn.write(b"\n")
  output += tn.read_until(b"^end", timeout=7).decode('utf-8')
  # Adjust the timeout for your environment, increase the time if you
  are not getting the full output.
print(output)
tn.close()
```
Chapter 9 of the Part I book covers Regular Expressions and here a Regular Expression pattern ^end was used to signify the start of the line ^, followed by the word end. From experience, I know that most of the Cisco switch show run command's output usually ends with the line ending with the word end. You can also replace this with another keyword; for example, line con 0 or banner exec.

7-2. Now run the python telnet_lab2_check.py command from u22s1 and check the VLAN and switchport change on SW1 (see Figure 2-12).

jdoe@u22s1:~/ch2_telnet$ **python telnet_lab2_check.py** |

*(continued)*
```

CHAPTER 2   PYTHON NETWORK AUTOMATION LABS - BASIC TELNET

Task

```
jdoe@u22s1:~/ch2_telnet$ python telnet_lab2_check.py
***********************************************************
* IOSv is strictly limited to use for evaluation, demonstration and IOS *
* education. IOSv is provided as-is and is not supported by Cisco's *
* Technical Advisory Center. Any use or disclosure, in whole or in part, *
* of the IOSv Software or Documentation to any third party for any *
* purposes is expressly prohibited except as otherwise authorized by *
* Cisco in writing.                                        *
***********************************************************
SW1#show vlan

VLAN Name                             Status    Ports
---- -------------------------------- --------- -------------------------------
1    default                          active    Gi0/0, Gi0/1, Gi0/2, Gi0/3
                                                Gi3/0, Gi3/1, Gi3/2, Gi3/3
2    Data_vlan_2                      active    Gi2/0, Gi2/1, Gi2/2, Gi2/3
3    Data_vlan_3                      active
4    Voice_vlan_4                     active    Gi2/0, Gi2/1, Gi2/2, Gi2/3
5    Wireless_vlan_5                  active    Gi1/0, Gi1/1, Gi1/2, Gi1/3
101  PYTHON_VLAN_101                  active
202  PYTHON_VLAN_202                  active
303  PYTHON_VLAN_303                  active
404  PYTHON_VLAN_404                  active
505  PYTHON_VLAN_505                  active
```

Figure 2-12. Telnet Lab 2, u22s1 - run telnet_lab2_check.py script

You can run your Cisco show commands using the Python script from your Python server. The whole point of this exercise is to keep you away from the switch's command line and also get you more familiar with writing and running the script from the your server.

You have successfully added four VLANs in sequential order and configured eight switch ports in their respective VLANs. Next, let's explore the use of for loops to add multiple random VLANs in Labs 3 and 4.

Telnet Lab 3: Configure Random VLANs Using a for Loop

In the previous lab, VLANs 2 to 5 were configured with multiple lines of code. Here, you'll practice adding VLANs with a more concise approach, using a for loop. Although you'll only be adding five random VLANs in this lab, the for loop method will significantly reduce the lines of code needed and save you time.

CHAPTER 2   PYTHON NETWORK AUTOMATION LABS - BASIC TELNET

Follow these steps to configure random VLANs—101, 202, 303, 404, and 505—on SW1 via Telnet. Utilize the u22s1 Linux server (192.168.127.20) and SW1 (192.168.127.101) for this lab (see Figure 2-13). Copy the add_vlans_single.py file and rename it add_vlans_for_loop.py for this lab. By the end of this exercise, SW1 should be configured with the VLANs specified in Lab 3 objectives.

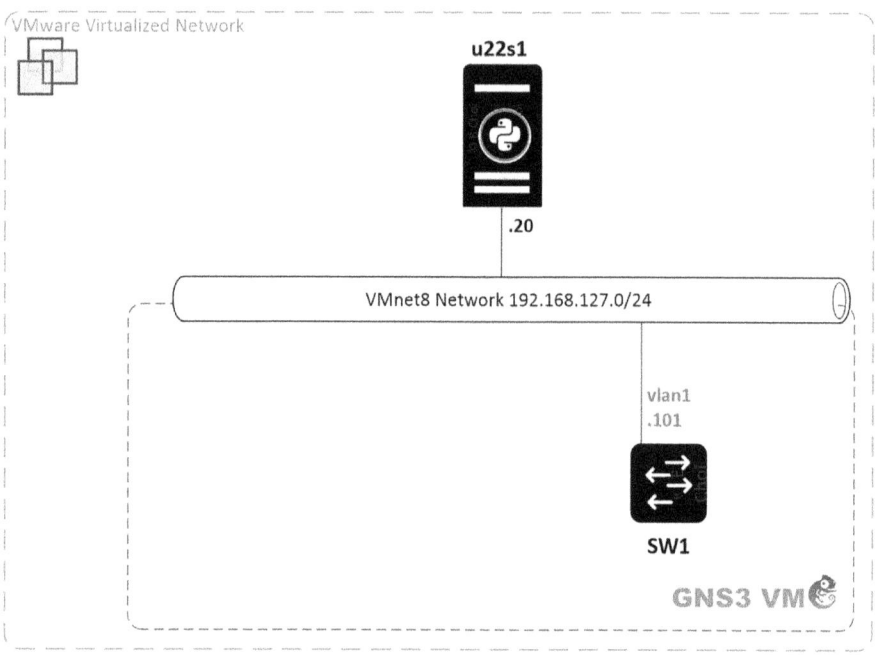

Figure 2-13.   *Telnet Lab 3: devices in use*

Telnet Lab 3 objectives are to configure the following on SW1:

- VLAN 101, Data_vlan_101
- VLAN 202, Data_vlan_202
- VLAN 303, Voice_vlan_303
- VLAN 404, Wireless_vlan_404
- VLAN 505, Wireless_vlan_505

Tip

The art of repetition: embracing for loops for effective programming

In this lab, focus on understanding the functionality of a for loop and its application within this context. Loops are specifically engineered for executing the same task iteratively, and by delving into this, you'll harness the true potential of programming. Learn Python concepts and them effectively use them at work to achieve your goal. Best of luck!

Time to write some more Python code.

#	Task

①	First, move to the telnet_labs directory, use the cp command to copy add_vlans_single.py, and rename it add_vlans_for_loop.py. The commands are shown here:
	
	 ```jdoe@u22s1:~$ cd ch2_telnet```
	 ```jdoe@u22s1:~/ch2_telnet$ ls add_vlans_*```
	 ```add_vlans_single.py```
	 ```jdoe@u22s1:~/ch2_telnet$ cp add_vlans_single.py add_vlans_for_loop.py```
	 ```jdoe@u22s1:~/ch2_telnet$ nano add_vlans_for_loop.py```
	 ```jdoe@u22s1:~/ch2_telnet$ ls add_vlans_*```
	 ```add_vlans_for_loop.py add_vlans_single.py```

②	Update the script to resemble the provided format. Use nano's Ctrl+K option to cut/copy the information and Ctrl+U to paste it back. Pay close attention to syntax and whitespace throughout the update. Every comma and space holds significance, particularly focusing on proper indentation using the requisite four, eight, and twelve spaces for code blocks. Additionally, for better code readability, aim to keep your code blocks within three levels of nesting whenever feasible.

(continued)

#	Task

	#!/usr/bin/env python3 # This is the shebang line, ignored by Python but read by the Linux Operating System to determine the application version for running the .py file.
	import getpass
	import telnetlib
	HOST = "**192.168.127.101**"
	user = input("Enter your username: ")
	password = getpass.getpass()
	tn = telnetlib.Telnet(HOST)
	tn.read_until(b"Username: ")
	tn.write(user.encode('ascii') + b"\n")
	if password:
	    tn.read_until(b"Password: ")
	    tn.write(password.encode('ascii') + b"\n")
	# Get into config mode
	tn.write(b"conf t\n")
	# Adds 5 vlans to the list with for loop
	vlans = [101, 202, 303, 404, 505] # vlans to add in a list
	for i in vlans: # call (index) each item from list vlans
	    command_1 = "vlan " + str(i) + "\n" # concatenate first command
	    tn.write(command_1.encode('ascii')) # send command_1 with ASCII encoding
	    command_2 = "name PYTHON_VLAN_" + str(i) + "\n" # concatenate second command

(*continued*)

#	Task

> **tn.write(command_2.encode('ascii'))** # send command_2 with ASCII encoding
>
> tn.write(b"end\n")
> tn.write(b"exit\n")
> print(tn.read_all().decode('ascii'))

Telnetlib's encoding and decoding process may seem convoluted. **Encoding in telnetlib involves converting a string into bytes suitable for network transmission using the .encode('encoding_type') method.** This process ensures that the string is converted into a format that can be sent across the network. **Decoding, on the other hand, takes the received bytes and converts them back into a string using the .decode('encoding_type') method.** This step is essential for interpreting the received data into a readable format within the Python environment after it has been transmitted over the network.

③ To execute this script, use either the python or the python3 command. Recall that you included an alias in the user's /.bashrc file to run the script without appending the 3 at the end of the word python. Here's an example:

jdoe@u22s1:/ch2_telnet$ **python3 add_vlans_for_loop.py**
OR
jdoe@u22s1:/ch2_telnet$ **python add_vlans_for_loop.py**

Enter the username and password and press the Enter key. The displayed output from your script should match Figure 2-14.

(continued)

CHAPTER 2   PYTHON NETWORK AUTOMATION LABS - BASIC TELNET

#	Task

```
jdoe@u22s1: ~/ch2_telnet                                    —   □   ×

*************************************************************
* IOSv is strictly limited to use for evaluation, demonstration and IOS *
* education. IOSv is provided as-is and is not supported by Cisco's    *
* Technical Advisory Center. Any use or disclosure, in whole or in part, *
* of the IOSv Software or Documentation to any third party for any    *
* purposes is expressly prohibited except as otherwise authorized by  *
* Cisco in writing.                                                    *
*************************************************************
SW1#conf t
Enter configuration commands, one per line.  End with CNTL/Z.
SW1(config)#vlan 101
SW1(config-vlan)#name PYTHON_VLAN_101
SW1(config-vlan)#vlan 202
SW1(config-vlan)#name PYTHON_VLAN_202
SW1(config-vlan)#vlan 303
SW1(config-vlan)#name PYTHON_VLAN_303
SW1(config-vlan)#vlan 404
SW1(config-vlan)#name PYTHON_VLAN_404
SW1(config-vlan)#vlan 505
SW1(config-vlan)#name PYTHON_VLAN_505
SW1(config-vlan)#end
SW1#exit

jdoe@u22s1:~/ch2_telnet$
```

Figure 2-14. Telnet Lab 3, u22s1 - add_vlans_for_loop.py output

④ Run the show vlan command on SW1 to confirm newly configured VLANs. If you can see the new VLANs in the switch's VLAN table, then your Python script used a for loop to add random VLANs to the switch. Once again, use your Python script to check the VLANs on your switch.

```
jdoe@u22s1:~/ch2_telnet$ cp undebug_all.py sh_vlan1.py
doe@u22s1:~/ch2_telnet$ nano sh_vlan1.py
jdoe@u22s1:~/ch2_telnet$ cat sh_vlan1.py
import getpass
import telnetlib

HOST = "192.168.127.101"
user = input("Enter your username: ")
password = getpass.getpass()

tn = telnetlib.Telnet(HOST)

tn.read_until(b"Username: ")
tn.write(user.encode('ascii') + b"\n")
```

(continued)

```
if password:
    tn.read_until(b"Password: ")
    tn.write(password.encode('ascii') + b"\n")
```

tn.write(b"terminal length 0\n")
tn.write(b"show vlan\n")
```
tn.write(b"exit\n")
print(tn.read_all().decode('ascii'))
```

jdoe@u22s1:~/ch2_telnet$ **python sh_vlan1.py**

As a result of executing this script, you should get output like Figure 2-15 shows.

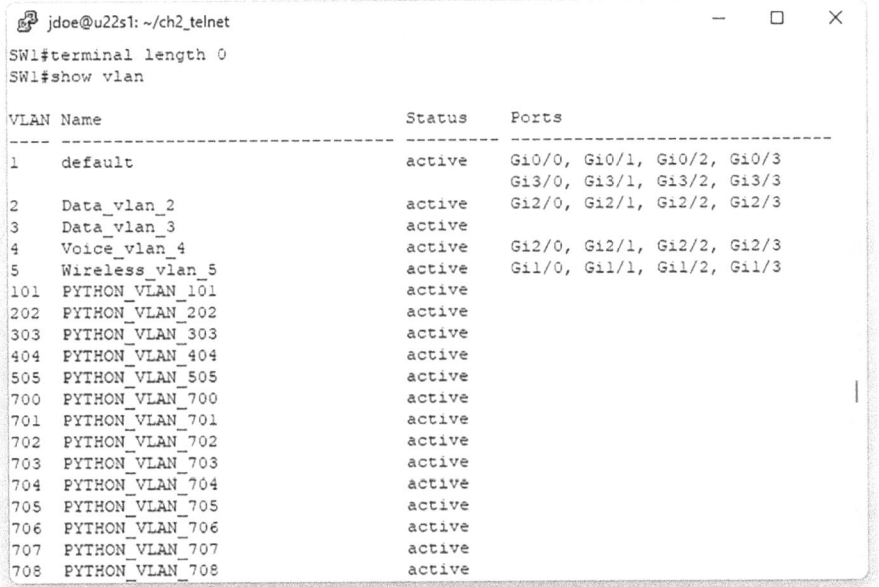

Figure 2-15. *Telnet Lab 3, u22s1 – show vlan output*

Now that you know about `for` loops and `while` loops, the next section delves into the application of the `while` loop to achieve the same outcome on another switch.

CHAPTER 2 PYTHON NETWORK AUTOMATION LABS - BASIC TELNET

Telnet Lab 4: Configure Random VLANs Using a while Loop

Similar to Lab 3, you'll replicate the creation of VLANs 101, 202, 303, 404, and 505 on sw2 via Telnet. You'll continue working from u22s1 for this lab, but sw2 (192.168.127.102) must be powered on. Keep in mind that the SW1 should be kept powered on all the times, as it is the uplink switch for sw2. Also, you will need it for a for loop verification command in the last task. To begin, copy the add_vlans_for_loop.py file and create a new script named add_vlans_while_loop.py (see Figure 2-16).

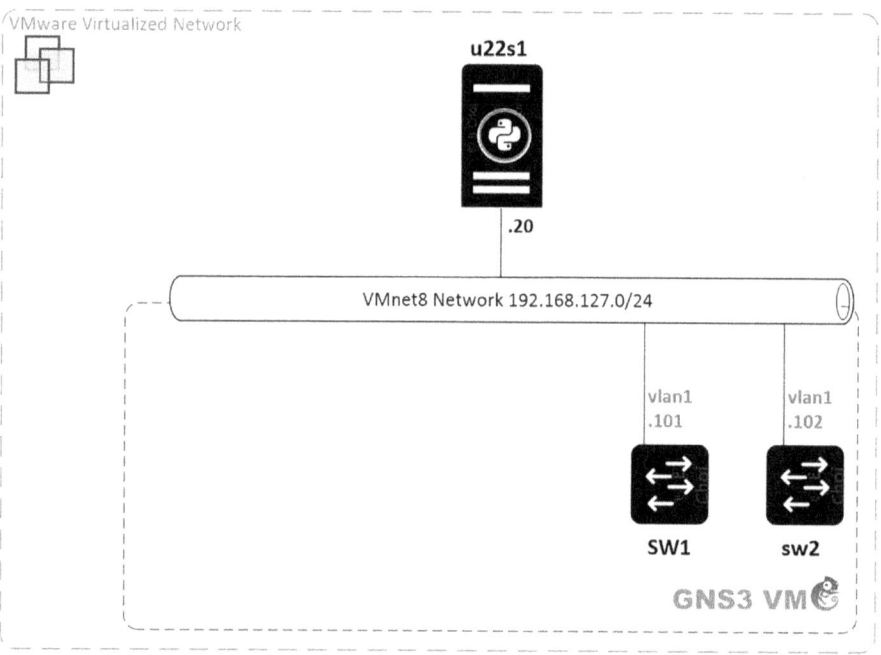

Figure 2-16. Telnet Lab 4, devices in use

At the end of this lab, sw2 should be configured with the same VLAN set as SW1, aligning with the objectives outlined next.

Telnet Lab 4 objectives: Configure the following on sw2 and perform config change verification.

- VLAN 101, Data_vlan_101
- VLAN 202, Data_vlan_202
- VLAN 303, Voice_vlan_303

CHAPTER 2   PYTHON NETWORK AUTOMATION LABS - BASIC TELNET

- VLAN 404, Wireless_vlan_404
- VLAN 505, Wireless_vlan_505
- Use the show vlan command from u22s1 to inspect the VLAN configurations on sw2 and SW1. Write a single script to perform Telnet connections into both devices, executing commands using a for loop.

It's time to delve into more coding.

#	Task
①	Given that you're currently working within the telnet_labs directory on u22s1, use the cp command to duplicate add_vlans_for_loop.py and create add_vlans_while_loop.py. Execute the following commands to generate the new script: jdoe@u22s1:~/ch2_telnet$ **cp add_vlans_for_loop.py add_vlans_while_loop.py** jdoe@u22s1:~/ch2_telnet$ **nano add_vlans_while_loop.py**

②	Alter your new script to resemble the following script. This script demonstrates an example of a while loop used to add the same VLANs as in the previous for loop example. Focus on grasping the working logic of the while loop in this script, without worrying excessively about the specific login protocol's security concerns. Refer to the embedded explanation for further insights.
	
	``python
	#!/usr/bin/env python3
	import getpass
	import telnetlib
	HOST = "192.168.127.102" # Update this to the IP address of sw2 switch
	user = input("Enter your username: ")
	password = getpass.getpass()
	
	tn = telnetlib.Telnet(HOST)
	tn.read_until(b"Username: ")
	tn.write(user.encode('ascii') + b"\n")
	``

(continued)

#	Task
| | ```
if password:
    tn.read_until(b"Password: ")
    tn.write(password.encode('ascii') + b"\n")
# Switch to configuration mode
tn.write(b"conf t\n")
# Create a list of 5 VLANs and use a while loop to configure them on
the switch
vlans = [101, 202, 303, 404, 505] # VLANs to add to the list
i = 0 # Initialize the index value
# Send commands to configure VLANs using a while loop
while i < len(vlans): # While 'i' is less than the length of 'vlans'
    # Concatenate and send the first command to create the VLAN
    command_1 = "vlan " + str(vlans[i]) + "\n"
    tn.write(command_1.encode('ascii')) # Send 'command_1' with ASCII
    encoding
    # Concatenate and send the second command to name the VLAN
    command_2 = "name PYTHON_VLAN_" + str(vlans[i]) + "\n"
    tn.write(command_2.encode('ascii'))  # Send 'command_2' with
    ASCII encoding
    i += 1 # Increment 'i' to move to the next VLAN in the list
# End configuration mode and exit Telnet session
tn.write(b"end\n")
tn.write(b"exit\n")
print(tn.read_all().decode('ascii'))
``` |

(continued)
```

#	Task
③	After you have rewritten your `while` loop script, before running your script, let's check the connection to sw2 (192.168.127.102). This time you will modify the previous os Python ping script to send four packets and check the reliability of the communication by calculating the returned ping responses. Type these commands or use the cut-and-paste method to complete the following `ping_os_percentage.py` script.

```python
#ping_os_percentage.py
import os

def get_ping_success_rate(host):
    response = os.system("ping -c 4 " + host)
    return response # Return the response from the ping command

def calculate_success_rate(response):
    return (4 - response) / 4 * 100 # Calculate success rate as a
                                    percentage

HOST = "192.168.127.102"

ping_response = get_ping_success_rate(HOST) # Check the server's
                                              responsiveness through
                                              ping

if ping_response > 3:
    print("Server is not responsive.")
else:
    print("Server is responsive.")

success_rate = calculate_success_rate(ping_response)
print(f"Ping success rate: {success_rate:.2f}%") # Print the success
                                                   rate of the ping
```

Execute this script. If you see the `Server is responsive` message with a 100 percent success rate, then the connection to sw2 is reliable at this point. If your response is `Server is not responsive` and you receive a minus value, you could have a network connectivity issue within your network, or the switch might be offline. Figure 2-17 shows both messages for your reference. |

(continued)

CHAPTER 2   PYTHON NETWORK AUTOMATION LABS - BASIC TELNET

#	Task
	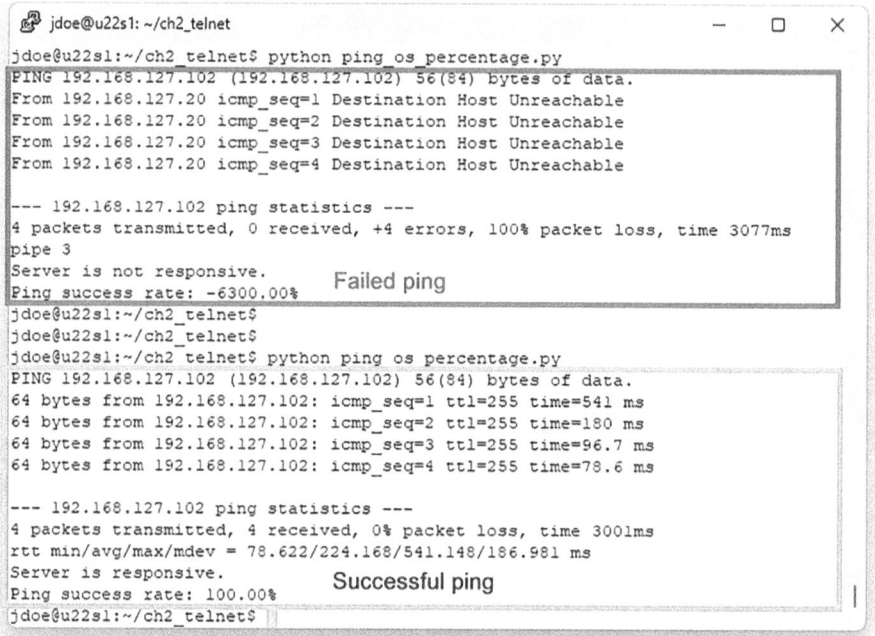\n\n*Figure 2-17. Telnet Lab 4, ping responses*
(4)	Now it is time to run your VLAN creating a `while` loop script. From the `u22s1` server, execute the script using the following command (see Figure 2-18).\n\n`jdoe@u22s1:~/ch2_telnet$` **`python add_vlans_while_loop.py`**

(continued)

#	Task

```
jdoe@u22s1:~/ch2_telnet$ python add_vlans_while_loop.py
Enter your username: jdoe
Password:

*************************************************************
* IOSv is strictly limited to use for evaluation, demonstration and IOS *
* education. IOSv is provided as-is and is not supported by Cisco's     *
* Technical Advisory Center. Any use or disclosure, in whole or in part, *
* of the IOSv Software or Documentation to any third party for any      *
* purposes is expressly prohibited except as otherwise authorized by    *
* Cisco in writing.                                                     *
*************************************************************
sw2#conf t
Enter configuration commands, one per line.  End with CNTL/Z.
sw2(config)#vlan 101
sw2(config-vlan)#name PYTHON_VLAN_101
sw2(config-vlan)#vlan 202
sw2(config-vlan)#name PYTHON_VLAN_202
sw2(config-vlan)#vlan 303
sw2(config-vlan)#name PYTHON_VLAN_303
sw2(config-vlan)#vlan 404
sw2(config-vlan)#name PYTHON_VLAN_404
sw2(config-vlan)#vlan 505
sw2(config-vlan)#name PYTHON_VLAN_505
sw2(config-vlan)#end
sw2#exit

jdoe@u22s1:~/ch2_telnet$
```

Figure 2-18. Telnet Lab 4, u22s1 - execute add_vlans_while_loop.py

⑤ Now make a copy of sh_vlan1.py and name the file sh_vlan2.py. Then modify it like the following script, so it uses the for loop to run the command on both switches. To achieve this, you will have to place the username and password collection at the beginning of the code to make these variables global. This way, the username and password can be reused by the for loop command.

```
jdoe@u22s1:~/ch2_telnet$ cp sh_vlan1.py sh_vlan2.py
jdoe@u22s1:~/ch2_telnet$ nano sh_vlan2.py
jdoe@u22s1:~/ch2_telnet$ cat sh_vlan2.py
# sh_vlan2.py
import getpass # Importing getpass for password input
import telnetlib # Importing telnetlib for Telnet connections

# Prompt for credentials
user = input("Enter your username: ") # Input username
password = getpass.getpass() # Securely input password
```

(*continued*)

#	Task
	``
# List of hosts
hosts = ["192.168.127.102", "192.168.127.101"] # Define hosts

for host in hosts: # Iterate through hosts
    tn = telnetlib.Telnet(host) # Establish Telnet connection
    tn.read_until(b"Username: ") # Read 'Username' prompt
    tn.write(user.encode('ascii') + b"\n") # Send encoded username
if password: # Check if a password is provided

    tn.read_until(b"Password: ") # Read 'Password' prompt
    tn.write(password.encode('ascii') + b"\n") # Send encoded password
tn.write(b"terminal length 0\n") # Set terminal length
tn.write(b"show vlan\n") # Show VLAN details
tn.write(b"exit\n") # Exit Telnet session
output = tn.read_all().decode('ascii') # Read and decode output
print(f"Output for {host}:\n{output}") # Display output per host
```
Execute the Python script. When you get the same output as shown in Figure 2-19, you have completed the Telnet Lab 4.

```
jdoe@u22s1:~/ch2_telnet$ python sh_vlan2.py
Enter your username: jdoe
Password: ********
``` |

(continued)
```

#	Task

```
jdoe@u22s1: ~/ch2_telnet                                    −   □   ×
jdoe@u22s1:~/ch2_telnet$ python sh_vlan2.py
Enter your username: jdoe
Password:
Output for 192.168.127.102:

************************************************************
* IOSv is strictly limited to use for evaluation, demonstration and IOS *
* education. IOSv is provided as-is and is not supported by Cisco's     *
* Technical Advisory Center. Any use or disclosure, in whole or in part,*
* of the IOSv Software or Documentation to any third party for any      *
* purposes is expressly prohibited except as otherwise authorized by    *
* Cisco in writing.                                                     *
************************************************************
sw2#terminal length 0
sw2#show vlan

VLAN Name                             Status    Ports
---- -------------------------------- --------- -------------------------------
1    default                          active    Gi0/0, Gi0/1, Gi0/2, Gi0/3
                                                Gi1/0, Gi1/1, Gi1/2, Gi1/3
                                                Gi2/0, Gi2/1, Gi2/2, Gi2/3
                                                Gi3/0, Gi3/1, Gi3/2, Gi3/3
101  PYTHON_VLAN_101                  active
202  PYTHON_VLAN_202                  active
303  PYTHON_VLAN_303                  active
404  PYTHON_VLAN_404                  active
505  PYTHON_VLAN_505                  active
1002 fddi-default                     act/unsup
```

Figure 2-19. Telnet Lab 4, u22s1 - execute sh_vlan2.py

Expand your knowledge

Until ping (death) do us part, amen

Many network engineers often dedicate more time to networking devices and colleagues than to their families. An earlier Cisco TAC (Technical Assistance Center) survey revealed higher divorce rates among married TAC engineers compared to other groups. This humorously implies that engineers are married to their ping tools.

From junior-level IT technicians to top-tier professionals in the ICT industry, the ping tool is a ubiquitous utility used daily for troubleshooting. It holds immense significance, so much so that it's considered the tool a network engineer might take to their proverbial grave due `to the fundamental role it plays in a network engineer's life. However, leveraging the power of Python, there exists an array of diverse ping tools available. These encompass the built-in modules, `os.system()` and `subprocess.run()`, which offer fundamental and advanced command execution capabilities. Additionally, there are external Python modules like ping3, pythonping, and scapy, each presenting a unique set of ICMP functionalities.

Exploring each tool provides an invaluable understanding of their distinctive features and methodologies. While all focus on ICMP tests, their individual mileage varies. Among these, although scapy is not NMAP, it notably emerges as a powerful and versatile Python ICMP tool, capable of more extensive network interactions and manipulations. This enhances its utility beyond standard ping operations.

Ping method 1 - os: Executes commands in a subshell for basic system operations; suitable for simple command execution like ping.

```
# ping_os_m.py
import os

response = os.system("ping -c 4 192.168.127.102")
print(response)
```

Ping method 2 - subprocess: Provides advanced control over subprocesses, allowing input/output capture; useful for executing external commands.

```
# ping_subprocess_m.py
import subprocess

ping_command = ["ping", "-c", "4", "192.168.127.102"]
completed_process = subprocess.run(ping_command, stdout=subprocess.PIPE, stderr=subprocess.PIPE, text=True)
print(completed_process.stdout)
```

Ping method 3 - ping3: Simplifies ICMP ping operation, enabling host reachability check, with additional control over parameters.

```
# ping_ping3_m.py
from ping3 import ping

response = ping('192.168.127.102')
print(response)
```

Ping method 4 - pythonping: Offers a high-level interface for executing ping commands, providing an easy-to-use Pythonic approach.

```
"""
```

```
ping_pythonping_m.py
Execute the script as a root user.
Log in as the root user first using 'su -' command.
Install pythonping module using 'pip3 install pythonping'
"""
```

from pythonping import ping

response = ping('192.168.127.102', count=4)
print(response)
print(response.rtt_avg_ms) # more features than normal ping

Ping method 5 - scapy: Empowers packet crafting, manipulation, and network scanning, including ICMP echo requests, with a flexible and interactive approach.

```
"""
# ping_scapy_m.py
Execute the script as a root user.
Log in as the root user first using 'su -' command.
Install pythonping module using 'pip3 install scapy'
More powerful and more detailed ping output than the other tools.
"""
```

from scapy.layers.inet import IP, ICMP
from scapy.sendrecv import sr1

packet = IP(dst="192.168.127.102")/ICMP()
response = sr1(packet, timeout=2)
response.show()

Over the years, I've been experimenting with a variety of ICMP and NMAP tools. For further insights into the ICMP and NMAP tools, visit the author's WordPress site: https://italchemy.wordpress.com/

CHAPTER 2   PYTHON NETWORK AUTOMATION LABS - BASIC TELNET

Telnet Lab 5: Configure 100 VLANs Using the for ~ in range Loop

In this lab, you will create 100 VLANs using the for ~ in range loop method on the SW1 (192.168.127.101) and sw2 (192.168.127.102) switches. You will create the script from the same Python server, the u22s1 Linux server (192.168.127.20). Copy the add_vlans_while_loop.py file and create a new script called add_vlans_for_range.py. At the end of this lab, both SW1 and sw2 should be configured with VLANs 700 to 799. See Figure 2-20.

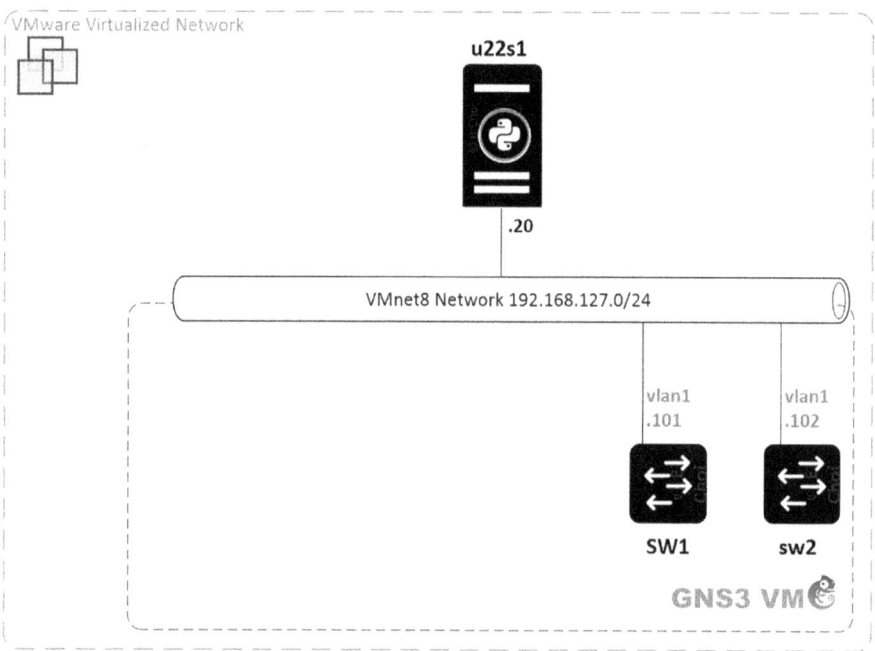

Figure 2-20.   Telnet Lab 5, devices in use

Telnet Lab 5 objectives: Configure VLANs 700-799.
VLANs 700-799 (i.e., PYTHON_VLAN_700, PYTHON_VLAN_701..., PYTHON_VLAN_799)

Ready to code again? Let's start without further ado.

#	Task
①	Continuing from the last lab, copy the previous while loop script and create the add_100_vlans.py script. jdoe@u22s1:~/ch2_telnet$ **cp add_vlans_while_loop.py add_100_vlans.py**

②	This time you will be configuring 100 VLANs using a for ~ range loop on both switches using a single script. You must be extremely careful with the leading whitespace and the code blocks for this script. Be consistent with the whitespace and the script's code blocks. Refer to the embedded explanation as you modify your script. Use # or """ """ to add as many comments as you want. At the end of the modification, your Python script should look like the following code.
	
	This script is available for download, but you should try to modify the script on Linux's nano text editor to get familiar with text editing on Linux.
	
	jdoe@u22s1:~/ch2_telnet$ **cp add_vlans_while_loop.py add_100_vlans.py**
	jdoe@u22s1:~/ch2_telnet$ **nano add_100_vlans.py**
	#!/usr/bin/env python3.10
	import getpass # Import for password input
	import telnetlib # Import for Telnet connections
	
	HOSTS = ["192.168.127.101", "192.168.127.102"] # List of switch IP addresses
	user = input("Enter your username: ") # Prompt for username
	password = getpass.getpass() # Securely prompt for password
	
	for HOST in HOSTS: # Iterate through switch IPs
	    print("SWITCH IP : " + HOST) # Print device IP
	    tn = telnetlib.Telnet(HOST) # Establish Telnet connection
	    tn.read_until(b"Username: ") # Read 'Username' prompt
	    tn.write(user.encode('ascii') + b"\n") # Send encoded username
	
	    if password: # Check for password
	        tn.read_until(b"Password: ") # Read 'Password' prompt
	        tn.write(password.encode('ascii') + b"\n") # Send encoded password
	    tn.write(b"conf t\n") # Enter configuration mode

(continued)

#	Task

> ```
> for n in range(700, 800):  # Configure VLANs
>     tn.write(f"vlan {n}\n".encode('utf-8'))  # Send VLAN commands
>     tn.write(f"name PYTHON_VLAN_{n}\n".encode('utf-8'))  # Send
>     VLAN names
>
> tn.write(b"end\n")  # Exit configuration mode
> tn.write(b"exit\n")  # Exit Telnet session
> print(tn.read_all().decode('ascii'))  # Print output
> ```

③ | Your Python server needs reliable network connectivity to both switches, SW1 and sw2. From u22s1, perform a quick ICMP test.

```
jdoe@u22s1:~/ch2_telnet$ cp ping_sw1.py ping_sw1_2.py
jdoe@u22s1:~/ch2_telnet$ nano ping_sw1_2.py
jdoe@u22s1:~/ch2_telnet$ cat ping_sw1_2.py
import os
os.system("ping -c 4 192.168.127.101")
os.system("ping -c 4 192.168.127.102")
jdoe@u22s1:~/ch2_telnet$ python ping_sw1_2.py
PING 192.168.127.101 (192.168.127.101) 56(84) bytes of data.
64 bytes from 192.168.127.101: icmp_seq=1 ttl=255 time=16.1 ms
[...omitted for brevity]
rtt min/avg/max/mdev = 8.310/14.403/23.064/5.765 ms
PING 192.168.127.102 (192.168.127.102) 56(84) bytes of data.
64 bytes from 192.168.127.102: icmp_seq=1 ttl=255 time=17.4 ms
[...omitted for brevity]
```

(continued)

#	Task
④	The network connection seems to be fine. Now, run the add_100_vlans.py script to add 100 VLANs to both switches. This script will Telnet into the first switch to add 100 VLANs and then Telnet into the second switch to add another 100 VLANs. This script may take a few minutes to complete, so be patient. Also, if your lab configuration is on an older computer like mine, consider reducing the number range from 100 to 10 so you speed up the process. ``` jdoe@u22s1:~/ch2_telnet$ python add_100_vlans.py Enter your username: jdoe Password: ******* SWITCH IP : 192.168.127.101 [...omitted for brevity] SW1#conf t Enter configuration commands, one per line. End with CNTL/Z. SW1(config)#vlan 700 SW1(config-vlan)#name PYTHON_VLAN_700 SW1(config-vlan)#vlan 701 SW1(config-vlan)#name PYTHON_VLAN_701 [...omitted for brevity] SWITCH IP : 192.168.127.102 [...omitted for brevity] sw2(config-vlan)#vlan 798 sw2(config-vlan)#name PYTHON_VLAN_798 sw2(config-vlan)#vlan 799 sw2(config-vlan)#name PYTHON_VLAN_799 sw2(config-vlan)#end sw2#exit ```

(continued)

CHAPTER 2   PYTHON NETWORK AUTOMATION LABS - BASIC TELNET

#	Task
| ⑤ | After the script execution is complete, use the `sh_vlan2.py` script from the previous lab to validate the configuration changes. You should see newly configured VLANs, 700 to 799, on both switches. Refer to Figure 2-21.<br><br>jdoe@u22s1:~/ch2_telnet$ **python sh_vlan2.py**<br>Enter your username: **jdoe**<br>Password: **\*\*\*\*\*\*\*\***<br><br>```
jdoe@u22s1: ~/ch2_telnet                                         —    □    ×
1005  trnet-default                          act/unsup

VLAN Type  SAID       MTU   Parent RingNo BridgeNo Stp  BrdgMode Trans1 Trans2
---- ----- ---------- ----- ------ ------ -------- ---- -------- ------ ------
1    enet  100001     1500  -      -      -        -    -        0      0
2    enet  100002     1500  -      -      -        -    -        0      0
3    enet  100003     1500  -      -      -        -    -        0      0
4    enet  100004     1500  -      -      -        -    -        0      0
5    enet  100005     1500  -      -      -        -    -        0      0
101  enet  100101     1500  -      -      -        -    -        0      0
202  enet  100202     1500  -      -      -        -    -        0      0
303  enet  100303     1500  -      -      -        -    -        0      0
404  enet  100404     1500  -      -      -        -    -        0      0
505  enet  100505     1500  -      -      -        -    -        0      0
700  enet  100700     1500  -      -      -        -    -        0      0
701  enet  100701     1500  -      -      -        -    -        0      0
702  enet  100702     1500  -      -      -        -    -        0      0
703  enet  100703     1500  -      -      -        -    -        0      0
704  enet  100704     1500  -      -      -        -    -        0      0
705  enet  100705     1500  -      -      -        -    -        0      0
706  enet  100706     1500  -      -      -        -    -        0      0
707  enet  100707     1500  -      -      -        -    -        0      0
708  enet  100708     1500  -      -      -        -    -        0      0
709  enet  100709     1500  -      -      -        -    -        0      0
710  enet  100710     1500  -      -      -        -    -        0      0
711  enet  100711     1500  -      -      -        -    -        0      0
712  enet  100712     1500  -      -      -        -    -        0      0
713  enet  100713     1500  -      -      -        -    -        0      0
```<br><br>*Figure 2-21. Telnet Lab 5, u22s1 - sh_vlan2.py partial output* |

(continued)
```

#	Task
6	To remove (reverse) the 100 VLANs from your switches, you only need to modify two lines of code in the original script. The following example will copy the original script and rename it reverse_100_vlans.py. Open this new file and modify two lines of code, as shown here. Add no in front of vlan, so it becomes no vlan. Then, on the next line, add a # to ignore this line. jdoe@u22s1:~/ch2_telnet$ **cp add_100_vlans.py reverse_100_vlans.py** jdoe@u22s1:~/ch2_telnet$ **nano reverse_100_vlans.py** [...omitted for brevity] for n in range (700, 800): tn.write(b"**no** vlan " + str(n).encode('UTF-8') + b"\n") **#**tn.write(b"name PYTHON_VLAN_" + str(n).encode('UTF-8') + b"\n") [...omitted for brevity] From your server, run python reverse_100_vlans.py to completely remove the vlans. jdoe@u22s1:~/ch2_telnet$ **python reverse_100_vlans.py** Enter your username: **jdoe** Password: ********
7	Lastly, use sh_vlan2.py to check the deletion of 100 VLANs on SW1 and sw2. jdoe@u22s1:~/ch2_telnet$ **python sh_vlan2.py** Enter your username: **jdoe** Password: ******* [...omitted for brevity]

Now you have completed Telnet Lab 5. Next, you'll configure a new user account across your managed network devices.

CHAPTER 2   PYTHON NETWORK AUTOMATION LABS - BASIC TELNET

Telnet Lab 6: Add a Privilege 3 User on Multiple Devices Using IP Addresses from an External File

In this lab, your task involves setting up a new user with restricted privileges across all routers and switches within your network, requiring the devices displayed in Figure 2-22 to be powered on. You will create a distinct file containing IP addresses. This file will enable your script to sequentially read IP addresses line by line and configure each device.

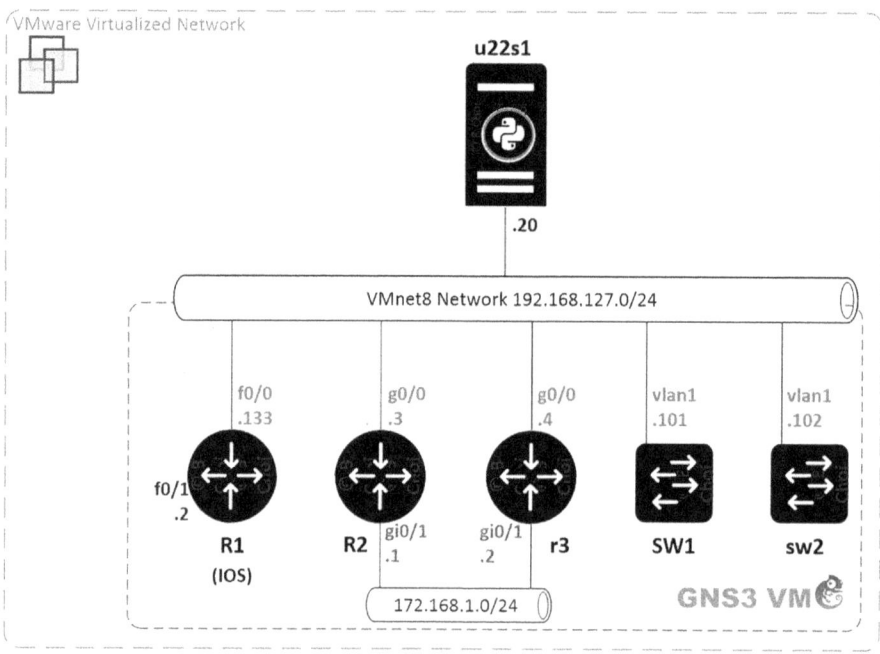

Figure 2-22. *Telnet Lab 6, devices in use*

The new user should be given a junior network administrator privilege to run show commands. You will be assigning this new user a privilege 3 and allowing this user to view the network devices' running configuration and interface status. Also, you'll change the file mode to the executable to run your script without typing python or python3.

At the end of this lab, you will have a local account created on each device, so the new user can run the show running-config view full, show ip interface brief, and other show commands.

102

CHAPTER 2   PYTHON NETWORK AUTOMATION LABS - BASIC TELNET

Telnet Lab 6 objectives: Configure the `junioradmin` accounts on all five devices.

- New username: `junioradmin`
- Password: `cisco321`
- Privilege level: 3
- Command 1: `privilege exec all level 3 show running-config`
- Command 2: `file privilege 3`

It's coding time once more—time to commence your work.

#	Task
①	First, go to your GNS3 user's main window and power on all of the devices involved, if you have not yet powered them on. The following devices must be up and running for this lab: R1, R2, r3, SW1, sw2, and u22s1.
②	Next, go to the u22s1 Python server and create a text file containing the IP addresses of your network devices. Be sure to list IP addresses on individual lines, separated by a new line for each distinct address. Follow these steps to create and save the `ip_addresses.txt` file. Place the `ip_addresses.txt` file in the current directory. jdoe@u22s1:~/ch2_telnet$ **touch ip_addresses.txt** jdoe@u22s1:~/ch2_telnet$ **nano ip_addresses.txt** **192.168.127.3** # R2 **192.168.127.4** # r3 **192.168.127.101** # SW1 **192.168.127.102** # sw2 **192.168.127.133** # R1

(continued)

#	Task
| ③ | To save time again, you'll reuse the old script to create a new one. Copy the script from Lab 5 and create a new script called add_junioradmin.py. There is not a lot you have to modify but carefully update the required line referring to the lab objectives.<br><br>jdoe@u22s1:~/ch2_telnet$ **cp add_100_vlans.py add_junioradmin.py**<br>jdoe@u22s1:~/ch2_telnet$ **nano add_junioradmin.py**<br>jdoe@u22s1:~/ch2_telnet$ **cat add_junioradmin.py**<br><br>```
#!/usr/bin/env python3.10 # Shebang for Python script, required for this lab!

import getpass  # Import password input
import telnetlib  # Import Telnet library
import time # Import time module

user = input("Enter your username: ")  # Prompt for username
password = getpass.getpass()  # Securely prompt for password

# Read IP addresses file
with open("ip_addresses.txt") as file: # Open file for reading
for ip in file: # Iterate through IP addresses
   print("Now configuring : " + ip) # Display configuring message
   HOST = ip.strip() # Trim IP address

tn = telnetlib.Telnet(HOST)  # Connect to device via Telnet
tn.read_until(b"Username: ")  # Read 'Username' prompt
tn.write(user.encode('ascii') + b"\n")  # Send username

if password:  # Check for password existence
   tn.read_until(b"Password: ")  # Read 'Password' prompt
   tn.write(password.encode('ascii') + b"\n")  # Send password

time.sleep(1) # Introduce a 1-second delay

# Configure new user with privileges
tn.write(b"conf t\n") # Enter configuration mode
tn.write(b"username junioradmin privilege 3 password cisco321\n")
# Create new user
``` |

(continued)
```

#	Task

```
tn.write(b"privilege exec all level 3 show running-config\n")
# Permit 'show running-config'
print("Added a new privilege 3 user") # Display completion message

tn.write(b"end\n") # Exit configuration mode
tn.write(b"exit\n") # Terminate Telnet session
print(tn.read_all().decode('ascii')) # Show configuration output
```

④	Now create a new ping Python script by combining the os ping method, but use the open method to read the IP addresses from the ip_addresses.txt file. When your script is complete, it should look like the following code.

```
doe@u22s1:~/ch2_telnet$ nano ping_read_ips.py
doe@u22s1:~/ch2_telnet$ cat ping_read_ips.py

import os # Import os module

with open("ip_addresses.txt", "r") as file: # Open and read the 'ip_
                                              addresses.txt' file
    for line in file: # Loop through each line in the file
        ip_address = line.split()[0] # Extract the IP address
                                       from the line

        # Execute the ping command for each IP address
        response = os.system(f"ping -c 4 {ip_address}")

        # Check and print the ping result
        if response == 0: # Check if the ping was successful
            print(f"Ping to {ip_address} successful.")
            # Print successful ping
        else:
            print(f"Ping to {ip_address} unsuccessful.")
            # Print unsuccessful ping
```

When you execute the Python code, it will read each IP address from the text file and ping each IP. If your ping script works, the output should look like Figure 2-23.

(continued)

#	Task

```
64 bytes from 192.168.127.101: icmp_seq=3 ttl=255 time=11.3 ms
64 bytes from 192.168.127.101: icmp_seq=4 ttl=255 time=28.1 ms

--- 192.168.127.101 ping statistics ---
4 packets transmitted, 4 received, 0% packet loss, time 3005ms
rtt min/avg/max/mdev = 9.940/17.515/28.075/7.371 ms
Ping to 192.168.127.101 successful.
PING 192.168.127.102 (192.168.127.102) 56(84) bytes of data.
64 bytes from 192.168.127.102: icmp_seq=1 ttl=255 time=36.4 ms
64 bytes from 192.168.127.102: icmp_seq=2 ttl=255 time=21.8 ms
64 bytes from 192.168.127.102: icmp_seq=3 ttl=255 time=30.0 ms
64 bytes from 192.168.127.102: icmp_seq=4 ttl=255 time=41.2 ms

--- 192.168.127.102 ping statistics ---
4 packets transmitted, 4 received, 0% packet loss, time 3006ms
rtt min/avg/max/mdev = 21.825/32.350/41.181/7.252 ms
Ping to 192.168.127.102 successful.
PING 192.168.127.133 (192.168.127.133) 56(84) bytes of data.
64 bytes from 192.168.127.133: icmp_seq=1 ttl=255 time=4.09 ms
64 bytes from 192.168.127.133: icmp_seq=2 ttl=255 time=2.27 ms
64 bytes from 192.168.127.133: icmp_seq=3 ttl=255 time=4.74 ms
64 bytes from 192.168.127.133: icmp_seq=4 ttl=255 time=8.75 ms

--- 192.168.127.133 ping statistics ---
4 packets transmitted, 4 received, 0% packet loss, time 3005ms
rtt min/avg/max/mdev = 2.269/4.962/8.752/2.368 ms
Ping to 192.168.127.133 successful.
jdoe@u22s1:~/ch2_telnet$
```

Figure 2-23. *Telnet Lab 6, u22s1 - ping_read_ips.py execution output*

(5) Alternatively, if you don't want to use a script but want to check the connectivity using a single Linux command from u22s1, then use fping. First, install it on the server before running the command. If you get ... is alive, the IP address is reachable.

jdoe@u22s1:~/ch2_telnet$ **fping 192.168.127.3 192.168.127.4 192.168.127.101 192.168.127.102 192.168.127.133**
Command 'fping' not found, but can be installed with:
sudo apt install fping
jdoe@u22s1:~/ch2_telnet$ **sudo apt install fping**
[sudo] password for jdoe: **********
[...omitted for brevity]
jdoe@u22s1:~/ch2_telnet$ **fping 192.168.127.3 192.168.127.4 192.168.127.101 192.168.127.102 192.168.127.133**

(*continued*)

#	Task
	192.168.127.3 is alive 192.168.127.4 is alive 192.168.127.133 is alive 192.168.127.101 is alive 192.168.127.102 is alive
(6)	6-1. You've included #!/usr/bin/env python3.10 at the beginning of your script. This line informs the Linux system that the application requires Python 3 to run. Initially, the script isn't executable, as observed when listing the file: jdoe@u22s1:~/ch2_telnet$ **ls -l add_junioradmin.py** -rw-rw-r-- 1 jdoe jdoe 789 Oct 31 02:02 add_junioradmin.py To execute the script without using the python command, you need to modify the file's mode using the chmod +x command: Notice the color change on your file. jdoe@u22s1:~/ch2_telnet$ **chmod +x add_junioradmin.py** jdoe@u22s1:~/ch2_telnet$ **ls -l add_junioradmin.py** -rwxrwxr-x 1 jdoe jdoe 789 Oct 31 02:02 add_junioradmin.py 6-2. If you aim to simplify script execution to solely using the script name, you can add the following PATH variable to your Linux server. This command provides temporary session-based functionality while you're logged in. There are methods to set this permanently, but this will be covered shortly. At this stage, your Python script can be run using only the script's name. Press Ctrl+Z to pause and exit if you want to resume running the script later. jdoe@u22s1:~/ch2_telnet$ **PATH="$(pwd):$PATH"** jdoe@u22s1:~/ch2_telnet$ **add_junioradmin.py** Enter your username: ^Z # Press Ctrl+Z [2]+ Stopped add_junioradmin.py 6-3. For a more streamlined and permanent execution, you can add the directory containing the script to the PATH variable on your Linux server. To achieve this:

(continued)

#	Task
	6-3a. Identify and edit the user's profile configuration file (e.g., ~/.bashrc for the Ubuntu 22 Server). The location of this profile may differ from one Linux distribution to another. 6-3b. Append the following line to the end of the file: **export PATH="/home/jdoe/ch2_telnet:$PATH"** 6-3c. Apply the changes using this command: jdoe@u22s1:~/ch2_telnet$ **source ~/.bashrc** Following this, you can log out of your SSH connection and reconnect. Once in the /home/jdoe/ch2_telnet directory, execute add_junioradmin.py. It should run without the need for python or ./ preceding the script.
⑦	Assuming that all your devices are reachable on the network, now run the script only using the script name, that is add_junioradmin.py. The script will add the junior admin to all five devices, as shown here: jdoe@u22s1:~/ch2_telnet$ **add_junioradmin.py** Enter your username: **jdoe** Password:******** When your junior user creation script runs, it should configure the new account on all five devices—three routers and two switches. Imagine if you need to update a local username or password on hundreds of devices. Now you can relate to a good use case for such a tool (see Figure 2-24).

(continued)

#	Task

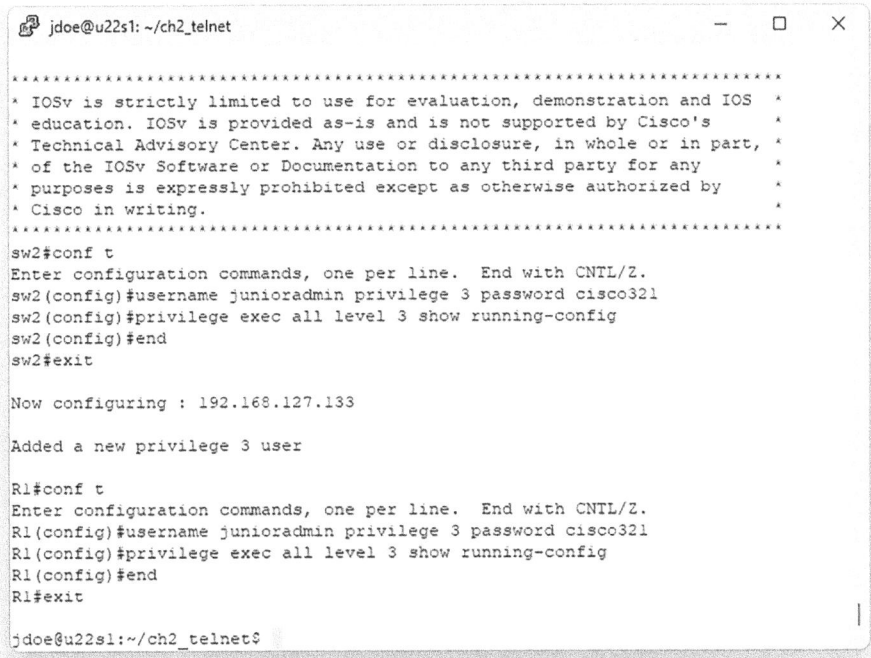

Figure 2-24. Telnet Lab 6, u22s1 - add_junioradmin.py output example

⑧	From u22s1, write another script, check_junioradmin.py, to read the IP addresses from the ip_addresses.txt file and verify the configuration of the user. Use the show run | in username junioradmin and show run | in level 3 commands. jdoe@u22s1:~/ch2_telnet$ **nano check_junioradmin.py** #!/usr/bin/env python3.10 import getpass import telnetlib import time user = input("Enter your username: ") password = getpass.getpass()

(*continued*)

#	Task

```
with open("ip_addresses.txt") as file:
    for ip in file:
        print("-"*79) # separates output and helps readability
        print(f"Now checking : {ip}")
        HOST = ip.strip()

        tn = telnetlib.Telnet(HOST)
        tn.read_until(b"Username: ")
        tn.write(user.encode('ascii') + b"\n")
        if password:
            tn.read_until(b"Password: ")
            tn.write(password.encode('ascii') + b"\n")

        tn.write(b"show run | in username junioradmin\n")
        # show command 1
        tn.write(b"show run | in level 3\n") # show command 2
        tn.write(b"exit\n")

        print(tn.read_all().decode('ascii'))
```

When your script execution completes, the end output of the code should look similar to Figure 2-25.

(*continued*)

CHAPTER 2   PYTHON NETWORK AUTOMATION LABS - BASIC TELNET

#	Task
	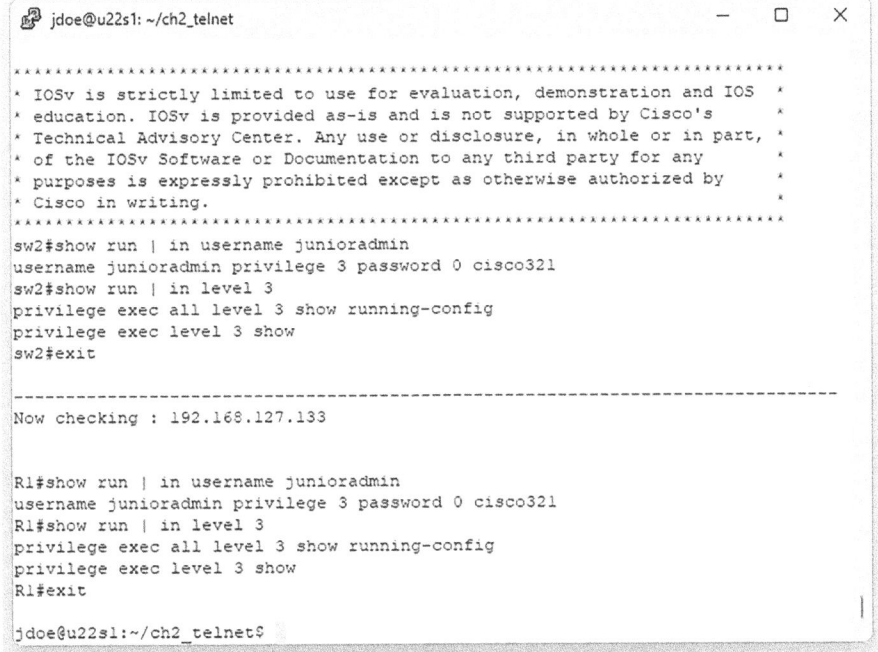 *Figure 2-25. Telnet Lab 6, u22s1 – verify junioradmin user configuration*
⑨	Now use PuTTY from your Windows host PC to Telnet into SW1 (192.168.127.101) with the junioradmin username and password `cisco321`. Then run the `show ip interface brief` or `show running-config view full` command. If you want to log in to another device, use 192.168.127.X, where X is the last octet of the device you want to Telnet. The example in Figure 2-26 displays the logged in user(s) and shows the command output of SW1 from the host PC's PuTTY.

(continued)

CHAPTER 2   PYTHON NETWORK AUTOMATION LABS - BASIC TELNET

Figure 2-26. Telnet Lab 6, Windows host PC – log in and run show command example

You have effectively established the junior admin user accounts across multiple devices by utilizing IP addresses read from an external file. It is essential to note that the IP addresses need not be contiguous; they can be disparate and drawn from lists or external files. With minor adjustments, the Python concepts acquired from these Telnet exercises can be easily applied to SSH labs.

Telnet Lab 7: Taking Backups of running-config (or startup-config) to Local Server Storage

In the last Telnet lab of this book, you will copy and modify the previous Telnet script to capture the current running configuration of each device in your network. The backed-up `running-config` files will be saved on `u22s1`'s local directory. In this lab, you will be using the `datetime` module to get the timestamp and add the name of the files with this timestamp so you know when the backups were taken. You will require all routers and switches to be powered on, as in Lab 6 (see Figure 2-27).

CHAPTER 2   PYTHON NETWORK AUTOMATION LABS - BASIC TELNET

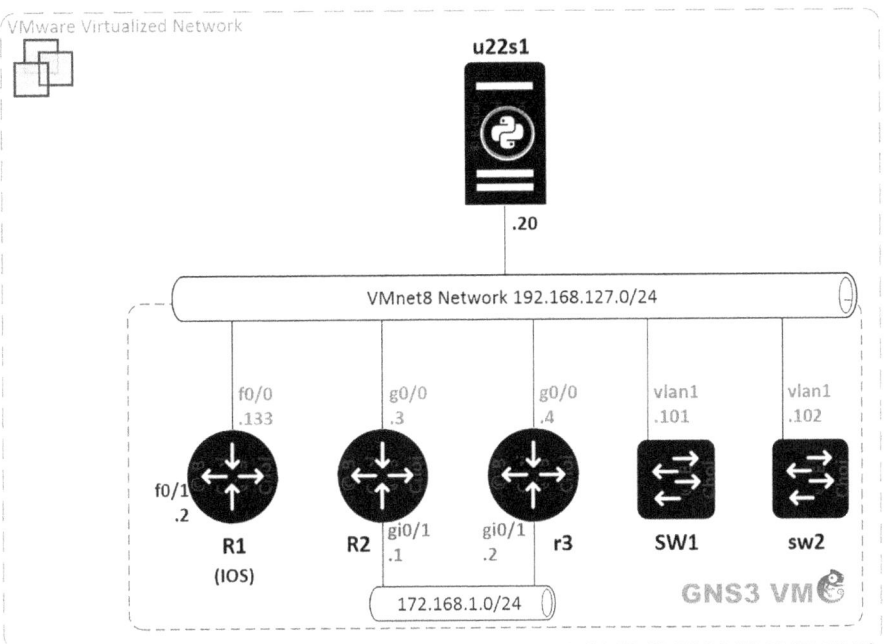

Figure 2-27. *Telnet Lab 7, devices in use*

At the end of this lab, you will have the backups of each device's running configuration with the timestamp on your Python server's local storage.

Telnet Lab 7 objectives:
Back up all devices' running configurations in a folder on u22s1

Time to get the coding wheels turning for one last time in this chapter.

#	Task
①	Once again, begin the lab by copying the script from the previous lab. The file name given here is take_backups.py, but you do not have to follow this naming convention. Provide a more meaningful name to this file so you can remember it easily. jdoe@u22s1:~/ch2_telnet$ **cp add_junioradmin.py take_backups.py** jdoe@u22s1:~/ch2_telnet$ **nano take_backups.py**

(*continued*)

113

#	Task
②	Now, open the file in the nano (or vi) text editor and make the modifications highlighted in the following source code:

```
#!/usr/bin/env python3.10
import getpass  # Import getpass module
import telnetlib  # Import Telnet library
from datetime import datetime # Import datetime module

# Get current timestamp
saved_time = datetime.now().strftime("%Y%m%d_%H%M%S")

# Input for username
user = input("Enter your username: ")
password = getpass.getpass()  # Input for password

file = open("ip_addresses.txt")  # Open file containing IP addresses

for ip in file:  # Loop through IP addresses
    print("Back up running-config of " + (ip))

    HOST = ip.strip()  # Trim IP address
    tn = telnetlib.Telnet(HOST)  # Connect to the device via Telnet
    tn.read_until(b"Username: ")  # Wait for 'Username' prompt
    tn.write(user.encode('ascii') + b"\n")  # Send username

if password:
    tn.read_until(b"Password: ")  # Wait for 'Password' prompt
    tn.write(password.encode('ascii') + b"\n")  # Send password

# Set terminal length & fetch config
tn.write(("terminal length 0\n").encode('ascii'))
tn.write(("show clock\n").encode('ascii'))
tn.write(("show running-config\n").encode('ascii'))
tn.write(("exit\n").encode('ascii'))
```

(continued)

CHAPTER 2 PYTHON NETWORK AUTOMATION LABS - BASIC TELNET

#	Task

	`readoutput = tn.read_all()` # Read output
	# Save output to file with timestamp
	`saveoutput = open(str(saved_time) + "_running_config_" + HOST, "wb")`
	`saveoutput.write(readoutput)`
	`saveoutput.close()` # Close file

③	Use the previous Linux `fping` command again to check the network connectivity on your network. If any of your nodes (network devices) are not reachable, you must troubleshoot the connectivity issue first. jdoe@u22s1:~/ch2_telnet$ **fping 192.168.127.3 192.168.127.4 192.168.127.101 192.168.127.102 192.168.127.133** 192.168.127.3 is alive 192.168.127.4 is alive 192.168.127.133 is alive 192.168.127.102 is alive 192.168.127.101 is alive
④	Now, it is time to run the code one last time in this chapter. Run the take_backup.py script from your u22s1 server (see Figure 2-28). jdoe@u22s1:~/ch2_telnet$ **python take_backups.py** Enter your username: **jdoe** Password: *********** 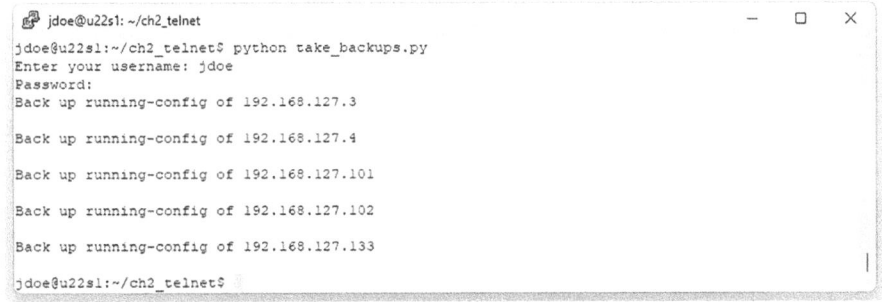 *Figure 2-28. Telnet Lab 7, u22s1 - take_backups.py output*

(continued)

CHAPTER 2   PYTHON NETWORK AUTOMATION LABS - BASIC TELNET

#	Task
⑤	Upon successful execution of the script, use the `ls -lh 20*` command to inspect the locally stored backup files containing the running configurations of all network devices. These files should commence with the year, month, and day, followed by the precise time the backup was created (see Figure 2-29).

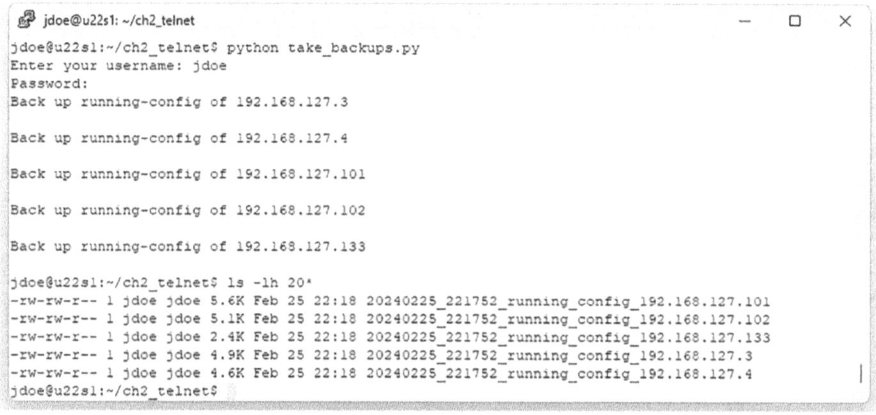

Figure 2-29. Telnet Lab 7, u22s1 – ls -lh 20* output after first execution

| ⑥ | Executing the script again will result in five additional files saved with distinct names based on their respective timestamps. This functionality ensures that existing files are preserved, as new backup files are created each time the script runs (see Figure 2-30). |

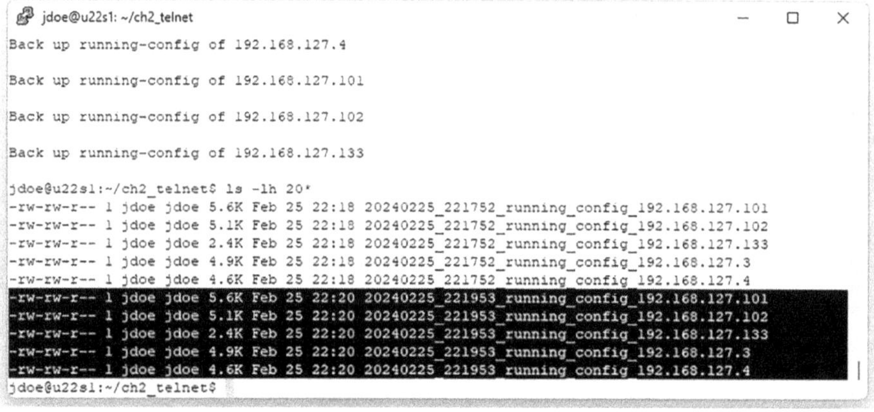

Figure 2-30. Telnet Lab 7, u22s1 – ls -lh 20* output after second execution

(*continued*)

#	Task
| ⑦ | Enhance the script by adding the `write memory` or `copy running-config startup-config` commands to retain configuration changes. Duplicate the `check_junioradmin.py` file and make slight modifications to capture the running configurations of all five devices. The configuration remains almost identical, except for the parts in bold and highlighted. For simplicity, the `write memory` method is used in this case.<br><br>```
jdoe@u22s1:~/ch2_telnet$ cp check_junioradmin.py save_run.py
jdoe@u22s1:~/ch2_telnet$ nano save_run.py
jdoe@u22s1:~/ch2_telnet$ cat save_run.py
[...omitted for brevity]
with open("ip_addresses.txt") as file:
for ip in file:
   print("-"*79)
   print(f"Saving config on : {ip}")
   HOST = ip.strip()
[...omitted for brevity]
   tn.write(b"write memory\n")
   tn.write(b"exit\n")
   print(tn.read_all().decode('ascii'))
```<br>After the execution of the `save_run.py` file, you should see a screenshot similar to Figure 2-31.<br><br>```
jdoe@u22s1:~/ch2_telnet$ python save_run.py
Enter your username: jdoe
Password: ************
``` |

CHAPTER 2   PYTHON NETWORK AUTOMATION LABS - BASIC TELNET

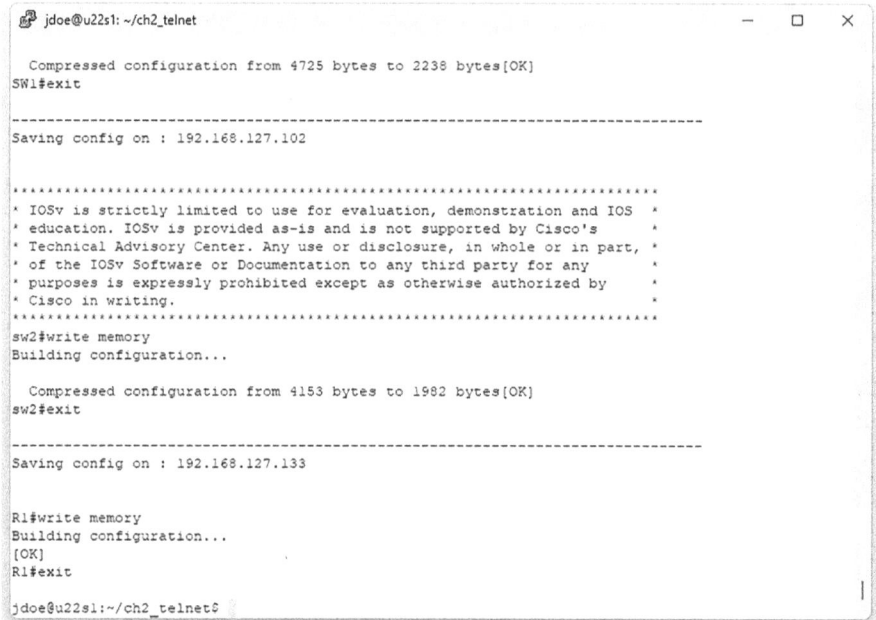

Figure 2-31. *Telnet Lab 7, u22s1 – save configuration script output*

Your Python script on the Linux server efficiently backs up the current running configurations of both routers and switches to the local directory. Although the process of backing up to an S/FTP server is also on the "topics to cover" wish list, this is reserved for upcoming SSH labs. Looking ahead, your next challenge involves SSH labs, using the `paramiko` and `netmiko` libraries, and building upon the various concepts you have learned from the Telnet labs.

While Telnet is becoming less common in production environments, the focus of this chapter was on writing seven straightforward Telnet applications and experiencing what writing working applications is like in real life. These exercises offered a hands-on understanding of how Python code directly correlates with practical work scenarios. Acquiring a grasp of Python concepts and coding proficiency is crucial, yet transcending theory to application—to effectively link these skills to real-world tasks—is paramount once you've mastered the basics. It's this bridge between theory and application that truly solidifies the value of your coding experience.

Summary

This chapter focused on seven basic Telnet labs to draw upon and apply the knowledge gained from the chapters in the Part I book. You explored writing Telnet scripts, copying, rewriting, and expanding ideas, and performing script executions and configuration validations from the Python Linux server, without needing to console or directly Telnet/SSH into the devices yourself. Throughout these labs, you learned to use some new Python modules such as the `ping` and `datetime` modules. I also introduced Linux's `fping` to check connections to multiple devices using a single Linux command. Initially, you executed simple configuration changes and then transitioned to making changes using Python scripts, leveraging the power of `for` and `while` loops and `range` commands. Furthermore, you mastered the process of reading IP addresses from an external file and slowly improved your scripts. The culmination involved making configuration backups for all five network devices using Python's `telnetlib`, with timestamps integrated into each file name. **You have experienced the robustness of the current lab setup and now you will systematically learn to write your actions and thoughts into lines of code.** Looking ahead to Chapter 3, you will continue your journey by delving into Python Network Automation via SSH connections. This exploration will be conducted using the `paramiko` and `netmiko` SSH libraries, building upon the foundational understanding gained in these Telnet automation labs.

CHAPTER 3

Python Network Automation Labs: SSH in Action, paramiko and netmiko Labs

Welcome to Chapter 3, an immersive journey into Python's SSH libraries—`paramiko` and `netmiko`—as integral tools for managing networking devices. This chapter delves into the enduring role of SSH in network infrastructure, tracing the evolution of remote management tools, and explores the robust functionalities of the `paramiko` and `netmiko` libraries. It's a hands-on experience that weaves the intricate tapestry of networking and Python programming concepts into practical application. Get ready to traverse through a series of labs that offer real-world exposure, starting with understanding SSH functionalities on Linux servers and Cisco devices, followed by Python SSH labs utilizing `paramiko`. You'll engage in interactive labs focusing on configuring device settings such as clock, time zones, and NTP servers on Cisco devices. Additionally, exploratory labs will demonstrate interactive `paramiko` scripts that preserve running configurations. Transitioning to the `netmiko` library, you learn to install and use it in practical labs, including using a dictionary for device information, crafting a `netmiko` Disable Telnet script, conducting configuration comparisons, and examining SSH key exchanges between devices—network device to Linux and network device to network device. The hands-on approach here equips you with the practical insights to immediately apply these SSH modules in your work. Furthermore, these labs form the cornerstone for building an IOS XE upgrade application and offer an in-depth exploration of automating device upgrades.

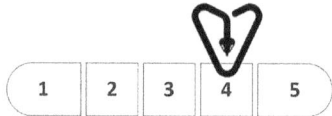

The Enduring Role of SSH in Network Infrastructure: Navigating the Evolution of Remote Management Tools

Since its introduction in 1995, SSH (Secure Shell) has remained a vital tool in modern enterprise network management. Despite the advancements in technology, SSH continues to hold a prominent position in remote enterprise device management, serving as a more secure alternative to the dated Telnet protocol introduced in 1969. Network and systems engineers today continue to favor SSH due to its robust security features and functionality, distinguishing it from GUI-based tools.

Despite the emergence of newer deployment methods, like plug-n-play and zero-touch provisioning, alongside the rise of software-based enterprise network device management such as APIs and Webhooks in the 2000s, SSH remains an indispensable fallback option when GUIs, APIs, or operating systems encounter issues. Its enduring appeal makes it the preferred method for connecting to remote enterprise network devices, likely retaining its dominance in the foreseeable future.

For instance, the initialization and provisioning of cloud-based solution physical switches heavily depend on SSH access, both in the factory and during staging, as well as in deploying physical infrastructure in production. Although many software developers, DevOps, and SRE engineers are distanced from hardware, having visibility into the underlying physical infrastructure remains critical for genuine infrastructure (including network) engineers. Cloud systems, whether public, private, or hybrid, extensively rely on hardware-based servers in data centers. They still rely on conventional methods like SSH and console connections.

CHAPTER 3   PYTHON NETWORK AUTOMATION LABS: SSH IN ACTION, PARAMIKO AND NETMIKO LABS

The shift toward software-driven networking devices residing in the cloud is evident; however, the foundational networking concepts persist despite the change in management approaches. Today, crucial infrastructure functions as software on remote data center servers, emphasizing the relevance of the "somebody else's computer" concept, particularly in infrastructure engineering.

While the focus for most developers and engineers primarily revolves around software and applications, disregarding in-depth networking and infrastructure understanding, preserving a foundational understanding of infrastructure management remains crucial. Despite the ever-evolving landscape of computing and application consumption, the core infrastructure managed by network and infrastructure experts continues to rely on traditional practices. At this infrastructure level, the reliability and familiarity of the "good old" console and SSH connections remain fundamental to keep the lights on for everyone.

Python Network Automation Labs Using the paramiko and netmiko Libraries

In the previous chapter, you delved into fundamental Python networking concepts using Telnet labs. Whether using Telnet or SSH, these Python networking principles can be seamlessly applied to your scenarios. However, SSH stands out as a significantly more secure remote login protocol compared to Telnet, requiring a specialized SSH library to access networking devices.

As you're well aware, Telnet operates as a plain-text network protocol, rendering it less secure in comparison to SSH. In most secure networks, Telnet usage is generally prohibited, allowing only encrypted SSH connections. When establishing an SSH connection between devices, it's not merely a matter of connecting with a pre-shared key, as in a wireless network. Devices must agree on the communication protocol, key exchange algorithms, host key algorithms, and ciphers for encoding and decoding the communication.

Buckle up, as in this chapter, you delve into practical SSH implementations, exploring SSH key exchange methods used between various devices, including network device to Linux and network device to network device. Understanding these SSH foundations is crucial to harnessing the full potential of this protocol within your Python scripts.

Connecting to Cisco devices using SSH in Python mandates the initial installation of the Python paramiko library on your Python server, along with the configuration of encrypted RSA keys on the Cisco devices. To kick start this chapter, let's dive deep into exploring the SSH protocol, which will be extensively used throughout the remainder of this book. The focus centers on the practical facets of the SSH protocol, uncovering the specific version of SSH running on each system and comprehending the SSH information hidden within your Python server and network devices. Following this exploration, you'll progress to the installation of the Python SSH library, paramiko, initiating your SSH lab journey within this chapter.

> **Tip**
>
> **Accessing the source code**
>
> While I encourage you to type each line of code for a genuine coding experience, if you need to save time and expedite your learning, you can use the cut-and-paste method to run the code in your environment. All the Python code (scripts) used in this book is accessible through the author's or publisher's GitHub repositories.
>
> Author's GitHub: https://github.com/pynetauto/apress_pynetauto_ed2.0/tree/main/source_codes/Part2
>
> Apress GitHub: https://github.com/Apress

Checking the SSH in Action on a Linux Server

As a precaution, it is best practice to check the SSH versions and the SSH authentication methods used on your devices. You'll learn how to discover SSH-related information in your Python server, u22s1.

First, you can check the authentication methods allowed on the server by looking at the sshd_config file, but it does not have a lot of details to reveal here and you know this information is in another file somewhere on your Linux server. This is sort of good news and bad news, because it allows you to customize the authentication method—this is good news for users like us.

```
jdoe@u22s1:~$ sudo cat /etc/ssh/sshd_config
[...omitted for brevity]
```

Second, check the SSH version on your Ubuntu Python server, u22s1, using the ssh -V command. Now that you checked the OpenSSH in action, let's look at the way to dig up more details on this application.

```
jdoe@u22s1:~$ ssh -V
OpenSSH_8.9p1 Ubuntu-3ubuntu0.4, OpenSSL 3.0.2 15 Mar 2022
```

Next, run the ssh -v command. The output will provide you with the usage information. Highlighted are more commonly used options. You can use these options to learn more about the SSH service running on your Linux server.

```
jdoe@u22s1:~$ ssh -v
usage: ssh [-46AaCfGgKkMNnqsTtVvXxYy] [-B bind_interface]
           [-b bind_address] [-c cipher_spec] [-D [bind_address:]port]
           [-E log_file] [-e escape_char] [-F configfile] [-I pkcs11]
           [-i identity_file] [-J [user@]host[:port]] [-L address]
           [-l login_name] [-m mac_spec] [-O ctl_cmd] [-o option] [-p port]
           [-Q query_option] [-R address] [-S ctl_path] [-W host:port]
           [-w local_tun[:remote_tun]] destination [command [argument ...]]
```

Now, use the ssh -Q commands to verify the supported authentication methods on your server. These commands offer a wealth of detailed SSH information supported by your Linux server. Following the initial lab, you'll also discover a more convenient method to obtain this information, using nmap. The nmap command will be used on the server, routers, and switches.

Table 3-1. *ssh -Q Query Options to Examine OpenSSH-Supported Authentication Method*

SSH -Q Command	Description
ssh -Q cipher	**Displays the list of supported ciphers.**
ssh -Q cipher-auth	Displays supported authenticated encryption ciphers.
ssh -Q mac	Shows available message authentication code (MAC) algorithms.
ssh -Q kex	**Lists the supported key exchange algorithms.**
ssh -Q key	**Displays supported key types.**
ssh -Q key-cert	Shows supported key types for certified keys.
ssh -Q key-sig	Lists the supported signature algorithms for keys.
ssh -Q help	Provides help with querying options.

For this lab, you are primarily focused on the outputs of the ssh -Q cipher, ssh -Q kex, and ssh -Q key commands, but you can try all of the commands suggested in Table 3-1. To aid in your comprehension, I have extracted the complete output and highlighted the significant ciphers, key exchanges, and keys used in this lab.

```
jdoe@u22s1:~$ ssh -Q cipher
3des-cbc
aes128-cbc
aes192-cbc
aes256-cbc
aes128-ctr
aes192-ctr
aes256-ctr
aes128-gcm@openssh.com
aes256-gcm@openssh.com
chacha20-poly1305@openssh.com
jdoe@u22s1:~$ ssh -Q kex
diffie-hellman-group1-sha1
diffie-hellman-group14-sha1
diffie-hellman-group14-sha256
diffie-hellman-group16-sha512
```

diffie-hellman-group18-sha512
diffie-hellman-group-exchange-sha1
diffie-hellman-group-exchange-sha256
ecdh-sha2-nistp256
ecdh-sha2-nistp384
ecdh-sha2-nistp521
curve25519-sha256
curve25519-sha256@libssh.org
sntrup761x25519-sha512@openssh.com
jdoe@u22s1:~$ **ssh -Q key**
ssh-ed25519
ssh-ed25519-cert-v01@openssh.com
sk-ssh-ed25519@openssh.com
sk-ssh-ed25519-cert-v01@openssh.com
ssh-rsa
ssh-dss
ecdsa-sha2-nistp256
ecdsa-sha2-nistp384
ecdsa-sha2-nistp521
sk-ecdsa-sha2-nistp256@openssh.com
ssh-rsa-cert-v01@openssh.com
ssh-dss-cert-v01@openssh.com
ecdsa-sha2-nistp256-cert-v01@openssh.com
ecdsa-sha2-nistp384-cert-v01@openssh.com
ecdsa-sha2-nistp521-cert-v01@openssh.com
sk-ecdsa-sha2-nistp256-cert-v01@openssh.com

Checking the SSH in Action on Cisco Devices

Let's check the SSH details on your Cisco devices. Use the show ip ssh command on R2, SW1 and R1. The outputs are similar to r3 and sw2. On R1, this command only reveals limited information, as this is an older IOS device. In these examples, you can compare the different authentication methods between a router and a switch, but they are almost identical. This information will come in handy in your first SSH lab.

CHAPTER 3   PYTHON NETWORK AUTOMATION LABS: SSH IN ACTION, PARAMIKO AND NETMIKO LABS

R2#**show ip ssh**
SSH Enabled - version 1.99
Authentication methods:publickey,keyboard-interactive,password
Authentication Publickey Algorithms:x509v3-ssh-rsa,ssh-rsa
Hostkey Algorithms:x509v3-ssh-rsa,ssh-rsa
Encryption Algorithms:aes128-ctr,aes192-ctr,aes256-ctr
MAC Algorithms:hmac-sha2-256,hmac-sha2-512,hmac-sha1,hmac-sha1-96
KEX Algorithms:diffie-hellman-group-exchange-sha1,diffie-hellman-group14-sha1
Authentication timeout: 120 secs; Authentication retries: 3
Minimum expected Diffie Hellman key size : 2048 bits
IOS Keys in SECSH format(ssh-rsa, base64 encoded): R2.pynetauto.local
ssh-rsa AAAAB3NzaC1yc2EAAAADAQABAAAAgQDDTCIMdKtzNOc/
OLIMpBn6Tr8Xv5dcYhDIzEAIqJmK
hO8N4COB68EYEgUrXAXM+6bngiIzAt7ppoyaXAwSab+P5OhAO7CCZCSaLZUJrD9u26Vg7q+
+vnsbXrcM
7UXoTytx3sylem4Tym55B6SlLjThdUlNlOlRzkXrKkrPv+k/tQ==

SW1#**show ip ssh**
SSH Enabled - version 1.99
Authentication methods:publickey,keyboard-interactive,password
Encryption Algorithms:aes128-ctr,aes192-ctr,aes256-ctr,aes128-cbc,3des-cbc,aes192-cbc,aes256-cbc
MAC Algorithms:hmac-sha1,hmac-sha1-96
Authentication timeout: 120 secs; Authentication retries: 3
Minimum expected Diffie Hellman key size : 1024 bits
IOS Keys in SECSH format(ssh-rsa, base64 encoded): SW1.pynetauto.local
ssh-rsa
AAAAB3NzaC1yc2EAAAADAQABAAAAgQCcB2EEDR4qu4jiTfORe9pOm7QhBHLk6O57c2GsLDdx
9slXy8PxOdmHOJ/xa1cwZBK3RlaoOPHKHmCVm8wOzPJ+hNdpmvc6pvRMPOcpCa4DCykbP2
bqS26LOgkp
33Bgn+SIwiFKNY6xZX7XwON+fNtdnAlm2uNNEwz7Cv8STbPkUQ==

R1#**show ip ssh**
SSH Enabled - version 1.99
Authentication timeout: 120 secs; Authentication retries: 3

CHAPTER 3   PYTHON NETWORK AUTOMATION LABS: SSH IN ACTION, PARAMIKO AND NETMIKO LABS

You've had a quick review of the SSH settings on both your server and network devices, so it is time to go back to the Python server and install the paramiko library to get the first lab going.

Python SSH Labs: paramiko

On your Python automation server, u22s1, you can install paramiko using the pip command shown here. This command works on Debian-based or Red Hat-based Linux distributions. If you are using Python 3 with dist-packages 2.9.3 on your Linux installation, you may not have to install this SSH library separately. Also, use the pip freeze command with the grep option to check the version of paramiko.

```
jdoe@u22s1:~$ pip3 install paramiko
[...omitted for brevity]
jdoe@u22s1:~$ pip freeze | grep paramiko
paramiko==2.9.3
```

After you have confirmed that paramiko is on your server, quickly run the Python interpreter session, and run import paramiko to check that the library can be imported without error. When you are ready, you can begin the first SSH lab after going over the devices in use for SSH Lab 1.

```
jdoe@u22s1:~$ python
Python 3.10.12 (main, Jun 11 2023, 05:26:28) [GCC 11.4.0] on linux
Type "help", "copyright", "credits" or "license" for more information.
>>> import paramiko
>>> # If no error, then you are ready to move to Lab 1. Exit Python using 'exit()'.
```

paramiko Lab 1: Configure the Clock and Time Zone of All Devices Interactively in the Python Interpreter

You will partake in an interactive SSH session to manage your network devices from the Python interpreter shell. Before connecting to the Cisco devices via SSH, the initial step involves generating RSA keys for each device. As you won't be using the AAA RADIUS server for authentication in your lab, you'll create local certificates for each device.

CHAPTER 3   PYTHON NETWORK AUTOMATION LABS: SSH IN ACTION, PARAMIKO AND NETMIKO LABS

Follow the outlined tasks to prepare for the first SSH lab. In this lab, you'll log in to the Python interpreter on your Python server and create a real-time script to interact with all routers and switches, adjusting their time and time zone settings. I expect you to power on all three routers and two switches alongside your server. However, if your PC operates on an older, low-performance CPU or has limited memory, you're only required to power on one router and one switch. This approach aims to prevent potential high CPU utilization and memory contention issues during the lab. Refer to Figure 3-1 for guidance.

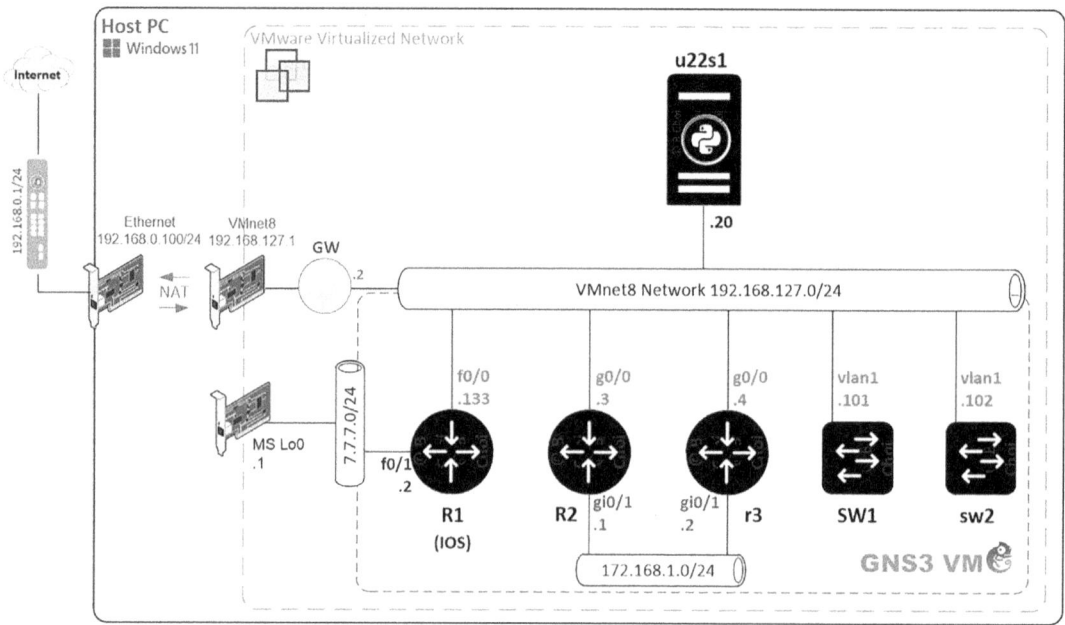

Figure 3-1. *SSH paramiko Lab 1, devices in use*

Additionally, after the first SSH lab, I've included an in-depth practical exploration of SSH between network devices and the Linux server to further expand your knowledge.

#	Task

①	1-1. Use the following Cisco IOS commands to create 1,024-bit local RSA keys on each device for SSH connection, but leave the device to use both SSH and Telnet for now. The following example shows the configuration on only the R2 router, but the same configuration must be applied to all other devices.
	
	R2#**configure terminal**
	R2(config)#**ip domain-name pynetauto.local**
	R2(config)#**crypto key generate rsa**
	The name for the keys will be: R2.pynetauto.local
	Choose the size of the key modulus in the range of 360 to 4096 for your
	  General Purpose Keys. Choosing a key modulus greater than 512 may take
	  a few minutes.
	How many bits in the modulus [512]: **1024**
	% Generating 1024 bit RSA keys, keys will be non-exportable...
	[OK] (elapsed time was 3 seconds)
	R2(config)#
	*Aug 9 13:04:41.381: %SSH-5-ENABLED: SSH 1.99 has been enabled
	R2(config)#**line vty 0 15**
	R2(config-line)#**transport input all** # allow both SSH and Telnet
	                                          remote connection
	R2(config-line)#**end**
	R2#**write memory**
	
	1-2a. Now apply this configuration to R2, SW1, sw2, and R1. You can copy the following commands once and paste them into each device's console:
	
	conf t
	ip domain-name pynetauto.local
	crypto key generate rsa
	1024
	!

(continued)

#	Task
	``
line vty 0 15
transport input all
end
!
write memory
```
1-2b. **Optionally**, you can configure devices using a telnetlib Python script. However, as this chapter is dedicated to SSH, your task is limited to configuring five devices. For this purpose, **you are excused for using the traditional cut-and-paste method**. However, I have provided the config_rsa_telnet.py file in the ch3 folder. This file can be used instead of the cut-and-paste method. You will be prompted for a username and password for each device. Feel free to explore and experiment with this Python script at your leisure. **For this specific section, I highly recommend sticking to the manual method to ensure the correct configuration entry. This step is critical for the success of all subsequent labs in this chapter as well as in the remaining chapters.** |
| ② | Furthermore, if you want to test the SSH connection from a Linux server to your network devices, you can use the ssh user_name@IP_address command. Nevertheless, due to the older SSH key exchange methods used by Cisco devices, you might encounter certain SSH challenges when exchanging keys between your Linux server and Cisco network devices.<br><br>2-1. Let's see how you can solve this problem with R1, and IOS devices first. When you run the standard SSH command, the IOS router offers an unsupported key exchanging method on port 22.<br><br>jdoe@u22s1:~$ **ssh jdoe@192.168.127.133**<br><br>Unable to negotiate with 192.168.127.133 port 22: no matching key exchange method found. Their offer: diffie-hellman-group1-sha1<br><br>2-2. Now use the -oKexAlgorithms=+ command and append the diffie-helman offer from Step 2-1. This time another message is prompted that the old IOS router is offering ssh-rsa for the host key type.<br><br>jdoe@u22s1:~$ **ssh jdoe@192.168.127.133 -oKexAlgorithms=+diffie-hellman-group1-sha1** |

*(continued)*
```

#	Task
	Unable to negotiate with 192.168.127.133 port 22: no matching host key type found. Their offer: ssh-rsa
	2-3. To accept the `ssh-rsa` host key type offer, append -oHostkeyAlgorithms=ssh-rsa to your command. Now, it is complaining about not matching the cipher.
	jdoe@u22s1:~$ **ssh jdoe@192.168.127.133 -oKexAlgorithms=+diffie-hellman-group1-sha1 -oHostkeyAlgorithms=ssh-rsa**
	Unable to negotiate with 192.168.127.133 port 22: no matching cipher found. Their offer: aes128-cbc,3des-cbc,aes192-cbc,aes256-cbc
	2-4. To address this issue, you can enhance the command by adding the -c aes256-cbc option and checking if it enables you to proceed further. If everything works as intended, you'll receive an RAA key fingerprint offer. Fortunately, this time it does work, providing a great sense of relief and sparing you from further scrutiny of additional SSH key-related messages. Additionally, **do not accept the key; instead, use the break sequence to decline the offer and proceed to the next step.**
	jdoe@u22s1:~$ **ssh jdoe@192.168.127.133 -oKexAlgorithms=+diffie-hellman-group1-sha1 -oHostkeyAlgorithms=ssh-rsa -c aes256-cbc**
	The authenticity of host '192.168.127.133 (192.168.127.133)' can't be established.
	RSA key fingerprint is SHA256:ULiBawN9ozpSo9C1ERp8cuCdSdeZjOHLIGpcEiYlrOw.
	This key is not known by any other names
	Are you sure you want to continue connecting (yes/no/[fingerprint])? ^C #Ctrl+C to break
	jdoe@u22s1:~$
	2-5. So, you have learned that for a Linux server to SSH into an older Cisco IOS-based router, it has to match three SSH algorithms supported by R1, namely kex, Host Key, and Cipher (see Figure 3-2).

(*continued*)

CHAPTER 3   PYTHON NETWORK AUTOMATION LABS: SSH IN ACTION, PARAMIKO AND NETMIKO LABS

#	Task

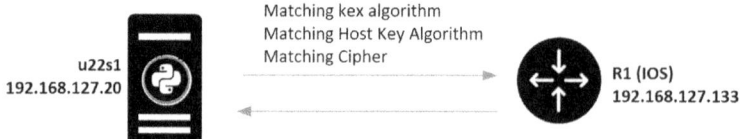

Figure 3-2. SSH paramiko Lab 1, Linux to IOS router SSH key exchange

2-6. Now you do not want to write a long command to SSH into R1, so you can simplify this and use the standard SSH command. One way to permanently fix this issue is to add these settings in the user's SSH config file. Now modify the config file under the /home/user_name/.ssh/config file, as shown here:

jdoe@u22s1:~$ **sudo nano ~/.ssh/config**
Host 192.168.127.133
    KexAlgorithms diffie-hellman-group1-sha1
    HostkeyAlgorithms ssh-rsa
    Ciphers aes128-cbc,3des-cbc,aes192-cbc,aes256-cbc

2-7. Now, SSH into R1 to accept the RSA key fingerprint and log in. After you confirm your login, return to u22s1 using the exit command.

jdoe@u22s1:~$ **ssh jdoe@192.168.127.133**
The authenticity of host '192.168.127.133 (192.168.127.133)' can't be established.
RSA key fingerprint is SHA256:ULiBawN9ozpSo9C1ERp8cuCdSdeZjOHLIGpcEiYlrOw.
This key is not known by any other names
Are you sure you want to continue connecting (yes/no/[fingerprint])?
yes
Warning: Permanently added '192.168.127.133' (RSA) to the list of known hosts.
(jdoe@192.168.127.133) Password: ********
R1#**exit**

Connection to 192.168.127.133 closed.

jdoe@u22s1:~$ # returned to Python server

(continued)

#	Task

2-8. Log in for the second time. The RSA key fingerprint is now in your Linux server's SSH config; it no longer prompts to accept a new key.

jdoe@u22s1:~$ **ssh jdoe@192.168.127.133**
(jdoe@192.168.127.133) Password: **********
R1#**exit**
Connection to 192.168.127.133 closed.
jdoe@u22s1:~$

2-9. The accepted RSA key fingerprint for SSH connections is typically stored in the known_hosts file on your local machine. This file maintains records of the keys of the remote hosts you've connected to. In Ubuntu or any Linux-based system, the known_hosts file is usually located in the ~/.ssh/ directory in the user's home directory. When you check the known_hosts file, all the entries are hashed and less human-readable, and you are unable to tell which entries belong to which hosts.

jdoe@u22s1:~$ **cat ~/.ssh/known_hosts**
[...omitted for brevity]
|1|5p4Y/Aj9Ja89VHoASHRrCbTmJJg=|65rSS+lUJBNXiF4yKbLZIro7Ep8= ecdsa-sha2-nistp256 AAAAE2VjZHNhLXNoYTItbmlzdHAyNTYAAAAIbmlzdHAy
                                NTYAAABBBDZU1oJVsS8V1l7WGAep+HLdGgHfMSg
                                nVJolF/ATd4v1WBZ16gj8xJsDJwS8f9iqLPLp1u
                                SvpAcFAFOxtA+8kjs=
|1|1zQSIIRrBCPjvHKS/oUBcUn5N/w=|gCwKXeMwBuHzDEVFDLSVjvzgXlc= ssh-rsa AAAAB3NzaC1yc2EAAAADAQABAAAAgQCbwwhbBfBAKvH8R/sYk/cY1BWN /O4LOMAhYeubxIC87aDMHmgAkAiBjkQQTIYYqHFrrSzeKu7mIOaDW8HxOy NngQ2RpyhjPPrM2z5/WTJpjy4uM8KRPHmHmmvfK/FVq+dootmsSVySVpz1nwdo8hGpZ thrM3ETNyQoxKZOLG6XTw==

(continued)

CHAPTER 3   PYTHON NETWORK AUTOMATION LABS: SSH IN ACTION, PARAMIKO AND NETMIKO LABS

#	Task

	2-10. You can view the entry for the specific host (192.168.127.133) and its associated key fingerprint by using the `ssh-keygen` command. Here's how you can do it:
	
	`jdoe@u22s1:~$ `**`ssh-keygen -F 192.168.127.133`**
	`# Host 192.168.127.133 found: line 4`
	`|1|1zQSIIRrBCPjvHKS/oUBcUn5N/w=|gCwKXeMwBuHzDEVFDLSVjvzgXlc= ssh-rsa`
	`AAAAB3NzaC1yc2EAAAADAQABAAAAgQCbwwhbBfBAKvH8R/sYk/cY1BWN`
	`/O4LOMAhYeubxIC87aDMHmgAkAiBjkQQTIYYqHFrrSzeKu7mIOaDW8HxO`
	`yNngQ2RpyhjPPrM2z5/WTJpjy4uM8KRPHmHmmvfK/FVq+dootmsSVySVpz1nwdo8hGpZ`
	`thrM3ETNyQoxKZOLG6XTw==`
	
	You delved extensively into troubleshooting the SSH key exchange between a Linux and a Cisco IOS router. I trust you found the in-depth exploration and resolution process valuable. Now, let's leverage this foundational learning to address similar issues that might arise with CML (Cisco Modeling Labs) devices.

③	Let's proceed with the same procedures to troubleshoot the SSH key exchange with R2, r3, SW1, and sw2, the CML network devices. For demonstration purposes, you'll use R2 with the IP address 192.168.127.3. After troubleshooting, to resolve the issue comprehensively, you'll hard code the settings into the same `~/.ssh/known_hosts` file. 3-1. First, try the standard command to SSH into R2. You'll get slightly different `diffie-hellman-group` offers from a CML device. `jdoe@u22s1:~$ `**`ssh jdoe@192.168.127.3`** `Unable to negotiate with 192.168.127.3 port 22: no matching key exchange method found. Their offer: diffie-hellman-group-exchange-sha1,diffie-hellman-group14-sha1` For your information, CML switches (SW1 and sw2) will present three `diffie-hellman-group` offers: `diffie-hellman-group-exchange-sha1`, `diffie-hellman-group14-sha1`, and `diffie-hellman-group1-sha1`. However, if one of these matches, you can handle these devices in the same manner.

(*continued*)

#	Task

3-2. Now use the -oKexAlgorithms=+ command and append the first diffi-helman-group offer. Once again, no matching host key type found error is given and it is offering ssh-rsa.

jdoe@u22s1:~$ **ssh jdoe@192.168.127.3 -oKexAlgorithms=+diffie-hellman-group-exchange-sha1**
Unable to negotiate with 192.168.127.3 port 22: no matching host key type found. Their offer: ssh-rsa

3-3. Add the -oHostkeyAlgorithms=ssh-rsa to the end of the last command, keeping your fingers crossed. You've been offered the RSA key fingerprint, which is great news. **Once again, refrain from accepting the key and use the breakout sequence to proceed to the next step.**

jdoe@u22s1:~$ **ssh jdoe@192.168.127.3 -oKexAlgorithms=+diffie-hellman-group-exchange-sha1 -oHostkeyAlgorithms=ssh-rsa**
The authenticity of host '192.168.127.3 (192.168.127.3)' can't be established.
RSA key fingerprint is SHA256:OvBXP/mWproeIUw4GgHUuNBdWjpkG1nvlvE4hB8/pYc.
This key is not known by any other names
Are you sure you want to continue connecting (yes/no/[fingerprint])?
^C

3-4. Now, reopen the user's .ssh/config file and append the information highlighted here. Be sure to replace the IP with an asterisk (*) to indicate that all devices (except R1) should follow these settings. This method ensures that the existing CML devices or any future devices will adhere to the key exchange settings specified under Host *.

jdoe@u22s1:~$ **sudo nano ~/.ssh/config**
Host 192.168.127.133
    KexAlgorithms diffie-hellman-group1-sha1
    HostkeyAlgorithms ssh-rsa
    Ciphers aes128-cbc,3des-cbc,aes192-cbc,aes256-cbc

(continued)

#	Task
	Host * **KexAlgorithms +diffie-hellman-group-exchange-sha1,diffie-hellman-group14-sha1,diffie-hellman-group1-sha1** **HostkeyAlgorithms +ssh-rsa** 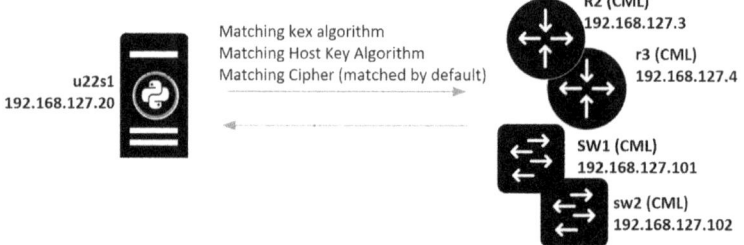 *Figure 3-3. SSH paramiko Lab 1, Linux to CML Router/Switch SSH key exchange* Here you have learned that the CML routers and switches require the correct key exchange algorithm and host key algorithm on your Linux server for SSH to work (see Figure 3-3). 3-5. Now try the standard SSH command; you should be prompted with the RSA key offer again. **This time accept the key by answering 'yes'.** jdoe@u22s1:~$ **ssh jdoe@192.168.127.3** The authenticity of host '192.168.127.3 (192.168.127.3)' can't be established. RSA key fingerprint is SHA256:OvBXP/mWproeIUw4GgHUuNBdWjpkG1nvlvE4hB8/pYc. This key is not known by any other names Are you sure you want to continue connecting (yes/no/[fingerprint])? **yes** Warning: Permanently added '192.168.127.3' (RSA) to the list of known hosts. [...omitted for brevity]

(continued)

#	Task
	Repeat the same process on r3, SW1, and sw2. jdoe@u22s1:~$ **ssh jdoe@192.168.127.4** *[...omitted for brevity]* jdoe@u22s1:~$ **ssh jdoe@192.168.127.101** *[...omitted for brevity]* jdoe@u22s1:~$ **ssh jdoe@192.168.127.102** *[...omitted for brevity]* 3-6. At this juncture, you can execute the following command to verify the key entries within the user's ~/.ssh/known_hosts file. However, it appears that this task is repetitive. To streamline and automate this process, you can leverage your Python skills. **While it's just one command across four devices, this is the inception of network automation—recognizing opportunities and seizing the moment.** **ssh-keygen -F 192.168.127.3** **ssh-keygen -F 192.168.127.4** **ssh-keygen -F 192.168.127.101** **ssh-keygen -F 192.168.127.102** Create a new directory for this chapter and navigate to it. Subsequently, generate a new Python file titled check_rsa_key_entries.py. Import the subprocess module and commence writing your code using a for loop to automate the four command lines mentioned with a Python script. jdoe@u22s1:~$ **mkdir ch3_ssh && cd ch3_ssh** jdoe@u22s1:~/ch3_ssh$ **nano check_rsa_key_entries.py** jdoe@u22s1:~/ch3_ssh$ **cat check_rsa_key_entries.py** **import subprocess** # List of IP addresses **ip_addresses = ['192.168.127.3', '192.168.127.4', '192.168.127.101', '192.168.127.102']**

(continued)

#	Task
	``python
# Loop through the IP addresses and run the ssh-keygen command for each
for ip in ip_addresses:
    command = f"ssh-keygen -F {ip}"
    process = subprocess. Open(command, shell=True,
    stdout=subprocess.PIPE, stderr=subprocess.PIPE)
stdout, stderr = process.communicate()
# Print the output
print(stdout.decode())
if stderr:
    print(stderr.decode())
```

3-7. Execute your Python code to check the key entries in your known_hosts file. Confirm that all entries exist for each respective IP address.

```
jdoe@u22s1:~/ch3_ssh$ python check_rsa_key_entries.py
Host 192.168.127.3 found: line 5
|1|4LwuTBOuo+U/Vuf72Dd9LUTkbQs=|jv+KdWAIycsFemmfk1Y6CSOktko=
ssh-rsa AAAAB3NzaC1yc2EAAAADAQABAAAAgQDDTCIMdKtzNOc/OLIMpBn6Tr8X
v5dcYhDIzEAIqJmKhO8N4COB68EYEgUrXAXM+6bngiIzAt7ppo
yaXAwSab+P5OhAO7CCZCSaLZUJrD9u26Vg7q++vnsbXrcM7UXoTytx3sylem4Tym55
B6SlLjThdUlNlOlRzkXrKkrPv+k/tQ==
Host 192.168.127.4 found: line 6
[...omitted for brevity]
Host 192.168.127.101 found: line 7
[...omitted for brevity]
Host 192.168.127.102 found: line 8
[...omitted for brevity]
``` |

(continued)
```

#	Task
④	4-1. Now that you've confirmed your Python server can SSH into all network devices and have the `paramiko` library installed, let's create a simple SSH authentication application to test your login using `paramiko` and the `show clock` command. As this is an initial code development phase, you can write basic code without following the PEP 8 convention. You'll later refactor it and turn it into a Python script. ```\njdoe@u22s1:~/ch3_ssh$ python\nPython 3.10.12 (main, Jun 11 2023, 05:26:28) [GCC 11.4.0] on linux\nType "help", "copyright", "credits" or "license" for more information.\n>>> import paramiko # Importing the paramiko library\n>>> import time # Importing the time library\n>>> ssh_client = paramiko.SSHClient() # Create an SSH client object\n>>> ssh_client.set_missing_host_key_policy(paramiko.AutoAddPolicy())\n# Set a policy to auto-accept unknown SSH host keys\n>>> ssh_client.connect("192.168.127.133", 22, username="jdoe", password="cisco123", look_for_keys=False)\n# Establish an SSH connection using IP, port, username, password, and disabling key lookup\n>>> conn = ssh_client.invoke_shell()\n# Start an interactive shell session\n>>> conn.send("show clock\n")\n# Send a command to the remote shell (here, showing the clock)\n11 # Unknown code snippet (should be in Python code)\n>>> time.sleep(1) # Introduce a 1-second delay\n>>> output = conn.recv(65535)\n# Receive the output from the remote shell\n>>> print(output.decode()) # Print the decoded output\n R1#show clock\n00:35:25.115 UTC Fri Mar 1 2002\nR1#\n>>> ssh_client.close()\n>>>\n```

(continued)

#	Task

As you can see, the login was successful—the show clock command ran and you received the expected output as the confirmation. Now, this can be converted into a properly structured code. This code is saved as checkAuth_1.py and shared in the ch3 folder for your reference.

Note: remote_connection.recv(65535), why 65535? The number 65535 is the maximum number of bytes that recv will attempt to receive.

4-2. Develop this code into a Python script to test SSH authentication for all of your devices in your current network.

```
jdoe@u22s1:~/ch3_ssh$ nano checkAuth_2.py
jdoe@u22s1:~/ch3_ssh$ cat checkAuth_2.py
import paramiko
import time

username = 'jdoe'
password = 'cisco123'
devices = ['192.168.127.3', '192.168.127.4', '192.168.127.101',
'192.168.127.102', '192.168.127.133']

ssh_client = paramiko.SSHClient()
ssh_client.set_missing_host_key_policy(paramiko.AutoAddPolicy())

for ip in devices:
    ssh_client.connect(ip, 22, username=username, password=password,
    look_for_keys=False) # this is a mandatory setting
    print(f"Connected to {ip}")
    conn = ssh_client.invoke_shell()
    conn.send("show clock\n")
    time.sleep(2)
    output = conn.recv(65535)
    print(output.decode())
    print("-"*79)

ssh_client.close()
```

(continued)

#	Task
	Once you have written this code, execute the application. If you see the time on all network devices, your application is working as designed.
| | ```
jdoe@u22s1:~/ch3_ssh$ python checkAuth_2.py
Connected to 192.168.127.133

R1#show clock
00:45:24.175 UTC Fri Mar 1 2002
R1#
[...omitted for brevity]
-----------------------------------------------------------------
Connected to 192.168.127.102
[...omitted for brevity]

sw2#show clock
00:43:12.245 UTC Fri Mar 1 2002
sw2#
-----------------------------------------------------------------
``` |
| ⑤ | Now copy the last Python script, checkAuth_2.py, and develop the code to reconfigure your time and time zone on your routers and switches. |
| | ```
jdoe@u22s1:~/ch3_ssh$ cp checkAuth_2.py update_time_tz_1.py
jdoe@u22s1:~/ch3_ssh$ nano update_time_tz_1.py
jdoe@u22s1:~/ch3_ssh$ cat update_time_tz_1.py
import paramiko # Import paramiko library
import time # Import Time module

username = 'jdoe' # Define the username
password = 'cisco123' # Define the password
devices = ['192.168.127.3', '192.168.127.4', '192.168.127.101',
'192.168.127.102', '192.168.127.133'] # List of device IP addresses

ssh_client = paramiko.SSHClient() # Create an SSHClient object
ssh_client.set_missing_host_key_policy(paramiko.AutoAddPolicy())
# Set policy to automatically add missing host keys
``` |

(continued)

#	Task
	``
for ip in devices: # Iterate through the list of devices
    ssh_client.connect(hostname=ip, username=username,
    password=password, look_for_keys=False) # Connect to the device
    print(f"Connected and configuring {ip}") # Indicate connection status
    conn = ssh_client.invoke_shell() # Invoke the shell
    conn.send("conf t\n") # Send commands to configure the device
    conn.send("clock timezone AEST +10\n")
    conn.send("clock summer-time AEST recurring\n")
    conn.send("exit\n")
    conn.send("clock set 15:30:00 04 Nov 2023\n")
    conn.send("write memory\n")
    conn.send("exit\n")
    time.sleep(2) # Add delay
    output = conn.recv(65535) # Receive the command output
    print(output.decode()) # Print the output
    print(f"Time and Time Zone configured on {ip}") # Indicate
    configuration completion
ssh_client.close() # Close the SSH connection
```
Once you are happy with your code, run the script to update the time and time zones to your local time.
```
jdoe@u22s1:~/ch3_ssh$ python update_time_tz_1.py
Connected and configuring 192.168.127.3
[...omitted for brevity]
R1#conf t
Enter configuration commands, one per line. End with CNTL/Z.
R1(config)#clock timezone AEST +10
R1(config)#clock summer-time AEST recurring
R1(config)#exit
R1#clock set 15:30:00 04 Nov 2023
R1#write memory
Building configuration...

Time and Time Zone configured on 192.168.127.133
``` |

(continued)
```

CHAPTER 3   PYTHON NETWORK AUTOMATION LABS: SSH IN ACTION, PARAMIKO AND NETMIKO LABS

#	Task
⑥	You created and used checkAuth_2.py in Steps 4-2, which was the time-checking script. It doubles as the SSH authentication verification tool. Run the script and confirm the time. If the time has changed to your time zone, you have successfully updated the time settings. jdoe@u22s1:~/ch3_ssh$ **python checkAuth_2.py** [...omitted for brevity] Connected to 192.168.127.102 ** * IOSv is strictly limited to use for evaluation, demonstration and IOS * * education. IOSv is provided as-is and is not supported by Cisco's * * Technical Advisory Center. Any use or disclosure, in whole or in part, * * of the IOSv Software or Documentation to any third party for any * * purposes is expressly prohibited except as otherwise authorized by * * Cisco in writing. * --- sw2#show clock 19:42:09.672 AEST Sat Nov 4 2023 sw2#
⑦	The previous script represents a simplified solution that effectively handles the task with a basic data structure. However, the data used can be more effectively organized using the JSON data format, often termed the "Pythonic" way of coding. I've named this script update_time_tz_1.py for comparison and stored it in the ch3 folder, along with update_time_tz_2.py and update_time_tz_3.py. These scripts are identical in function and configure the time and time zone on Cisco devices but showcase distinct methods of structuring variables in Python scripts. More often than not with paramiko Python scripts, programmers prefer to use a Python list containing multiple dictionaries. Although it looks similar to JSON format, it is a Python list. The transition of your variables in these scripts demonstrates the benefits of organizing your data, particularly from the perspective of Python code readability.

(*continued*)

#	Task

	Quick and dirty:
	``python
	username = 'jdoe'
	password = 'cisco123'
	devices = ['192.168.127.3', '192.168.127.4', '192.168.127.101', '192.168.127.102', '192.168.127.133']
	``
	
	Quick and PEP 8 guided:
	``python
	username = 'jdoe'
	password = 'cisco123'
	devices = [
	    '192.168.127.133', # R1
	    '192.168.127.3', # R2
	    '192.168.127.4', # r3
	    '192.168.127.101', # SW1
	    '192.168.127.102' # sw2
	]
	``
	
	The Pythonic way:
	``python
	devices = [
	    {'ip_address': '192.168.127.3', 'username': 'jdoe', 'password': 'cisco123'},
	    {'ip_address': '192.168.127.4', 'username': 'jdoe', 'password': 'cisco123'},
	    {'ip_address': '192.168.127.101', 'username': 'jdoe', 'password': 'cisco123'},
	    {'ip_address': '192.168.127.102', 'username': 'jdoe', 'password': 'cisco123'},
	    {'ip_address': '192.168.127.133', 'username': 'jdoe', 'password': 'cisco123'}
	]
	``

(continued)

#	Task

	A Python list inspired by JSON/JavaScript, but not in JSON format:
	``
	devices = [
	   {
	      'ip_address' : '192.168.127.3',
	      'vendor' : 'cisco',
	      'username' : 'jdoe',
	      'password' : 'cisco123'
	   },
	   {
	      'ip_address' : '192.168.127.4',
	      'vendor' : 'cisco',
	      'username' : 'jdoe',
	      'password' : 'cisco123'
	   },
	   {
	      'ip_address' : '192.168.127.101',
	      'vendor' : 'cisco',
	      'username' : 'jdoe',
	      'password' : 'cisco123'
	   },
	   {
	      'ip_address' : '192.168.127.102',
	      'vendor' : 'cisco',
	      'username' : 'jdoe',
	      'password' : 'cisco123'
	   },
	   {
	      'ip_address' : '192.168.127.133',
	      'vendor' : 'cisco',
	      'username' : 'jdoe',
	      'password' : 'cisco123'
	   }
	]
	``

(continued)

CHAPTER 3   PYTHON NETWORK AUTOMATION LABS: SSH IN ACTION, PARAMIKO AND NETMIKO LABS

#	Task
⑧	The last example's data structure is a Python list containing multiple dictionaries. Each dictionary within the list represents a device and its associated details such as IP address, vendor, username, and password. This structure is specific to Python and is not in JSON format yet; however, it can be easily transformed into JSON format using the json library in Python. Let's convert the Python list of dictionaries into JSON format using the json library and its dumps method: ```\njdoe@u22s1:~/ch3_ssh$ nano list_to_json1.py\njdoe@u22s1:~/ch3_ssh$ cat list_to_json1.py\nimport json\n\ndevices = [\n {\n 'ip_address': '192.168.127.3',\n 'vendor': 'cisco',\n 'username': 'jdoe',\n 'password': 'cisco123'\n },\n {\n 'ip_address': '192.168.127.4',\n 'vendor': 'cisco',\n 'username': 'jdoe',\n 'password': 'cisco123'\n },\n {\n 'ip_address': '192.168.127.101',\n 'vendor': 'cisco',\n 'username': 'jdoe',\n 'password': 'cisco123'\n },\n```

#	Task

```
    {
        'ip_address': '192.168.127.102',
        'vendor': 'cisco',
        'username': 'jdoe',
        'password': 'cisco123'
    },
    {
        'ip_address': '192.168.127.133',
        'vendor': 'cisco',
        'username': 'jdoe',
        'password': 'cisco123'
    }
]
json_devices = json.dumps(devices, indent=4)
print(json_devices)
```

Run the code provided to observe the difference. The `json.dumps()` method converted the Python list into a JSON-formatted string. Using the `indent=4` argument improves the readability of the printed output. Have you noticed the subtle aesthetic difference? **The Python list used single quotes for key and value pairs in the dictionary, while the converted JSON data used double quotation marks.** You'll learn more about this topic later, but it's important to remember the distinction between Python lists in dictionary format and JSON-formatted data.

jdoe@u22s1:~/ch3_ssh$ **python list_to_json1.py**

```
[
    {
        "ip_address": "192.168.127.3",
        "vendor": "cisco",
        "username": "jdoe",
        "password": "cisco123"
    },
```

#	Task
| | ```
    {
       "ip_address": "192.168.127.4",
       "vendor": "cisco",
       "username": "jdoe",
       "password": "cisco123"
    },
    {
       "ip_address": "192.168.127.101",
       "vendor": "cisco",
       "username": "jdoe",
       "password": "cisco123"
    },
    {
       "ip_address": "192.168.127.102",
       "vendor": "cisco",
        "username": "jdoe",
        "password": "cisco123"
    },
    {
       "ip_address": "192.168.127.133",
       "vendor": "cisco",
       "username": "jdoe",
       "password": "cisco123"
    }
]
``` |
| ⑨ | Now open the Python in interactive mode. While you are in interactive mode, you can use dir(library_name) to display the modules available to each library after importing a library. You have used the SSHClient and AutoAddPolicy modules from the paramiko library in the previous example. Can you locate the modules?

`>>> import paramiko`
`>>> dir(paramiko)` |

(*continued*)
```

#	Task
	`['AUTH_FAILED', 'AUTH_PARTIALLY_SUCCESSFUL', 'AUTH_SUCCESSFUL', 'Agent', 'AgentKey', 'AuthHandler', 'AuthenticationException', 'AutoAddPolicy', 'BadAuthenticationType', 'BadHostKeyException', 'BaseSFTP', 'BufferedFile', 'Channel', 'ChannelException', 'ChannelFile', 'ChannelStderrFile', 'ChannelStdinFile', 'ConfigParseError', 'CouldNotCanonicalize', 'DSSKey', 'ECDSAKey', 'Ed25519Key', 'GSSAuth', 'GSS_AUTH_AVAILABLE', 'GSS_EXCEPTIONS', 'HostKeys', 'IncompatiblePeer', 'InteractiveQuery', 'Message', 'MissingHostKeyPolicy', 'OPEN_FAILED_ADMINISTRATIVELY_PROHIBITED', 'OPEN_FAILED_CONNECT_FAILED', 'OPEN_FAILED_RESOURCE_SHORTAGE', 'OPEN_FAILED_UNKNOWN_CHANNEL_TYPE', 'OPEN_SUCCEEDED', 'PKey', 'Packetizer', 'PasswordRequiredException', 'ProxyCommand', 'ProxyCommandFailure', 'PublicBlob', 'RSAKey', 'RejectPolicy', 'SFTP', 'SFTPAttributes', 'SFTPClient', 'SFTPError', 'SFTPFile', 'SFTPHandle', 'SFTPServer', 'SFTPServerInterface', 'SFTP_BAD_MESSAGE', 'SFTP_CONNECTION_LOST', 'SFTP_EOF', 'SFTP_FAILURE', 'SFTP_NO_CONNECTION', 'SFTP_NO_SUCH_FILE', 'SFTP_OK', 'SFTP_OP_UNSUPPORTED', 'SFTP_PERMISSION_DENIED', 'SSHClient', 'SSHConfig', 'SSHConfigDict', 'SSHException', 'SecurityOptions', 'ServerInterface', 'SubsystemHandler', 'Transport', 'WarningPolicy', '__all__', '__author__', '__builtins__', '__cached__', '__doc__', '__file__', '__license__', '__loader__', '__name__', '__package__', '__path__', '__spec__', '__version__', '__version_info__', '_version', 'agent', 'auth_handler', 'ber', 'buffered_pipe', 'channel', 'client', 'common', 'compress', 'config', 'dsskey', 'ecdsakey', 'ed25519key', 'file', 'hostkeys', 'io_sleep', 'kex_curve25519', 'kex_ecdh_nist', 'kex_gex', 'kex_group1', 'kex_group14', 'kex_group16', 'kex_gss', 'message', 'packet', 'pipe', 'pkey', 'primes', 'proxy', 'py3compat', 'rsakey', 'server', 'sftp', 'sftp_attr', 'sftp_client', 'sftp_file', 'sftp_handle', 'sftp_server', 'sftp_si', 'ssh_exception', 'ssh_gss', 'sys', 'transport', 'util']`

#	Task

⑩	To view each module's packages in `paramiko`, you can first import each module and again use `dir(module_name)` to view some details of this module.
	
	``
	>>> from paramiko import SSHClient
	>>> dir(SSHClient)
	['__class__', '__delattr__', '__dict__', '__dir__', '__doc__', '__enter__', '__eq__', '__exit__', '__format__', '__ge__', '__getattribute__', '__gt__', '__hash__', '__init__', '__init_subclass__', '__le__', '__lt__', '__module__', '__ne__', '__new__', '__reduce__', '__reduce_ex__', '__repr__', '__setattr__', '__sizeof__', '__str__', '__subclasshook__', '__weakref__', '_auth', '_families_and_addresses', '_key_from_filepath', '_log', 'close', 'connect', 'exec_command', 'get_host_keys', 'get_transport', 'invoke_shell', 'load_host_keys', 'load_system_host_keys', 'open_sftp', 'save_host_keys', 'set_log_channel', 'set_missing_host_key_policy']
	``

Wow, that was an intensive lab session. While I haven't covered all aspects of the SSH protocol, typically a dedicated chapter is given to SSH topics in networking books, I've taken a practical approach here. You have written a simple authentication application to test SSH connectivity and expanded it to create an application for modifying the time and time zone of your network devices. Additionally, this section examined the minor differences between a Python list with dictionaries versus JSON-formatted data. You also saw how to convert the Python list data into JSON-formatted data to understand the data structure better. Finally, you explored how to inspect the modules and packages in the `paramiko` module using the Python `dir()` module. Keeping my promise, next is a quick way to discover more about the network using NMAP, an exceptional network tool highly regarded by both network and security engineers. For an in-depth understanding, I've provided a detailed dive into NMAP on my blog.

You can access NMAP by following the URL provided for more insights. Useful NMAP port scan commands: https://wordpress.com/post/italchemy.wordpress.com/2807 and Cent OS 8 – Add SFTP service [2/2], install NMAP on Ubuntu 20 to scan port and test SFTP: https://wordpress.com/post/italchemy.wordpress.com/3714

Expand your knowledge

NMAP to the rescue to complete your SSH query questions

Here's a simpler method to scan port 22 and retrieve the SSH authentication methods from your Linux server, Cisco routers, and switches. NMAP, a powerful port-scanning tool, offers various options to probe any open port on the network. However, it's crucial to use this tool with caution and adhere to the port-scanning policies in your organization. In the hands of a skilled network engineer, NMAP is an extremely powerful asset, but in the wrong hands of a bad actor, it can be a malicious tool.

Let's see what NAMP can offer you about SSH authentication details on your server and Cisco devices.

Begin by verifying if NMAP is installed. If not, proceed to install the software.

```
jdoe@u22s1:~$ nmap -p 22 --script ssh-auth-methods localhost
Command 'nmap' not found, but can be installed with:
sudo snap install nmap  # version 7.94, or
sudo apt  install nmap  # version 7.91+dfsg1+really7.80+dfsg1-2ubuntu0.1
See 'snap info nmap' for additional versions.
jdoe@u22s1:~$ sudo apt install nmap -y # install nmap, on Fedora, replace apt with dnf/yum
```

[...omitted for brevity]

To check the key exchange algorithms supported by a local SSH server using NMAP, you can use the `--script ssh2-enum-algos` option along with the `-p` flag to specify the port. You can replace the localhost with an external IP to probe this information on a remote system.

u22s1 (localhost, 127.0.0.1, 192.168.127.20), 22.04.3 LTS (Jammy Jellyfish)

CHAPTER 3   PYTHON NETWORK AUTOMATION LABS: SSH IN ACTION, PARAMIKO AND NETMIKO LABS

```
jdoe@u22s1:~$ nmap -p 22 --script ssh2-enum-algos localhost
Starting Nmap 7.80 ( https://nmap.org ) at 2023-11-04 10:13 AEDT
Nmap scan report for localhost (127.0.0.1)
Host is up (0.00021s latency).

PORT   STATE SERVICE
22/tcp open  ssh
| ssh2-enum-algos:
|   kex_algorithms: (10)
[...omitted for brevity]
|       diffie-hellman-group-exchange-sha256
|       diffie-hellman-group16-sha512
|       diffie-hellman-group18-sha512
|       diffie-hellman-group14-sha256
|   server_host_key_algorithms: (4)
|       rsa-sha2-512
|       rsa-sha2-256
|       ecdsa-sha2-nistp256
|       ssh-ed25519
|   encryption_algorithms: (6)
|       chacha20-poly1305@openssh.com
|       aes128-ctr
|       aes192-ctr
|       aes256-ctr
[...omitted for brevity]
```

R1 (192.168.127.133), Cisco IOS Software, 3700 Software (C3745-ADVENTERPRISEK9-M), Version 12.4(25d)

```
jdoe@u22s1:~$ nmap -p 22 --script ssh2-enum-algos 192.168.127.133
Starting Nmap 7.80 ( https://nmap.org ) at 2023-11-04 10:22 AEDT
Nmap scan report for 192.168.127.133
Host is up (0.020s latency).

PORT   STATE SERVICE
22/tcp open  ssh
| ssh2-enum-algos:
|   kex_algorithms: (1)
```

CHAPTER 3   PYTHON NETWORK AUTOMATION LABS: SSH IN ACTION, PARAMIKO AND NETMIKO LABS

```
|        diffie-hellman-group1-sha1
|    server_host_key_algorithms: (1)
|        ssh-rsa
|    encryption_algorithms: (4)
|        aes128-cbc
|        3des-cbc
|        aes192-cbc
|        aes256-cbc
|    mac_algorithms: (4)
|        hmac-sha1
|        hmac-sha1-96
|        hmac-md5
|        hmac-md5-96
|    compression_algorithms: (1)
|_       none

Nmap done: 1 IP address (1 host up) scanned in 0.36 seconds
```

R2 (192.168.127.3), Cisco IOS Software, IOSv Software (VIOS-ADVENTERPRISEK9-M), Version 15.7(3)M3. Note, r3 (192.168.127.4) is using the same software.

```
jdoe@u22s1:~$ nmap -p 22 --script ssh2-enum-algos 192.168.127.3
Starting Nmap 7.80 ( https://nmap.org ) at 2023-11-04 10:25 AEDT
Nmap scan report for 192.168.127.3
Host is up (0.29s latency).

PORT   STATE SERVICE
22/tcp open  ssh
| ssh2-enum-algos:
|   kex_algorithms: (2)
|        diffie-hellman-group-exchange-sha1
|        diffie-hellman-group14-sha1
|    server_host_key_algorithms: (1)
|        ssh-rsa
|    encryption_algorithms: (3)
|        aes128-ctr
```

CHAPTER 3   PYTHON NETWORK AUTOMATION LABS: SSH IN ACTION, PARAMIKO AND NETMIKO LABS

```
|         aes192-ctr
|         aes256-ctr
|     mac_algorithms: (4)
|         hmac-sha2-256
|         hmac-sha2-512
|         hmac-sha1
|         hmac-sha1-96
|     compression_algorithms: (1)
|_        none

Nmap done: 1 IP address (1 host up) scanned in 1.77 seconds
```

SW1 (192.168.127.101), Cisco IOS Software, vios_l2 Software (vios_l2-ADVENTERPRISEK9-M), Version 15.2(4.0.55)E. Note, sw2 (192.168.127.102) is using the same software.

```
jdoe@u22s1:~$ nmap -p 22 --script ssh2-enum-algos 192.168.127.101
Starting Nmap 7.80 ( https://nmap.org ) at 2023-11-04 10:24 AEDT
Nmap scan report for 192.168.127.101
Host is up (0.029s latency).

PORT   STATE SERVICE
22/tcp open  ssh
| ssh2-enum-algos:
|     kex_algorithms: (3)
|         diffie-hellman-group-exchange-sha1
|         diffie-hellman-group14-sha1
|         diffie-hellman-group1-sha1
|     server_host_key_algorithms: (1)
|         ssh-rsa
|     encryption_algorithms: (7)
|         aes128-ctr
|         aes192-ctr
|         aes256-ctr
|         aes128-cbc
|         3des-cbc
|         aes192-cbc
|         aes256-cbc
```

```
|    mac_algorithms: (2)
|        hmac-sha1
|        hmac-sha1-96
|    compression_algorithms: (1)
|_       none

Nmap done: 1 IP address (1 host up) scanned in 0.57 seconds
```

I can assure you that you are almost finished with the first lab topic. You've already covered logging into Cisco devices from your Linux server and touched on the implications. However, in the enterprise networking environment, SSH commands serve to navigate between network devices. Let's quickly explore this process and understand some of the caveats associated with this feature within your lab.

Expand your knowledge

SSH connection between Cisco routers and switches

You can use the `ssh -l user_name IP_address` command to establish an SSH connection between any two Cisco devices. From the R2 console, execute the following command to SSH into r3, SW1, or sw2.

R2#**ssh -l jdoe 192.168.127.4** # SSH to r3

```
**********************************************************************
* IOSv is strictly limited to use for evaluation, demonstration and IOS *
* education. IOSv is provided as-is and is not supported by Cisco's    *
* Technical Advisory Center. Any use or disclosure, in whole or in part,*
* of the IOSv Software or Documentation to any third party for any     *
* purposes is expressly prohibited except as otherwise authorized by   *
* Cisco in writing.                                                    *
**********************************************************************
Password: ******** # enter your password

**********************************************************************
* IOSv is strictly limited to use for evaluation, demonstration and IOS *
* education. IOSv is provided as-is and is not supported by Cisco's    *
* Technical Advisory Center. Any use or disclosure, in whole or in part,*
```

CHAPTER 3   PYTHON NETWORK AUTOMATION LABS: SSH IN ACTION, PARAMIKO AND NETMIKO LABS

```
 * of the IOSv Software or Documentation to any third party for any      *
 * purposes is expressly prohibited except as otherwise authorized by    *
 * Cisco in writing.                                                     *
 *************************************************************************
r3#exit # logout to get back to R1
```

As seen, R2, r3, SW1, and sw2 are all CML devices utilizing identical SSH key exchange and host key algorithms, with minimal SSH algorithm-related issues (see Figure 3-4).

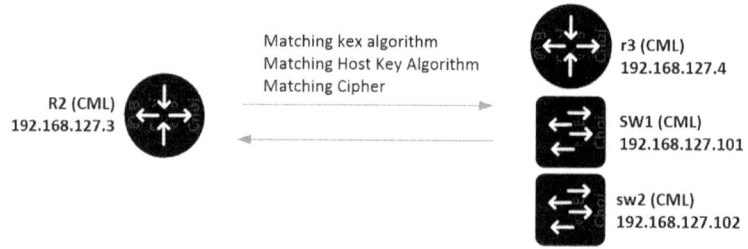

Figure 3-4. SSH paramiko Lab 1, CML to CML SSH key exchange

```
[Connection to 192.168.127.4 closed by foreign host]

R2#ssh -l jdoe 192.168.127.101 # SSH to SW1
[...omitted for brevity]
R2#ssh -l jdoe 192.168.127.102 # SSh to sw2
[...omitted for brevity]
```

You might be asking what about the R1? Try to SSH into R1 from R2. You will receive the No matching cipher found message, as shown here.

```
R2#ssh -l jdoe 192.168.127.133
[Connection to 192.168.127.133 aborted: error status 0]
R2#
*Nov  2 11:45:38.260: %SSH-3-NO_MATCH: No matching cipher found: client
aes128-ctr,aes192-ctr,aes256-ctr server aes128-cbc,3des-cbc,aes192-
cbc,aes256-cbc
```

What about the other direction? Try to SSH into R2 from R1.

158

R1#**ssh -l jdoe 192.168.127.3**

[Connection to 192.168.127.3 aborted: error status 0]
R1#
Nov 4 06:07:49.071: SSH2 CLIENT 0: no matching cipher found: client aes128-cbc,3des-cbc,aes192-cbc,aes256-cbc server aes128-ctr,aes192-ctr,aes256-c

Essentially, this demonstrates a two-way issue, not a one-way problem. Next, execute the ssh -c ? command on both R2 and R1. You will observe three matching ciphers available for use.

```
R2#ssh -c ?
  3des        triple des
SSHv2 only cipher list:
  aes128-cbc  AES 128 bits
  aes128-ctr  AES-CTR 128 bits
  aes192-cbc  AES 192 bits
  aes192-ctr  AES-CTR 192 bits
  aes256-cbc  AES 256 bits
  aes256-ctr  AES-CTR 256 bits

R1#ssh -c ?
  3des        triple des
SSHv2 only cipher list:
  aes128-cbc  AES 128 bits
  aes192-cbc  AES 192 bits
  aes256-cbc  AES 256 bits
```

You can address the first error by using the ssh command with the -c option. However, this time, the two routers do not agree on the matching kex (Key Exchange) algorithm, so it produces another error. When attempting to SSH between R2 and R1 (see Figure 3-5), you'll notice the absence of matching kex algorithms between the CML-based device (R2) and the legacy IOS-based device (R1).

CHAPTER 3   PYTHON NETWORK AUTOMATION LABS: SSH IN ACTION, PARAMIKO AND NETMIKO LABS

Figure 3-5. *SSH paramiko Lab 1, CML to IOS SSH key exchange*

```
From R2 to R1:
R2#ssh -c aes256-cbc -l jdoe 192.168.127.133
[Connection to 192.168.127.133 aborted: error status 0]
R2#
*Nov  2 11:51:38.835: %SSH-3-NO_MATCH: No matching kex algorithm found:
client diffie-hellman-group-exchange-sha1,diffie-hellman-group14-sha1
server diffie-hellman-group1-sha1

From R1 to R2:
R1#ssh -c aes256-cbc -l jdoe 192.168.127.3
[Connection to 192.168.127.3 aborted: error status 0]
R1#
*Mar  1 03:44:01.111: SSH2 CLIENT 0: kex algo not supported:
client diffie-
hellman-group1-sha1, server diffie-hellman-group14-sha1
```

On R1 (IOS), there isn't a command to configure the `ssh server kex` algorithm. On the contrary, the CML router has the `ip ssh server algorithm kex` option, but it lacks the kex used by R1. There is no kex algorithm match between these two operating systems. In simpler terms, there isn't a method to SSH into each other using the SSH protocol. It's important to note that Telnet logins function seamlessly between the two routers.

```
R2(config)#ip ssh server algorithm kex?
  diffie-hellman-group-exchange-sha1  DH_GRPX_SHA1 diffie-hellman key
  exchange algorithm
  diffie-hellman-group14-sha1         DH_GRP14_SHA1 diffie-hellman key
  exchange algorithm
```

I can only speculate that this might be by design due to the security vulnerabilities associated with the old IOS's `diffie-hellman-group1-sha1` SSH algorithm. At the time of writing this book, I couldn't find a comprehensive explanation as to why Cisco would prevent SSH communications between the new and old routers. Therefore, let's leave this issue on the backburner and save time.

Here is a quick troubleshooting tip for more stubborn SSH key problems.

Tip

Dealing with SSH key problems: tips for Cisco router and switch settings

If you encounter known-hosts issues due to multiple attempts to add the keys, you can refer to the instructions in the following forum to remove the records from your Linux server and attempt to re-establish the SSH connection to Cisco networking devices.

Link: `https://superuser.com/questions/30087/remove-key-from-known-hosts`

Moreover, if you encounter RSA key issues on your Cisco routers or switches, you can "zerorize" the RSA to re-create the RSA key. To clear the previous key, use the following command:

R1 (config)# **crypto key zeroize rsa**

After issuing this command, you should reconfigure the crypto RSA key.

```
R1(config)# hostname <name>
R1(config)# ip domain-name <domain>
R1(config)# crypto key generate rsa
R1(config)# ip ssh version 2
```

If your GNS3 devices are encountering Diffie-Hellman key exchange issues and you haven't found a solution, save the running configuration to a notepad, remove the node from the topology, and then reconfigure your device. This problem might be associated with the GNS3, VMware, and Cisco image file incompatibility issue. If the problem persists, it's advisable to seek guidance from articles and resources in the Cisco/GNS3/VMware community for a resolution before proceeding. Good luck with your troubleshooting!

CHAPTER 3   PYTHON NETWORK AUTOMATION LABS: SSH IN ACTION, PARAMIKO AND NETMIKO LABS

paramiko Lab 2: Configuring an NTP Server on Cisco Devices Without User Interaction (NTP Lab)

In this lab, you'll configure two routers, R2 and r3, using information from two external files. The first file contains each router's IP address per line, and the other file contains the username and password. Using external files removes the necessity for a network administrator to interactively input information while sitting in front of the device's remote console.

In the upcoming chapter, you'll learn to use Linux cron as a scheduler to run Python applications autonomously, effectively turning your application into an automated tool. Although a Python script with Linux cron might seem modest (cost-free and lacking a polished GUI) compared to advanced features offered by tools like Cisco ACI and Red Hat Ansible Tower, understanding how to customize your environment and work with cron is essential. Networks vary in size and complexity, and the scheduler's core function remains consistent. Before you begin the lab, take a look at the devices in use (see Figure 3-6).

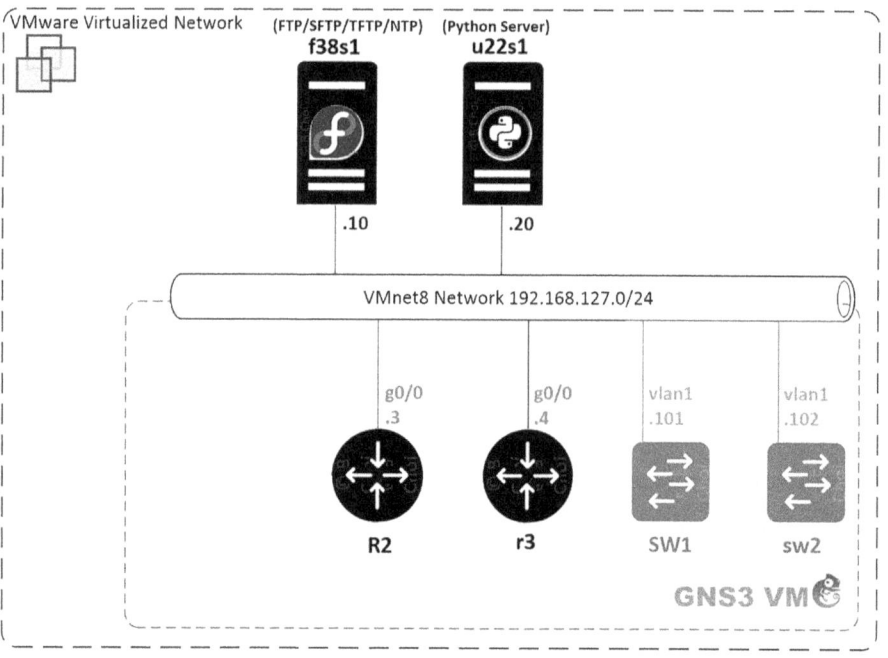

Figure 3-6. SSH paramiko Lab 2, devices in use

To facilitate NTP services for the routers, start the IP Services server, f38s1. Throughout this lab, you'll need to power on u22s1, f38s1, R2, and r3 while keeping sw1 and sw2 running in the background to establish connections with the CML routers. This lab encompasses several tasks, so it's advisable to write the script and execute it within the lab environment.

#	Task
①	On your Python server, u22s1, continue to work in the ch3_ssh directory and follow these steps to create two external files in the same directory: jdoe@u22s1:~/ch3_ssh$ **nano routerlist** jdoe@u22s1:~/ch3_ssh$ **cat routerlist** **192.168.127.3** **192.168.127.4** jdoe@u22s1:~/ch3_ssh$ **nano adminpass** jdoe@u22s1:~/ch3_ssh$ **cat adminpass** **jdoe** **cisco123**
②	I conducted a Google search, found an example ssh_client.py script, and adapted its content to suit my requirements. You can visit the following site to access the base template SSH script and commence your SSH scripting. URL: https://gist.github.com/ghawkgu/944017 You will be using the ssh_client.py script as a base to build your Python script. **#!/usr/bin/env python** **import paramiko** **hostname = 'localhost'** **port = 22** **username = 'foo'** **password = 'xxxYYYxxx'**

(continued)

#	Task
	``
if __name__ == "__main__":
    paramiko.util.log_to_file('paramiko.log')
    s = paramiko.SSHClient()
    s.load_system_host_keys()
    s.connect(hostname, port, username, password)
    stdin, stdout, stderr = s.exec_command('ifconfig')
    print stdout.read()
    s.close()
```
The script you are going to write will set up the NTP server on both routers, R2 and r3, thus ensuring synchronized time across the network. **NTP (Network Time Protocol) holds significant importance in enterprise networks to maintain accurate timestamps on syslog, monitoring, and reporting servers down to the millisecond**. While this isn't a networking book and won't delve deeply into stratum and NTP functionalities, a few key points are worth noting. First, Cisco and many other vendor devices don't rely on Microsoft W32tm services from Microsoft Windows-based NTP servers. Secondly, for certain Cisco networking and server devices, time must be set to a minimum stratum to be considered a reliable time source. Typically, this value is equal to or less than stratum 5, where a lower stratum implies higher trustworthiness in timekeeping.

It's time to modify the base `paramiko` login configuration. Once completed, it should resemble the following script:

```
jdoe@u22s1:~/ch3_ssh$ nano ssh_ntp_lab.py
jdoe@u22s1:~/ch3_ssh$ cat ssh_ntp_lab.py
#!/usr/bin/env python3
import paramiko # Importing paramiko library
import time # Importing time module
from datetime import datetime # Importing datetime module

t_ref = datetime.now().strftime("%Y-%m-%d_%H-%M-%S") # Time reference in desired time format
file1 = open("routerlist") # Open routerlist as file1
``` |

(continued)
```

#	Task
| | ```python
for line in file1: # For loop for router IP address
    print(t_ref) # Print time reference
    print("Now logging into " + (line)) # Print statement
    ip_address = line.strip() # Remove any whitespace
    file2 = open("adminpass") # Open adminpass as file2
    for line1 in file2: # Read the first line(admin ID) in file2
        username = line1.strip() # Remove any whitespace

    for line2 in file2: # Read the second line (password) in file2
        password = line2.strip() # Remove any whitespace
    ssh_client = paramiko.SSHClient() # Create paramiko SSH client object
    ssh_client.set_missing_host_key_policy(paramiko.AutoAddPolicy())
    # Automatically accept host key policy
    ssh_client.connect(hostname=ip_address, username=username,
    password=password, look_for_keys=False) # SSH connection objects
    print("Successful connection to " + (ip_address) + "\n") # Print
                                                              statement
    print("Now completing following tasks : " + "\n") # Print statement
    remote_connection = ssh_client.invoke_shell() # Invoke shell session
    output1 = remote_connection.recv(65535) # Catches and removes the
                                              login prompt output
    remote_connection.send("configure terminal\n") # Move to
                                                    configuration mode
    print("Configuring NTP Server") # Print statement
    remote_connection.send("ntp server 192.168.127.10\n") # Configure NTP
                                                           server
    remote_connection.send("end\n") # Go back to exec privilege mode
    remote_connection.send("copy running-config start-config\n")
                                                   # Send save command
``` |

(*continued*)
```

#	Task
	print() # Print remote_connections to add a new line **time.sleep(2)** # Pause for 2 seconds **output2 = remote_connection.recv(65535)** # Capture session in output variable **print((output2).decode('ascii'))** # Print output using ASCII decoding **print("Successfully configured your device & Disconnecting from " + (ip_address))** # Print statement **ssh_client.close** # Close SSH connection **time.sleep(2)** # Pause for 2 seconds **file1.close() # Close file1** **file2.close() # Close file2**
| ③ | Since you have had NMAP installed and it can check the open port status with ease, let's use it to check if your R2 and r3 are listening on TCP port 22 for SSH and UDP port 123 for NTP. First, check that the TCP port 22 is open on R2 and r3 for the Python server to log in.<br>**SSH TCP port 22 check on R2 and r3:**<br><br>```
jdoe@u22s1:~/ch3_ssh$ nmap -p 22 192.168.127.3 192.168.127.4
Starting Nmap 7.80 ( https://nmap.org ) at 2023-11-05 00:09 AEDT
Nmap scan report for 192.168.127.3
Host is up (0.050s latency).

PORT   STATE SERVICE
22/tcp open  ssh # OK

Nmap scan report for 192.168.127.4
Host is up (0.053s latency).

PORT   STATE SERVICE
22/tcp open  ssh # OK

Nmap done: 2 IP addresses (2 hosts up) scanned in 0.15 seconds
``` |

(continued)
```

#	Task
	Next, check that UDP port 123 is open on f38s1, R2, and r3. If any of the STATE is showing as closed, you will have to troubleshoot the issue. Usually with the closed status, it could be misconfiguration or a firewall issue. If the STATE is showing `filtered`, it is usually a firewall-related issue. **Check UDP 123 port for NTP is open on NTP server, R2 and r3:** ``` jdoe@u22s1:~/ch3_ssh$ sudo nmap -p 123 -sU 192.168.127.3 192.168.127.4 192.168.127.10 # with UDP port, use -sU option as shown Starting Nmap 7.80 (https://nmap.org) at 2023-11-05 01:48 AEDT Nmap scan report for 192.168.127.3 Host is up (0.037s latency). PORT STATE SERVICE 123/udp open ntp # OK MAC Address: 0C:AE:5C:A3:00:00 (Unknown) Nmap scan report for 192.168.127.4 Host is up (0.039s latency). PORT STATE SERVICE 123/udp open ntp # OK MAC Address: 0C:3C:CB:05:00:01 (Unknown) Nmap scan report for 192.168.127.10 Host is up (0.00031s latency). PORT STATE SERVICE 123/udp open ntp # OK MAC Address: 00:0C:29:05:BC:8C (VMware) Nmap done: 3 IP addresses (3 hosts up) scanned in 0.23 seconds ``` Alternatively, if you want to check the open ports manually on your router, you can log in to R2 and r3 and then run the `show control-plane host open-ports` command to check the open ports. Also, to avoid any NTP problems, run the `show run	in ntp server` command to confirm that the NTP server is configured correctly.

(continued)

#	Task

R2#**show run | in ntp server**
ntp server 192.168.127.10
R2#**show control-plane host open-ports**
Active internet connections (servers and established)
```
Prot        Local Address    Foreign Address       Service      State
tcp             *:22              *:0           SSH-Server     LISTEN
tcp             *:23              *:0              Telnet      LISTEN
udp            *:123              *:0                NTP       LISTEN
```
r3#**show run | in ntp server**
ntp server 192.168.127.10
r3#**show control-plane host open-ports**
Active internet connections (servers and established)
```
Prot        Local Address    Foreign Address       Service      State
tcp             *:22              *:0           SSH-Server     LISTEN
tcp             *:23              *:0              Telnet      LISTEN
udp            *:123              *:0                NTP       LISTEN
```

(4) Now, before running your script, check that the NTP service is running correctly on f38s1. If you find that the NTP service is not running, you must troubleshoot the issue before running the script. On Fedora 38, you previously installed chronyd in Chapter 8 of the Part I book, as an NTP server. Here are commands to troubleshoot common issues with chronyd. In the next step, you have to complete the server configuration by modifying the chrony.conf file to make it a functional server.

[jdoe@f38s1 ~]$ **systemctl status chronyd**
[jdoe@f38s1 ~]$ **systemctl restart chronyd**
[jdoe@f38s1 ~]$ **systemctl enable chronyd**
[jdoe@f38s1 ~]$ **sudo firewall-cmd --zone=public --add-port=123/udp --permanent**
[jdoe@f38s1 ~]$ **sudo firewall-cmd -reload**

If you forgot to install chrony, use the dnf install command to install it.
[jdoe@f38s1 ~]$ **dnf install chrony**

(continued)

#	Task

	Also, make sure that the current time on your NTP server is correct.
	`[jdoe@f38s1 ~]$ `**`date`**
	`Sun 05 Nov 2023 00:18:59 AEDT`
	
	For chronyd, you must get a list of NTP servers closest to your home region from the URL specified here, and you have to update the NTP server pool according to your time zone. Make sure you comment out the first default line with a hash (#). Ensure that you choose the public NTP server closest to your location from the following site:
	
	URL: https://www.pool.ntp.org/en/
	
	Open the file in sudo mode to modify the `chrony.conf file` under /etc/. This was configured in Chapter 8 of the Part I book, but I am cross-checking to make sure that the correct NTP servers have been specified for my location.
	
	`[jdoe@f38s1 ~]$ `**`sudo nano /etc/chrony.conf`**
	`GNU nano 7.2`
	`/etc/chrony.conf`
	`# Use public servers from the pool.ntp.org project.`
	`# Please consider joining the pool (https://www.pool.ntp.org/join.html).`
	`#pool 2.fedora.pool.ntp.org iburst`
	`server 0.au.pool.ntp.org iburst`
	`server 1.au.pool.ntp.org iburst`
	`server 2.au.pool.ntp.org iburst`
	`server 3.au.pool.ntp.org iburst`
	`[...omitted for brevity]`
	
	While you are on the same config file, go to the end of the line and confirm that your network has been allowed to access this NTP server from the local network. If you configured this in Chapter 8 of Book I, it should be already there.
	`[...omitted for brevity]`
	`# Allow NTP client access from local network.`
	`#allow 192.168.0.0/16`
	`allow 192.168.127.0/24`

(continued)

#	Task
	As the final check, ensure that the `chronyd` service is correctly running on `f38s1`. `[jdoe@f38s1 ~]$ `**`systemctl status chronyd`** ● chronyd.service - NTP client/server Loaded: loaded (/usr/lib/systemd/system/chronyd.service; enabled; preset: enabled) Drop-In: /usr/lib/systemd/system/service.d └─10-timeout-abort.conf Active: active (running) since Sun 2023-11-05 00:04:53 AEDT; 2h 3min ago Docs: man:chronyd(8) man:chrony.conf(5) Process: 699 ExecStart=/usr/sbin/chronyd $OPTIONS (code=exited, status=0/SUCCESS) Main PID: 715 (chronyd) Tasks: 1 (limit: 4631) Memory: 5.4M CPU: 141ms CGroup: /system.slice/chronyd.service └─715 /usr/sbin/chronyd -F *[...omitted for brevity]* If you are happy with all the pre-checks, go to the next step and execute the NTP updating Python script.
⑤	My `chronyd` time server is running and the pre-check has been completed successfully. All looks to be in good working order in your environment, so go back to the `u22s1` server and run the `ssh_ntp_lab.py` script. `jdoe@u22s1:~/ch3_ssh$ `**`python ssh_ntp_lab.py`** 2023-11-05_00-33-58 Now logging into 192.168.127.3 Successful connection to 192.168.127.3 Now completing following tasks : Configuring NTP Server

CHAPTER 3   PYTHON NETWORK AUTOMATION LABS: SSH IN ACTION, PARAMIKO AND NETMIKO LABS

#	Task
	configure terminal Enter configuration commands, one per line. End with CNTL/Z. R2(config)#ntp server 192.168.127.10 R2(config)#end R2#copy running-config start-config Destination filename [start-config]? Successfully configured your device & Disconnecting from 192.168.127.3 2023-11-05_00-33-58 *[...omitted for brevity]*
⑥	Once the NTP script runs successfully, it may take ten minutes or so for the NTP synchronization to take place between the NTP server, f38s1 and R2, and r3. Note that the NTP protocol is a rather very slow protocol for a full synchronization so you have to be patient for the times to be synchronized. 6-1. Meanwhile, make a copy of checkAuth_2.py and modify it so it runs the show commands and you can check the NTP association and synchronization from the comfort of your Python server. ```\njdoe@u22s1:~/ch3_ssh$ cp checkAuth_2.py display_show.py\njdoe@u22s1:~/ch3_ssh$ nano display_show.py\njdoe@u22s1:~/ch3_ssh$ cat display_show.py\nimport paramiko\nimport time\n\nusername = 'jdoe'\npassword = 'cisco123'\ndevices = [\n#'192.168.127.133', # Hash out this line\n'192.168.127.3',\n'192.168.127.4',\n#'192.168.127.101', # Hash out this line\n#'192.168.127.102' # Hash out this line\n]\n```

#	Task
	``
ssh_client = paramiko.SSHClient()
ssh_client.set_missing_host_key_policy(paramiko.AutoAddPolicy())

for ip in devices:
    ssh_client.connect(ip, 22, username=username, password=password,
    look_for_keys=False)
    print(f"Connected to {ip}")
    conn = ssh_client.invoke_shell()
    conn.send("show ntp associations\n")
    conn.send("show ntp status\n")
    time.sleep(2)
    output = conn.recv(65535)
    print(output.decode())
    print("-"*79)

ssh_client.close()
```
6-2. Run this Python script and you should be able to confirm that R2 and r3 synchronized to the NTP server, f38s1, 192.168.127.10.

```
jdoe@u22s1:~/ch3_ssh$ python display_show.py
Connected to 192.168.127.3
****
* IOSv is strictly limited to use for evaluation, demonstration and IOS *
* education. IOSv is provided as-is and is not supported by Cisco's   *
* Technical Advisory Center. Any use or disclosure, in whole or in part,*
* of the IOSv Software or Documentation to any third party for any   *
* purposes is expressly prohibited except as otherwise authorized by   *
* Cisco in writing.   *
****
``` |

(continued)
```

#	Task
| | ```
R2#show ntp associations

address         ref clock        st    when   poll reach delay   offset    disp
*~192.168.127.10  162.159.200.123  4     56     64   1    15.810  2342.01 189.79
* sys.peer, # selected, + candidate, - outlyer, x falseticker, ~ configured
R2#show ntp status
Clock is synchronized, stratum 5, reference is 192.168.127.10
nominal freq is 1000.0003 Hz, actual freq is 999.5003 Hz, precision is 2**14
ntp uptime is 649600 (1/100 of seconds), resolution is 1001
reference time is E8F0DEDC.AAED26A8 (01:16:44.667 AEST Sun Nov 5 2023)
clock offset is 2342.0104 msec, root delay is 43.42 msec
root dispersion is 7956.45 msec, peer dispersion is 189.79 msec
loopfilter state is 'CTRL' (Normal Controlled Loop), drift is 0.000499999 s/s
system poll interval is 64, last update was 428 sec ago.
R2#
------------------------------------------------------------------
Connected to 192.168.127.4
[...omitted for brevity]
``` |

You have successfully configured the NTP server on the R2 and r3 routers. The synchronization of time with the Linux NTP server in your network is now achieved. Moreover, by moving the user credentials and router IP addresses to two distinct external files, you've streamlined the process, eliminating the need to manually input each IP address and the admin credentials. These preloaded files will prove beneficial in the forthcoming cron lab in Chapter 4. While critical, the topics of password security and password vaults extend beyond the scope of this book. Now, let's swiftly proceed to the next SSH lab, focusing on backing up your devices to your TFTP server.
```

CHAPTER 3   PYTHON NETWORK AUTOMATION LABS: SSH IN ACTION, PARAMIKO AND NETMIKO LABS

Warning

Managing your host PC hardware resources

In the upcoming lab, you'll be ramping up your activities with two Linux VMs and five routers and switches. As you boot each VM, you'll notice a spike in CPU and memory usage, so it's essential to manage your host PC's resources wisely for a smooth lab operation. The accompanying screenshot illustrates the surge in Task Manager when all GNS3 network devices are simultaneously powered on. Exercise patience with your computer's resources and allow your host PC's performance to stabilize before commencing your next lab.

Learning a comprehensive enterprise technology on a single laptop does have its limitations. It might be slower than a production environment yet running such labs costs less than a cup of your daily coffee and its power button is within easy reach of your fingertips (see Figure 3-7).

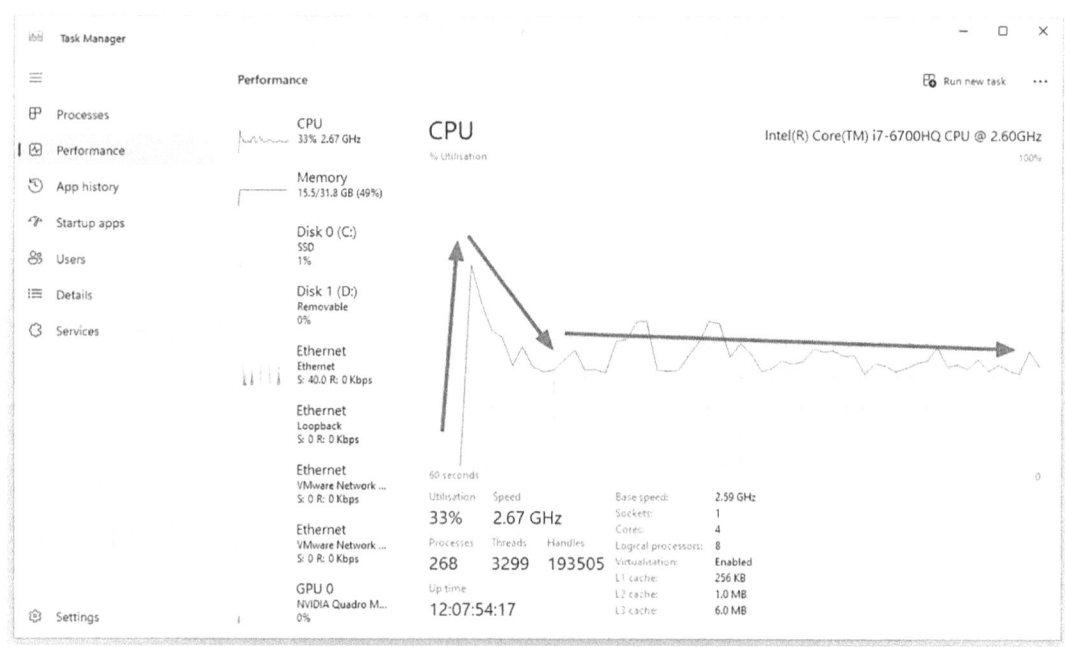

Figure 3-7. Windows host PC: Task Manager Performance

You have to be knowledgeable about the performance of VMs running in different environments. The VMs can be somewhat affected by Type-2 hypervisors like VMware Workstation 17 Pro in your lab settings. These hypervisors run atop an existing operating system, adding a small layer between the VMs and hardware. This results in a slight performance overhead compared to Type-1 (bare-metal) hypervisors directly integrated with the system hardware.

paramiko Lab 3: Create an Interactive paramiko SSH Script to Save Running Configurations to the TFTP Server

In this lab, you'll be creating an interactive tool using Python's input function along with the getpass module to collect the network administrator's ID and password. To maintain simplicity, the script will use a list containing all IP addresses, and the backup files will be stored on the TFTP server operating on your IP services server, f38s1 (192.168.127.130). **With minor adaptations, you can personalize and develop operational tools to save time and resources for your team and organization.** This lab focuses on creating device configuration backups for only five devices. In a development environment, if you have more than two of the same devices, you can use the Python loops, so you can test the basic functions. The only limitation is the testing of the scale, which may require testing on hundreds of devices operating in all sorts of network conditions. Consider the workload when these tasks need to be conducted across tens or hundreds of network devices within customers' networks. If you aim to run such scripts routinely using an automated scheduler, engineers no longer need to be tied to their computers performing backups manually. While enterprise-level network device configuration backup solutions like SolarWinds Network Configuration Manager (NCM), Cisco Secure ACS (Access Control System), ManageEngine Firewall Analyzer, Tripwise Enterprise, and Infoblox NetMRI exist, you will delve deeper into this subject, learning about the working mechanisms of these tools by re-creating and executing them in your lab. Often, a product's popularity is attributed to its marketing and user-friendly graphical interface. However, many tools function similarly under the hood, and the main difference is on the surface (see Figure 3-8).

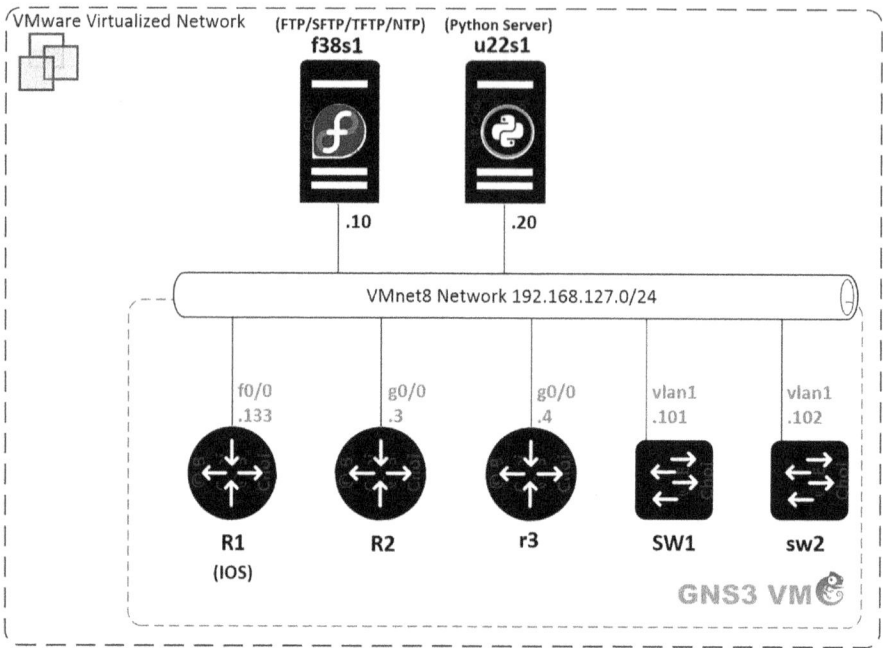

Figure 3-8. SSH paramiko lab 3, devices in use

Vendor solutions offer exceptional features and usually suit standardized customer environments well. However, they often require some customization to enhance their versatility and utility in real production environments. Without further ado, let's proceed to create an interactive tool for SSHing into your routers and switches to perform backups.

Tip

Configuring IP services server port 69 for TFTP in Fedora using the ufw command

In a previous Telnet lab, you tested the TFTP and, if you've been continuing your work from that lab, the port should already be open and prepared for further workload. On your IP services server, if port 69 is not listed (in open state), you can enable it using the ufw command with the following syntax:

On f38s1 server:
```
[jdoe@f38s1 ~]$ sudo ufw allow 69/udp
[jdoe@f38s1 ~]$ netstat -tuln | grep 69 # Check if port 69 is open
```

```
udp6       0      0 :::69                    :::*    # Prioritize IPv6 by
```
default, but Ipv4 port 69 is open

Also, here are quick reminder commands to get your TFTP services running on f38s1.

```
[jdoe@f38s1 ~]$ systemctl status tftp-server # Check the status of the
TFTP service
[jdoe@f38s1 ~]$ systemctl start tftp-server # Start the TFTP service
[jdoe@f38s1 ~]$ systemctl enable tftp-server # Set the TFTP service to
start on bootup
```

If in doubt, you can always check if port 69 is in an open or service state using the NMAP.

```
On u22s1 server:
jdoe@u22s1:~$ sudo nmap -p 69 -sU 192.168.127.10 # Prob the IP server
from u22s1
Starting Nmap 7.80 ( https://nmap.org ) at 2023-11-05 10:21 AEDT
Nmap scan report for 192.168.127.10
Host is up (0.00032s latency).

PORT    STATE          SERVICE
69/udp  open|filtered  tftp
MAC Address: 00:0C:29:05:BC:8C (VMware)

Nmap done: 1 IP address (1 host up) scanned in 0.38 seconds
```

Let's get started on this SSH lab—lots of fun ahead!

#	Task
①	As the first step, working from your Python server, u22a1, check if you can ping all network devices from your Python automation server. This time you are going to use fping again and check the connectivity from a single command. If you still have not installed fping, use the following command to do so now: jdoe@u22s1:~/ch3_ssh$ **sudo apt-get install fping** After fping has been installed successfully, run the command like this to confirm your server's network connectivity to all six nodes: jdoe@u22s1:~/ch3_ssh$ **fping 192.168.127.3 192.168.127.4 192.168.127.101 192.168.127.102 192.168.127.133 192.168.127.10** 192.168.127.10 is alive 192.168.127.101 is alive 192.168.127.102 is alive 192.168.127.3 is alive 192.168.127.4 is alive 192.168.127.133 is alive
②	Now check the TFTP services on f38s1 as a precaution. [jdoe@f38s1 tftpboot]$ sudo systemctl status tftp-server.service ● tftp-server.service - Tftp Server Loaded: loaded (/etc/systemd/system/tftp-server.service; enabled; preset: disabled) Drop-In: /usr/lib/systemd/system/service.d └─10-timeout-abort.conf Active: active (running) since Sun 2023-11-05 14:42:06 AEDT; 14min ago TriggeredBy: ● tftp-server.socket Docs: man:in.tftpd Main PID: 3492 (in.tftpd) Tasks: 1 (limit: 4632) Memory: 196.0K CPU: 24ms

(continued)

#	Task
|   | ```
CGroup: /system.slice/tftp-server.service
└─3492 /usr/sbin/in.tftpd -c -p -s /var/lib/tftpboot # Storage
directory

Nov 05 14:42:06 f38s1.pynetauto.com systemd[1]: Started tftp-server.
service - Tftp Server.
``` |
| ③ | To make this lab more interesting, you'll aim to measure the end-to-end task duration from initiation to completion. This will help you measure your script's speed and tweak sleep timers as necessary. You might have noticed in both the `paramiko` and Telnet scripts, that `time.sleep(s)` code has been used to allow sufficient response time for the virtual routers and switches. Similarly, with `netmiko` modules, many timers are already built into the library. However, they might not always be entirely accurate, necessitating the use of sleep timers to cater to different devices' needs in dissimilar operational conditions. To clear things up, in Python, a module represents a file that holds Python code and definitions. It can include functions, classes, and variables. On the other hand, a library comprises a collection of modules using Python code. A script is a Python file or program that uses modules and libraries to carry out specific tasks tailored to the user's requirements. Essentially, a Python script relies heavily on various modules and libraries to accomplish its functions in the Python ecosystem.

In developing Python scripts, the creation of small tools and ideas enhances the value of Python scripts. The development timer tool script presented here measures the time it takes to count from 1 to 10 million. By adjusting the numbers, you can assess your computer's counting speed. You can also place your function or tasks in between the script to gauge its interval execution times. In Python, one function might process data more rapidly than another. It's worthwhile to test the speed of your scripts, considering that Python isn't notably renowned for its speed. **Often, it's the Python coder's choice of modules or tools that may add lag to a script. In the realm of network automation, while speed matters, precision often takes precedence over it**. For instances that demand speed, the C programming language, closest to machine language, can be used. In May 2023, Mojo, a superset of Python was introduced to general users to overcome Python's sluggish performance problems and a supercharged version of new Python. |

(continued)
```

#	Task

	I encourage you to try the following code on your server and observe how this simple script performs. While it may seem basic, many simple scripts collectively build the foundation of your application.
	
	``
	jdoe@u22s1:~/ch3_ssh$ nano timer_tool.py
	import time # Import the time module
	
	start_timer = time.mktime(time.localtime()) # Start time
	
	# Run a function or task here
	big_number = range(10000000)
	for i in big_number:
	print(i, end=" ") # Print the numbers
	
	# Uncomment the line below if you want to print the numbers
	# differently
	# print(" ".join(str(i) for i in big_number))
	
	total_time = time.mktime(time.localtime()) - start_timer
	# Calculate total time
	
	print("Total time:", total_time, "seconds") # Display total time
	``

4	Guess how long your computer will take to count from 1 to 10 million. From your u22s1 Python server, run the timer tool by typing python timer_tool.py. On my Ubuntu VM, it took a full 26 seconds to count from 1 to 10 million. The speed will be relative to the resources assigned to your VM to the performance of your host PC as the Type-2 hypervisor. ``` jdoe@u22s1:~/ch3_ssh$ python timer_tool.py [...omitted for brevity] 9999974 9999975 9999976 9999977 9999978 9999979 9999980 9999981 9999982 9999983 9999984 9999985 9999986 9999987 9999988 9999989 9999990 9999991 9999992 9999993 9999994 9999995 9999996 9999997 9999998 9999999 Total time : 26.0 seconds ```

(continued)

#	Task
| ⑤ | To create an interactive script for communication with the network administrator, you'll develop a username and password collector tool that leverages the getpass module from the getpass library. This module allows you to mask the passwords during input, and with certain adjustments, you can prompt the user to re-enter the password to verify it. This helps address issues arising from potential mistyped passwords, which you can manage by integrating if statements into your password script.<br><br>```
jdoe@u22s1:~/ch3_ssh$ nano password_tool.py
from getpass import getpass  # Import getpass module
def get_credentials():  # Define function
    username = input("*Enter Network Admin ID : ")  # Get username input
    password = None  # Set password
    while not password:  # Loop until a password is entered
        password = getpass("*Enter Network Admin PWD : ")  # Get password
        password_verify = getpass("**Confirm Network Admin PWD : ")
        # Verify password
        if password != password_verify:  # Check password match
            print("! Network Admin Passwords do not match. Please try again.")
            # Notify mismatch
            password = None  # Reset password
    print(username, password)  # Display entered credentials
    return username, password  # Return username and password
get_credentials()  # Execute the function
``` |
| ⑥ | Create and execute the password tool script from your u22s1 server. For testing purposes, you'll print the username and password, but remember to remove the print(username, password) code from the function and comment it out after pressing the Enter key. Silence the print statement as needed.<br><br>```
jdoe@u22s1:~/ch3_ssh$ python password_tool.py
*Enter Network Admin ID : jdoe
*Enter Network Admin PWD : ********
**Confirm Network Admin PWD : ********
jdoe cisco123
``` |

(continued)

CHAPTER 3　PYTHON NETWORK AUTOMATION LABS: SSH IN ACTION, PARAMIKO AND NETMIKO LABS

#	Task
⑦	Now that you have developed two mini-tools, let's proceed to write the interactive code for creating `running-config` backups of your lab network devices. Once the `paramiko` script is complete, your script should resemble the following. Detailed explanations are provided within the code, so carefully review it before coding. **If you encounter issues, refer to the source code but try to type most of the code manually instead of copying and pasting.** jdoe@u22s1:~/ch3_ssh$ **nano interactive_backup.py** jdoe@u22s1:~/ch3_ssh$ **cat interactive_backup.py** ```#!/usr/bin/env python3 # Shebang for Python script
import time # Importing the time module
import paramiko # Importing paramiko for SSH connections
from datetime import datetime # Importing datetime module
from getpass import getpass # Importing getpass for secure password
                              input

t_ref = datetime.now().strftime("%Y-%m-%d_%H-%M-%S") # Get current
                                                       timestamp
device_list = ["192.168.127.3", "192.168.127.4", "192.168.127.101",
"192.168.127.102", "192.168.127.133"] # List of device IPs
start_timer = time.mktime(time.localtime()) # Start time for the script

def get_credentials():
global username, password
username = input("*Enter Network Admin ID : ")   # Asking for
                                                   username
password = None
while not password:
password = getpass("*Enter Network Admin PWD : ") # Asking for
                                                    password
password_verify = getpass("**Confirm Network Admin PWD : ")
# Confirming password
if password != password_verify:
``` |

(continued)

182
```

#	Task
| | ```python
print("! Network Admin Passwords do not match. Please try again.")
# Password mismatch prompt
password = None
return username, password

get_credentials() # Get user credentials

for ip in device_list:
print(t_ref) # Print timestamp
print("Now logging into " + (ip)) # Logging into device
ssh_client = paramiko.SSHClient() # Creating SSH client object
ssh_client.set_missing_host_key_policy(paramiko.AutoAddPolicy())
# Auto-accepting unknown SSH keys
ssh_client.connect(hostname=ip, username=username,
password=password, look_for_keys=False) # SSH connection
print("Successful connection to " + (ip) + "\n") # Successful
                                                   connection message
print("Now making running-config backup of " + (ip) + "\n")
# Creating backup message
remote_connection = ssh_client.invoke_shell() # Start an interactive
                                                 shell
time.sleep(3)
remote_connection.send("copy running-config tftp\n")
# Command to copy running config to TFTP
remote_connection.send("192.168.127.10\n") # TFTP server IP
remote_connection.send((ip) + ".bak@" + (t_ref) + "\n") # Backup
                                                          filename
time.sleep(3)
print() # Empty line
time.sleep(3)
``` |

(continued)
```

#	Task
| | ```
output = remote_connection.recv(65535) # Receiving output from the
                                         SSH server
print((output).decode('ascii')) # Printing received output
print(("Successfully backed-up running-config to TFTP &
Disconnecting from ") + (ip) + "\n") # Backup success message
print("-" * 80) # Separator line
ssh_client.close  # Closing the SSH connection
time.sleep(1)
total_time = time.mktime(time.localtime()) - start_timer #
Calculating total script execution time
print("Total time : ", total_time, "seconds") # Displaying total
                                                time taken
``` |
| ⑧ | In Chapter 8 of the Part I book, you used the /var/lib/tftpboot/ directory as your file-sharing location on the f38s1 server. If you quickly run the mandatory ls command, you will see the old files from your previous labs.<br><br>```
[jdoe@f38s1 ~]$ cd /var/lib/tftpboot
[jdoe@f38s1 tftpboot]$ ls -lh
total 17K
-rw-r--r--. 1 jdoe    jdoe      37 Sep 30 10:06 f38s1_file06.txt
-rw-r--r--. 1 nobody  nobody  1.9K Sep 30 14:42 r1-confg
-rw-r--r--. 1 nobody  nobody    36 Sep 30 10:28 u22s1_file09.txt
``` |
| ⑨ | Return to the Python automation server (u22s1). Now, initiate the interactive backup script to SSH into each device and create backups of their running configurations on the TFTP server. Execute the python interactive_backup.py command, provide the network administrator credentials, and then relax while observing the process.<br><br>```
jdoe@u22s1:~/ch3_ssh$ python interactive_backup.py
*Enter Network Admin ID : jdoe
*Enter Network Admin PWD : ********
**Confirm Network Admin PWD : ********
2023-11-05_15-01-43
Now logging into 192.168.127.3
Successful connection to 192.168.127.3
``` |

(continued)
```

#	Task
	Now making running-config backup of 192.168.127.3 *[...omitted for brevity]* --- Total time : 53.0 seconds 2023-11-05_15-01-43 Now logging into 192.168.127.133 Successful connection to 192.168.127.133 Now making running-config backup of 192.168.127.133 R1#copy running-config tftp Address or name of remote host []? 192.168.127.10 Destination filename [r1-confg]? 192.168.127.133.bak@2023-11-05_15-01-43 !! 1929 bytes copied in 2.184 secs (883 bytes/sec) R1# Successfully backed-up running-config to TFTP & Disconnecting from 192.168.127.133 --- Total time : 63.0 seconds It took 63 seconds to make backups of five lab devices' running configuration to the TFTP server. In the upcoming netmiko lab, you'll delve into file management using FTP. However, the principles you've acquired thus far are transferable and can be applied to FTP or SFTP scenarios.
⑩	Reconnect to your f38s1 server. Next, navigate to the /var/lib/tftpboot directory and execute the ls -lh command. You'll locate all your device's running-config backups, each labeled with the date and timestamp, as defined in your code. With this, you've finished running the SSH interactive script to capture the running configurations of all devices. While you're using only five devices here, as the count of devices grows into tens and hundreds, handling them in production will be effortless. Sweet!

(continued)

#	Task

	`[jdoe@f38s1 tftpboot]$ `**`ls -lh 192*`**
	`-rw-r--r--. 1 nobody nobody 4.5K Nov  5 15:02 192.168.127.101.bak@2023-11-05_15-01-43`
	`-rw-r--r--. 1 nobody nobody 4.0K Nov  5 15:02 192.168.127.102.bak@2023-11-05_15-01-43`
	`-rw-r--r--. 1 nobody nobody 1.9K Nov  5 15:02 192.168.127.133.bak@2023-11-05_15-01-43`
	`-rw-r--r--. 1 nobody nobody 3.7K Nov  5 15:02 192.168.127.3.bak@2023-11-05_15-01-43`
	`-rw-r--r--. 1 nobody nobody 3.5K Nov  5 15:02 192.168.127.4.bak@2023-11-05_15-01-43`

You've reached the halfway point of this chapter. You've finished the `paramiko` labs by completing the running configuration backups through an SSH connection and TFTP file transfer. You can conveniently modify the configuration to use either FTP or SFTP file transfer for additional testing and development. In the latter half of this chapter, you'll engage with the versatile `netmiko` library in various scenarios.

Python SSH Labs: netmiko

`paramiko` is a robust SSH library used to manage network devices through an SSH connection. However, fine-tuning certain aspects of your script is essential for smooth operation. The precise timing of connections and execution of various SSH commands require special attention to prompts and timing. Fortunately, a CCIE emeritus engineer, Kirk Byers, has dedicated considerable effort to address the potential issues one might encounter when using `paramiko`. His solution is `netmiko`, an impressive Python library designed to simplify `paramiko` SSH connections to various vendor networking devices. In the Python community, many skilled individuals generously share their expertise and insights.

After using `netmiko` for approximately five years, I cannot express enough gratitude to Kirk Byers for helping me navigate through many of the pitfalls associated with `paramiko`. He has preemptively addressed several issues related to `paramiko`'s SSH connections, allowing you to benefit from one engineer's ingenuity and hard work in pioneering Python Network Automation.

For further reading, here are Kirk's official netmiko links and GitHub sites:

Kirk Byers' Python Home: https://pynet.twb-tech.com/

Getting started with netmiko: https://ktbyers.github.io/netmiko/#tutorialsexamplesgetting-started

netmiko supported device list: https://github.com/ktbyers/netmiko/blob/master/netmiko/ssh_dispatcher.py

Installing the netmiko Library

Before writing your first netmiko Python script, make sure that you have netmiko libraries installed using pip install netmiko, as shown here:

```
jdoe@u22s1:~/ch3_ssh$ pip install netmiko
Defaulting to user installation because normal site-packages is not writeable
Collecting netmiko
  Downloading netmiko-4.2.0-py3-none-any.whl (213 kB)
[...omitted for brevity]
Requirement already satisfied: pycparser in /usr/local/lib/python3.10/dist-packages (from cffi>=1.4.1->pynacl>=1.5->paramiko>=2.9.5->netmiko) (2.21)
Installing collected packages: netmiko
Successfully installed netmiko-4.2.0
```

After installing netmiko, launch your Python interpreter and import the netmiko module. You can verify the installed libraries using the pip freeze | grep netmiko command. However, at times, even if the library is listed, you might encounter issues using it, especially in environments running multiple versions of Python. If no errors are returned in the interpreter, netmiko has been successfully installed, and you're prepared to write the following:

```
jdoe@u22s1:~/ch3_ssh$ pip freeze | grep netmiko
netmiko==4.2.0
jdoe@u22s1:~/ch3_ssh$ python
Python 3.10.12 (main, Jun 11 2023, 05:26:28) [GCC 11.4.0] on linux
```

```
Type "help", "copyright", "credits" or "license" for more information.
>>> import netmiko
>>>
```

After verification, press Ctrl+Z, type 'exit()', or use 'quit()' to exit/close your Python 3 interpreter.

netmiko Lab 1: netmiko Uses a Dictionary for Device Information, Not a JSON Object

When utilizing netmiko, **you need to provide device information in a dictionary format, containing basic SSH login details such as the device type, IP (hostname), administrator ID, and password as the essential elements.** Additionally, you can include other optional information like the secret password, port number, and logging status. At first glance, this dictionary format resembles a data set in JavaScript Object Notation (JSON) format. However, it's essential to note that the netmiko device information is a built-in Python dictionary, differing from a JSON data set used by browsers and servers for data transmission. A typical netmiko dictionary usually takes the form as follows:

```
c9300x = {
    'device_type': 'cisco_ios',
    'host':    '10.20.30.40',
    'username': 'jdoe',
    'password': 'password123',
    'port' : 8022,          # optional, if not using defaults to 22
    'secret': 'secret123',    # optional
}
```

If you want to flatten out the previous dictionary as in the earlier paramiko examples, you can write the dictionary like this. Still, the previous structured information looks nicer on the eyes.

```
c9300x = {"device_type":"cisco_ios", "ip":"10.20.30.40", "username":"jdoe", "password":"cisco123" }
```

You can also enhance the netmiko dictionary by mixing it with some input statements and getpass functions to request user inputs, which looks like this:

```
from netmiko import ConnectHandler
from getpass import getpass
device1 = {
'device_type': 'cisco_ios',
'ip': input('IP Address : '),
'username': input('Enter username : '),
'password': getpass('SSH password : '),
}
```

I covered this briefly as part of the previous paramiko lab. Before moving onto netmiko Lab 1, confirm that the netmiko dictionary is not a JSON object. Let's quickly type this code into the Python interpreter to confirm this fact.

Open your Python interpreter and type the following into the interpreter window. Make sure that you use the double quotation marks when creating the dictionary, device1. Python will automatically convert them to single quotation marks, and you will find out why. First, check the attributes of a netmiko dictionary.

```
>>> device1 = {"device_type":"cisco_ios", "ip":"192.168.127.3", "username":"jdoe", "password":"cisco123"}
>>> print(device1)
{'device_type': 'cisco_ios', 'ip': '192.168.127.3', 'username': 'jdoe', 'password': 'cisco123'}
>>> print(type(device1))
<class 'dict'>
>>> dir(device1)
['__class__', '__class_getitem__', '__contains__', '__delattr__', '__delitem__', '__dir__', '__doc__', '__eq__', '__format__', '__ge__', '__getattribute__', '__getitem__', '__gt__', '__hash__', '__init__', '__init_subclass__', '__ior__', '__iter__', '__le__', '__len__', '__lt__', '__ne__', '__new__', '__or__', '__reduce__', '__reduce_ex__', '__repr__', '__reversed__', '__ror__', '__setattr__', '__setitem__', '__sizeof__', '__str__', '__subclasshook__', 'clear', 'copy', 'fromkeys', 'get', 'items', 'keys', 'pop', 'popitem', 'setdefault', 'update', 'values']
```

Convert the `device1` dictionary into the JSON format using the `dumps` method. When you print the converted JSON objects, they look like a Python dictionary, but the key difference is that all key and value sets are enclosed in double quotation marks. At first glance, you would think that the JSON object is the same as the `netmiko` dictionary object, but it is a string object, as revealed by the `dir()` method. The JSON object belongs to Python's built-in `str` class, and of course, the `netmiko` dictionary belongs to the built-in `dict` class.

```
>>> import json
>>> device2 = json.dumps(device1)
>>> print(device2)
{"device_type": "cisco_ios", "ip": "192.168.127.3", "username": "jdoe",
"password": "cisco123"}
>>> print(type(device2))
<class 'str'>
>>> dir(device2)
['__add__', '__class__', '__contains__', '__delattr__', '__dir__',
'__doc__', '__eq__', '__format__', '__ge__', '__getattribute__', '__
getitem__', '__getnewargs__', '__gt__', '__hash__', '__init__', '__init_
subclass__', '__iter__', '__le__', '__len__', '__lt__', '__mod__', '__
mul__', '__ne__', '__new__', '__reduce__', '__reduce_ex__', '__repr__',
'__rmod__', '__rmul__', '__setattr__', '__sizeof__', '__str__', '__
subclasshook__', 'capitalize', 'casefold', 'center', 'count', 'encode',
'endswith', 'expandtabs', 'find', 'format', 'format_map', 'index',
'isalnum', 'isalpha', 'isascii', 'isdecimal', 'isdigit', 'isidentifier',
'islower', 'isnumeric', 'isprintable', 'isspace', 'istitle', 'isupper',
'join', 'ljust', 'lower', 'lstrip', 'maketrans', 'partition',
'removeprefix', 'removesuffix', 'replace', 'rfind', 'rindex', 'rjust',
'rpartition', 'rsplit', 'rstrip', 'split', 'splitlines', 'startswith',
'strip', 'swapcase', 'title', 'translate', 'upper', 'zfill']
```

Let's quickly check if device1 is the same as device2.

```
>>> if device1 == device2:
...     print(True)
```

CHAPTER 3   PYTHON NETWORK AUTOMATION LABS: SSH IN ACTION, PARAMIKO AND NETMIKO LABS

```
... else:
...     print(False)
...
False # Confirmed that netmiko dictionary is not the same as JSON-
formatted data
```

In netmiko Lab 1, you'll delve into creating a directory and generating a file using Tcl shell commands on a Cisco router. This exercise aims to provide practice in creating dummy files for subsequent deletion from the router's flash memory. For simplicity, the exercise uses an IOSv router, R2, to ease the learning process, particularly for those from non-Cisco networking backgrounds. The focus here is on understanding the concepts rather than dwelling on the age of the IOSv image.

As previously discussed at the beginning of this chapter, while using the latest tools is beneficial, they aren't imperative for grasping the fundamental networking concepts or Python Network Automation. Refer to Figure 3-9 for further guidance.

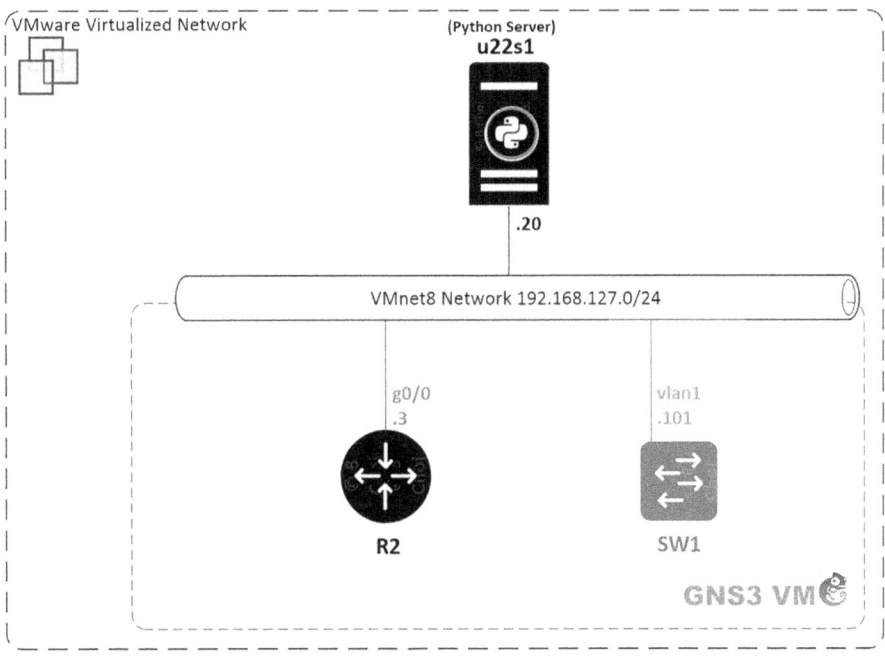

Figure 3-9. *SSH netmiko Lab 1, devices in use*

CHAPTER 3   PYTHON NETWORK AUTOMATION LABS: SSH IN ACTION, PARAMIKO AND NETMIKO LABS

The objective of this lab is to provide practice in removing files from the Cisco device's flash memory using a Python script, eliminating the need for direct console interaction with a Cisco device. The process starts with formatting the flash memory of R2, creating a directory using the `mkdir` command, and utilizing traditional Tcl shell commands to generate files for deletion.

You aim to develop Python code capable of removing outdated files and navigating to specific directories to locate targeted files for removal. Although this functionality might not immediately seem relevant to your work, it's extremely valuable in the development of IOS upgrade applications. The ability of your Python scripts to delete outdated IOS and other files is critical due to the limited storage space in routers, switches, firewalls, and most other purpose-built computing devices. This same limitation also extends to virtualized network devices, where storage space for network device VMs is constrained to manage costs and enable vendors to safeguard their profit margins through licensing. Essentially, they use different vCPU, memory, and storage sizes to apply varied pricing strategies to customers. When I was a young network engineer, facing this challenge was a daily occurrence. Now, as a seasoned network engineer, it remains an ongoing challenge. With age and experience, I've gained wisdom and understanding that storage doesn't simply appear out of thin air. This is where vendors leverage their investments in research and marketing to capitalize on the significance of bytes.

It's time to put theory into action.

#	Task
①	Open the console window of R2 and format disk 0 so there are no files. **R2#format flash:** Format operation may take a while. Continue? [confirm] # Press <Enter> Format operation will destroy all data in "flash:". Continue? [confirm] # Press <Enter> Format: All system sectors written. OK... Format: Total sectors in formatted partition: 4193217 Format: Total bytes in formatted partition: 2146927104 Format: Operation completed successfully.

#	Task
	Format of flash0: complete R2#**show flash:** -#- --length-- -----date/time------ path 2142711808 bytes available (8192 bytes used)
②	Use `mkdir flash:/directory_name` to create a directory under `flash:`. R2#**mkdir flash:/old_files** Create directory filename [old_files]? # Press <Enter> Created dir flash0:/old_files R2#**dir** Directory of flash0:/ 2 drw- 0 Nov 5 2023 20:29:00 +10:00 old_files 2142720000 bytes total (2142707712 bytes free)
③	Now use the good old Tcl shell to create four files, as shown here. Two files will be created under `flash:` and the other two files will be created under `flash:old_files/`. R2#**Tclsh** R2(tcl)#**puts [open "flash:Delete_me_1.bin" w+]** file0 R2(tcl)#**puts [open "flash:Don't_delete_me_1.txt" w+]** file1 R2(tcl)#**puts [open "flash:old_files/Delete_me_2.bin" w+]** # Created in old_files directory file2 R2(tcl)#**puts [open "flash:old_files/Don't_delete_me_2.txt" w+]** # Created in old_files directory file3 R2(tcl)#**dir** Directory of flash0:/

(continued)

#	Task

```
       2  drw-    0    Nov 5 2023 20:29:00 +10:00   old_files # a directory
       3  -rw-    0    Nov 5 2023 20:30:18 +10:00   Delete_me_1.bin # a file
       4  -rw-    0    Nov 5 2023 20:30:26 +10:00   Don't_delete_me_1.txt
                                                    #a file

2142720000 bytes total (2142707712 bytes free)
R2(tcl)#dir flash:/old_files/
Directory of flash0:/old_files/

       5  -rw-    0    Nov 5 2023 20:30:34 +10:00   Delete_me_2.bin #a file
       6  -rw-    0    Nov 5 2023 20:30:42 +10:00   Don't_delete_me_2.txt
                                                    #a file

2142720000 bytes total (2142707712 bytes free)
R2(tcl)#^Z # Ctrl+Z break sequence
R2#
```

④ You've successfully created a dummy directory and generated dummy IOS (.bin) files on R2's flash. Let's now write an amazing SSH script using the `netmiko` library to log in to R2 and execute a Python script for deleting specific .bin files.

You might wonder why go through the trouble of creating this in a Python script when manual file deletion from the router's CLI is an option? The answer lies in the benefits. Firstly, as a reader interested in network automation coding, putting in the effort is essential to reap the rewards. Second, the script can communicate and collaborate effectively, unlike your team members who spend more time downstairs chitchatting, smoking, or drinking coffee all day, following the 80/20 rule (the Pareto Principle) in the IT industry. You are reading this book because you are or want to become that 20 percent. Moreover, envision dealing with not just one, but a hundred routers or switches and needing to repeat this process regularly as part of your job. The real power of Python Network Automation lies in its loops—the `for` and `while` loops. In the subsequent lab, you'll improve this script to a newer version for deployment on multiple devices. This iterative process is called improvement in general terms, but in programming, it's referred to as "refactoring".

(continued)

#	Task

Review the script with the embedded comments and start writing the code. Once you are happy with your work, proceed to Step 5 to execute your script. I've named this Python application file netmiko_delete_me.py.

```
jdoe@u22s1:~/ch3_ssh$ nano netmiko_delete_me.py
jdoe@u22s1:~/ch3_ssh$ cat netmiko_delete_me.py
#!/usr/bin/env python3
import re
from netmiko import ConnectHandler
from getpass import getpass
import time

device1 = {
    'device_type': 'cisco_ios',
    'ip': input('IP Address : '), # User input for IP address
    'username': input('Enter username : '), # User input for username
    'password': getpass('SSH password : '), # Securely prompts user
                                            # for SSH password
}

net_connect = ConnectHandler(**device1) # Establishing SSH connection
net_connect.send_command("terminal length 0\n") # Setting terminal
                                                # length to display all
                                                # output

time.sleep(1) # Pause for 1 second

dir_flash = net_connect.send_command("dir flash:\n") # Command to
list files in flash memory
print(dir_flash) # Display flash directory content

p30 = re.compile(r'D[0-9a-zA-Z]{4}.*.bin') # Regular Expression for
                                           # .bin files
m30 = p30.search(dir_flash) # Searching for .bin files in flash

time.sleep(1) # Pause for 1 second
```

(*continued*)

#	Task
| | ```python
# Performing file deletion if conditions are met
if bool(m30) == True:
    # Prompt user for file deletion
    print("If you can see 'Delete_me.bin' file, select it and press
    Enter.")

    del_cmd = "del flash:/" # Partial command for file deletion
    old_ios = input("*Old IOS (bin) file to delete : ")
    # User input for file to delete

    while not p30.match(old_ios) or old_ios not in dir_flash:
        old_ios = input("**Old IOS (bin) file to delete : ")
        # Request correct file name

    command = del_cmd + old_ios # Complete deletion command

    # Executing deletion command with netmiko send_command_timing
    output = net_connect.send_command_timing(
        command_string=command,
        strip_prompt=False,
        strip_command=False
    )

    # Handling confirmation and disconnecting SSH session
    if "Delete filename" in output:
        output += net_connect.send_command_timing(command_
        string="\n",
            strip_prompt=False,
            strip_command=False)
    if "confirm" in output:
        output += net_connect.send_command_timing(command_string="y",
            strip_prompt=False,
            strip_command=False)
    net_connect.disconnect # Disconnecting from SSH session
    print(output) # Displaying deletion output
``` |

(continued)
```

#	Task
| | ```python
# Additional file deletion process if the previous conditions are not met
elif bool(m30) == False:
    print("No IOS file under 'flash:/', select the directory to
    view.") # Informational message
    open_dir = input("*Enter Directory name : ") # User input for
                                                   directory name

    # Requesting user input until the correct response is received
    while not open_dir in dir_flash:
        open_dir = input("** Enter Directory name : ")

    open_dir_cmd = (r"dir flash:/" + open_dir) # Complete command for
                                                 directory listing
    send_open_dir = net_connect.send_command(open_dir_cmd)
    # Sending the directory listing command
print(send_open_dir) # Displaying directory content

    p31 = re.compile(r'D[0-9a-zA-Z]{4}.*.bin')
# Regular Expression for .bin files in a specific directory
    m31 = p31.search(send_open_dir) # Searching for .bin files in the
                                      specified directory

    if bool(m31) == True:
        # Prompt user for file deletion in the specified directory
        print("If you see old IOS (bin) in the directory. Select it
        and press Enter.")

        del_cmd = "del flash:/" + open_dir + "/" # Command for
                                                   file deletion in
                                                   specified directory
        old_ios = input("*Old IOS (bin) file to delete : ")
# User input for file to delete
``` |

(continued)
```

#	Task
| | ```python
    while not p30.match(old_ios) or old_ios not in send_open_dir:
        old_ios = input("**Old IOS (bin) file to delete : ")
# Request correct file name

    command = del_cmd + old_ios # Complete deletion command

    # Executing deletion command with netmiko send_command_timing
    output = net_connect.send_command_timing(
        command_string=command,
        strip_prompt=False,
        strip_command=False
    )

    # Handling confirmation and disconnecting SSH session
    if "Delete filename" in output:
        output += net_connect.send_command_timing(command_string="\n",
            strip_prompt=False,
            strip_command=False)
    if "confirm" in output:
        output += net_connect.send_command_timing(command_string="y",
            strip_prompt=False,
            strip_command=False)
    net_connect.disconnect # Disconnecting from SSH session
    print(output) # Displaying deletion output
else:
    ("No IOS found.")
    exit() # Exiting script if conditions aren't met
    net_connect.disconnect # Disconnecting from SSH session
``` |
| ⑤ | Now, on your u22s1 (192.168.127.20) Python server and fping/ping R2 (192.168.127.3) to check the connectivity. You are still working with the SSH protocol, and the netmiko lab is still part of the SSH labs, so you should be working within the /home/jdoe/ch3_ssh directory.
jdoe@u22s1:~/ch3_ssh$ **fping 192.168.127.3**
192.168.127.3 is alive |

(continued)
```

#	Task

⑥	Now check the file under your directory and execute the netmiko_delete_me.py script using python netmiko_delete_me.py. When the script returns the output, select the .bin file and press the Enter key to remove it from R2's flash.
	jdoe@u22s1:~/ch3_ssh$ **pwd**
	/home/jdoe/ch3_ssh
	jdoe@u22s1:~/ch3_ssh$ **ll netmiko***
	-rw-rw-r-- 1 jdoe jdoe 2982 Nov 5 22:02 netmiko_delete_me.py
	jdoe@u22s1:~/ch3_ssh$ **python netmiko_delete_me.py**
	IP Address : **192.168.127.3**
	Enter username : **jdoe**
	SSH password : ********
	Directory of flash0:/
	
	    2 drw- 0 Nov 5 2023 20:29:00 +10:00 old_files
	    3 -rw- 0 Nov 5 2023 20:30:18 +10:00 Delete_me_1.bin # your target file
	    4 -rw- 0 Nov 5 2023 20:30:26 +10:00 Don't_delete_me_1.txt
	
	2142720000 bytes total (2142707712 bytes free)
	If you can see the Delete_me.bin file, select it and press Enter.
	*Old IOS (bin) file to delete : **Delete_me_1.bin**
	del flash:/Delete_me_1.bin
	Delete filename [Delete_me_1.bin]?
	Delete flash0:/Delete_me_1.bin? [confirm]y
	R2#
	R2#
	jdoe@u22s1:~/ch3_ssh$

(continued)

CHAPTER 3   PYTHON NETWORK AUTOMATION LABS: SSH IN ACTION, PARAMIKO AND NETMIKO LABS

#	Task
7	Let's verify whether the Delete_me file has been removed from the flash:. While you could access R2's console and execute show flash:, let's opt to rerun the same script to confirm your earlier task. If you observe only one file and one directory, you've completed the initial task. Additionally, in this instance, as there are no .bin files within flash:, a distinct prompt will be displayed, asking you to choose a directory for the .bin file search. Use Ctrl+Z to practice exiting the script. jdoe@u22s1:~/ch3_ssh$ **python netmiko_delete_me.py** IP Address : **192.168.127.3** Enter username : **jdoe** SSH password : ******** Directory of flash0:/ 2 drw- 0 Nov 5 2023 20:29:00 +10:00 old_files # a directory 4 -rw- 0 Nov 5 2023 20:30:26 +10:00 Don't_delete_me_1.txt # a file 2142720000 bytes total (2142707712 bytes free) No IOS file under 'flash:/', select the directory to view. *Enter Directory name : **^Z** # Ctrl+Z break sequence [5]+ Stopped python3 netmiko_delete_me.py jdoe@u22s1:~/ch3_ssh$
8	Now, run the same script for the third time. When prompted to select a directory, cut and paste, or simply type in the directory name from the output. This action then runs the script to look for the elusive .bin file and display the result. When the script finds the .bin file, go ahead and remove the Delete_me_2.bin file. jdoe@u22s1:~/ch3_ssh$ **python netmiko_delete_me.py** IP Address : **192.168.127.3** Enter username : **jdoe** SSH password : ******** Directory of flash0:/ 2 drw- 0 Nov 5 2023 20:29:00 +10:00 old_files 4 -rw- 0 Nov 5 2023 20:30:26 +10:00 Don't_delete_me_1.txt

(*continued*)

#	Task
| | ```
2142720000 bytes total (2142707712 bytes free)
No IOS file under 'flash:/', select the directory to view.
*Enter Directory name : old_files
Directory of flash0:/old_files/

    5  -rw-    0   Nov 5 2023 20:30:34 +10:00  Delete_me_2.bin # your
target file
    6  -rw-    0   Nov 5 2023 20:30:42 +10:00  Don't_delete_me_2.txt

2142720000 bytes total (2142707712 bytes free)
If you see old IOS (bin) in the directory. Select it and press Enter.
*Old IOS (bin) file to delete : Delete_me_2.bin
del flash:/old_files/Delete_me_2.bin
Delete filename [/old_files/Delete_me_2.bin]?
Delete flash0:/old_files/Delete_me_2.bin? [confirm]y
R2#
R2#
jdoe@u22s1:~/ch3_ssh$
``` |
| 9 | Next, run the script one last time and validate the last deletion. If no files with the .bin extension are found, the script will exit and you will return to your Python server console screen.<br>```
jdoe@u22s1:~/ch3_ssh$ python netmiko_delete_me.py
IP Address : 192.168.127.3
Enter username : jdoe
SSH password : ********
Directory of flash0:/

    2  drw-    0   Nov 5 2023 20:29:00 +10:00  old_files
    4  -rw-    0   Nov 5 2023 20:30:26 +10:00  Don't_delete_me_1.txt
``` |

(continued)

#	Task
| | ```
2142720000 bytes total (2142707712 bytes free)
No IOS file under 'flash:/', select the directory to view.
*Enter Directory name : old_files
Directory of flash0:/old_files/

    6  -rw-         0  Nov 5 2023 20:30:42 +10:00  Don't_delete_me_2.txt

2142720000 bytes total (2142707712 bytes free)
jdoe@u22s1:~/ch3_ssh$
``` |
| ⑩ | Now go to R2 and delete the working directory and file, as you have completed this lab.<br>```
R2#dir
Directory of flash0:/

    2  drw-         0  Nov 5 2023 20:29:00 +10:00  old_files
    4  -rw-         0  Nov 5 2023 20:30:26 +10:00  Don't_delete_me_1.txt

2142720000 bytes total (2142707712 bytes free)
R2#delete /recursive /force flash:old_files
R2#dir
Directory of flash0:/

    4  -rw-         0  Nov 5 2023 20:30:26 +10:00  Don't_delete_me_1.txt

2142720000 bytes total (2142711808 bytes free)
R2#delete flash:Don't_delete_me_1.txt
Delete filename [Don't_delete_me_1.txt]? # Press <Enter>
Delete flash0:/Don't_delete_me_1.txt? [confirm] # Press <Enter>
R2#dir
Directory of flash0:/

No files in directory

2142720000 bytes total (2142711808 bytes free)
``` |

CHAPTER 3   PYTHON NETWORK AUTOMATION LABS: SSH IN ACTION, PARAMIKO AND NETMIKO LABS

Congratulations! You've acquired the skills to create Python code using the netmiko library, which is capable of deleting and searching for files in a router's flash memory. This marks the inception of Python Network Automation, where simple tasks evolve into script snippets—paving the way for more intricate applications. You've learned how to develop and assemble code snippets, culminating in more comprehensive scripts, or what some term a Python application. In this context, the terms "script" and "application" are used interchangeably, despite their subtle distinctions and resemblances. The journey continues as you move forward to craft a port-scanning tool—a valuable asset for inspecting open ports. This tool will not only enrich your skillset but will also find practical application in your ongoing work, offering enhanced capabilities in network analysis and security.

netmiko Lab 2: Develop a Simple Port Scanner Using a Socket Module and Then Develop a netmiko Disable Telnet Script

In this SSH lab, you'll delve into writing a simple port-scanner tool using the socket module. The aim is to develop this scanning tool within your netmiko Lab 2 scripts (e.g., netmiko_disable_telnet.py) to examine opened port 23 (Telnet) and subsequently deactivate these ports using a Python script using the netmiko library. Previously, although you were aware that Telnet is an insecure protocol for remote device management, you configured all routers and switches with transport input all under line vty 0 15, enabling both Telnet and SSH connections for device management. To disable Telnet access, the task involves reconfiguring the Virtual Teletype (vty) lines using the transport input ssh command. To begin this lab, ensure all routers and switches are powered on for the lab's execution.

As a result of this lab, you'll gain proficiency in scanning network devices for open ports, verifying if port 23 (or any other selected ports) is active. You'll secure your devices by deactivating the Telnet port 23 across all devices, restricting device management to SSH connections exclusively. Try to focus on the core concepts and develop the necessary tools to accomplish the lab objectives. Refer to Figure 3-10 for devices in use before you begin this lab.

CHAPTER 3   PYTHON NETWORK AUTOMATION LABS: SSH IN ACTION, PARAMIKO AND NETMIKO LABS

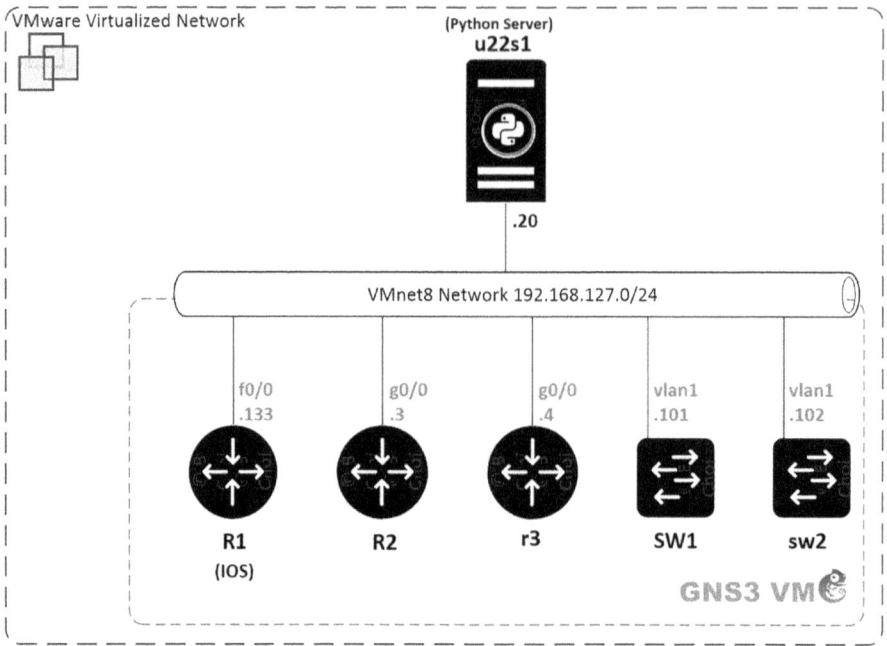

Figure 3-10. *SSH netmiko Lab 2, devices in use*

Let's first develop a mini-scanning tool and verify the open ports on your routers and switches. Then you'll incorporate the scanning tool into your script, netmiko_disable_ telnet.py. Eager to write more code? You can start your coding journey now.

#	Task
①	First, you have to create a socket object using a socket.socket(family, type) with the family set to the socket.AF_INET and the type set to socket.SOCK_STREAM. According to Python socket programming, socket.AF_INET specifies the IP address family for IPv4 and socket.SOCK_STREAM specifies the socket type for TCP. To check a port status, use socket.connect_ex(dest) with the socket as the socket object and dest as a tuple containing the IP address and desired port number. If the port is open, socket.connect_ex() returns 0, but if the port is closed, it will return different digits based on the port being scanned. At the end of the port scan, you must close the socket using socket.close().

(continued)

#	Task
| | A simple port scanner template you are going to use looks like this, and it allows you to scan only a single device with a single port. If you put this onto your Linux server, create a new file called scan_open_port.py and run the code. You will be able to check the port status on a single destination.<br>jdoe@u22s1:~/ch3_ssh$ **nano scan_open_port.py**<br><br>```python
import socket # Import socket module
sock = socket.socket(socket.AF_INET, socket.SOCK_STREAM)
# Create a socket object
dest = ("192.168.127.3", 23) # IP and Port to scan
port_open = sock.connect_ex(dest) # Create socket connect object

if port_open == 0: # If a port is opened, it returns 0
    print(port_open) # print result = 0
    print(f"On {dest[0]}, port {dest[1]} is open.")# Informational
else:
    print(port_open) # print result, an integer other than 0
    print(f"On {dest[0]}, port {dest[1]} is closed.") # Informational

sock.close()   # close socket object
```<br><br>When you run the script in your lab on the R2 (192.168.127.3) router, the result will look like this:<br>jdoe@u22s1:~/ch3_ssh$ **python scan_open_port.py**<br>0 # returns 0, so the port is open<br>On 192.168.127.3, port 23 is open.<br><br>If you change port 23 (Telnet) to 80 (HTTP) and scan the port again, it returns 111 and the port 80 is closed message. Looking pretty good so far.<br>jdoe@u22s1:~/ch3_ssh$ **python scan_open_port.py**<br>111<br>On 192.168.127.3, port 80 is closed. |

(continued)
```

#	Task

②	The previous script works well for a single port on one device but lacks scalability for your diverse use cases. To enhance its adaptability within your environment, modifications are required. Leveraging the existing script as a basis, let's reimagine it to scan ports across multiple devices concurrently. My version of the script resembles the following, yet in coding, there's no definitive right or wrong approach. Feel free to add your personal touch by making modifications as you see fit. Given that this script scans multiple ports, it has been named scan_open_ports.py (with the plural ports).
	jdoe@u22s1:~/ch3_ssh$ **nano scan_open_ports.py**
	import socket # Import socket module
	ip_addresses = ["192.168.127.3", "192.168.127.4", "192.168.127.101", "192.168.127.102", "192.168.127.133"] # IP address list
	
	for ip in ip_addresses: # Iterate through IP addresses
	    for port in range(22, 24): # Iterate over port range 22-23
	        dest = (ip, port) # Define IP and port
	        try: # Try connecting
	            with socket.socket(socket.AF_INET, socket.SOCK_STREAM)
	as sock: # Create socket object
	                sock.settimeout(3) # Set timeout
	                connection = sock.connect(dest)
	# Connect to a destination on a specified port
	                print(f"On {ip}, port {port} is open!")
	# Print if the port is open
	        except: # If the connection fails
	            print(f"On {ip}, port {port} is closed.")
	# Print if the port is closed
	
	After you've completed the script, execute it. The outcome should resemble the following if both Telnet (port 22) and SSH (port 23) are open across your network devices. Alternatively, you can specify a different range of ports for another port-scanning task.

(continued)

CHAPTER 3   PYTHON NETWORK AUTOMATION LABS: SSH IN ACTION, PARAMIKO AND NETMIKO LABS

#	Task
	jdoe@u22s1:~/ch3_ssh$ **python scan_open_ports.py** On 192.168.127.3, port 22 is open! On 192.168.127.3, port 23 is open! On 192.168.127.4, port 22 is open! On 192.168.127.4, port 23 is open! On 192.168.127.101, port 22 is open! On 192.168.127.101, port 23 is open! On 192.168.127.102, port 22 is open! On 192.168.127.102, port 23 is open! On 192.168.127.133, port 22 is open! On 192.168.127.133, port 23 is open!

③	You've created a Python port scanner in two steps, which is now available for your use. Now that you've added another tool to your toolkit, let's build on the previous script. You'll proceed to create code that inspects the port status and, if Telnet is enabled on your devices, allows the Python application to SSH in, disable the Telnet on port 23, and then save the altered settings.
	jdoe@u22s1:~/ch3_ssh$ **nano netmiko_disable_telnet.py**
	#!/usr/bin/env python3
	import re
	from netmiko import ConnectHandler
	from getpass import getpass
	import time
	import socket
	
	# Enhanced User ID and password collection tool
	def get_credentials():
	    global username # Global username variable
	    global password # Global password variable
	    username = input("Enter your username : ") # User input for username
	    password = None
	    while not password:
	        password = getpass() # Securely input password

(continued)

#	Task
| | ```python
            password_verify = getpass("Retype your password : ")
# Verify password
            if password != password_verify:
                print("Passwords do not match. Please try again.")
                password = None
    return username, password

get_credentials()  # Get user credentials

device1 = {'device_type': 'cisco_ios', 'ip': '192.168.127.3',
'username': username, 'password': password}
device2 = {'device_type': 'cisco_ios', 'ip': '192.168.127.4',
'username': username, 'password': password}
device3 = {'device_type': 'cisco_ios', 'ip': '192.168.127.101',
'username': username, 'password': password}
device4 = {'device_type': 'cisco_ios', 'ip': '192.168.127.102',
'username': username, 'password': password}
device5 = {'device_type': 'cisco_ios', 'ip': '192.168.127.133',
'username': username, 'password': password}
devices = [device1, device2, device3, device4, device5]

# Loop through devices
for device in devices:
    ip = device.get("ip", "")
    for port in range(23, 24):  # Check port 23
        dest = (ip, port)  # Create destination IP and port
        try:
            with socket.socket(socket.AF_INET, socket.SOCK_STREAM)
            as sock:   # Create socket object
                sock.settimeout(5)  # Set timeout
                connection = sock.connect(dest)
                # Connect to the destination on the specified port
                print(f"On {ip}, port {port} is open!")
                # Inform if port is open
``` |

(continued)
```

#	Task
| | ```python
            net_connect = ConnectHandler(**device)
            # Connect to network device
            show_clock = net_connect.send_command("show
            clock\n") # Show clock on the device
            print(show_clock) # Display time
            config_commands = ['line vty 0 15', 'transport input
            ssh'] # Configurations for SSH
            net_connect.send_config_set(config_commands)
            # Apply configuration
            output = net_connect.send_command("show run |
            b line vty") # Display line configurations
            print() # Empty line for spacing
            print('-' * 79) # Separator line
            print(output) # Display configurations
            print('-' * 79) # Separator line
            print() # Empty line for spacing
            net_connect.disconnect() # Close connection
    except:
        print(f"On {ip}, port {port} is closed.") # Inform if
        port is closed
``` |
| (4) | Run the netmiko_disable_telnet.py script and update the configuration to disable Telnet logins.

jdoe@u22s1:~/ch3_ssh$ **python netmiko_disable_telnet.py**
Enter your username : **jdoe**
Password: ********
Retype your password : ********
On 192.168.127.3, port 23 is open!
23:34:15.592 AEST Sun Nov 5 2023

----
line vty 0 4
 logging synchronous
 login local |

(continued)
```

#	Task
	``
 transport input ssh # 'all' changed to 'ssh', disabled Telnet.
line vty 5 15
 logging synchronous
 login local
 transport input ssh # 'all' changed to 'ssh', disabled Telnet.
!
no scheduler allocate
ntp server 192.168.127.10
!
end
```
---
On 192.168.127.4, port 23 is open!
23:34:24.399 AEST Sun Nov 5 2023
*[...omitted for brevity]* |
| ⑤ | When you run the scan_open_ports.py script, Telnet will be disabled for all network devices. Consequently, the expected output should indicate a closed status for Telnet ports. This change signifies an enhancement in your network's security posture.<br>jdoe@u22s1:~/ch3_ssh$ **python scan_open_ports.py**<br>On 192.168.127.3, port 22 is open!<br>On 192.168.127.3, port 23 is closed.<br>On 192.168.127.4, port 22 is open!<br>On 192.168.127.4, port 23 is closed.<br>On 192.168.127.101, port 22 is open!<br>On 192.168.127.101, port 23 is closed.<br>On 192.168.127.102, port 22 is open!<br>On 192.168.127.102, port 23 is closed.<br>On 192.168.127.133, port 22 is open!<br>On 192.168.127.133, port 23 is closed. |

*(continued)*
```

CHAPTER 3   PYTHON NETWORK AUTOMATION LABS: SSH IN ACTION, PARAMIKO AND NETMIKO LABS

#	Task
⑥	Once you have completed the previous task successfully, modify the main code and your except line, so that if port 23 is closed, the configuration is saved. In other words, after port 23 is closed, the `running-config` will be saved to the `running-config`, also if the port is closed, the configuration will be saved anyway. This prevents the loss of the good `running-config`. You always want to save the last known good configuration to the `startup-config` so you can recover from outages quicker. jdoe@u22s1:~/ch3_ssh$ **cp netmiko_disable_telnet.py netmiko_disable_telnet_save.py** jdoe@u22s1:~/ch3_ssh$ **nano netmiko_disable_telnet_save.py** `[...omitted for brevity]` <pre> print('-' * 79)
print(output)
# Saving running-config to start-up config for open
port devices
net_connect = ConnectHandler(**device)
write_mem = net_connect.send_command("write
memory\n")
print(write_mem)
print('-' * 79)
print()
net_connect.disconnect()
except:
print(f"On {ip}, port {port} is closed.")
# This is for saving the configuration after the
configuration change.
net_connect = ConnectHandler(**device)
write_mem = net_connect.send_command("write memory\n")
print(write_mem)
print('-' *79)
net_connect.disconnect()</pre>

(continued)

211

CHAPTER 3   PYTHON NETWORK AUTOMATION LABS: SSH IN ACTION, PARAMIKO AND NETMIKO LABS

#	Task
(7)	Run the script one more time to save the currently running configuration. jdoe@u22s1:~/ch3_ssh$ **python netmiko_disable_telnet_save.py** Enter your username : **jdoe** Password: ******** Retype your password : ******** On 192.168.127.3, port 23 is closed. Building configuration... [OK] --- On 192.168.127.4, port 23 is closed. Building configuration... [OK] --- On 192.168.127.101, port 23 is closed. Building configuration... Compressed configuration from 4587 bytes to 2137 bytes[OK] --- On 192.168.127.102, port 23 is closed. Building configuration... Compressed configuration from 4078 bytes to 1926 bytes[OK] --- On 192.168.127.133, port 23 is closed. Building configuration... [OK] ---
(8)	**Optionally**, if you want to conduct another round of code validation, re-enable the transport input all configuration under line vty 0 15 for any router or switch. Then, proceed to rerun the validation test until you are satisfied with the code quality.

> **Warning**
>
> **Considerations when using port-scanning tools**
>
> Exercise caution when using scanning tools to avoid violating company security policies or client regulations. Refrain from executing port-scanning tasks without appropriate change control and official approval at your workplace. For production use, ensure that SSH serves as the primary remote console management protocol. By adhering to security best practices, you can responsibly remove Telnet services using an automated script.

You've created a miniature port-scanning tool through Python's socket module, which served as the groundwork for developing an SSH (netmiko) tool to disable the insecure Telnet protocol across your lab devices. This enhancement results in a more secure lab topology, where access to routers and switches is limited to SSH connections for device management. The use case extends beyond the lab environment into production, enabling you to independently construct the tool and automate the task previously performed manually. There's room for innovation by expanding the tool's capabilities and potential use cases.

netmiko Lab 3: config compare

In this netmiko lab, you'll write a quick network device configuration comparison tool using the difflib library. This script will integrate components from a previous lab, leveraging the port scanner to verify communication prechecks by examining the open status of port 22 for SSH connections. The aim is to compare the running configurations of two similar devices side by side. This also can be extended to the pre- and post-configuration change comparison tool for the same device. If you've worked as a network engineer, you might be acquainted with comparison tools like the Notepad++ compare module.

Traditionally, manual intervention is required to log in to each device, gather configurations by running show commands, save them as separate logs or text files, and then use a tool like Notepad++ for comparison. However, this tool streamlines the process for legacy devices accessible through SSH connections. For devices supporting REST API and Webhook, this comparison method can extend to collecting running configurations via API calls/Webhook. The script can be applied to the acquired dataset.

Furthermore, to enhance this script, powerful data modules like pandas and xlsxwriter can be integrated to interpret lines as data frames and extract discrepancies between configurations, automating various manual tasks. Note that you could also use the NAPALM library to yield similar outcomes. NAPALM primarily abstracts network device configurations, while netmiko focuses specifically on SSH connections and automation. Despite NAPALM's presence in network automation, its weird market positioning between Python and Ansible is not convincing and falls beyond the scope of this book. To those readers keen on discovering the napalm library, I suggest doing so at your own convenience. I may consider adding a chapter on the napalm library in the subsequent edition of this book.

To undertake this lab, access to R2 (192.168.127.3), r2 (192.168.127.4), SW1 (192.168.127.101), and sw2 (192.168.127.102) is required. Using WinSCP for accessing data stored on the u22s1 Python server is recommended. Preparing by downloading and installing WinSCP or FileZilla in advance for file retrieval will speed up the process. See Figure 3-11 for reference.

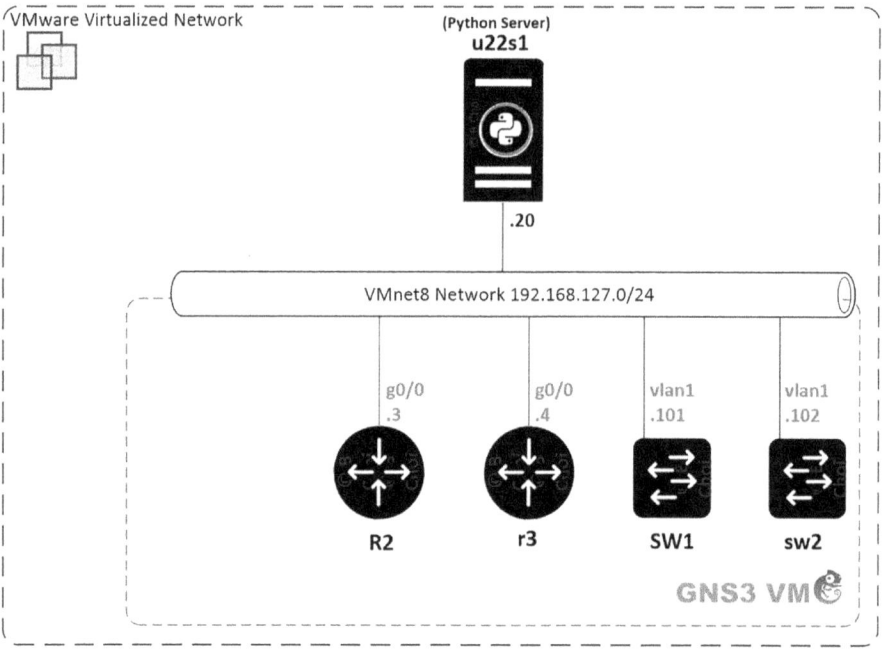

Figure 3-11. SSH netmiko Lab 3, devices in use

CHAPTER 3   PYTHON NETWORK AUTOMATION LABS: SSH IN ACTION, PARAMIKO AND NETMIKO LABS

Upon the conclusion of this lab, your script will generate three files in its running directory: two text files containing the running configuration, and an XML file detailing the comparison results. This tool is a modern alternative to the traditional method of comparing two files and can accurately spotlight disparities between the two text files. This functionality is vital for automating configuration comparisons before and after making changes. In the future, integration with an AI bot might enable it to determine the success or failure of a change based on the outcome.

Time for a practical application. Follow these steps to finish this lab:

#	Task
| ① | First, copy the port-scanner tool from the previous lab and make the necessary modifications so that SSH port 22's open status is probed by this tool. After you have made changes, the script should look like the following:<br>```python
for device in devices:  # Loop through netmiko devices list
    ip = device.get("ip", "")  # Get the value of the key "ip" from
                                 the device dictionary
    for port in range (22, 23):  # For only port 22
        dest = (ip, port)  # Combine ip and port number to form dest
                             object
        try:
            with socket.socket(socket.AF_INET, socket.SOCK_STREAM)
            as sock:
                sock.settimeout(3)
                connection = sock.connect(dest)  # Send socket request
                print (f"on {ip}, port {port} is open!")
                # Informational, pass object using f string
        except:
            print (f"On {ip}, port {port} is closed. Check the
            connectivity to {ip} again.")
            exit()
``` |

(continued)
```

CHAPTER 3   PYTHON NETWORK AUTOMATION LABS: SSH IN ACTION, PARAMIKO AND NETMIKO LABS

#	Task

(2)	Here's the complete source code of the netmiko_compare_config.py script. This application uses various modules: the getpass module for password retrieval, the time module to create pauses, the socket module for probing open ports, difflib for configuration file comparison, and, notably, netmiko's ConnectHandler to establish SSH connections to two devices. Try typing the code to familiarize yourself with Python coding. When coding, spending hours in front of your screen and keyboard is inevitable. For more in-depth explanations, refer to the comments provided within the code.
	
	``python
	jdoe@u22s1:~/ch3_ssh$ nano netmiko_compare_config.py
	#!/usr/bin/env python3
	#---#
	Import required modules
	import time
	import socket
	import difflib
	from getpass import getpass
	from netmiko import ConnectHandler
	#---#
	Borrowed from previous labs
	# Functions to collect credentials and IP addresses of devices
	def get_input(prompt=''):
	    try:
	        line = input(prompt)
	    except NameError:
	        line = input(prompt)
	    return line
	def get_credentials():
	    #Prompts for, and returns a username and password
	    username= get_input("Enter Network Admin ID : ")
	    password = None
	``

(continued)

#	Task

```
        while not password:
            password = getpass("Enter Network Admin PWD    : ")
            password_verify = getpass("Confirm Network Admin PWD   : ")
            if password != password_verify:
                print("Passwords do not match. Please try again.")
                password = None
        return username, password
    # For IP addresses of comparing devices
    def get_device_ip():
        #Prompts for, and returns a first_ip and second_ip
        first_ip = get_input("Enter primary device IP      : ")
        while not first_ip:
            first_ip = get_input("* Enter primary device IP     : ")
        second_ip = get_input("Enter secondary device IP    : ")
        while not second_ip:
            second_ip = get_input("* Enter secondary device IP : ")
        return first_ip, second_ip
    #-----------------------------------------------------------------
    # Run the functions to collect credentials and ip addresses
    print("-"*79)
    username, password = get_credentials()
    first_ip, second_ip = get_device_ip()
    print("-"*79)
    #-----------------------------------------------------------------
    # netmiko device dictionaries
    device1 = {            # netmiko dictionary for device1
    'device_type': 'cisco_ios',
    'ip': first_ip,
    'username': username,
    'password': password,
    }
```

(continued)

#	Task
| | ```python
device2 = {              # netmiko dictionary for device2
    'device_type': 'cisco_ios',
    'ip': second_ip,
    'username': username,
    'password': password,
}
devices = [device1, device2]
#----------------------------------------------------------------
# Re-use port scanner as a pre-check tool for reachability
verification tools.
# If an IP is not reachable, the application will exit due to a
communication problem.
for device in devices: # Loop through    devices
    ip = device.get("ip", "") # get value of the key "ip"
    for port in range (22, 23): # only port 22
        dest = (ip, port) # Combine ip and port number to form dest
                          object
        try:
            with socket.socket(socket.AF_INET, socket.SOCK_STREAM)
            as sock:
                sock.settimeout(3)
                connection = sock.connect(dest)
                print(f"on {ip}, port {port} is open!")
        except:
            print(f"On {ip}, port {port} is closed. Check the
            connectivity to {ip} again.")
            exit()
# Prompt the user to make a decision to run the tool
response = input(f"Make a comparison of {first_ip} and {second_ip}
now? [Yes/No]")
response = response.lower()
if response == 'yes':
``` |

(continued)
```

#	Task

```python
        print(f"* Now making a comparison : {first_ip} vs {second_ip}")
        # Informational
        for device in devices: # Loop through    devices
            ip = device.get("ip", "") # Get value of the key "ip"
            try:
                net_connect = ConnectHandler(**device) # Create netmiko
                                                        connection object
                net_connect.send_command("terminal length 0\n")
                output = net_connect.send_command("show running-
                config\n") # Run show running-config
                show_run_file = open(f"{ip}_show_run.txt", "w+") # Create
                                                                    a file
                show_run_file.write(output) # Write output to file
                show_run_file.close() # Close-out the file
                time.sleep(1)
                net_connect.disconnect() # Disconnect SSH connection
            except KeyboardInterrupt: # Keyboard Interrupt
                print("-"*79)
    else:
        print("You have selected No. Exiting the application.")
        # Informational
        exit()
    #--------------------------------------------------------------------
    # Compare the two show running-config files and display them in html
    file format.
    # Prepare for comparison of the text files
    device1_run = f"./{first_ip}_show_run.txt"
    # Create device1_run object, ./ is present working folder
    device2_run = f"./{second_ip}_show_run.txt" # Create device2_run
                                                    object
```

(*continued*)

#	Task
| | ```
device1_run_lines = open(device1_run).readlines()
# Convert into strings first for comparison
time.sleep(1)
device2_run_lines = open(device2_run).readlines()
# Convert into strings first for comparison
time.sleep(1)
# Four arguments required in HtmlDiff function
difference = difflib.HtmlDiff(wrapcolumn=60).make_file(device1_run_lines, device2_run_lines, device1_run, device2_run)
difference_report = open(first_ip + "_vs_" + second_ip + "_compared.html", "w") # Create html file to write the difference
difference_report.write(difference) # Writes the differences to the difference_report
difference_report.close()
print("** Device configuration comparison completed. Please Check the html file to check the differences.")
print("-"*79)
time.sleep(1)
``` |
| ③ | After completing writing the previous script, use the `python` command to run it under the ch3_ssh directory. After running this interactive application, you should have three files automatically created in your Linux Python server's present working directory (pwd).<br><br>```
jdoe@u22s1:~/ch3_ssh$ python netmiko_compare_config.py
-----------------------------------------------------------------
Enter Network Admin ID      : jdoe
Enter Network Admin PWD     : ********
Confirm Network Admin PWD   : ********
Enter primary device IP     : 192.168.127.3 # R2
Enter secondary device IP   : 192.168.127.4 # r3
-----------------------------------------------------------------
``` |

(*continued*)

#	Task
	on 192.168.127.3, port 22 is open! on 192.168.127.4, port 22 is open! Make a comparison of 192.168.127.3 and 192.168.127.4 now? [Yes/No]**yes** * Now making a comparison : 192.168.127.3 vs 192.168.127.4 ** Device configuration comparison completed. Please Check the html file to check the differences. --
④	If you are running into errors or issues, download the source code from my GitHub repository and carefully re-examine your code and settings. Source code downloads URL: https://github.com/pynetauto/apress_pynetauto_ed2.0/tree/main/source_codes/Part2/ch3_ssh Run the `ls` command on the Python Network Automation server, and you should find the three files under the current working directory. jdoe@u22s1:~/ch3_ssh$ **ls -lh 192.168*** -rw-rw-r-- 1 jdoe jdoe 3.9K Nov 6 02:31 192.168.127.3_show_run.txt -rw-rw-r-- 1 jdoe jdoe 59K Nov 6 02:31 192.168.127.3_vs_192.168.127.4_compared.html -rw-rw-r-- 1 jdoe jdoe 3.7K Nov 6 02:31 192.168.127.4_show_run.txt
⑤	Now, using WinSCP or FileZilla, log in to the Ubuntu Python automation server with your credentials and move the files to your Windows host. If you have not installed the software, install it here before logging in. I am using the SCP protocol with port 22. URL: https://winscp.net/eng/download.php Figure 3-12 shows a WinSCP example with an SCP connection (port 22) to the server. Locate the three files under /home/jdoe/ch3_ssh/ and drag and drop them to your Windows host PC folder.

(*continued*)

CHAPTER 3   PYTHON NETWORK AUTOMATION LABS: SSH IN ACTION, PARAMIKO AND NETMIKO LABS

#	Task

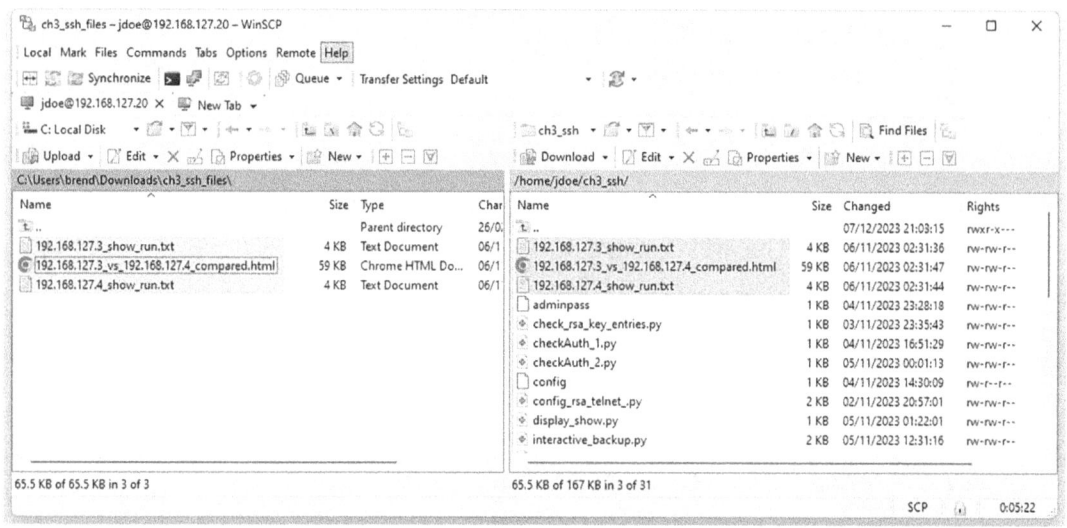

Figure 3-12. Win-SCP, copying compared and running-config backups to Windows host

Now, go to the downloaded folder and open the .html file to review the script's result. You can compare any similar devices using this interactive tool. This tool can be modified slightly to compare the difference between firewall rules and configuration, and one practical example is to use such a tool to compare Palo Alto firewalls. Using its simple XML API call feature, you can compare the primary to standby configurations. The same script can also be used to compare the configuration before and after a change has been performed on a device. How you want to use the tool is totally up to your imagination. See Figure 3-13.

(*continued*)

CHAPTER 3 PYTHON NETWORK AUTOMATION LABS: SSH IN ACTION, PARAMIKO AND NETMIKO LABS

#	Task

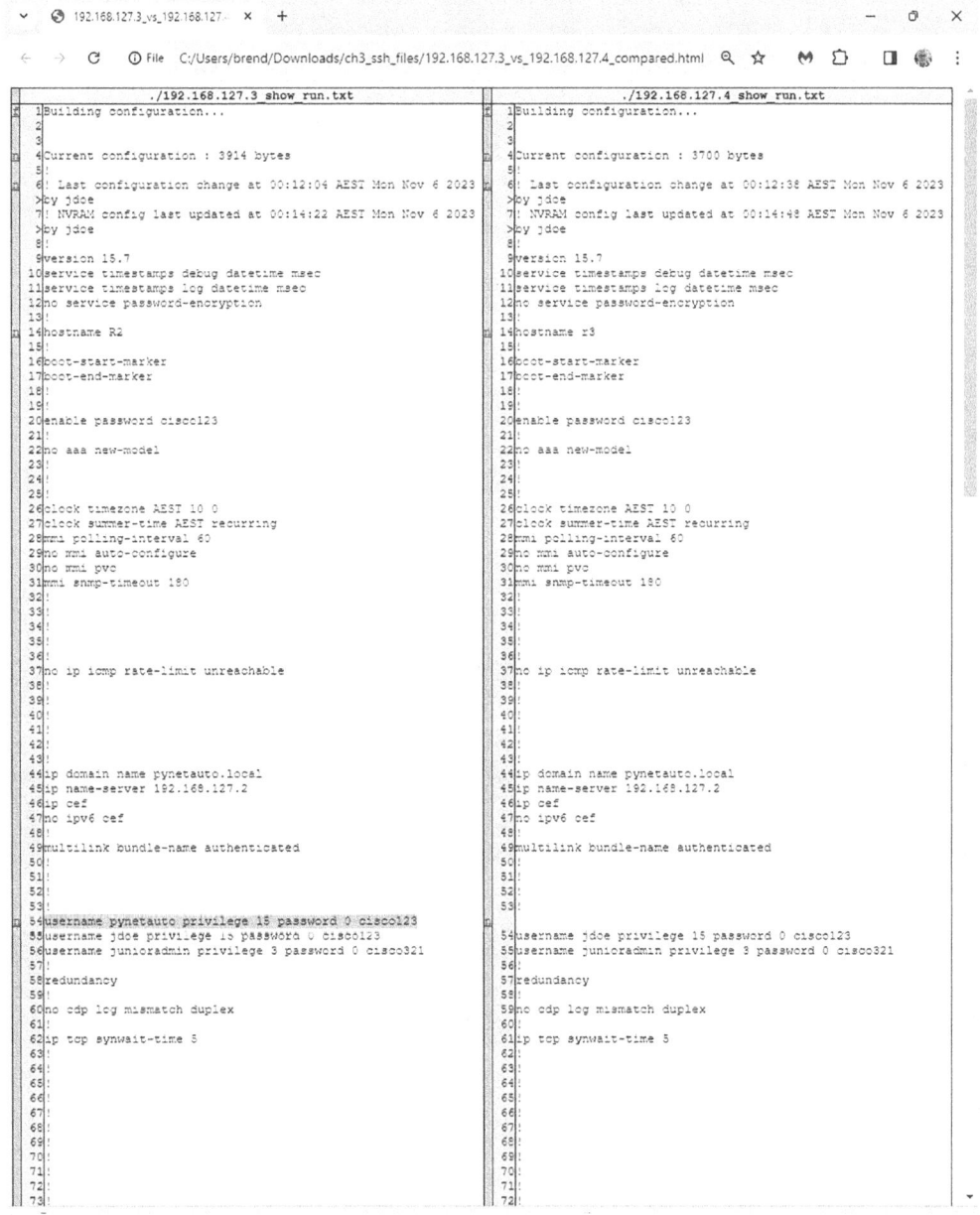

Figure 3-13. *Web browser, opening an HTML file for a review, R2 vs. r3*

(*continued*)

#	Task
6	Now rerun the same script and this time compare the two switches, SW1 and sw2.

jdoe@u22s1:~/ch3_ssh$ **python netmiko_compare_config.py**

Enter Network Admin ID   : **jdoe**
Enter Network Admin PWD   : ************
Confirm Network Admin PWD   : ************
Enter primary device IP   : **192.168.127.101** # SW1
Enter secondary device IP   : **192.168.127.102** # sw2
----
on 192.168.127.101, port 22 is open!
on 192.168.127.102, port 22 is open!
Make a comparison of 192.168.127.101 and 192.168.127.102 now? [Yes/No]**yes**
* Now making a comparison : 192.168.127.101 vs 192.168.127.102
** Device configuration comparison completed. Please Check the html file to check the differences.
----
jdoe@u22s1:~/ch3_ssh$ **ls -lh 192.168.***
-rw-rw-r-- 1 jdoe jdoe 4.6K Nov 6 02:41 192.168.127.101_show_run.txt
-rw-rw-r-- 1 jdoe jdoe 67K Nov 6 02:41 192.168.127.101_vs_192.168.127.102_compared.html
-rw-rw-r-- 1 jdoe jdoe 4.1K Nov 6 02:41 192.168.127.102_show_run.txt
-rw-rw-r-- 1 jdoe jdoe 3.9K Nov 6 02:31 192.168.127.3_show_run.txt
-rw-rw-r-- 1 jdoe jdoe 59K Nov 6 02:31 192.168.127.3_vs_192.168.127.4_compared.html
-rw-rw-r-- 1 jdoe jdoe 3.7K Nov 6 02:31 192.168.127.4_show_run.txt

Access the file through WinSCP and open the 192.168.127.101_vs_192.168.127.102_compared.html file as shown in Figure 3-14. |

CHAPTER 3 PYTHON NETWORK AUTOMATION LABS: SSH IN ACTION, PARAMIKO AND NETMIKO LABS

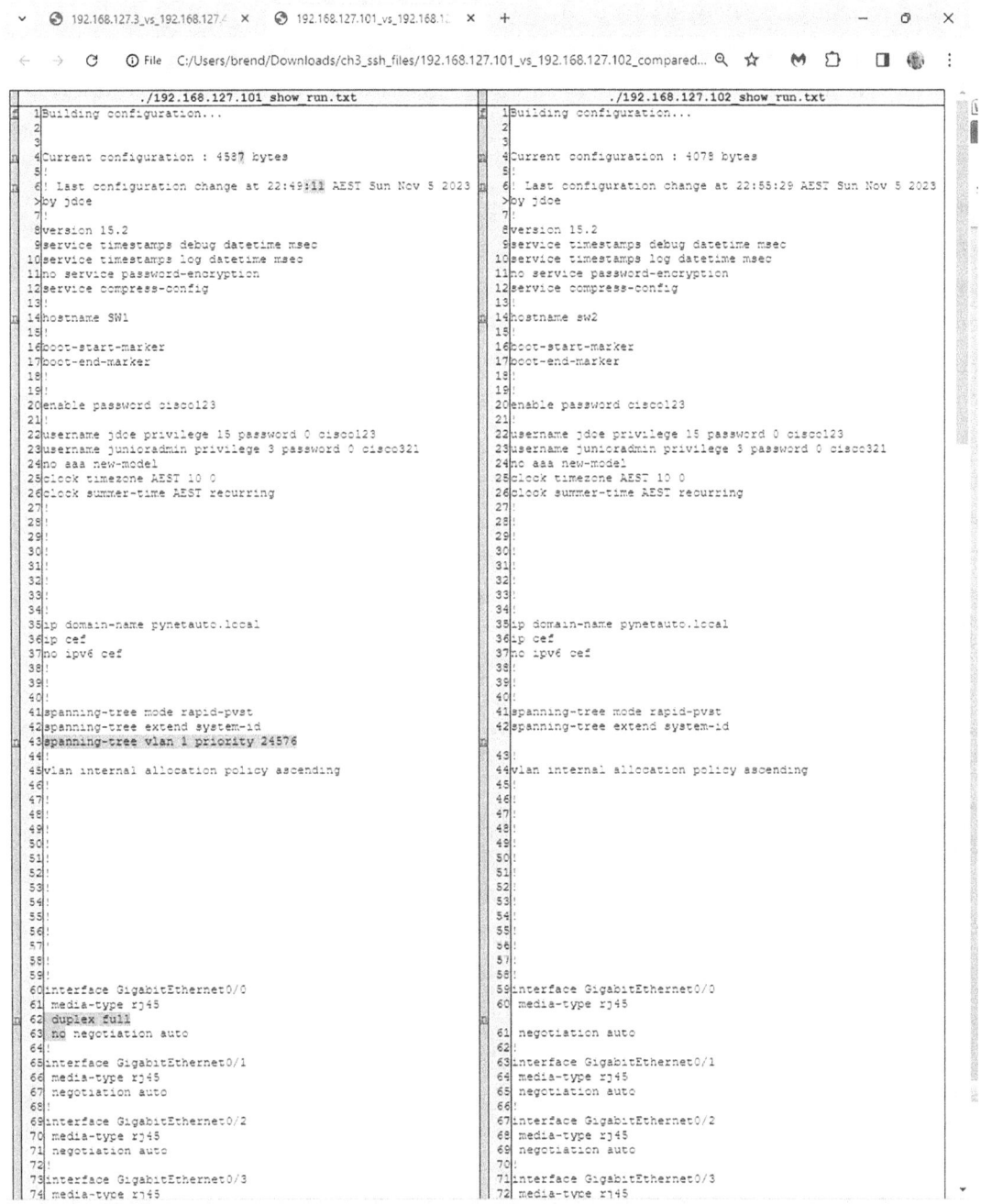

Figure 3-14. Web browser, opening an HTML file for a review, SW1 vs sw2

Today, many new networking devices offer support for REST/XML APIs and Webhooks. However, countless companies have heavily invested in enterprise network devices that only accommodate Telnet, SSH, and SNMP, with active technical support expected for several more years. Network engineers typically begin their journey facing a black command-line console connected via SSH or Telnet. **Your knowledge expansion into network programmability starts by studying the diverse behaviors of network engineers and comprehending their decision-making processes while they stare into the dark CLI screen.** Network automation transcends the creation of robots, bots, or software designed to work harder. It involves understanding behavior as network engineers and identifying areas where automation can significantly aid technicians. Given that nearly all enterprise networking devices support SSH, harnessing the power of Python, along with Python modules, in automating mundane tasks is an essential tool, especially during this transitional phase. However, the crux in this new software-defined network (SDN) and "cloud for anything" era is to understand the fundamental elements that enhance resilience and stability within your managed networks. Exploring how software programming, such as Python, can propel you to the next level becomes crucial in this context.

Summary

This chapter delved into the realm of network automation, immersing readers in Python's essential SSH libraries—`paramiko` and `netmiko`—and their pivotal role in managing network infrastructure. This exploration vividly illustrated the enduring significance of SSH in the evolution of remote management tools, offering profound insights into their practical applications for network engineers. Through a series of detailed labs, the chapter navigated SSH functionality on Linux servers and Cisco devices while examining `paramiko`'s varied capabilities. Ranging from configuring device settings like clocks and NTP servers to executing sophisticated tasks like storing configurations on a TFTP server, these labs enabled interactive `paramiko` script creation. Notably, SSH key exchanges between devices were extensively explored, emphasizing secure data transmission protocols. `netmiko`, a key Python library for vendor networking devices, was thoroughly showcased, from installation to multifaceted functionalities throughout the labs. Demonstrating its versatility, these labs used dictionaries for device information, script creation for disabling Telnet, and conducting configuration comparisons. The chapter's hands-on approach emphasized the immediate practicality of these SSH modules, merging networking and programming principles into real-world applications. As the chapter concluded, it integrated acquired knowledge, emphasizing the pivotal role of SSH libraries in constructing an IOS XE upgrade application. The chapter equipped you with practical insights and skills to adeptly navigate these libraries, strengthening your understanding of networking environments and core programming concepts. Next, you'll explore scheduling Python applications and dive into Python SNMPv3, introduced through cron and SNMP exploration labs. This will further highlight Python's role in network automation within workplace scenarios.

CHAPTER 4

Python Network Automation Labs cron

In this chapter, the focus shifts toward harnessing the power of Python for network automation through practical labs and the exploration of cron. As you aspire to streamline work processes using Python, this chapter guides you through hands-on exercises, providing a foundational understanding of how cron functions and how you can put your Python applications to work with little human interaction. Starting with the intricacies of the Linux Scheduler, cron, you'll learn about Ubuntu Server's task scheduler (crontab) and Fedora's task scheduler (crond). The journey progresses with insights into cloning GNS3 projects, simulating router failures, and optimizing network recovery. Lab pre-tasks, such as deleting VPCS and adding L2 extra switches to emulate a new switch installation, set the stage for a comprehensive cron learning experience. Alongside, you'll master cron job definitions with examples to bolster your understanding. This chapter culminates with a summary, offering a concise recap of the key takeaways and reinforcing your grasp of Python's role in network automation and the task scheduling of developed Python applications.

Unveiling the Power of cron in Python Networking

Wonderful to see you progressing through the initial chapters of this book! By now, you've mastered the ability to develop Python scripts for various networking tasks. Your goal, as you continue your journey, is to harness Python for automating mundane tasks at work. After crafting your code and building scripted applications, the next challenge is to execute these scripts seamlessly, independent of your direct involvement. Windows systems provide the Windows Task Scheduler, and on Linux, cron allows you to schedule scripts to run autonomously at specified times or intervals. As your comfort with Python and Linux deepens, writing and scheduling scripts to run automatically will become second nature.

Advancing further, you'll find the need to execute these scripted applications without direct user interaction. This is where the role of task schedulers, particularly Linux cron, becomes pivotal. By leveraging cron jobs, you can automate the execution of Python scripts, streamlining repetitive tasks and fostering a seamless workflow without manual intervention.

In this chapter, your exploration focuses on the intricacies of cron, delving into the foundational concepts and practical applications within Python-based network automation lab settings. You will initiate your lab by creating a full clone of the last GNS3 project, setting the stage for in-depth exploration. The journey continues with a deep dive into the use of cron for task scheduling, providing you with hands-on experiences that will not only expand your knowledge but also deepen your understanding of enterprise-level commercial tools. This exploration serves to bridge the gap between theoretical concepts and their real-world application within a sophisticated IP services environment. Get ready for a hands-on journey into the dynamic world of cron and its pivotal role in network automation.

Cloning a GNS3 Project for the Next Lab

To prepare for your lab, you will make a true clone of the GNS3 project from Chapter 14 of Book I. The GNS3 project export cloning method helped you to make a new project, but it compromised the original lab's configuration files, so in this sense, it is not a true clone of a GNS3 project. Sometimes this may not be the result you want for your labs. You might want to keep the original files intact and create a true clone of an old GNS3 project. With the project export method of cloning, the original lab becomes unusable as the Dynamips files also get exported as the new project is created. A better way to create a new project without compromising the original one is to make a real duplicate of the original project folder and then reimport it with another name. You want to keep the `pynetauto-lab` project as a fully working project, but at the same time, you do not want to create a new GNS3 project from the ground up. Let's learn how you can achieve this.

CHAPTER 4 PYTHON NETWORK AUTOMATION LABS CRON

#	Task
①	If you are continuing from Chapter 3, you have to first save all the running configurations using the simple `save_config.py` script provided here. (Optionally, you could console in and run `write memory` or `copy running-config startup-config`, but this would be too cumbersome, so I recommend you use the `save_configs.py` script given here.) Once the configurations of all devices have been saved, power off all network devices gracefully. This script uses the `netmiko` library and most parts are borrowed from previous labs, so the code should be familiar to you; no comments are included to clear the clutter.

```
jdoe@u22s1:~$ mkdir ch15_cron && cd ch15_cron
jdoe@u22s1:~/ch15_cron$ nano save_configs.py
jdoe@u22s1:~/ch15_cron$ cat save_configs.py
from netmiko import ConnectHandler
import time

start_time = time.time() # Start timer

username = 'jdoe' # For lab purpose only
password = 'cisco123' # For lab purpose only
devices = [
    '192.168.127.3',
    '192.168.127.4',
    '192.168.127.101',
    '192.168.127.102',
    '192.168.127.133'
]

for ip in devices:
    device = {
        'device_type': 'cisco_ios',
        'ip': ip,
        'username': username,
        'password': password
    }
```

(continued)

#	Task
	``
try:
    ssh_conn = ConnectHandler(**device)
    print(f"Connected to {ip}")
    ssh_conn.send_command("write memory")
    print("saved running-config")
    time.sleep(2)
    output = ssh_conn.send_command("")
    print(output)
    print("-" * 79)
    ssh_conn.disconnect()
except Exception as e:
    print(f"{ip} is offline")
    continue
```
end_time = time.time() # End timer<br>execution_time = end_time - start_time # Calculate execution time<br>print(f"Total execution time: {execution_time:.2f} seconds") # seconds in two decimal points |
| ② | Once all network devices are powered off, exit GNS3 so the GNS3 and GNS3 VMs are closed completely (see Figures 4-1 and 4-2).  *Figure 4-1. pynetauto-lab GNS3 project, stop all devices* |

*(continued)*
```

#	Task
	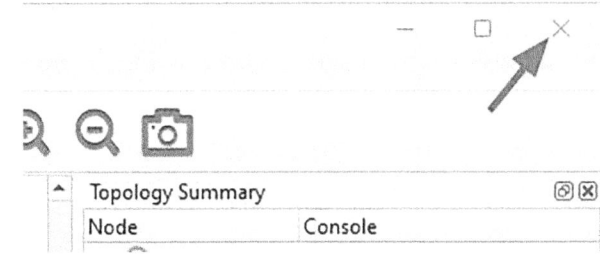 *Figure 4-2. pynetauto-lab GNS3 project, exiting GNS3 and GNS3 VM*
③	Go to the C:\Users\<user_name>\GNS3\projects folder and make a copy of the pynetauto-lab folder so it looks like Figure 4-3. Replace <your_name> with your name. 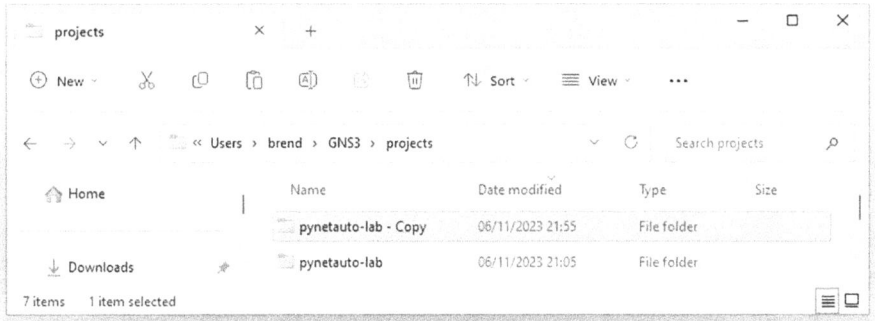 *Figure 4-3. GNS3 project, making a copy of an existing project folder*
④	Go to the desktop of your Windows host PC and double-click the GNS3 icon on the desktop. Wait until GNS3 and GNS3 VM both become fully operational (see Figure 4-4).  *Figure 4-4. GNS3 start desktop icon*

(continued)

CHAPTER 4   PYTHON NETWORK AUTOMATION LABS CRON

#	Task
⑤	When the project window opens, select Open a Project from Disk (see Figure 4-5) and navigate to the C:\Users\<user_name>\GNS3\projects\pynetauto-lab - Copy folder. Then select the pynetauto-lab GNS3 project file and then click the Open button.

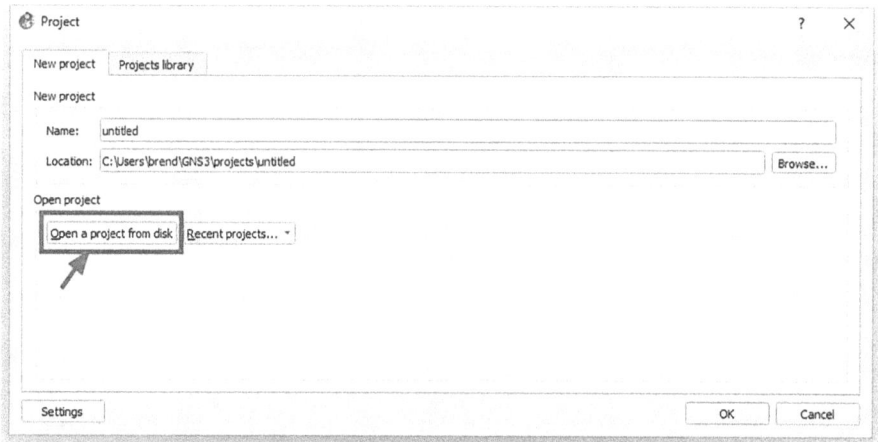

Figure 4-5. GNS3, opening a project from disk

⑥	Once the copy of the pynetauto-lab project is fully launched, wait for the lab to open correctly and then select GNS3's File ➤ Save Project As option (see Figure 4-6).

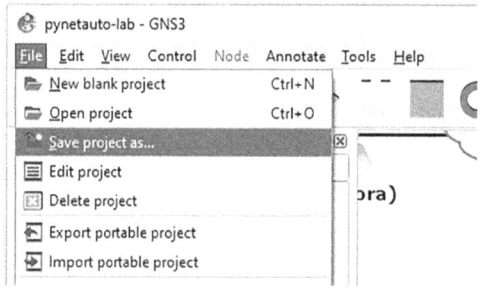

Figure 4-6. GNS3, the Save Project As menu item

(*continued*)

CHAPTER 4   PYTHON NETWORK AUTOMATION LABS CRON

#	Task
7	Give your project a name. I named it `pynetauto-devops` (see Figure 4-7).  *Figure 4-7. GNS3, saving as a new project*
8	Now, go back to the `C:\Users\<user_name>\GNS3\projects` folder, and you will find a new project folder called `pynetauto-devops`. The temporary folder has served its purpose and has now become redundant; you can go ahead and delete the `pynetauto-lab - Copy` folder permanently (see Figure 4-8). 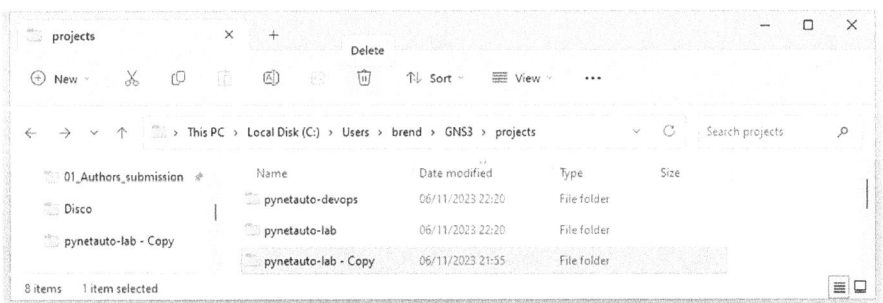 *Figure 4-8. GNS3, deleting the redundant project copy*

(continued)

CHAPTER 4   PYTHON NETWORK AUTOMATION LABS CRON

#	Task
⑨	You have just made a true clone of the last GNS3 project. When you go back to the GNS3 main window and the new `pynetauto-devops` project is opened, you will notice that your new GNS3 project file has the correct name. Now power up all devices and start checking their running configurations (see Figure 4-9).

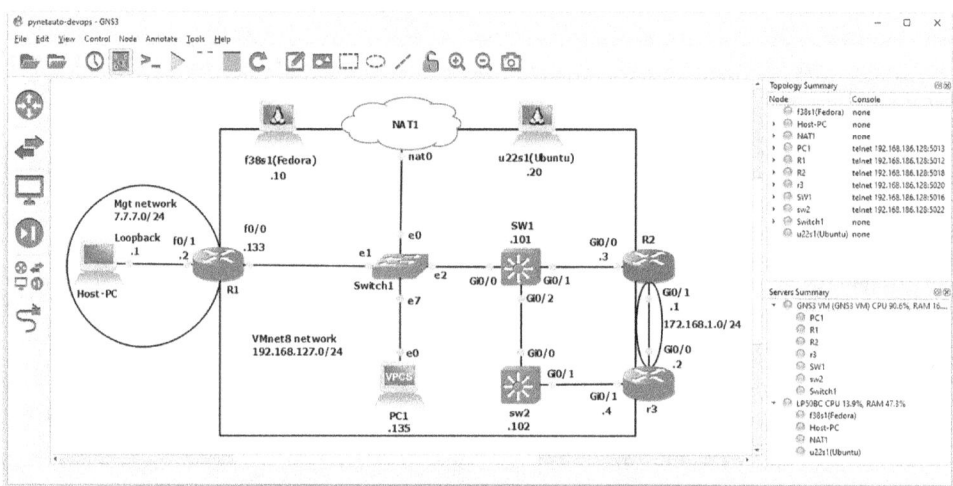

Figure 4-9. *GNS3, fully cloned pynetauto-devops project*

Tip

Improving GNS3 lab efficiency: adjusting memory for enhanced lab operations

In the upcoming lab, you'll include two additional IOSv L2 switches to GNS3 VM. If you plan to run all devices and both Linux servers simultaneously, it's advisable to consider adjusting the memory allocation. If your system currently operates with 16GB of memory, maintain it at 8192MB. However, if your host has 32GB of memory or more, boosting the allocation to 12288MB is recommended for optimal performance (see Figure 4-10).

CHAPTER 4   PYTHON NETWORK AUTOMATION LABS CRON

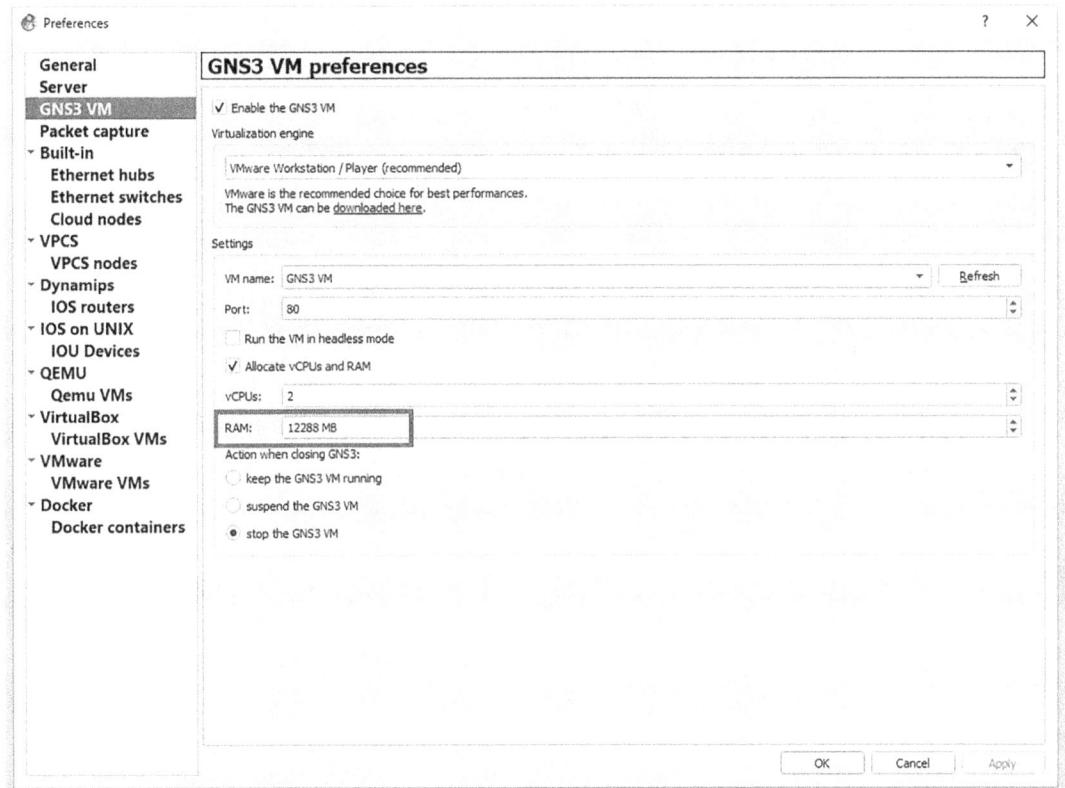

Figure 4-10.   GNS3, increase RAM for GNS3 VM

Warning

Managing network failures: simulated hardware replacement and recovery

In your netmiko Lab 1 from Chapter 3, you formatted the flash drive and cleared its configuration for testing purposes. If you're continuing from the previous chapter in one go, you are about to encounter a problem. However, don't worry—you've preserved the running configurations of all your devices in the netmiko TFTP configuration backup lab. You'll simulate a failed router (R2) replacement by using the last backed-up running configuration.

In real-world scenarios, network engineers frequently conduct hardware replacements, offering a hands-on experience akin to what you're simulating. If you're using a virtual router in your private cloud, you'll swiftly deploy another vRouter and restore its configurations along with the services. Similarly, in a public cloud environment, you'll use the default vRouter from the service providers' marketplace and restore the configuration to reinstate the services. Whether dealing with hardware, a public cloud, or a private cloud, the fundamental concept of replacing a faulty enterprise network device remains alike and consistently demands reliable human intervention. When computers encounter malfunctions, bugs, or unidentified issues, a self-healing bot won't be capable of rectifying network issues—human intervention remains essential. Although it's feasible that core network equipment may eventually transition to the cloud, accompanied by using AI and ML for deploying a real self-healing bot to substitute the engineers, this remains a possibility for the future. I hope that this advancement does not happen within our lifetime or the lifetime of next generation of engineers.

Router Failure Simulation, P1 Network Recovery

All devices should boot up, except for R2. The reason for this failure is already known, and it's related to the netmiko Lab 1 in Chapter 3. During that lab, you formatted the flash drive of R2, leading to the deletion of all files, including the critical OS files (refer to Figure 4-11).

When routers boot up, they load programs into their running memory. Under normal operations, everything runs smoothly. However, if crucial system files are wiped and the router is reloaded, it won't boot up properly due to the missing system files. This situation replicates common issues that network engineers encounter in their day-to-day operations, such as bugs, change failures, hardware malfunctions, or human errors. I intentionally caused this outage for this exercise to provide you with an unexpected educational detour! This opportunity offers an invaluable hands-on experience, allowing you to learn and grasp the intricate tasks and challenges faced by network engineers, all without the usual weight of business-related pressures.

CHAPTER 4 PYTHON NETWORK AUTOMATION LABS CRON

#	Task
①	**If this were a real production environment, what would you do in this situation?** Because you lost the console connection and have no Out-of-bound (OOB) connection to this device for recovery, the fastest recovery path to shorten the service outage is to replace this device with another. If this was a physical router, you would RMA (Return Material Authorization) the device and replace it onsite. If it is a virtual device like this, you simply deploy another router with the same spec and perform emergency recovery. Roll up your sleeves so you can recover from this situation (see Figure 4-11). 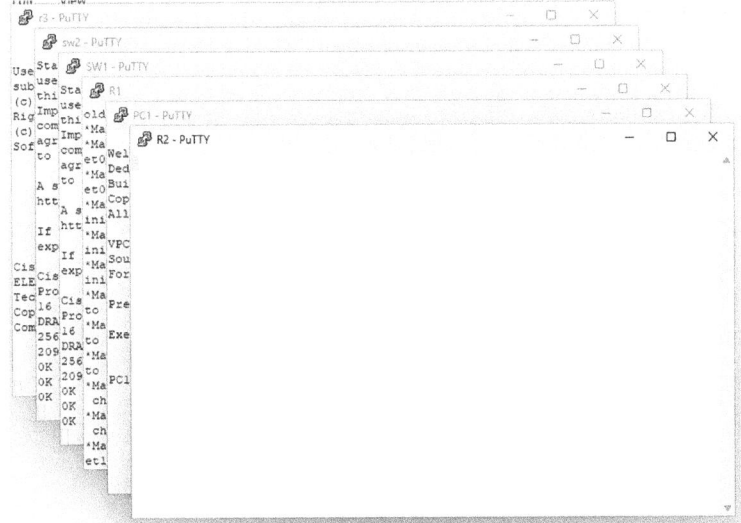 ***Figure 4-11.*** *GNS3, identifying non-working device*
②	Envisage that you are working on a P1 case and performing an emergency router replacement for a failed IP service router. First, power down R2 and then delete R2 from the GNS3 canvas. When prompted with the following message, select Yes (see Figure 4-12). 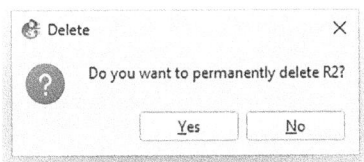 ***Figure 4-12.*** *GNS3, Delete R2 from Topology canvas* After deleting R2, your GNS3 topology should look similar to Figure 4-13.

(*continued*)

CHAPTER 4 PYTHON NETWORK AUTOMATION LABS CRON

#	Task
	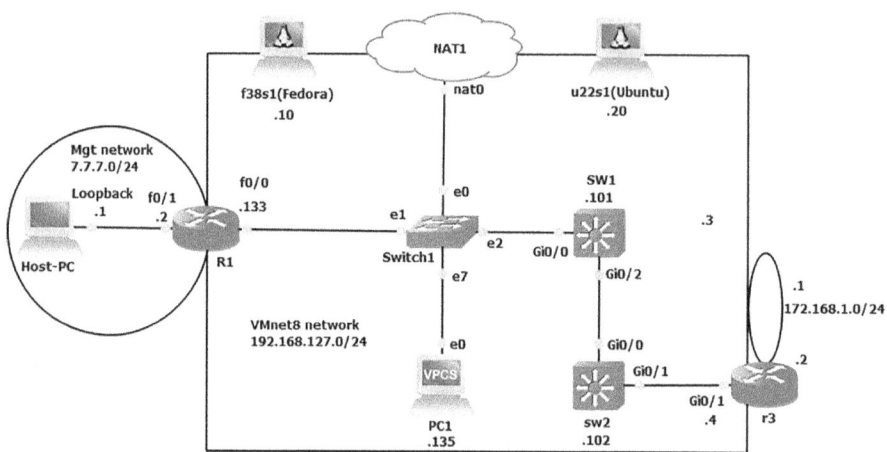
Figure 4-13. GNS3, R2 removed from Topology canvas
③	Replace with another Cisco CML router by drag and drop and make connections as before (see Figure 4-14).
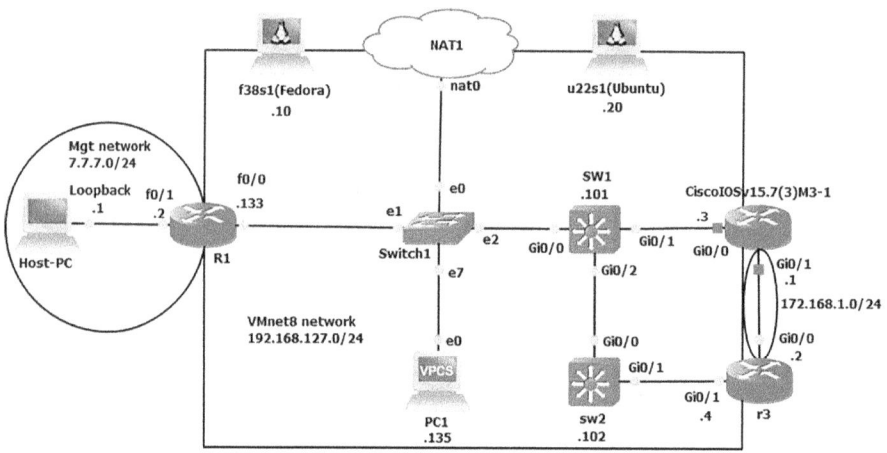
Figure 4-14. GNS3, emergency router replacement

(continued)

#	Task
④	Rename the new router R2 in the topology and power on your new R2 router (see Figure 4-15). 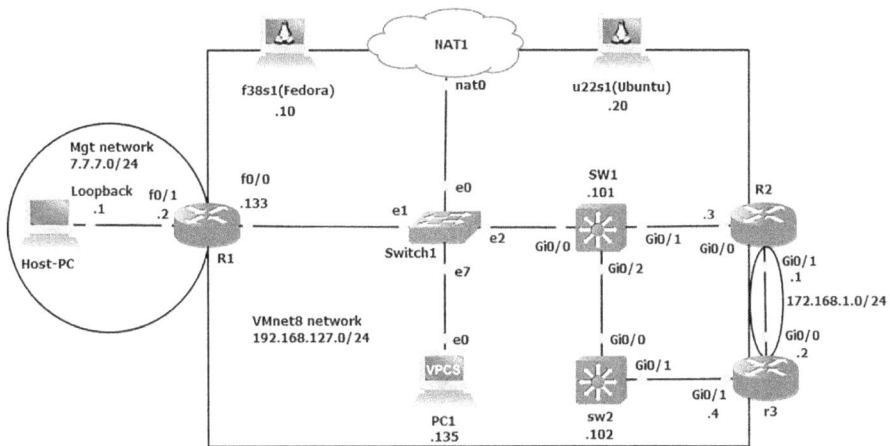 *Figure 4-15. GNS3, new router powered-on*

⑤	Open the console to the new R2 router and let it fully load up the CML OS. Then configure the router as shown to put it on the network.
	`Router(config)#`**`int g0/0`**
	`Router(config-if)#`**`ip address 192.168.127.3 255.255.255.0`**
	`Router(config-if)#`**`no shut`**
	
	Test the connectivity to the TFTP server where your backup configuration is held. Remember the TFTP network device backup lab? You used each device's IP address and saved the configuration.
	`Router#`**`ping 192.168.127.10`**
	`Type escape sequence to abort.`
	`Sending 5, 100-byte ICMP Echos to 192.168.127.10, timeout is 2 seconds:`
	`.!!!!`
	`Success rate is 80 percent (4/5), round-trip min/avg/max = 14/14/16 ms`

(continued)

CHAPTER 4   PYTHON NETWORK AUTOMATION LABS CRON

#	Task

⑥	Now follow the instructions to copy R2's `running-config` file from the TFTP server at 192.168.127.10. The target file you need to copy to `startup-config` is 192.168.127.3.bak@2023-11-05_15-01-43. Your file name will be different, so you will have to grab the file name from your IP services server, f38s1.
	
	`Router#`**`copy tftp: startup-config`**
	`Address or name of remote host []?` **`192.168.127.10`**
	`Source filename []?` **`192.168.127.3.bak@2023-11-05_15-01-43`** `# Update to your file name`
	`Destination filename [startup-config]? Press <Enter>`
	`Accessing tftp://192.168.127.10/192.168.127.3.b`
	`ak@2023-11-05_15-01-43...`
	`Loading 192.168.127.3.bak@2023-11-05_15-01-43 from 192.168.127.10 (via GigabitEthernet0/0): !`
	`[OK - 3776 bytes]`
	`[OK]`
	`3776 bytes copied in 1.912 secs (1975 bytes/sec)`
	
	Run the show `startup-config` command to confirm that the correct file has been downloaded. This step is extremely important as you may have downloaded the wrong file or there are some configuration modifications you have to make before saving the `startup-config` file to the `running-config` file.
	
	`Router#`**`show startup-config`**
	`[...omitted for brevity]`

⑦	My `startup-config` file looks good, so I am committing the changes by copying the `startup-config` file to the `running-config` file, which will restore the R2 configurations.
	`Router#`**`copy startup-config running-config`**
	`Destination filename [running-config]? Press <Enter>`
	`[...omitted for brevity]`
	
	Once you observe that the router name is correct, check your IP address and the status of R2's GigabitEthernet 0/0 interface, which is the uplink to your network.

(*continued*)

#	Task
	R2#**show ip int brief**
	Interface IP-Address OK? Method Status Protocol
	GigabitEthernet0/0 192.168.127.3 YES manual up up
	GigabitEthernet0/1 172.168.1.1 YES TFTP administratively down down
	GigabitEthernet0/2 unassigned YES unset administratively down down
	GigabitEthernet0/3 unassigned YES unset administratively down down
	Loopback0 2.2.2.2 YES TFTP up up
	Loopback1 4.4.4.4 YES TFTP up up

(8)	Now, use the following tcl shell script to test the end-to-end network connectivity to confirm the completion of R2 restoration. If you can communicate to other devices, your emergency router replacement has been completed successfully. Well done!
	R2#
	tclsh
	foreach ip {
	    192.168.127.4
	    192.168.127.10
	    192.168.127.20
	    192.168.127.101
	    192.168.127.102
	    192.168.127.133
	    7.7.7.2
	    8.8.8.8
	} {

(continued)

#	Task
	``
        if {[catch { ping $ip repeat 3 timeout 1 } result]} {
            puts "$ip is offline"
        } else {
            puts "$ip is online"
        }
    }
```
Type escape sequence to abort.
Sending 3, 100-byte ICMP Echos to 192.168.127.4, timeout is 1 seconds:
.!!
Success rate is 100 percent (2/3), round-trip min/avg/max = 39/54/68 ms192.168.127.4 is online
*[...omitted for brevity]*
Type escape sequence to abort.
Sending 3, 100-byte ICMP Echos to 8.8.8.8, timeout is 1 seconds:
.!!
Success rate is 100 percent (2/3), round-trip min/avg/max = 34/56/89 ms8.8.8.8 is online |

Great job replacing and restoring R2 configuration in a hurry! You have completed an emergency router swap and now R2 is back on the network. Now it is time to install two more switches in preparation for the next lab.

# Lab Pre-Task, Deleting VPCS and Adding More L2 Switches

Now let's add a couple more switches to make your lab a little more interesting.
```

CHAPTER 4   PYTHON NETWORK AUTOMATION LABS CRON

#	Task
①	To prepare for the next part of your exploration, you will now add two more switches, as shown in Figure 4-16. First, delete PC1 (VPCS) and replace it with sw3 with a random IP address of 192.168.127.153/24 and with SW4 with a random IP address of 192.168.127.244/24. Power on the new switches and let the CPU and memory settle down. Also shown is a Cisco c8000v router to be installed on VMware for a later IOS upgrade lab; this icon is only a placeholder, so you do not have to worry about the c8000v router until the next chapter (see Figure 4-16). 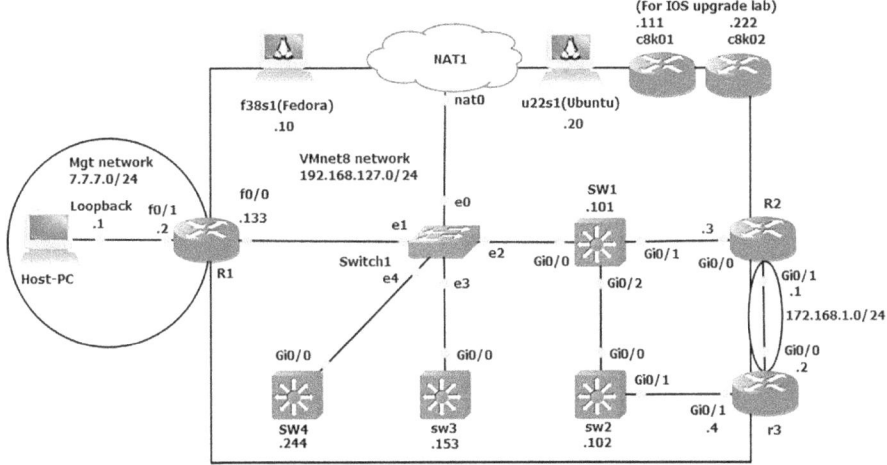 *Figure 4-16. GNS3, adding two new switches*
②	While configuring the new switches, initiate an extended ping from R2, `ping ip 192.168.127.153 repeat 10000`. Also, from r3, execute `ping ip 192.168.127.244 repeat 10000`. Ensure these pings continue to run continuously in the background. R2#**ping ip 192.168.127.153 repeat 10000** Type escape sequence to abort. Sending 10000, 100-byte ICMP Echos to 192.168.127.153, timeout is 2 seconds: .... *[…omitted for brevity]* r3#**ping ip 192.168.127.244 repeat 10000** Type escape sequence to abort. Sending 10000, 100-byte ICMP Echos to 192.168.127.244, timeout is 2 seconds: .... *[…omitted for brevity]*

(continued)

#	Task
③	After powering on the switches, proceed to configure both switches using the specified configurations. Ensure the disabling of IP routing on both switches and set up the last-resort gateway. For sw3, direct outbound traffic through R2, and for SW4, route unknown traffic through R1 initially. The following configuration is provided in the ch4_cron folder, so you can simply cut and paste the commands into your router. ! sw3 initial configuration: **enable** **configure terminal** **hostname sw3** **no ip routing** **ip default-gateway 192.168.127.3** ! Route unknown destination to R2 **enable password cisco123** **username jdoe privilege 15 secret cisco123** **line vty 0 15** **login local** **transport input all** **exit** **ip domain name pynetauto.local** **crypto key generate rsa** **1024** **interface vlan 1** **description "Native interface"** ! For later SNMP lab **ip address 192.168.127.153 255.255.255.0** **no shutdown** **interface GigabitEthernet0/0** **description "1GB Main Connection"** ! For later SNMP lab **no negotiation auto** **duplex full** **end**

(continued)

#	Task

	`copy running-config startup-config`
	`!`
	
	! SW4 initial configuration:
	`enable`
	`configure terminal`
	`hostname SW4`
	`no ip routing`
	`ip default-gateway 192.168.127.133` ! Route unknown destination to R1
	`enable password cisco123`
	`username jdoe privilege 15 secret cisco123`
	`line vty 0 15`
	`login local`
	`transport input all`
	`exit`
	`ip domain name pynetauto.local`
	`crypto key generate rsa` ! use 1024 bits
	`1024`
	`interface vlan 1`
	`ip address 192.168.127.244 255.255.255.0`
	`no shutdown`
	`interface GigabitEthernet0/0`
	`no negotiation auto`
	`duplex full`
	`end`
	`copy running-config startup-config`
	`!`

(continued)

#	Task
④	Once the configuration is finalized, you'll receive the ping responses from the new switches. To cease and exit the ongoing extended ping or traceroute request, use the Ctrl+Shift+6 key combination. In Cisco routers, Ctrl+Shift+6 is vital to interrupt active commands like extended pings or traceroutes. It stops continuous tests, preventing network congestion and resource overload. This action is crucial to managing network diagnostics and to ensuring smooth operation without overwhelming devices. R2#**ping ip 192.168.127.153 repeat 10000** Type escape sequence to abort. Sending 10000, 100-byte ICMP Echos to 192.168.127.153, timeout is 2 seconds: ....!!!!!!!!!!!!!!!!!!!!! # Ctrl+Shift+6, use break sequence Success rate is 85 percent (102/115), round-trip min/avg/max = 17/25/55 ms r3#**ping ip 192.168.127.244 repeat 10000** Type escape sequence to abort. Sending 10000, 100-byte ICMP Echos to 192.168.127.244, timeout is 2 seconds: .... !!!!!!!!!!!!!!!!!!!!!!!!!!!!!!!!!!!!!!! # Ctrl+Shift+6, use break sequence Success rate is 88 percent (102/115), round-trip min/avg/max = 12/21/44 ms **Remember the Ctrl+Shift+6 key combination. If your router or switch hangs or you want to interrupt an operation, you can use this key combination to exit from the operation.**

(continued)

#	Task
⑤	To verify that all unknown outbound traffic from the new switches is correctly routed through R1 for sw3 and R2 for SW4, execute the `clear arp` command and perform traceroutes to Google's public DNS, such as 8.8.8.8 or 8.8.4.4.

Encountering issues while tracerouting the Google DNS could indicate problems with your Internet's DNS or incorrect device configurations. Alternatively, a potential cause might be the recent Windows update affecting GNS3, disrupting the loopback interface used by GNS3. If pinging the DNS IP address presents challenges, set it aside for now and proceed to the next section.

```
sw3#clear arp
sw3#traceroute 8.8.8.8
Type escape sequence to abort.
Tracing the route to 8.8.8.8
VRF info: (vrf in name/id, vrf out name/id)
  1 192.168.127.3 19 msec 37 msec 7 msec # confirmed that the traffic
    routes via R1
  2 192.168.127.2 15 msec *  *
3  *  *  *
[omitted for brevity]
30 *  *  *
sw3#ping 8.8.8.8
Type escape sequence to abort.
Sending 5, 100-byte ICMP Echos to 8.8.8.8, timeout is 2 seconds:
!!!!!
Success rate is 100 percent (5/5), round-trip min/avg/max = 14/31/89 ms

SW4#clear arp
SW4#traceroute 8.8.4.4
Type escape sequence to abort.
Tracing the route to 8.8.4.4
```
|

(continued)

#	Task

	VRF info: (vrf in name/id, vrf out name/id)
	  1 192.168.127.133 19 msec 19 msec 25 msec # confirmed that the traffic routes via R2
	  2 192.168.127.2 16 msec * *
	3 * * *
	[omitted for brevity]
	30 * * *
	SW4#**ping 8.8.8.8**
	Type escape sequence to abort.
	Sending 5, 100-byte ICMP Echos to 8.8.8.8, timeout is 2 seconds:
	!!!!!
	Success rate is 100 percent (5/5), round-trip min/avg/max = 7/32/79 ms

If you've successfully finished these tasks without drama, you're ready to proceed with more labs. By the end of the previous task, your topology should now include two additional IOSvL2 switches. In this next section, you access your Linux servers and become proficient in utilizing the Linux task scheduler, cron.

Quick-Start to the Linux Scheduler, cron

In the earlier chapters of the Part I book, you were introduced to fundamental Linux concepts and even learned to build a Linux server capable of handling multiple IP services. As you delve deeper into Linux, you'll quickly recognize that tasks that might have been challenging or restricted within a Windows environment become effortlessly achievable in a Linux system. Linux enables swift installation of features and services, getting your IP services up and running promptly.

For those accustomed to Windows, the Windows Task Scheduler is a familiar tool. Similarly, Linux has its task-scheduling counterpart known as cron, operating without a graphical interface. Unlike the Windows Task Scheduler, cron is simpler, more dependable, and light due to its lack of a graphical user interface. It serves as Linux's default tool for task scheduling, allowing the execution of native Linux commands and setting up applications like Python scripts to run at specified intervals.

Proficient Linux system administrators extensively utilize automated shell-based scripting or other programming languages for tasks related to system monitoring and backups. The indispensable tool for this purpose in Linux is cron, which serves the equivalent function of the Windows Task Scheduler. As you finish writing Python applications (scripts), understanding how to schedule their execution at specific times and intervals becomes pivotal. Thus, mastering the use of cron for managing scheduled tasks is crucial. While off-the-shelf task schedulers are available for those with ample IT budgets, leveraging free and open-source tools becomes a logical choice, especially when companies can afford to hire smart and skilled professionals like you.

This section guides you through scheduling tasks using cron in both Ubuntu and Fedora systems. Although the basic syntax is almost identical, there are enough variations between them to treat each system uniquely.

Ubuntu LTS Task Scheduler: crontab

On an Ubuntu Server, the task scheduler is known as crontab. You will create a basic Python greeting application to familiarize yourself with task scheduling using Ubuntu's crontab. These instructions will be executed on your u22s1 virtual server (see Figure 4-17). What you learn in scheduling a simple script can easily be applied to schedule more complicated ones. The complexity of the script does not significantly impact the task-scheduling process. Focus on mastering the inner workings of cron; once you understand the basic cron syntax, setting up a script schedule remains a relatively straightforward task. You only need access to your Ubuntu Server for the task in this section.

Figure 4-17. Ubuntu Server, u22s1

#	Task
①	Create a new directory to place your script in, write the three-line Python code for crontab testing, and save it as print_hello_friend.py. You can change the greeting in your script. Here you will import the datatime module from the datetime library and put the print (datetime.now()) statement before printing hello. Because I am from Australia, my favorite greeting is "G'day mate!" ``` jdoe@u22s1:~$ cd ch4_cron jdoe@u22s1:~/ch4_cron$ nano print_hello_friend.py jdoe@u22s1:~/ch4_cron$ cat print_hello_friend.py from datetime import datetime print(datetime.now()) print("G'day Mate!") ```
②	Next, run the code once to check that it prints out the time and your version of the "hello" statement. ``` jdoe@u22s1:~/ch4_cron$ python print_hello_friend.py 2023-11-08 22:01:47.024040 G'day Mate! ```
③	You'll now use crontab to schedule the execution of this Python code. Before creating a schedule, you'll quickly learn how to use crontab. First, you'll need to decide whether to run the script as a standard user or as a root user. Second, determine if you want to execute your Python code as a non-executable file or change it to an executable format before running it. You've already learned how to convert a non-executable file into an executable on a Linux server. If you've chosen to run crontab as a non-root Linux user, you must execute the crontab -e command using sudo as a prefix. This means typing sudo crontab -e. However, if you've decided to run crontab as the root user, you won't need to use the sudo command. ``` 3a. non-root user crontab scheduler execution jdoe@u22s1:~/ch4_cron$ sudo crontab -e # DO NOT run this command yet. [sudo] password for jdoe: ********* OR 3b. root user crontab scheduler execution jdoe@u22s1:~/ch4_cron$ crontab -e # DO NOT run this command yet. ```

(continued)

#	Task
④	To ensure that cron executes your script accurately, it's crucial to verify the location of the Python executable file. Begin by running the `which python3` command to confirm the directory. Initially, execute the `python -V` command to reveal your Python version. Subsequently, use `which python3`; the output will display the symbolic link pointing to /usr/bin/python3 connected to python3.10. This signifies that whether you use /usr/bin/python3 or usr/bin/python3.10, both will direct to /usr/bin/python3.10. Here is an example of the command outputs and soft links: ``` jdoe@u22s1:~/ch4_cron$ python -V Python 3.10.12 jdoe@u22s1:~/ch4_cron$ which python3 /usr/bin/python3.10 jdoe@u22s1:~/ch4_cron$ sudo ls -lh /usr/bin/python* lrwxrwxrwx 1 root root 10 Aug 18 2022 /usr/bin/python3 -> python3.10 #Softlinked to Python3.10 lrwxrwxrwx 1 root root 17 Aug 18 2022 /usr/bin/python3-config -> python3.10-config -rwxr-xr-x 1 root root 960 Jan 25 2023 /usr/bin/python3-futurize -rwxr-xr-x 1 root root 964 Jan 25 2023 /usr/bin/python3-pasteurize -rwxr-xr-x 1 root root 5.7M Jun 11 15:26 /usr/bin/python3.10 lrwxrwxrwx 1 root root 34 Jun 11 15:26 /usr/bin/python3.10-config -> x86_64-linux-gnu-python3.10-config ``` Furthermore, use the `pwd` command to check your working directory for your Python script, as you'll need this information soon to schedule your code. ``` jdoe@u22s1:~/ch4_cron$ pwd /home/jdoe/ch4_cron ``` Ensuring these details are accurate will help you effectively schedule your Python code using cron.

(continued)

#	Task

⑤	To proceed, execute the crontab -e command to access the crontab scheduler. The jdoe user has sudo privileges to run this command without using sudo. When using crontab -e for the first time, a prompt will appear to select your preferred text editor. Opt for the simplest or the recommended choice, which is the nano text editor (Option 1).
	
	jdoe@u22s1:~/ch4_cron$ **crontab -e** # Run this command.
	no crontab for jdoe - using an empty one
	
	Select an editor. To change later, run 'select-editor'.
	  1. /bin/nano <---- easiest
	  2. /usr/bin/vim.basic
	  3. /usr/bin/vim.tiny
	  4. /bin/ed
	
	Choose 1-4 [1]: **1**

⑥	Once you've selected nano as your text editor for the crontab scheduler, press the Enter key to open the crontab scheduler file. Swiftly navigate through the content and position your cursor at the bottom line of the file. Enter the exact text marked in bold at the end of your file and proceed to save the file using Ctrl+S followed by Ctrl+X to exit crontab. The specifics of cron time formats will be covered later, relieving you of the need to decipher each line or character at this point. For now, the best approach is to try it firsthand. Note that getting this right the first time can be a little challenging, but type the cron configuration exactly as shown here: [...omitted for brevity] *** * * * * /usr/bin/python3 /home/jdoe/ch4_cron/print_hello_friend.py** **>> /home/jdoe/ch4_cron/cron.log** This line will execute the Python script every minute and appends the output to cron.log in the current working directory.

(continued)

#	Task
⑦	Next, wait for one to two minutes. You should see that a logging file called cron.log is automatically created under your /home/jdoe/ch4_cron directory. ``` jdoe@u22s1:~/ch4_cron$ ls -lh total 12K -rw-rw-r-- 1 jdoe jdoe 39 Nov 8 22:35 cron.log -rw-rw-r-- 1 jdoe jdoe 73 Nov 8 22:01 print_hello_friend.py -rw-rw-r-- 1 jdoe jdoe 965 Nov 6 21:24 save_configs.py ``` Check the cron.log file with the cat command to check if the crontab is running normally. You will see that the cron job is running every minute, and the greeting is recorded in the cron.log file. So, now you have learned that the five stars (* * * * *) in the crontab -e meant every minute. ``` jdoe@u22s1:~/ch4_cron$ cat cron.log 2023-11-08 22:35:01.832670 G'day Mate! 2023-11-08 22:36:01.890324 G'day Mate! ```
⑧	To inspect your cron job, execute the crontab -l command, as demonstrated. If you're currently logged in as the user, specifying the user is optional. ``` jdoe@u22s1:~/ch4_cron$ crontab -l [...omitted for brevity] # m h dom mon dow command * * * * * /usr/bin/python3 /home/jdoe/ch4_cron/print_hello_friend.py >> /home/jdoe/ch4_cron/cron.log ``` Note that there might not be a cron job under the root user, therefore, use this command to examine another user's cron schedule. ``` jdoe@u22s1:~/ch4_cron$ sudo crontab -u root -l [sudo] password for jdoe:********* no crontab for root ```

(continued)

#	Task
⑨	If you need to oversee another user's cron job, use the sudo crontab -u user_name -l command. To adjust time intervals or disable another user's cron job, use sudo crontab -u user_name -e to make the necessary time adjustments. When the user doesn't have any cron jobs, the system will return a message stating no crontab for user_name. jdoe@u22s1:~/ch4_cron$ **sudo crontab -u pyadmin -e** jdoe@u22s1:~/ch4_cron$ **sudo crontab -u pyadmin -l** no crontab for pyadmin
⑩	Finally, to cancel (deactivate) the currently running crontab schedule, open the scheduler file with crontab -e, add # in front of a specific cron task line, or delete the line in question, and then save the file. Deactivate the print hello friend job by placing the # key. jdoe@u22s1:~/ch4_cron$ **crontab -e** [...omitted for brevity] # m h dom mon dow command **#** * * * * * /usr/bin/python3 /home/jdoe/ch4_cron/print_hello_friend.py >> /home/jdoe/ch4_cron/cron.log
⑪	Another way to control crontab's cron job is to stop/start/restart the crontab service by using the service or systemd (systemctl) commands. 11a. Using Service Command (SysV Init) –backward compatible method jdoe@u22s1:~/ch4_cron$ **sudo service cron** # Press <Enter> key without service status * Usage: /etc/init.d/cron {start|stop|status|restart|reload|force-reload} jdoe@u22s1:~/ch4_cron$ **sudo service cron stop** # To stop cron service OR 11b. Using Systemd (systemctl command) – modern method jdoe@u22s1:~/ch4_cron$ **sudo systemctl status** # Press <Enter> key without specific service name

(*continued*)

#	Task

[...omitted for brevity]
```
           ├─systemd-udevd.service
           │ └─554 /lib/systemd/systemd-udevd
           ├─cron.service
           │ └─2163 /usr/sbin/cron -f -P
           ├─nginx.service
```
[...omitted for brevity]
jdoe@u22s1:~/ch4_cron$ **sudo systemctl stop cron** # To stop cron service

In Ubuntu, service (SysV Init) manages services via /etc/init.d/, simple and backward compatible. systemctl (systemd) is more advanced, using /lib/systemd/system/ in modern Ubuntu (15.04 and later) for powerful and extensive service control and management with additional features and functionalities.

Tip

Understanding output redirection in cron: '>>' vs. '>': append vs. overwrite

In Step 6, every time the cron job runs and executes the script, it logs activity to the cron.log file. But what do the two right arrows or greater-than symbols (>>) mean? What happens when a single arrow (>) is used? Try it for yourself and check it out with your own eyes. Let's examine the difference.

Using two arrows appends logs to the next line, preserving previous content.

```
* * * * * /usr/bin/python3.8 /home/jdoe/my_cron/print_hello_
friend.py >> /home/jdoe/my_cron/cron.log
```

jdoe@u22s1:~/ch4_cron$ **cat cron.log**
2023-11-08 22:35:01.832670
G'day Mate!
2023-11-08 22:36:01.890324
G'day Mate!
[...omitted for brevity]
2023-11-09 08:00:01.936278

```
G'day Mate!
2023-11-09 08:01:02.042657
G'day Mate
```

Using a single arrow overwrites the file, capturing only the latest execution:

```
* * * * * /usr/bin/python3.8 /home/jdoe/my_cron/print_hello_
friend.py > /home/jdoe/my_cron/cron.log
```

```
jdoe@u22s1:~/ch4_cron$ cat cron.log
2023-11-09 08:01:02.042657
G'day Mate
```

Fedora Task Scheduler: crond

Minor differences exist between Fedora and Ubuntu's cron versions. Becoming adept at both Red Hat and Debian-derived Linux operating systems is crucial due to their prevalence in enterprise IT ecosystems. Fedora's crond, akin to Ubuntu's crontab, requires familiarization. Follow the instructions on your Fedora server to grasp crond's usage. To accomplish the tasks, log into f38s1 via SSH. Ensure the routers and switches are active in your lab topology, as shown in Figure 4-18.

CHAPTER 4   PYTHON NETWORK AUTOMATION LABS CRON

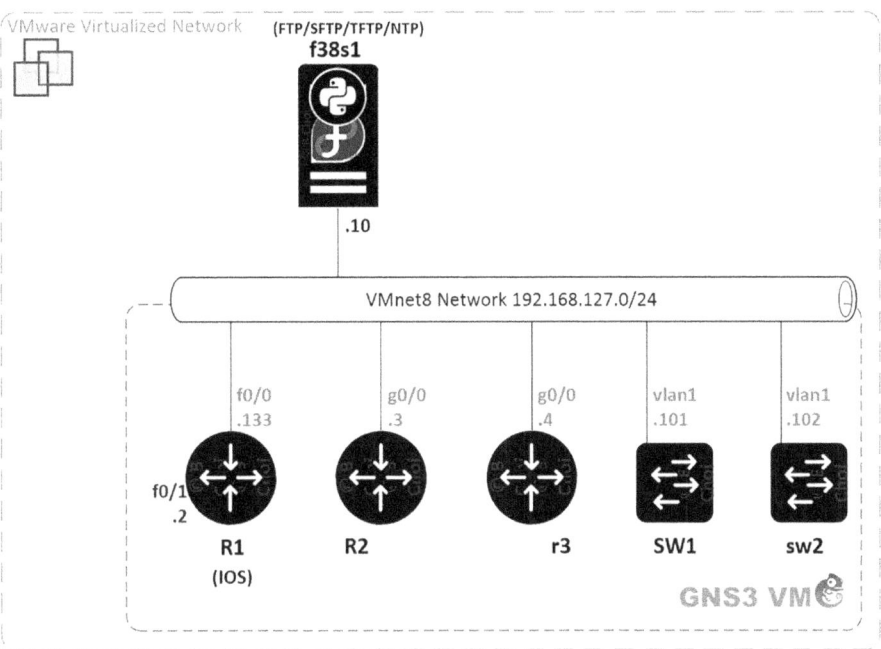

Figure 4-18. f38s1, crond lab required devices

#	Task
①	In a previous chapter, you learned about probing a specific port (port 22). Now, you'll explore how to ping an IP address from your Linux server using a Python script. Let's generate a basic ICMP pinging script to ping all IP addresses in your network and log the activity to the cron. log file. This time, you'll create an external text file named ip_addresses.txt to read the IP addresses of each device and subsequently perform three pings, recording the activity in the cron.log file. Create two files, ip_addresses.txt and ping_sweep1.py, by following these instructions: [jdoe@f38s1 ~]$ **mkdir ch4_ping_sweeper && cd ch4_ping_sweeper** [jdoe@f38s1 ch4_ping_sweeper]$ **nano ip_addresses.txt** [jdoe@f38s1 ch4_ping_sweeper]$ **cat ip_addresses.txt** 192.168.127.3 192.168.127.4 192.168.127.101 192.168.127.102 192.168.127.133

(continued)

#	Task
	Practicing coding extensively requires typing this Python code yourself. The script displays the current time and pings each IP address three times. If the IP is on the network, it prints IP is reachable; otherwise, it prints IP is either offline or ICMP is filtered.

```
[jdoe@f38s1 ch4_ping_sweeper]$ nano ping_sweep1.py
[jdoe@f38s1 ch4_ping_sweeper]$ cat ping_sweep1.py
#!/usr/bin/python3 # shebang line
import os # Import os module
import datetime # Import datetime module

with open("/home/jdoe/ch4_ping_sweeper/ip_addresses.txt", "r") as
ip_addresses: # Open and read file
    print("-" * 80)  # Print separator line
    print(datetime.datetime.now()) # Display current time

    for ip_add in ip_addresses: # Loop through IP addresses
        ip = ip_add.strip() # Remove whitespace
        rep = os.system('ping -c 3 ' + ip) # Send 3 pings
        if rep == 0: # Check ping response
            print(f"{ip} is reachable.") # Display reachable IP
        else:
            print(f"{ip} is either offline or ICMP is filtered.")
            # Display status
    print("-" * 80) # Print separator line
    print("All tasks completed.") # Display task completion
``` |

(continued)

#	Task

②	If you execute the script for verification purposes, if all devices are online, you should get responses from all network devices in the ip_addresses.txt file.
	
	`[jdoe@f38s1 ch4_ping_sweeper]$ `**`python ping_sweep1.py`**
	[...omitted for brevity]
	`192.168.127.102 is reachable.`
	`PING 192.168.127.133 (192.168.127.133) 56(84) bytes of data.`
	`64 bytes from 192.168.127.133: icmp_seq=1 ttl=255 time=4.43 ms`
	`64 bytes from 192.168.127.133: icmp_seq=2 ttl=255 time=9.79 ms`
	`64 bytes from 192.168.127.133: icmp_seq=3 ttl=255 time=7.13 ms`
	
	`--- 192.168.127.133 ping statistics ---`
	`3 packets transmitted, 3 received, 0% packet loss, time 2004ms`
	`rtt min/avg/max/mdev = 4.431/7.114/9.786/2.186 ms`
	`192.168.127.133 is reachable.`
	`----------------------------------------------------------------------`
	`All tasks completed.`

③	At this point, convert the Python script into an executable file for streamlined scheduling in crond. Initially, verify the script's user permissions. Use the chmod +x command to enable file execution and recheck the permissions afterward. Upon executing the chmod command, the file's color should change to green, with x added to the file attributes.
	
	`[jdoe@f38s1 ch4_ping_sweeper]$ `**`ls -lh`**
	`total 8.0K`
	`-rw-r--r--. 1 jdoe jdoe  76 Nov  9 09:22 ip_addresses.txt`
	`-rw-r--r--. 1 jdoe jdoe 458 Nov  9 09:23 ping_sweep1.py`
	`[jdoe@f38s1 ch4_ping_sweeper]$ `**`chmod +x ping_sweep1.py`**
	`[jdoe@f38s1 ch4_ping_sweeper]$ `**`ls -lh`**
	`total 8.0K`
	`-rw-r--r--. 1 jdoe jdoe  76 Nov  9 09:22 ip_addresses.txt`
	`-rwxr-xr-x. 1 jdoe jdoe 458 Nov  9 09:23 ping_sweep1.py`

(continued)

CHAPTER 4　PYTHON NETWORK AUTOMATION LABS CRON

#	Task
④	Fedora's crond has two noticeable differences from Ubuntu's crontab. The first obvious difference is the name, and, when you must reinstall cron on Fedora, you have to use the name `sudo yum install cronie` to install crond. You can use the `rpm -q cronie` command to check the crond version on Fedora.

```
[jdoe@f38s1 ch4_ping_sweeper]$ rpm -q cronie
cronie-1.6.1-4.fc38.x86_64
```

The second difference is where the cron job is scheduled. Unlike Ubuntu, Fedora's crond schedules tasks directly in the `/etc/crontab` file, as shown here (see Figure 4-19):

```
[jdoe@f38s1 ch4_ping_sweeper]$ ls -lh /etc/crontab
-rw-r--r--. 1 root root 451 Jan 19  2023 /etc/crontab
[jdoe@f38s1 ch4_ping_sweeper]$ cat /etc/crontab
```

```
SHELL=/bin/bash
PATH=/sbin:/bin:/usr/sbin:/usr/bin
MAILTO=root

# For details see man 4 crontabs

# Example of job definition:
# .---------------- minute (0 - 59)
# |  .------------- hour (0 - 23)
# |  |  .---------- day of month (1 - 31)
# |  |  |  .------- month (1 - 12) OR jan,feb,mar,apr ...
# |  |  |  |  .---- day of week (0 - 6) (Sunday=0 or 7) OR sun,mon,tue,wed,thu,fri,sat
# |  |  |  |  |
# *  *  *  *  * user-name  command to be executed
```

Figure 4-19. *Fedora, cat /etc/crontab output*

⑤	Subsequently, execute the `systemctl status crond` command to verify the operational status of the crond service.

```
[jdoe@f38s1 ch4_ping_sweeper]$ systemctl status crond
● crond.service - Command Scheduler
    Loaded: loaded (/usr/lib/systemd/system/crond.service; enabled;
    preset: enabled)
    Drop-In: /usr/lib/systemd/system/service.d
             └─10-timeout-abort.conf
```

(*continued*)

#	Task
	``
Active: active (running) since Thu 2023-11-09 09:17:44 AEDT;
16min ago
Main PID: 946 (crond)
    Tasks: 1 (limit: 4632)
    Memory: 1.0M
       CPU: 16ms
    CGroup: /system.slice/crond.service
            └─946 /usr/sbin/crond -n
```
*[...omitted for brevity]* |
| ⑥ | When scheduling a shell command or Python script to run on Fedora crond, you can configure and schedule cron commands in different formats. For example, just as in the Ubuntu example earlier, you could use a full path method to schedule a cron job. In Fedora, the same method will work, as shown here:<br><br>```
[jdoe@f38s1 ch4_ping_sweeper]$ ls -lh /usr/bin/python*
lrwxrwxrwx. 1 root root     9 Feb  8  2023 /usr/bin/python ->
./python3
lrwxrwxrwx. 1 root root    10 Feb  8  2023 /usr/bin/python3 ->
python3.11
-rwxr-xr-x. 1 root root   16K Feb  8  2023 /usr/bin/python3.11
-rwxr-xr-x. 1 root root  2.5K Jan 20  2023 /usr/bin/python-
argcomplete-check-easy-install-script
-rwxr-xr-x. 1 root root   387 Jan 20  2023 /usr/bin/python-
argcomplete-tcsh
[jdoe@f38s1 ch4_ping_sweeper]$ which python
/usr/bin/python
[jdoe@f38s1 ch4_ping_sweeper]$ python -V
Python 3.11.2
``` |

(continued)

CHAPTER 4   PYTHON NETWORK AUTOMATION LABS CRON

#	Task
	Here you are making the Python script an executable file and running it without specifying the application location. Your Python script must have the first shebang line (#!/usr/bin/python3) included in the script. Also, note that you have to specify a user to run the script; because /etc/crontab is a root user file, it is easiest to run the script as the root user, so when you schedule the task, make sure that the root user is specified for the cron job to run successfully. Now, open etc/crontab and schedule your ICMP script to send pings to all routers and switches in your lab every five minutes. */5 * * * * denotes every fifth minute (see Figure 4-20). The script will run at 0, 5, 10, 15, 20, 25, 30, 35, 40, 45, 50, 55, and again every 0 or 60 minutes until you pause or remove the cron job. [jdoe@f38s1 ch4_ping_sweeper]$ **pwd** /home/jdoe/ch4_ping_sweeper [jdoe@f38s1 ch4_ping_sweeper]$ **sudo nano /etc/crontab** [sudo] password for jdoe: ******** *[... omitted for brevity]* ***/5 * * * * root /usr/bin/python /home/jdoe/ch4_ping_sweeper/ping_sweep1.py >> /home/jdoe/ch4_ping_sweeper/cron.log 2>&1**  *Figure 4-20. Fedora, cat /etc/crontab task scheduling*

(continued)

#	Task
⑦	Finally, wait five minutes. When the scheduled job is executed, it will automatically create a file named cron.log. Once the file appears, open the cron.log file under the current working directory and check that the communication check task is taking place every five minutes. `[jdoe@f38s1 ch4_ping_sweeper]$ `**`ls -lh`** ``` total 12K -rw-r--r--. 1 root root 3.2K Jan 13 15:35 cron.log # Newly created file -rw-rw-r--. 1 jdoe jdoe 110 Jan 13 15:03 ip_addresses.txt -rwxrwxr-x. 1 jdoe jdoe 886 Jan 13 15:35 ping_sweep1.py ``` `[jdoe@f38s1 ch4_ping_sweeper]$ `**`cat cron.log`** ``` [... omitted for brevity] --- 192.168.127.102 ping statistics --- 3 packets transmitted, 3 received, 0% packet loss, time 2003ms rtt min/avg/max/mdev = 27.076/36.783/54.996/12.887 ms PING 192.168.127.133 (192.168.127.133) 56(84) bytes of data. 64 bytes from 192.168.127.133: icmp_seq=1 ttl=255 time=10.5 ms 64 bytes from 192.168.127.133: icmp_seq=2 ttl=255 time=10.6 ms 64 bytes from 192.168.127.133: icmp_seq=3 ttl=255 time=6.59 ms --- 192.168.127.133 ping statistics --- 3 packets transmitted, 3 received, 0% packet loss, time 2004ms rtt min/avg/max/mdev = 6.590/9.218/10.561/1.858 ms --- 2023-11-09 22:10:02.011644 192.168.127.3 is reachable. 192.168.127.4 is reachable. 192.168.127.101 is reachable. 192.168.127.102 is reachable. 192.168.127.133 is reachable. --- All tasks completed. ```

(continued)

#	Task
⑧	At the end of this lab, make sure you go back to the /etc/crontab file and add # to the beginning of the line to deactivate it. You can optionally delete the line and then save the changes. # */5 * * * * root /usr/bin/python /home/jdoe/ch4_ping_sweeper/ping_sweep1.py >> /home/jdoe/ch4_ping_sweeper/cron.log 2>&1

Tip

Installing crontab on Fedora is your choice

Out of the box, crontab is significantly more user-friendly than crond. Moreover, it offers the flexibility of scheduling cron jobs under various user accounts, making it more intuitive for many users. If you are like me, you can opt to install crontab on the Fedora server using the following Linux commands. However, it's important to note that in production environments, certain older Linux servers might lack Internet connectivity, therefore requiring the use of crond to schedule tasks. Therefore, it's beneficial to understand the distinctions between the two when operating on different Linux distributions. To install crontab on your Fedora VM, use these commands:

```
[jdoe@f38s1 ~]$ sudo dnf update
[jdoe@f38s1 ~]$ sudo dnf install crontabs
[jdoe@f38s1 ~]$ sudo systemctl enable crond.service
[jdoe@f38s1 ~]$ sudo systemctl start crond.service
```

Note: dnf (Dandified YUM), also known as the next-generation package management utility for RPM-based distributions, is replacing the older YUM (Yellowdog Updater, Modified) package management utilities in the Fedora, CentOS, and Red Hat distributions.

Next, let's review and unveil the cron job definitions or the five asterisks.

Learn cron Job Definitions with Examples

To utilize Linux cron effectively, it is essential to comprehend the significance of the five stars (asterisks) at the beginning of each cron job line. If you are already familiar with cron and understand the purpose of the five asterisks and how they are used, that is fantastic! You can proceed to the next section. However, if you are new to cron, continue reading this section to gain an understanding of the meaning behind the five asterisks. The excerpt shown in Figure 4-21 is directly from the /etc/crontab file in Fedora, providing all the information you need to comprehend how a command, script, or task is scheduled in cron.

```
# Example of job definition:
# .---------------- minute (0 - 59)
# |  .------------- hour (0 - 23)
# |  |  .---------- day of month (1 - 31)
# |  |  |  .------- month (1 - 12) OR jan,feb,mar,apr ...
# |  |  |  |  .---- day of week (0 - 6) (Sunday=0 or 7) OR sun,mon,tue,wed,thu,fri,sat
# |  |  |  |  |
# *  *  *  *  *  user-name  command to be executed
```

Figure 4-21. Crontab, example of job definition

For the sake of clarity, all exercises will be based on crontab (Ubuntu) rather than crond (Fedora), so open cron on your Ubuntu Server. Typically, a default cron job commences with five stars (asterisks). These five stars correspond to the frequency of task execution, or, in other words, the time and date interval of a specific task. Following these five values is the command (CMD) necessary to execute a particular task (see Figure 4-22).

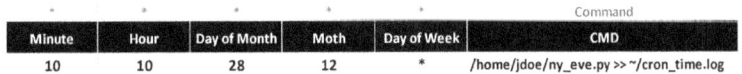

*	*	*	*	*	Command
Minute	Hour	Day of Month	Moth	Day of Week	CMD
10	10	28	12	*	/home/jdoe/ny_eve.py >> ~/cron_time.log

Figure 4-22. Cron job, an example

Table 4-1 shows the tabulated version of the example of a job definition shown earlier.

Table 4-1. Linux crontab: Time Unit

Unit	Available Values	Conversion
Minute of an hour	0–59 minutes	1 hour = 60 minutes
Hour of a day	0–23 hours	1 day = 24 hours
Day of month	1–31 days	1 month = 28–31 days
Month	1–12 or Jan, Feb, Mar, Apr…Dec	Jan, Feb, Mar, Apr…
Day of week	0–6	1 week = 7 days
CMD	Command to run	N/A

Depending on the type of Linux operating system you have, each command may have slight variations, but they are generally quite similar. If these values use a different format, you should consult the documentation provided by the OS distributor for alternative time unit variations. Next you'll examine more examples to gain a better understanding of the job definitions within the cron schedule. The following examples encompass the most common job definitions. If your requirements differ, you can adjust the examples provided in Table 4-2 and incorporate them into your cron job scheduling.

Table 4-2. Cron Examples with Explanations

#	Examples (Alternatives in Brackets)	Explanation
①	* * * * *	The five asterisks are the default value, which tells the Linux system to execute a task every minute. You have used the five asterisks to run a Python script already.
②	15 03 25 03 *	At 3:15 on the 25th day of March.
③	15 1,13 * * *	At 1:15 a.m. and 1:15 p.m.
④	00 07-11 * * *	On the hour between 7 a.m. to 11 a.m. (every day).

(*continued*)

Table 4-2. (*continued*)

#	Examples (Alternatives in Brackets)	Explanation
⑤	00 07-11 * * 1-5	Same as Example 4, but only runs on the weekdays (1 = Mon, to 5 = Fri).
⑥	0 2 * * 6,0	At 2 a.m. on Saturdays and Sundays only.
⑦	00 23 * * 1 (00 23 * * Mon)	At 11 p.m. on Mondays.
⑧	0/5 * * * *	At every 5th minute, every five minutes.
⑨	0/15 * * * *	At every 15th minute, every quarter of an hour.
⑩	0/30 * * * *	At every 30th minute, every half an hour.
⑪	0 0 * 1/1 *	At midnight on the first day of each month.
⑫	0/10 * * * 0 (0,10,20,30,40,50 * * * 0)	At every 10th minute; supported on newer cron versions. At every 10th minute; supported on older cron versions.
⑬	@reboot	After rebooting.
⑭	@hourly (0 * * * *)	At minute 0.
⑮	@daily (@midnight) (0 0 * * *)	At 00:00 (midnight).
⑯	@weekly (0 0 * * 0)	At midnight every Sunday.
⑰	@monthly (0 0 1 * *)	At midnight, the first day of the month.
⑱	@yearly (@annually) (0 0 1 1 *)	At midnight on January 1st of every year.

> **Expand your knowledge**
>
> **Unlocking cron secrets and practical applications**
>
> A more detailed description of cron can also be found on Wikipedia. You can learn more about cron and how to use it by visiting the following Wikipedia site.
>
> URL: https://en.wikipedia.org/wiki/Cron
>
> Spend some time on the following websites. You can learn more about how to use cron practically by using one of the cron maker sites listed here:
>
> URL: http://www.cronmaker.com/
>
> URL: https://crontab.guru/

In this section, you configured and completed the scheduling of simple Python scripts using Linux cron, and then you learned the meanings behind the five stars through examples. At this point, there is no need to memorize the cron definitions by heart; you can always return to this page as a reference. Additionally, in a later lab, you will apply what you have learned here and schedule your cron job to execute a Python script for your specific task.

In the upcoming chapter, you learn about the SNMP basics needed to prepare you for the development of Python and SNMPv3 applications in Chapter 6. This foundational understanding will pave the way for creating a straightforward SNMPv3 monitoring application in Python at a later stage.

Summary

As you wrap up Chapter 4, you have adeptly navigated the complexities of Python Network Automation labs and immersed yourself in the world of cron. Now armed with this knowledge, you can seamlessly schedule your Python applications to execute at specific times or intervals, providing the flexibility to run tasks even outside of business hours without compromising your well-deserved rest. The robust tool, cron, stands out as a powerful, free, and open-source task scheduler, emphasizing the value of integrating open-source tools whenever possible. From simulating router failures to refining network recovery strategies, each lab served as a reinforcement of your practical insights. The exploration of the Linux Scheduler, cron, was thorough, encompassing the Ubuntu LTS Task Scheduler (crontab) and Fedora Task Scheduler (crond). Your hands-on experiences, including the cloning of GNS3 projects and the execution of pre-tasks for lab optimization, laid a sturdy foundation. The mastery of cron job definitions with illustrative examples has further honed your skills. Looking ahead to Chapter 5, it is noteworthy that the next segment of your network automation journey will delve into the dynamic interplay between SNMP system monitoring and Python script interactions, promising a seamless transition into the next phase of your Python exploration in this evolving ICT field.

CHAPTER 5

Python Network Automation Labs: SNMP Discovery with Python

This informative chapter takes a closer look at SNMP, laying the groundwork for its integration into Python applications for network automation. The journey begins with a quick-start guide to SNMP, offering insights into its history, types of system monitoring, and a comparative analysis of SNMP vs. Syslog vs. APIs. The chapter then explores SNMP standard operations, distinguishing between SNMP polling and the trap (event reporting) method and explores various SNMP versions. The chapter provides a focused examination of the SNMP protocol within the TCP/IP stack, covering SNMP message types, SMI, MIB, and OID. It delves into SNMP access policies and synchronous vs. asynchronous communication in SNMP traps. It also demonstrates the practical use of Python for executing SNMPv3 queries. Hands-on experience includes understanding MIBs and OIDs, along with using SNMPwalk on a Linux server to enhance your skillset. The practical application becomes the focal point as I guide you through the installation and configuration of an SNMP server on Fedora. You'll gain insights into configuring Cisco routers and switches to support SNMPv3 using Python scripts, troubleshooting SSH connections between Fedora and Cisco devices, and setting up SNMP agents on Cisco devices. The chapter concludes with a comprehensive summary, consolidating your knowledge and providing a strong foundation for developing Python code for SNMPv3 applications. Explore the SNMP lab source code to understand how Python can be utilized to query Cisco device information and run queries for interface descriptions.

By the end of this chapter, you'll be well-equipped with the skills you need to effectively integrate Python into SNMPv3 application development for streamlined network automation.

Bridging Theory and Practice: Learning About SNMP with Python

Having journeyed through the narrative to this point in the book, you've refined your skills in developing Python scripts for diverse networking tasks. Simple Network Management Protocol (SNMP) has been the de facto protocol for systems and network monitoring and has been playing a pivotal role in gathering statistics and monitoring the health and performance of systems across the enterprise, private cloud, hybrid cloud, and public cloud networks. Leveraging Python's SNMP libraries empowers you to create custom scripts for querying network devices, tracking network traffic, and detecting issues. Despite the existence of numerous specialized commercial SNMP solution vendors, dedicating a chapter to learning about SNMP technologies and building tools yourself using Python will enhance your understanding of the protocol and its capabilities in the network automation field. It is crucial to note that SNMP primarily serves monitoring purposes, with more suitable protocols available for device configuration management. SNMP is rarely used for system or network provisioning and configuration. This focused chapter explores the foundational concepts of SNMP, providing practical applications within Python-based network automation lab settings. Continuing in the same GNS3 Network Project settings from the previous chapter saves time for an in-depth SNMP exploration. This chapter tables cover SNMP facts and assist you in approaching SNMP from various angles. Additionally there are some hands-on practical exercises at the end of the chapter. This chapter will help you apply your thinking to develop a tool using SNMP-based Python application tools.

You will delve deeply into SNMP for network device information and monitoring, enriching your understanding of the inner workings of enterprise-level tools. This chapter aims to bridge the gap between theoretical concepts and their real-world application within a sophisticated IP services environment. As always, you will be responsible for installing and configuring the SNMP server and clients yourself, running SNMP library-powered Python scripts from your Python/SNMP server to interact with your networking devices. Without hands-on installation and configuration, you're merely an end-user. Conversely, building everything yourself transforms you into the expert, granting you a deeper understanding of the system and its ecosystem.

Quick-Start Guide to SNMP

Despite being a cornerstone of enterprise network and systems monitoring, SNMP often remains undervalued, silently ensuring the health of critical network and systems infrastructure. Many network engineers overlook in-depth studies of SNMP, considering it less exciting than other networking topics. As professionals climb up the corporate ranks, their focus shifts toward client services, design, and implementation. SNMP quietly monitors vast enterprise networks but seldom takes center stage, perceived as a secondary service. **However, understanding Python's interaction with SNMP is crucial for application development in network monitoring.** In many commercial products, various programming languages have been used to drive cloud-based monitoring systems in recent years, and the vast majority of them use the Microsoft PowerShell behind the scenes and SNMPv2c code as a foundation, refactoring it to interact with network devices using SNMPv3. Developing full-fledged Python tools for SNMP is a specialized networking area that may take weeks, if not months, to master. Here, you'll dip your toes in to experience Python's interaction with network devices using SNMP.

To refresh your SNMP knowledge, you'll review some basic facts. If you're already familiar with SNMP, feel free to skip this section and proceed to the SNMP lab. For those new to SNMP, this serves as a quick start guide to understanding Python's interaction with SNMP.

Quick SNMP History

To gain a broader understanding of a technology, reviewing its history is a good icebreaker. SNMP originated as ICMP and evolved to become the preferred monitoring tool in the IT industry. Let's quickly go over its standardization in its early adoption period.

- **1981: ICMP's genesis.** The Internet Engineering Task Force (IETF) defined the Internet Control Management Protocol (ICMP). ICMP served as a protocol designed to assess the network communication status of devices. During this early phase, its primary purpose was to probe and monitor the health of networked devices.

- **1987: Transition to SGMP.** ICMP transformed and was redefined as the Simple Gateway Monitoring Protocol (SGMP), documented in RFC 1028. This shift marked a pivotal moment in the protocol's development, indicating a broader scope beyond basic network communication monitoring.

- **1988: Founding year of SNMP.**

- **1989 to 1990: Emergence of SNMP.** Building on the foundation laid by SGMP, the protocol underwent further refinement. During this period, SGMP evolved into the Simple Network Monitoring Protocol (SNMP). This transformation was formalized through the introduction of RFC 1155, RFC 1156, and RFC 1157. SNMP, in its new guise, extended its capabilities and became a standardized protocol for network monitoring.

- **1990 to the present: SNMP standardization.** From 1990 onwards, SNMP gained widespread acceptance among IT and networking vendors. It solidified its position as the standard protocol for monitoring IP devices. The ongoing development and adoption of SNMP was instrumental in enhancing its features and ensuring its compatibility with the evolving landscape of network technologies.

Source: https://www.ietf.org/standards/rfcs/ (RFC: 777, 1028, 1155, 1156, 1157)

Now that you've learned a bit of SNMP's history, let's take a brief look at what system monitoring aims to achieve through protocols like SNMP.

Types of System Monitoring

While SNMP, a comprehensive system monitoring tool, addresses various aspects of system health, specific monitoring areas exist. System monitoring, applicable to network devices (essentially purpose-built computers), spans physical, virtual, and cloud computing realms. The diversity of system monitoring types plays a pivotal role in sustaining the reliability, security, and efficiency of computer systems and applications. Organizations commonly deploy a mix of these monitoring types for holistic IT infrastructure management. **In recent times, the landscape of system monitoring has witnessed a notable shift toward cloud-based solutions, with vendors like Datadog, New Relic, Dynatrace, SolarWinds, LogicMonitor, and ScienceLogic offering holistic approaches to system monitoring.** Additionally, if your infrastructure resides and consists of major public cloud service providers, each vendor provides system monitoring as a service, such as Amazon CloudWatch, Microsoft Azure Monitor, and Google Cloud Monitoring, to name the big three. Most of these tools integrate with automated ticketing systems that generate tickets through APIs. They also page on-call engineers around the clock for immediate response and resolution to system problems. In a later Python lab, you will build a similar tool to gain a better understanding of how such tools are written and operate. Under the hood of each of these solutions, the basic monitoring concepts are universal and aim for the same or similar goal: to keep the systems up and running with as little downtime as possible. Therefore, understanding the types of system monitoring becomes crucial, as this information applies to the fundamental basics of all vendor-monitoring systems. Refer to Table 5-1 for insights into monitoring types, encompassing their definitions and purposes.

Table 5-1. *Types of System Monitoring*

#	Monitoring Type	Definition	Purpose
1	Performance Monitoring	System resources monitoring like CPU use, memory utilization, disk I/O, and network bandwidth.	Identify performance bottlenecks, prevent resource exhaustion, and optimize system responsiveness.
2	Application Performance Monitoring (APM)	Tracking the performance of specific applications, including response times, transaction success rates, and resource consumption.	Pinpoint issues affecting application performance, optimize code, and enhance user experience.
3	Network Monitoring	Observing network infrastructure to ensure connectivity, detect anomalies, and analyze traffic patterns.	Prevent network congestion, identify security threats, and ensure smooth data transmission.
4	Security Monitoring	Monitoring for security events, including intrusion detection, log analysis, and anomaly detection.	Identify and respond to security threats, ensure compliance with security policies, and protect sensitive data.
5	Log Monitoring	Analyzing log files generated by systems, applications, and network devices.	Detect and troubleshoot issues, track user activity, and comply with auditing requirements.
6	Uptime Monitoring	Ensuring that systems and services are available and operational.	Minimize downtime, maximize service reliability, and meet service level agreements (SLAs).
7	Capacity Planning	Forecasting future resource needs based on historical data and current usage trends.	Optimize resource allocation, avoid performance degradation, and plan for scalability.
8	Error Monitoring	Detecting and tracking errors, exceptions, and anomalies in applications.	Debug software issues, improve code quality, and enhance user experience.
9	End-User Experience Monitoring	Measuring and analyzing the experience of end-users interacting with applications.	Understand user satisfaction, identify areas for improvement, and ensure optimal user interactions.

(*continued*)

Table 5-1. (*continued*)

#	Monitoring Type	Definition	Purpose
10	**Cloud Monitoring**	Monitoring cloud-based infrastructure, services, and applications.	Ensure the availability and performance of cloud resources, optimize costs, and maintain security.
11	**Database Performance Monitoring**	Monitoring the performance of database servers, queries, and transactions.	Optimize database performance, identify slow queries, and ensure data integrity.
12	**Custom Metric Monitoring**	Tracking specific metrics or key performance indicators (KPIs) tailored to the needs of a particular system or application.	Gain insights into critical aspects of a system or application that are not covered by standard monitoring.

Now that you've explored various aspects of system monitoring, you will delve into the protocols used for effective system monitoring. While SNMP stands out as the most widely adopted protocol for this purpose today, newer contenders are emerging, such as APIs for system monitoring. Additionally, the reliable Syslog server continues to play a significant role in aggregating logs generated across production systems. This overview aims to provide insights into the current state of SNMP monitoring and sheds light on potential shifts in the enterprise system monitoring landscape, particularly with the evolving role of APIs in the future.

Comparison Table: SNMP vs. Syslog vs. APIs

SNMP (Simple Network Management Protocol), Syslog, and APIs (Application Programming Interfaces) represent distinct system monitoring approaches. **SNMP, founded in 1988, uses a pull-based model for network device data collection, often used for legacy devices.** Syslog, since 1980, has facilitated real-time insights by receiving log messages from devices across the network. APIs, in early development, offer versatility with pull and push models, supporting real-time updates and communication. SNMP has been the protocol choice for monitoring many network devices, Syslog for event tracking, and APIs for diverse applications like cloud integration, automation, and seamless communication between services and systems, emphasizing flexibility, security, and extensibility. The landscape of system monitoring is currently dynamic and subject to change.

Table 5-2 provides a comprehensive comparison of SNMP, Syslog, and APIs based on their key features and characteristics.

Table 5-2. SNMP, Syslog, and API Comparison

Feature	SNMP	Syslog	APIs
Founding Year	1988	1980	Early development; widespread use later
Major Release Versions	SNMPv1, SNMPv2c, SNMPv3	Syslog (RFC 3164), Syslog-NG, Rsyslog	Early APIs, RESTful APIs becoming common
Data Retrieval and Flexibility	Pull-based model; manager actively requests information from agents	Push-based model; devices send log messages to a central server	Offers versatility with both pull and push models; supports real-time updates
Ease of Use and Integration	Simple but configuration/setup can be complex	Commonly used for logging and auditing; standardized log collection	Developer-friendly; standard HTTP methods; often communicates in JSON
Real-time Monitoring	Operates on a periodic polling basis; may not provide real-time monitoring	Provides real-time insights into events and activities across the network	Supports real-time updates and notifications through webhooks
Security	SNMPv3 introduced security features; concerns with SNMPv2	Relies on network security measures; lacks encryption but is used in secure networks	Implements modern security standards like OAuth2 and API keys
Extensibility	Well-defined set of standard MIBs; extending/customizing can be complex	Supports severity levels for message categorization and prioritization	Supports versioning and can be extended more easily

(*continued*)

Table 5-2. (*continued*)

Feature	SNMP	Syslog	APIs
Device Support	Widely supported in legacy devices and network equipment	Widely supported across various network devices, operating systems, and applications	Varies; modern devices often come with RESTful APIs or other programmable interfaces
Use Cases	Traditionally used for monitoring network devices like routers, switches, and servers	Mainly used for tracking system events, debugging, and compliance monitoring	Versatile usage, including cloud integration, automation, and communication between applications and services

Note: Founding years and major release versions may differ, depending on specific implementations and software versions.

Having explored three system monitoring protocols prevalent in today's IT industry, you will now delve into the standard operations of SNMP and learn how SNMP works before proceeding with the creation of your SNMP Python scripts.

SNMP Standard Operation

SNMP is primarily known for its role in device monitoring, allowing administrators to collect and monitor data about networked devices. However, SNMP is not typically used to make direct system changes. Its primary functions include monitoring, querying, and collecting information from IP devices. Although SNMP can be used to make configuration changes, SNMP is not designed as a protocol for making configuration changes or altering system settings. Instead, it provides a standardized method for managing and monitoring network devices, ensuring their optimal performance and facilitating troubleshooting.

Much like many server and client systems, SNMP adheres to a similar model. The SNMP server is referred to as the manager, and it teams up with clients, known as agents. Even if you have completed your CCNA studies, you may not be familiar with the intricacies of standard SNMP operations. The CCNA curriculum touches on SNMP but does not delve deeply into the subject because Cisco has the Cisco Prime Infrastructure management platform as an SNMP monitoring product, but it still cannot compete with industry leaders such as SolarWinds, Paessler PRTG Network Monitor, Nagios, or ManageEngine OpManager. As a result, SNMP is often covered in CCNA as a courtesy topic. If this is your first-time using SNMP, it is recommended that you have a go at reading the rest of this chapter in full.

- **SNMP functions as a server (manager) and client (agent)**: Operates in a server-client model, where the SNMP server assumes the role of the manager, and devices being monitored act as clients or agents.

- **SNMP agent is installation on monitored devices**: Typically, the SNMP agent is installed on devices designated for monitoring, functioning as the interface between the manager and the monitored device.

- **SNMP follows preset rules for monitoring**: It adheres to predetermined rules to monitor network devices systematically. It collects and stores information crucial for monitoring and managing devices effectively.

- **Agent status reporting via polling or event reporting**: An SNMP agent can communicate its status to the manager using either polling, where the manager actively requests information, or event reporting, where the agent autonomously notifies the manager when specific events occur.

- **Management Information Base (MIB) as the information standard**: SNMP uses the Management Information Base (MIB) as the standardized framework for organizing and defining the various data elements that can be retrieved from a device. MIB facilitates the consistent collection of data.

- **Object Identification (OID) as MIB's primary key**: Serves as the primary key within MIB's management objects. It uniquely identifies and categorizes each data element within the MIB structure.

Understanding these fundamental aspects of SNMP's standard operation is crucial for effectively utilizing this protocol in network monitoring and management. If this is your initial encounter with SNMP, taking the time to familiarize yourself with these concepts will provide a solid foundation for your SNMP-related work.

SNMP Polling vs. the Trap (Event Reporting) Method

To collect data, the SNMP manager initiates the Get request message to the agent, which responds with its status. The SNMP server seeks specific information, and the SNMP client provides the necessary responses. In an environment where the SNMP server can modify the SNMP client's status, the server can utilize the SNMP Set request to affect changes. **In simpler terms, SNMP can be used to exert control over the client, although its primary function typically revolves around network monitoring.** The Polling method is more talkative on the network, generating more traffic compared to the Trap method. Also, note that in both methods, SNMP clients send information to the SNMP servers. In the Trap method, the SNMP client transmits information to the SNMP server's Trap receiver only if monitored information in its system exists. This method stands as the predominant approach in the industry today and generates a relatively modest amount of network traffic in a stable network. In short, SNMP Polling is where the SNMP server regularly requests information, whereas SNMP Trapping involves clients directly sending event data to the server without any prior request.

SNMP Versions

Table 5-3 presents an overview of three SNMP versions in use. Many legacy systems lack support for the more secure SNMPv3. **Consequently, SNMPv2c, often referred to as SNMPv2, remains the most utilized version in many enterprise networks and clouds today.** However, due to growing security concerns, an increasing number of enterprises are transitioning to SNMPv3. As aging devices are replaced and new devices with SNMPv3 compatibility become more prevalent, the deployment of SNMPv3 is gradually gaining momentum. Another influential factor prompting the shift away from SNMPv2 is the prevalence of virtualized and cloud-based infrastructures, now standard in enterprise network and server networking. **In the virtualized cloud environment, beyond the physical server farm and direct connections, nearly everything exists as files or software, liberating technologies from their hardware constraints.** Table 5-3 has a detailed comparison of the features associated with each SNMP version.

Table 5-3. Major SNMP Versions and Key Features

Version	Key Features
SNMPv1	• Uses a community string (key) • Lacks built-in security • Limited to 32-bit systems
SNMPv2c	• Most used version • Supports 32/64-bit systems • Improved performance for 64-bit systems • Almost identical features to SNMPv1 • Lower security compared to SNMPv3 • Offers cross-platform support
SNMPv3	• Manager and agent systems transformed into objects • Enhanced security for 64-bit systems • Provides authentication and privacy guarantees • Recognizes each SNMP entity with unique EngineID names

Understanding the SNMP Protocol in the TCP/IP Stack

SNMP functions as an application protocol within the TCP/IP protocol stack, residing on the transport layer in the OSI model. It uses UDP for communication between the server and the client. Here's a concise overview of SNMP communication port details and the interaction between managers and agents:

- The SNMP Get message uses UDP port 161 for communication.

- The SNMP Set message also uses UDP port 161 for communication.

- In the case of SNMP Trap usage, SNMP Trap communicates via UDP port 162. Notably, the agent sending the message does not need to confirm if the trap was sent successfully.

- When SNMP Inform is utilized, it follows a similar process as SNMP Trap, confirming the reception of the Inform message via UDP port 162.

This framework ensures effective communication between SNMP managers and agents, with specified port assignments for different SNMP operations.

About SNMP Message Types

Six types of messages are exchanged between the SNMP manager and SNMP agent, as summarized in Table 5-4.

Table 5-4. SNMP Message Types

Message Type	PDU Type	Explanation
get-request	0	Retrieves the value of the object specified in the agent management MIB.
get-next-request	1	Retrieves the next object value in the specified agent management MIB. Typically used for MIB tables.
get-bulk-request	1	Like get-next-request, retrieves the object's value in the agent management MIB multiple times.
set-request	2	Sets (changes) the value of the object specified in the agent management MIB.
get-response	3	Returns the result of the manager's request.
traps	4	Informs the manager when a specific event occurs in the agent management MIB.

*PDU-type values range between 0 and 4 only.

Table 5-4 outlines various SNMP message types and their corresponding Protocol Data Unit (PDU) types, providing a concise reference for understanding the purpose of each message in the SNMP manager-agent communication process. PDU-type values are limited to the range between 0 and 4.

Understanding SNMP's SMI, MIB, and OID

After providing a brief definition of SMI, MIB, and OID in SNMP, the following discussion looks at security methods, access policies, and the synchronization and asynchronization of traps. This section explores SNMP terminologies and their roles in monitored networks.

- **Structured Management Information (SMI)**: Serves as a tool for defining and configuring MIBs, playing a pivotal role in creating and managing MIB standards.

- **Management Information Base (MIB)**: Constitutes a collection of objects managed from the SNMP server's perspective. It acts as a database containing objects that administrators can view or set. Each object follows a tree-like structure, allowing administrators to check statuses or modify agent values based on specific MIB values. ASN.1 (Abstract Notation One), organized in a tree structure, serves as a scripting language for describing objects, including SNMP objects.

- **Object-Identifier (OID)**: An OID must be assigned to generate the MIB. Represented as a sequence of unique, specific numbers using a notation like IP addresses, OIDs function as identifiers (primary keys), designating administered objects within the MIB. Utilizing OID lookup tools can provide a detailed understanding of OIDs.

This overview clarifies the roles of SMI, MIB, and OID in SNMP-monitored networks, setting the stage for a deeper exploration of security methods, access policies, and the nuances of trap synchronization and asynchronization.

SNMP Access Policy

To uphold SNMP security, communication between the manager and agent relies on SNMP community strings. The manager must have the correct community string to access vital device information throughout the network. Typically, network vendors set the default public community string to "public" when shipping network equipment. However, many network administrators choose to modify this string to prevent potential intruders from seeking information about the network setup. SNMP extends different access levels to information based on distinct community strings, encompassing Read-Only, Read-Write, and Trap community strings.

- **SNMP Read-only community string**: Allows other SNMP managers to read information from the agent. InterMapper uses read-only information to map devices.

- **SNMP Read-Write community string**: Used for communication via requests, this string can both read and alter the status or settings of other devices. It is noteworthy that InterMapper operates solely in the read-only mode, foregoing the use of read-write string mode.

- **SNMP Trap community string**: The agent uses this community string to dispatch SNMP traps to InterMapper, which promptly receives and stores information, regardless of the specific SNMP Trap community string in use.

Understanding and managing these SNMP community strings ensures a layered approach to network security, enabling administrators to tailor access levels based on their network management requirements.

Synchronous vs. Asynchronous Communication While Using SNMP Traps

The following provides a concise overview of two SNMP Trap methods utilized by SNMP. Generally, when the Trap message is exchanged between the SNMP manager (server) and the agent (client) through SNMP, SNMP must use either synchronous or asynchronous mode. The distinctions between these two modes are explained here:

- **Synchronous mode**: In this mode, the sending and receiving sides operate with a shared reference clock independent of the data. Both the manager and the agent must synchronize their actions based on this common signal. Continuous data transmission in this mode requires the completion of one data transmission or reception between the sender and receiver before transmitting the next set of data.

- **Asynchronous mode**: In this mode, the time difference is determined with the received signal clock, and data is transmitted one character at a time, regardless of the sender's clock. The packet to be sent includes start and stop bits at the front. In SNMP agent operations using asynchronous mode, diverse data is exchanged when communicating with Get requests, allowing for irregular transmission regardless of time and order.

When SNMP Trap is used, the message is generated as either a general trap or an inform trap message, operating in an asynchronous communication method. The SNMP agent unilaterally transmits information without requiring confirmation from the SNMP manager, functioning asynchronously. Conversely, the SNMP agent operates in synchronous mode when the transmitted information requires acknowledgment from the receiver, which is the SNMP manager.

SNMP Tools and Python Integrations

As you can see in Table 5-5, there are plenty of SNMP/API system monitoring tools in the market. This section quickly reviews the list of the tools and languages supported by these SNMP tools. In the world of network monitoring, diverse tools cater to developers' needs, each with unique features. This section reviews the tools mentioned previously with Python integration and support. Cacti, with its core in PHP, supports SNMP and offers a web-based interface. Python developers can use libraries like PySNMP within Cacti. Nagios, known for extensibility, supports SNMP via plugins, with Python developers using libraries like pynag. Zabbix, mainly in C, supports SNMP and offers Python developers the Zabbix API for custom scripts. PRTG Network Monitor, in C#, uses SNMP, and Python developers can interact with the PRTG API. You do not need to memorize the information in Table 5-5; you should use it as a reference point.

Table 5-5. Monitoring Tools and SNMP Application Development

Monitoring Tool	Backend Language	Frontend Language	Monitoring Support
Cacti	PHP	JavaScript	SNMP, API
Nagios	C	PHP	SNMP
Zabbix	C	JavaScript	SNMP, API
PRTG Network Monitor	C# (.NET)	JavaScript	SNMP, API
Observium	PHP	JavaScript	SNMP, Syslog
LibreNMS	PHP	JavaScript	SNMP, API
MRTG (Multi Router Traffic Grapher)	Perl	PHP (GraphRRD)	SNMP
NetXMS	C++	JavaScript	SNMP
Grafana	Go	TypeScript (a superset of JavaScript)	API
Prometheus	Go	TypeScript (a superset of JavaScript)	API
Icinga	PHP, C++	PHP (Icinga Web 2)	SNMP, API
Checkmk	Python	Not specified	SNMP, API
SolarWinds Network Performance Monitor	Various (proprietary)	Not specified	SNMP, API

Observium, in PHP, relies on SNMP, and Python developers can engage through SNMP using Python. LibreNMS, built around SNMP, provides Python developers access to the LibreNMS API. MRTG, a classic SNMP tool, might not involve Python directly, but Python can be used for auxiliary scripting. NetXMS, in C++, uses SNMP, and Python developers can use scripts or the NetXMS API. Grafana, in Go, supports various data sources. While it's not SNMP-centric, Python developers can use scripts or apps for integration. Prometheus, in Go, supports a query language. Python developers use exporters for compatibility. Icinga supports SNMP and is extensible, with Python developers contributing SNMP-related plugins. Checkmk, in Python, supports SNMP, allowing Python developers to extend capabilities. SolarWinds Network Performance Monitor, despite proprietary languages, supports SNMP. Python developers can use libraries like `pysolarwinds` for customization.

Using Python to Run an SNMPv3 Query

Utilizing Python for SNMPv3 queries involves leveraging a Python SNMP library. Start the process by defining SNMPv3 parameters, encompassing authentication and privacy details. Specify the target device and context and designate the Object Identifier (OID) for the desired information. Execute the SNMP request through the `getCmd` function, scrutinize it for errors, and showcase the results, encompassing the OID and its corresponding value. Ensure the presence of a Python library such as `pysnmp` by using `pip install pysnmp`. Write the script according to your SNMPv3 credentials and target device specifications, adjusting the OID for specific data retrieval. The next section delves into various the Python SNMP libraries at your disposal.

SNMP-Related Python Libraries

For seamless communication between Python and other network devices via SNMP, it is crucial to install a suitable SNMP library. In this context, the recommended choice is the `pysnmp` module, proficient at interacting with routers and switches using Python. However, an array of SNMP Python modules is available, enabling you to make a selection based on your unique requirements for SNMP integration. Refer to Table 5-6 for an insightful overview of various Python SNMP-related modules.

Table 5-6. Python SNMP Libraries

SNMP Package Name	Features and Download Link
pysnmp	Python-based SNMP module Program execution speed is slow https://pypi.org/project/pysnmp/
python3-netsnmp	Uses default Python bindings for net-snmp Supports non-Pythonic interface https://pypi.org/project/python3-netsnmp/
snimpy	Python tool developed for SNMP query Developed based on PySNMP Program execution speed is slow https://snimpy.readthedocs.io/en/latest/
easysnmp	Library developed based on net-snmp Python-style interface support and operation Program execution speed is fast Simplified interface for SNMP operations https://easysnmp.readthedocs.io/en/latest/ https://github.com/easysnmp/easysnmp
fastsnmpy	Uses an asynchronous wrapper based on net-snmp Developed for SNMPwalk https://github.com/ajdvk/fastsnmpy

Choose the appropriate SNMP library based on your specific needs for program execution speed, interface support, and other considerations. Also, for the most up-to-date information on the latest Python SNMP libraries, I recommend all readers check online repositories such as PyPL (Python Package Index) or GitHub.

Understanding MIBs and OIDs and Browsing OIDs Directly

The SNMP agent plays a vital role in sharing extensive information with the SNMP manager. This section discusses the relationship between MIBs and OIDs. As discussed earlier, OID serves as an identifier designating an administered object in the MIB. This implies that the OID is an integral part of the MIB, utilizing a unique ID to represent each MIB. An example of an OID is 1.3.6.1.2.1.1.3.0; this OID functions as an identifier containing information about the uptime after a system has booted up. These distinctive numerical sequences for each SNMP can be accessed through the MIB library, with equipment vendors often providing this information on their official websites. When devices are connected to the network, systems and network devices can effectively communicate their system status via SNMP, utilizing MIBs and unique OIDs. In the following steps, an interactive Linux session will be used to execute SNMPwalk on Cisco CML routers and switches in your GNS3 lab. Subsequently, you will locate functional SNMPv2c Python code from the Internet, refactor it to support SNMPv3, and probe device information within your lab topology.

Learning to Use SNMPwalk on a Linux Server

To transform your Linux server into an SNMP manager, start by installing SNMP software. Next, establish a password for authentication, tailored to the SNMP version your agents use. Typically, the SNMP manager is set up as an independent server on a Linux or Windows server, with a program supporting the SNMP API being installed and utilized. Prominent SNMP server programs include SolarWinds Network Performance Monitor, Paessler PRTG Network Monitor, Nagios, and Zabbix. Many of these programs offer free distribution versions for testing purposes. For more comprehensive details on these applications, consult each vendor's website and available resources.

In the following task, I guide you through the installation of SNMP server software on a Linux server, configure SNMPv3 on all network devices, and execute the OID of virtual network devices' MIB using the command line. Begin by installing the SNMP manager software on your Linux server, following the steps outlined in the subsequent section. If you intend to monitor from another SNMP server, you can opt to install SNMP agent (client) software. For this tutorial, the Linux server in use is a Fedora server, so ensure you are logged into the f38s1 server to seamlessly follow the tasks step by step. The lab topology remains consistent with previous setups, requiring a Fedora server and all routers and switches powered on.

Installing and Configuring SNMP Server on Fedora

To equip your Fedora server (f38s1, 192.168.127.10) with SNMP services and configure user authentication and authorization settings, follow these steps. Enter the commands highlighted in bold directly into your f38s1 server.

#	Task

①	First, install the net-snmp-utils and net-snmp-devel packages on the Fedora server using the commands shown here. net-snmp-utils is a tool used for SNMPwalk.
	
	`[jdoe@f38s1 ~]$ `**`sudo dnf install -y net-snmp net-snmp-utils net-snmp-devel -y`**
	`[sudo] password for jdoe: **********`
	`[...omitted for brevity]`
	`Complete!`

②	Follow these simplified steps to configure an SNMPv3 user and password on your Fedora Server (f38s1, 192.168.127.10):
	
	2-1. Enable, start, and check SNMPd:
	`[jdoe@f38s1 ~]$ `**`sudo systemctl enable snmpd`**
	`[jdoe@f38s1 ~]$ `**`sudo systemctl start snmpd`**
	`[jdoe@f38s1 ~]$ `**`sudo systemctl status snmpd`**
	`[...omitted for brevity]`
	
	2-2. Stop the snmpd service to create your user in Steps 2-3. Stopping the service is important here.
	
	`[jdoe@f38s1 ~]$ `**`sudo systemctl stop snmpd`**` # An important step!`
	
	2-3. Configure SNMPv3 User (SNMPUser1) with an Authentication pass-phrase (AUTHPass1) and Encryption Pass-phrase (PRIVPass1) using the interactive method. Remember the user ID and both pass-phrases for later steps.
	
	Interactive Method: Enter the pass-phrases interactively.
	`[jdoe@f38s1 ~]$ `**`sudo net-snmp-create-v3-user SNMPUser1`**
	`Enter authentication pass-phrase:`
	AUTHPass1

(continued)

#	Task

	Enter encryption pass-phrase:
	  [press return to reuse the authentication pass-phrase]
	PRIVPass1
	adding the following line to /var/lib/net-snmp/snmpd.conf:
	    createUser SNMPUser1 MD5 "AUTHPass1" AES "PRIVPass1"
	adding the following line to /etc/snmp/snmpd.conf:
	    rwuser SNMPUser1
	Note: Replace AUTHPass1 and PRIVPass1 with your desired authentication and privacy passwords.
	
	In Steps 2-3, you have the option to configure the SNMPv3 user and password using interactive, manual, or command-line methods. **If you've used the interactive method, there is no need to use the following methods**; this is provided for informational purposes only.
	
	Manual Configuration: Open the SNMP configuration files and add user information directly to the files.
	[jdoe@f38s1 ~]$ **sudo nano /var/lib/net-snmp/snmpd.conf**
	[jdoe@f38s1 ~]$ **sudo nano /etc/snmp/snmpd.conf**
	Command-line Method: Use the command to add a new user.
	[jdoe@f38s1 ~]$ **sudo net-snmp-config --create-snmpv3-user -A AUTHPass1 -X PRIVPass1 -a MD5 -x DES SNMPUser1**

③	Issue the following commands to restart the SNMP service, and subsequently, inspect the status of the snmpd service. Ensure that the status reflects as Active: active (running).
	
	[jdoe@f38s1 ~]$ **sudo systemctl start snmpd** # An important step!
	[jdoe@f38s1 ~]$ **sudo systemctl status snmpd** # An important step!
	● snmpd.service - Simple Network Management Protocol (SNMP) Daemon.
	   Loaded: loaded (/usr/lib/systemd/system/snmpd.service; enabled;
	   preset: disabled)
	   Drop-In: /usr/lib/systemd/system/service.d
	        └─10-timeout-abort.conf

(continued)

#	Task
| | ```
    Active: active (running) since Fri 2023-11-10 20:50:09 AEDT; 6s ago
   Main PID: 1411 (snmpd)
      Tasks: 1 (limit: 4632)
     Memory: 4.3M
        CPU: 58ms
     CGroup: /system.slice/snmpd.service
             └─1411 /usr/sbin/snmpd -LS0-6d -f
[...omitted for brevity]
``` |
| ④ | Use the snmpwalk command to confirm the proper functioning of SNMP on the local server. Execute the snmpwalk command on the Fedora server, using the SNMPUser1 created in Steps 2-3. If the output displays an extensive set of MIB and OID information about your Fedora server, consider the task is completed. |

```
[jdoe@f38s1 ~]$ snmpwalk -v3 -u SNMPUser1 -l authPriv -a MD5 -A
AUTHPass1 -X PRIVPass1 localhost
SNMPv2-MIB::sysDescr.0 = STRING: Linux f38s1.pynetauto.com 6.2.9-300.
fc38.x86_64 #1 SMP PREEMPT_DYNAMIC Thu Mar 30 22:32:58 UTC 2023 x86_64
SNMPv2-MIB::sysObjectID.0 = OID: NET-SNMP-MIB::netSnmpAgentOIDs.10
DISMAN-EVENT-MIB::sysUpTimeInstance = Timeticks: (5433) 0:00:54.33
SNMPv2-MIB::sysContact.0 = STRING: Root <root@localhost> (configure /
etc/snmp/snmp.local.conf)
SNMPv2-MIB::sysName.0 = STRING: f38s1.pynetauto.com
SNMPv2-MIB::sysLocation.0 = STRING: Unknown (edit /etc/snmp/snmpd.conf)
SNMPv2-MIB::sysORLastChange.0 = Timeticks: (0) 0:00:00.00
SNMPv2-MIB::sysORID.1 = OID: SNMP-FRAMEWORK-MIB::snmpFrameworkMIBComp
liance
SNMPv2-MIB::sysORID.2 = OID: SNMP-MPD-MIB::snmpMPDCompliance
SNMPv2-MIB::sysORID.3 = OID: SNMP-USER-BASED-SM-MIB::usmMIBCompliance
[... omitted for brevity]
SCTP-MIB::sctpRtoInitial.0 = Gauge32: 0 milliseconds
SCTP-MIB::sctpMaxAssocs.0 = INTEGER: 0
SCTP-MIB::sctpValCookieLife.0 = Gauge32: 0 milliseconds
SCTP-MIB::sctpMaxInitRetr.0 = Gauge32: 0
```
```

The snmpwalk command output displays SNMP version 2 MIBs (SNMPv2-MIB), as SNMPv3 supports SNMPv2 MIBs for compatibility. Responses may include MIBs from various SNMP versions. Notably, SNMPv3 user authentication and privacy settings remain effective for SNMPv3 operations, ensuring security, despite the coexistence of SNMPv2 MIBs. This compatibility allows for interoperability while maintaining the integrity of SNMPv3's secure features.

Great job! You've successfully installed the SNMP server, created a new SNMP user, and executed your initial SNMPwalk on the local server. In the next steps, you'll leverage this newly configured tool to monitor your Cisco routers and switches.

> **Tip**
>
> **Overview of the option handles in the Linux snmpwalk command**
>
> The following provides a concise overview of the option handles utilized in the Linux snmpwalk command:
>
> ```
> snmpwalk -v3 -u SNMPUser1 -l authPriv -a MD5 -A AUTHPass1 -X PRIVPass1 127.0.0.1
> ```
> -v3: SNMP version
> -u: SNMP username
> -l: Security level
> -a: Authentication method
> -A: Passphrase

Configuring Cisco Routers and Switches to Support SNMPv3 Using a Python Script

The following settings pertain to SNMP on Cisco devices. Transitioning from SNMPv2c to SNMPv3 requires a deeper understanding of SNMP authentication. Despite SNMPv3's enhanced security, SNMPv2c remains widely used due to legacy compatibility issues. Unless your network consists of newer devices with SNMPv3 support, SNMPv2c is the default choice. However, as the network fleets age and retire, enterprise systems and network monitoring will lean toward SNMPv3. With the emergence of private, public, and hybrid clouds, moving a significant part of legacy infrastructure into software, and abstracting the management from end-users, most virtualized systems and network

CHAPTER 5   PYTHON NETWORK AUTOMATION LABS: SNMP DISCOVERY WITH PYTHON

devices are compatible with SNMPv3. There is little reason to stick with SNMPv2c. Often, it is the ignorance of existing systems and network administrators about SNMPv3 that still keeps SNMPv2c as the de facto monitoring protocol for many organizations. In many cases, technological challenges associated with SNMPv3 have been overcome. For security reasons and as a best practice, this book focuses on SNMPv3. A Python script friendly to SNMPv3 will be developed to streamline SNMP configuration across all Cisco devices.

To follow along with this lab, you can use more than two devices, but for this demo, you will be utilizing all six CML network devices (two routers and four switches) and your SNMP Python server, f38s1. The first lab will use R2 and four switches; later you will configure the SNMP-server configuration on r3 yourself. You will also configure the Fedora server to be a secondary Python server and revisit the SSH communication to consolidate your knowledge. If you are not using other devices, power them off for this lab to save computer resources (see Figure 5-1).

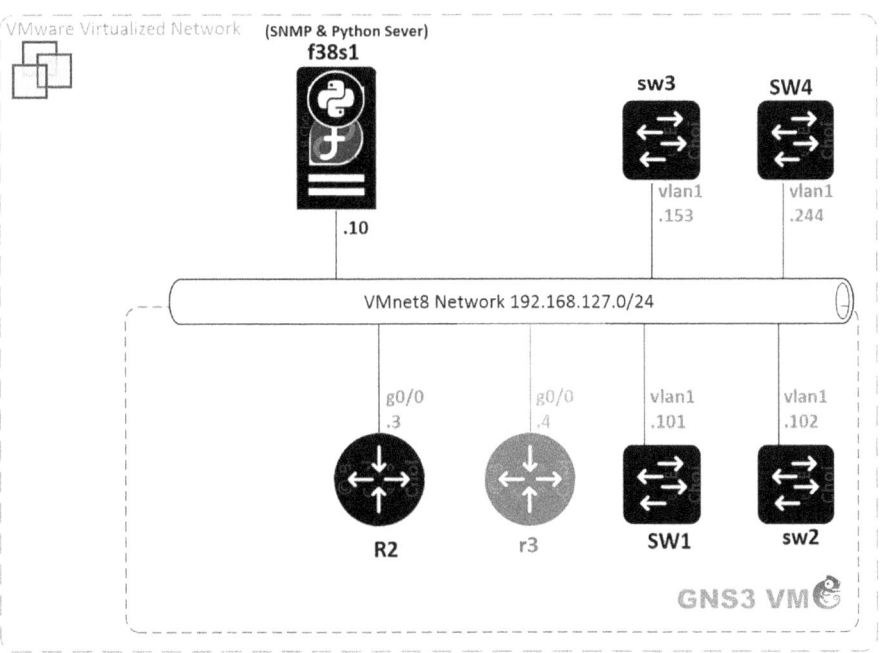

Figure 5-1. *SNMPv3 lab – devices in use*

Troubleshooting SSH Connections Between Fedora and Cisco Devices

Like the SSH connection challenges encountered in the paramiko and netmiko labs in Chapter 3, SSH-related issues are anticipated when attempting to connect to network devices from the Fedora server, f38s1. In Chapter 3, you exclusively relied on the Ubuntu Python Server to establish SSH connections with these devices and manage these endpoints. In this example, you'll delve into troubleshooting the actual problems as they manifest, addressing each issue methodically. Also, you will create testing tools to help you troubleshoot. Many students encounter setbacks with Python as a network tool due to such problems and the initial complexities associated with key exchange issues and Linux OS-related system settings. **This emphasizes the significance of attaining proficiency in the Linux OS, as merely possessing networking skills may not be sufficient to extend your horizon into the realm of network automation in general. A solid understanding of how Linux operates, the structure and location of specific directories and files, and adept system problem-solving are prerequisites**. Linux provides users with considerable freedom, but this can sometimes be challenging for user groups accustomed to Windows systems.

As you learned in an earlier chapter, there will be some SSH and key exchange challenges. Before moving on, you must troubleshoot and address these SSH communication issues between Fedora Linux and Cisco network devices. It's time to translate theory into practice.

CHAPTER 5   PYTHON NETWORK AUTOMATION LABS: SNMP DISCOVERY WITH PYTHON

#	Task

①	From your Fedora server, use the `ssh jdoe@IP_address` command to assess SSH login on each device. You may encounter SSH connection errors resembling the following. Analyzing the messages, I identify two potential issues. The `port 22: Connection refused` error might indicate an SSH problem linked to RSA key generation on the router/switch itself. On the other hand, `port 22: no matching key exchange method found` could indicate a key exchange problem related to key algorithms and host key algorithms on the Fedora server, especially because this server hasn't connected to any routers or switches via SSH before. You need to address both issues before progressing.
	
	`[jdoe@f38s1 ~]$ `**`mkdir ch5_snmp && cd ch5_snmp`**
	SSH login problem 1:
	`[jdoe@f38s1 ch5_snmp]$ `**`ssh jdoe@192.168.127.3`**
	`ssh: connect to host 192.168.127.3 port 22: Connection refused`
	`[jdoe@f38s1 ch5_snmp]$ `**`ssh jdoe@192.168.127.101`**
	`ssh: connect to host 192.168.127.101 port 22: Connection refused`
	
	SSH login problem 2:
	`[jdoe@f38s1 ch5_snmp]$ `**`ssh jdoe@192.168.127.102`**
	`Unable to negotiate with 192.168.127.102 port 22: no matching key exchange method found. Their offer: diffie-hellman-group-exchange-sha1,diffie-hellman-group14-sha1,diffie-hellman-group1-sha1`
	`[jdoe@f38s1 ch5_snmp]$ `**`ssh jdoe@192.168.127.153`**
	`Unable to negotiate with 192.168.127.153 port 22: no matching key exchange method found. Their offer: diffie-hellman-group-exchange-sha1,diffie-hellman-group14-sha1,diffie-hellman-group1-sha1`
	`[jdoe@f38s1 ch5_snmp]$ `**`ssh jdoe@192.168.127.244`**
	`Unable to negotiate with 192.168.127.244 port 22: no matching key exchange method found. Their offer: diffie-hellman-group-exchange-sha1,diffie-hellman-group14-sha1,diffie-hellman-group1-sha1`
	
	Note that your problem could be on another device, or you may only experience issues on R2 due to router replacement. Whatever the problem is, you must have a go at troubleshooting the issue, so you can continue with the rest of this chapter's activities.

(continued)

#	Task
②	If the `paramiko` and `netmiko` libraries are not installed, you need to install them before running your script. Execute the following instructions to install `paramiko` on the `f38s1` server. `[jdoe@f38s1 ch5_snmp]$ `**`pip freeze	grep paramiko`** `[jdoe@f38s1 ch5_snmp]$ `**`sudo dnf update -y`**` # Update the package manager ` ` (5-10 mins)` `[...omitted for brevity]` `[jdoe@f38s1 ch5_snmp]$ `**`sudo dnf install python3-pip -y`** `# install pip3 (1 min)` `[jdoe@f38s1 ch5_snmp]$ `**`pip3 --version`**` # check pip version` `pip 22.3.1 from /usr/lib/python3.11/site-packages/pip (python 3.11)` `[jdoe@f38s1 ch5_snmp]$ `**`sudo pip3 install paramiko`**` # install paramiko` `[...omitted for brevity]` `[jdoe@f38s1 ch5_snmp]$ `**`sudo pip3 install netmiko`**` # install netmiko` `[...omitted for brevity]`  `[jdoe@f38s1 ch5_snmp]$ `**`pip freeze	grep -E 'paramiko|netmiko'`** `# Check installed versions` `netmiko==4.2.0` `paramiko==3.3.1` `[jdoe@f38s1 ch5_snmp]$ `**`python`** `Python 3.11.6 (main, Oct 3 2023, 00:00:00) [GCC 13.2.1 20230728 (Red Hat 13.2.1-1)] on linux` `Type "help", "copyright", "credits" or "license" for more information.` `>>> `**`import paramiko`** `>>> `**`import netmiko`** `>>> `**`exit()`** `[jdoe@f38s1 ch5_snmp]$`

(continued)

CHAPTER 5   PYTHON NETWORK AUTOMATION LABS: SNMP DISCOVERY WITH PYTHON

#	Task

③	To help troubleshoot this problem, you'll create your own Python tool to test the SSH connection using the paramiko library.
	
	[jdoe@f38s1 ch5_snmp]$ **nano test_ssh1.py**
	[jdoe@f38s1 ch5_snmp]$ **cat test_ssh1.py**
	#Test SSH Connection to a single network device, only test the SSH connection via port 22.
	import paramiko
	from getpass import getpass
	
	def test_ssh_login(hostname, username, password):
	    try:
	        # Create an SSH client
	        ssh_client = paramiko.SSHClient()
	        # Automatically add the server's host key (this is insecure, see comments)
	        ssh_client.set_missing_host_key_policy(paramiko.AutoAddPolicy())
	        # Connect to the SSH server
	        ssh_client.connect(hostname, username=username, password=password, timeout=5)
	
	        print(f"SSH login to {hostname} successful.")
	    except paramiko.AuthenticationException:
	        print("Authentication failed. Please check your username and password.")
	    except paramiko.SSHException as e:
	        print(f"Error: Unable to establish SSH connection to {hostname}. Details: {str(e)}")
	    finally:
	        # Close the SSH connection
	        ssh_client.close()

(*continued*)

#	Task
| | ```
if __name__ == "__main__":
    # Get user input for SSH login details
    hostname = input("Enter the hostname or IP address: ")
    username = input("Enter the username: ")
    password = getpass("Enter the password: ")

    # Test SSH login
    test_ssh_login(hostname, username, password)
``` |
| ④ | To continue troubleshooting, execute the test_ssh1.py script to view the error messages.
jdoe@f38s1 ch5_snmp]$ **python test_ssh1.py**

Enter the hostname or IP address: **192.168.127.3**
Enter the username: **jdoe**
Enter the password:************
Traceback (most recent call last):
  File "/home/jdoe/ch5_snmp/test_ssh.py", line 29, in <module>
    test_ssh_login(hostname, username, password)
  File "/home/jdoe/ch5_snmp/test_ssh.py", line 11, in test_ssh_login
    ssh_client.connect(hostname, username=username,
    password=password, timeout=5)
  File "/usr/local/lib/python3.11/site-packages/paramiko/client.py",
  line 409, in connect
    raise NoValidConnectionsError(errors)
paramiko.ssh_exception.NoValidConnectionsError: [Errno None] Unable to connect to port 22 on 192.168.127.3 |

(*continued*)
```

#	Task

	The aforementioned issue is not Cisco device-specific; rather, it stems from the lack of configuration on your Fedora server with the necessary SSH settings for key exchanges. **As you learned in Chapter 3's `paramiko` labs, successful SSH key exchanges require proper configuration, a prerequisite for Python scripts to function seamlessly with libraries like `paramiko` or `netmiko`.** The same principle applies to the Fedora server; you must pre-configure these settings to enable SSH functionality with your network devices. Given that many network and security devices often run OSs that may not support the latest Python, the command executions take place on the server that has Python support. The client nodes do not support Python and in many cases are missing support for Python 3.6.5+. This aligns with the approach used by Ansible in its network modules, a topic you will explore in detail in one of the later Ansible labs.
	
	Now you can troubleshoot and resolve these SSH connection issues by adding the ~/.ssh/config file, drawing directly from the content covered in Chapter 3's `paramiko` SSH lab.
	
	`[jdoe@f38s1 ch5_snmp]$ `**`sudo ls ~/.ssh/config`**
	`[sudo] password for jdoe: *********`
	`ls: cannot access '/home/jdoe/.ssh/config': No such file or directory`
	`[jdoe@f38s1 ch5_snmp]$ `**`sudo nano ~/.ssh/config`**
	`[jdoe@f38s1 ch5_snmp]$ `**`sudo cat ~/.ssh/config`**
	`Host *`
	    `KexAlgorithms +diffie-hellman-group-exchange-sha1,diffie-hellman-group14-sha1,diffie-hellman-group1-sha1`
	    `HostkeyAlgorithms +ssh-rsa`
	
	This change should now fix the first issue faced by all network devices. Now you'll have a look at the second issue and troubleshoot it.

(continued)

CHAPTER 5 PYTHON NETWORK AUTOMATION LABS: SNMP DISCOVERY WITH PYTHON

#	Task

(5)	The issue at hand arises from an oversight: I neglected to generate the crypto key after replacing R2 in the preceding lab. Although you successfully replaced the device and reintegrated it into the network, you failed to regenerate the RSA key necessary for SSH connections. You can rectify this by re-creating the keys on the affected devices. Additionally, it is worth noting that you will be transitioning from SSH version 1.99 to version 2, and I'll provide an explanation of the differences between the two versions at the end of this section.
	
	`[jdoe@f38s1 ch5_snmp]$ `**`ssh jdoe@192.168.127.3`**
	`ssh: connect to host 192.168.127.3 port 22: Connection refused`
	
	`R2#`**`show ip ssh`**
	`SSH Disabled - version 1.99 # Read v1.99 vs v2.0 comparison at the end of this practical`
	`Authentication methods:publickey,keyboard-interactive,password`
	`Authentication Publickey Algorithms:x509v3-ssh-rsa,ssh-rsa`
	`Hostkey Algorithms:x509v3-ssh-rsa,ssh-rsa`
	`Encryption Algorithms:aes128-ctr,aes192-ctr,aes256-ctr`
	`MAC Algorithms:hmac-sha2-256,hmac-sha2-512,hmac-sha1,hmac-sha1-96`
	`KEX Algorithms:diffie-hellman-group-exchange-sha1,diffie-hellman-group14-sha1`
	`Authentication timeout: 120 secs; Authentication retries: 3`
	`Minimum expected Diffie Hellman key size : 2048 bits`
	`IOS Keys in SECSH format(ssh-rsa, base64 encoded):NONE`
	
	`R2(config)#`**`crypto key generate rsa`**
	`The name for the keys will be: R2.pynetauto.local`
	`Choose the size of the key modulus in the range of 360 to 4096 for your`
	`  General Purpose Keys. Choosing a key modulus greater than 512 may take`
	`  a few minutes.`
	
	`How many bits in the modulus [512]: `**`2048`**
	`% Generating 2048 bit RSA keys, keys will be non-exportable...`
	`[OK] (elapsed time was 10 seconds)`

(continued)

#	Task
| | ```
Nov 11 03:45:04.095: %SSH-5-ENABLED: SSH 1.99 has been enabled
R2(config)#**ip ssh version 2**

R2#**show ip ssh**
SSH Enabled - version 2.0 # Read v1.99 vs v2.0 comparison at the end
of this practical
Authentication methods:publickey,keyboard-interactive,password
Authentication Publickey Algorithms:x509v3-ssh-rsa,ssh-rsa
Hostkey Algorithms:x509v3-ssh-rsa,ssh-rsa
Encryption Algorithms:aes128-ctr,aes192-ctr,aes256-ctr
MAC Algorithms:hmac-sha2-256,hmac-sha2-512,hmac-sha1,hmac-sha1-96
KEX Algorithms:diffie-hellman-group-exchange-sha1,diffie-hellman-
group14-sha1
Authentication timeout: 120 secs; Authentication retries: 3
Minimum expected Diffie Hellman key size : 2048 bits
IOS Keys in SECSH format(ssh-rsa, base64 encoded): R2.pynetauto.local
ssh-rsa AAAAB3NzaC1yc2EAAAADAQABAAABAQC2xdfPWbyrl6chwwe5PFpCfzROIH+lF
+OyZMuvWW2E
/tXBSlFh/owjyHCVWO5/luM5QsIvIjvYi+At+KnhvMNz/1CzsM37KkBDsVoFiXHeq9xJB
OtNLm+OmOMj
HOGOBYFoC8cvErEAox+q1UryB7R72RstaUvPu+aKuV1GzCHR48OwhLXXSco1pWo+8hn7M
l76OQRY5bDa
rfPWVHAo1GnPI4OdGRo7UpY5p1sVhfPwkux/TOiQ/R/3iyrb9T48m6SalbhdONisVTm4o
XoO8NtzZsBm
2ZWFzN65Y9YWDic199Smv8QLOjhQf+iwCPG47OOnB7hSGsmHZFZ39H6pjVJV

Run the test_ssh1.py script again for your verification on R2.
[jdoe@f38s1 ch5_snmp]$ **python test_ssh1.py**
Enter the hostname or IP address: **192.168.127.3**
Enter the username: **jdoe**
Enter the password: *********
SSH login to 192.168.127.3 successful.
``` |

(continued)
```

#	Task

⑥	In the same way, fix the issue on SW2.
	
	[jdoe@f38s1 ch5_snmp]$ **ssh jdoe@192.168.127.101**
	ssh: connect to host 192.168.127.101 port 22: Connection refused
	
	SW1#**show ip ssh**
	SSH Disabled - version 1.99
	%Please create RSA keys to enable SSH (and of atleast 768 bits for SSH v2).
	Authentication methods:publickey,keyboard-interactive,password
	Encryption Algorithms:aes128-ctr,aes192-ctr,aes256-ctr,aes128-cbc,3des-cbc,aes192-cbc,aes256-cbc
	MAC Algorithms:hmac-sha1,hmac-sha1-96
	Authentication timeout: 120 secs; Authentication retries: 3
	Minimum expected Diffie Hellman key size : 1024 bits
	IOS Keys in SECSH format(ssh-rsa, base64 encoded): NONE
	
	SW1#**conf t**
	Enter configuration commands, one per line. End with CNTL/Z.
	SW1(config)#**crypto key generate rsa**
	The name for the keys will be: SW1.pynetauto.local
	Choose the size of the key modulus in the range of 360 to 4096 for your
	  General Purpose Keys. Choosing a key modulus greater than 512 may take
	  a few minutes.
	
	How many bits in the modulus [512]: **2048**
	% Generating 2048 bit RSA keys, keys will be non-exportable...
	[OK] (elapsed time was 4 seconds)
	
	*Nov 11 03:36:44.700: %SSH-5-ENABLED: SSH 1.99 has been enabled
	SW1(config)#**ip ssh version 2**
	SW1(config)#exit

(continued)

#	Task

SW1#**show ip ssh**
SSH Enabled - version 2.0
Authentication methods:publickey,keyboard-interactive,password
Encryption Algorithms:aes128-ctr,aes192-ctr,aes256-ctr,aes128-cbc,
3des-cbc,aes192-cbc,aes256-cbc
MAC Algorithms:hmac-sha1,hmac-sha1-96
Authentication timeout: 120 secs; Authentication retries: 3
Minimum expected Diffie Hellman key size : 1024 bits
IOS Keys in SECSH format(ssh-rsa, base64 encoded): SW1.pynetauto.
local
ssh-rsa AAAAB3NzaC1yc2EAAAADAQABAAABAQDFfu9BVkfX8724Rqo+N7DjA9XwRFJs7
q3nA+muJ9ti
p3dnerOLKf4TnveDAgwk1pCUDnSscS4qRWRxOaIxG6KwTDV2SLiCvzDA4Pby+wc4g7+p+
ZKzZXY/BdUe
zGyqo6e/
aCgHezTHTGZ6XhKmhNH8pSOVlu1lnmJPSftRKhsxxYUVDXWylZvnF7Xk5nXsCR
f9s9fgV18s
HAE3kngFh5J/ouPrf268gVHmxxo+VEUlQE6DyNXCmkYqle7dxWYVeRnIX7LFOkxw75SjX
Tw2L9mB9pwq
XuIki1L/XywD+zR9AUn2M/+cj8NOHa84rTO51jvlAwalPJkR21EsE4PgOET5

Use test_ssh2.py to validate the SSH login on SW1.
[jdoe@f38s1 ch5_snmp]$ python test_ssh2.py
Enter the hostname or IP address: **192.168.127.101**
Enter the username: **jdoe**
Enter the password: **********
SSH login to 192.168.127.101 successful.

(continued)

#	Task

⑦	To check the SSH connection on all devices, modify test_ssh1.py and create another file called test_ssh3.py. This new script will read all the IPs from the external file called ip_addresses.txt and test the SSH login at once. It will also run one of the simplest Cisco show commands, show clock. If you get the time from all your devices, that means your SSH connection is working between the Fedora server and the Cisco devices.
	
	[jdoe@f38s1 ch5_snmp]$ **nano test_ssh3.py**
	[jdoe@f38s1 ch5_snmp]$ **cat test_ssh3.py**
	``python
	# Reads IPs from the 'ip_addresses.txt' file and tests SSN connection for multiple devices.
	# Runs 'show clock' and output with the line separators.
	import paramiko
	from getpass import getpass
	
	def test_ssh_login(hostname, username, password):
	    try:
	        # Create an SSH client
	        ssh_client = paramiko.SSHClient()
	        # Automatically add the server's host key (this is insecure, see comments)
	        ssh_client.set_missing_host_key_policy(paramiko.AutoAddPolicy())
	        # Connect to the SSH server
	        ssh_client.connect(hostname, username=username, password=password, timeout=5)
	
	        print(f"SSH login to {hostname} successful.")
	
	        # Run 'show clock' command
	        stdin, stdout, stderr = ssh_client.exec_command("show clock")
	        output = stdout.read().decode("utf-8")
	``

(continued)

#	Task

```python
            # Print the time and hostname
            print(f"\nTime on {hostname}:\n{output}")
            print("-"*79)
    except paramiko.AuthenticationException:
        print("Authentication failed. Please check your username and
        password.")
    except paramiko.SSHException as e:
        print(f"Error: Unable to establish SSH connection to
        {hostname}. Details: {str(e)}")
    finally:
        # Close the SSH connection
        ssh_client.close()

if __name__ == "__main__":
    # Read IP addresses from the file
    file_path = 'ip_addresses.txt'
    try:
        with open(file_path, 'r') as file:
        ip_addresses = file.read().splitlines()
    except FileNotFoundError:
        print(f"Error: File '{file_path}' not found.")
    exit()

    # Get user input for SSH login details
    username = input("Enter the username: ")
    password = getpass("Enter the password: ")

    # Iterate over IP addresses and test SSH login
    for ip in ip_addresses:
        test_ssh_login(ip, username, password)
```

(*continued*)

CHAPTER 5 PYTHON NETWORK AUTOMATION LABS: SNMP DISCOVERY WITH PYTHON

#	Task
⑧	Confirm the IP addresses in your `ip_addresses.txt` file, execute the new SSH script, and confirm that the SSH login works for all five devices.
	``` [jdoe@f38s1 ch5_snmp]$ cat ip_addresses.txt 192.168.127.3 192.168.127.101 192.168.127.102 192.168.127.153 192.168.127.244 [jdoe@f38s1 ch5_snmp]$ python test_ssh3.py Enter the username: jdoe Enter the password: ******** SSH login to 192.168.127.3 successful. Time on 192.168.127.3: 14:27:42.490 AEST Sat Nov 11 2023 -- [...omitted for brevity] -- SSH login to 192.168.127.244 successful. Time on 192.168.127.244: *04:07:41.516 UTC Sat Nov 11 2023 -- ```

Well done! If you've reached this point, your Fedora server (acting as the SNMP server) is now equipped to control your Cisco network devices using Python scripts via an SSH connection.

Expand Your Knowledge

Transitioning safely: SSH 1.99 and SSH 2.0 coexistence strategies among key enterprise network equipment vendors

Leading enterprise network equipment vendors—such as Cisco, Juniper, Arista, Hewlett Packard Enterprise (HPE), Dell EMC, Extreme Networks, Brocade (now part of Broadcom), Huawei, Fortinet, and Palo Alto Networks—offer support for both SSH 1.99 and SSH 2.0. This provides flexibility for users with various security requirements and infrastructure configurations. In the case of Cisco, devices supporting SSH version 1.99 are often configured to maintain compatibility with legacy systems or networks still relying on SSH version 1.x, facilitating a smoother transition for organizations yet to fully migrate to SSH 2.0. While SSH 1.99 may be considered less secure due to vulnerabilities inherited from SSH version 1, its support in Cisco devices facilitates interoperability during transitional phases. In contrast, SSH 2.0 is the preferred and more secure option. This situation extends to other major enterprise network vendors that have not migrated all legacy platforms to only support SSH version 2.0. Therefore, the trend of supporting both SSH 1.99 and SSH 2.0 may persist for an extended period. Enterprise routers and switches supporting SSH 2.0 implement advanced security features, robust encryption algorithms, and improved key exchange mechanisms. SSH 2.0 is commonly used in modern network environments prioritizing security. Organizations are encouraged to configure Enterprise devices to use SSH 2.0 to benefit from these heightened security measures. In practice, administrators can configure the SSH version through the device's command-line interface (CLI) or graphical user interface (GUI), aligning with their organization's security policies and adapting to evolving standards. The coexistence of both SSH versions in Enterprise network devices underscores Cisco's commitment to providing versatile solutions for diverse networking requirements. Table 5-7 presents a quick comparison of the SSH protocol versions and their features.

Table 5-7. *Evolution of SSH Protocol Versions and Security Features*

Security Features	SSH 1.0	SSH 1.5	SSH 1.99	SSH 2.0
Version/Design	Initial version	Intermediate version	Transitional phase	A more secure and improved version
Compatibility	N/A	N/A	Supports SSH 1 and SSH 2. Intermediate step for compatibility	Not directly backward compatible with SSH 1.x
Key Exchange	Uses weaker key exchange	Improved from SSH 1.0	Relied on weaker protocols	Introduces improved key exchange protocols (e.g., Diffie-Hellman)
Encryption	Uses less secure algorithms	Improved from SSH 1.0	Relies on less secure algorithms	Uses stronger encryption techniques (e.g., AES)
Vulnerabilities	Had vulnerabilities	Addressed some vulnerabilities	Inherits vulnerabilities from SSH 1	Addresses security vulnerabilities found in SSH 1
Implementation	N/A	N/A	Supports SSH 1 and SSH 2	Many systems use SSH 2.0, with support for both protocols
Migration	N/A	Facilitates smoother transition	Facilitates smoother transition	Notable use in modern systems

Configuring SNMP Agents (engineID, group, and user) on Cisco Devices

Navigate to R2 to begin this lab. Each SNMP agent independently manages its system information, which becomes part of SNMPv3 messages. For SNMP agent devices like Cisco routers and switches, it is advisable to establish distinct SNMP engine IDs before SNMPv3 configuration. If an SNMP engine ID isn't explicitly defined during setup, a unique ID is generated using a combination of the enterprise number and the default MAC address. To properly configure SNMPv3 on Cisco devices, follow the recommended approach involving three `snmp-server` commands.

#	Task
①	Here you will configure R2 SNMP-server related configuration manually, so you understand how to configure the SNMPv3 configurations on Cisco devices. Later, you will write the Python script to do this task for you on all four Cisco switches. 1-1. The engine ID configuration command example looks like this: R2(config)# **snmp-server engineID local 1921681273** 1-2. The SNMP group name configuration command is shown next. GROUP1 is the SNMP group name in this case. This command will create an SNMPv3 group called GROUP1 with a private password. R2(config)# **snmp-server group GROUP1 v3 priv** 1-3. Using the same information as the manager-side SNMP user credentials, configure the SNMP user on a Cisco device using the following command: R2(config)# **snmp-server user SNMPUser1 GROUP1 v3 auth sha AUTHPass1 priv aes 128 PRIVPass1** *Nov 11 01:29:01.672: Configuring snmpv3 USM user, persisting snmpEngineBoots. Please Wait... You can use the `show snmp user` command to check the SNMP user information on Cisco devices. In the next step, based on the previous commands, you'll write a Python script to add this configuration to all devices at once. R2#**show snmp user** User name: SNMPUser1 Engine ID: 1921681273 storage-type: nonvolatile active Authentication Protocol: SHA Privacy Protocol: AES128 Group-name: GROUP1

(continued)

#	Task
②	You can enhance your network-testing capabilities by installing `fping` on your Fedora server to test multiple connections with a single command. However, to add an interesting twist to this lab, you'll utilize the previously created `ping` sweeper tool (from Chapter 4) as your pre-communication checking tool. Execute the following commands to duplicate the files and then rename the `ping_sweep1.py` script to `precheck_tool.py`. Creating this module is crucial, as you will later modify it for integration into your application, utilizing it as your custom module. ``` [jdoe@f38s1 ~]$ cd ch5_snmp [jdoe@f38s1 ch5_snmp]$ ls ../ch4_ping_sweeper cron.log ip_addresses.txt ping_sweep1.py [jdoe@f38s1 ch5_snmp]$ cp ../ch4_ping_sweeper/ping_sweep1.py ./precheck_tool.py [jdoe@f38s1 ch5_snmp]$ cp ../ch4_ping_sweeper/ip_addresses.txt ./ip_addresses.txt [jdoe@f38s1 ch5_snmp]$ nano ip_addresses.txt [jdoe@f38s1 ch5_snmp]$ cat ip_addresses.txt #192.168.127.3 # excluded as SNMP configuration is configured in step ① 192.168.127.101 192.168.127.102 **192.168.127.153** **192.168.127.244** [jdoe@f38s1 ch5_snmp]$ ls ip_addresses.txt precheck_tool.py ``` You should have two files under the `/home/jdoe/ch5_snmp` directory at the end of this task. The `ip_addresses.txt` file should contain all the IP addresses of CML routers and switches in your topology.

(continued)

#	Task

③	Open and modify precheck_tool.py using the nano text editor. Refactor the main part of the script into a function to make it callable from another Python script as a module. Make the following modifications to the file:
	
	[jdoe@f38s1 ch5_snmp]$ **nano precheck_tool.py**
	#!/usr/bin/python3
	import os
	
	def precheck_ping(ip):
	    response = os.system('ping -c 3 ' + ip)
	    if response == 0:
	        print(f"{ip} is reachable.")
	    else:
	        print(f"{ip} is either offline or ICMP is filtered. Continuing...")
	        # exit() # If you want the exit the application, remove '#'
	
	print("-" * 80)
	
	file_path = 'ip_addresses.txt' # Read IP addresses from the file
	
	try:
	    with open(file_path, 'r') as file:
	        ip_addresses = file.read().splitlines()
	except FileNotFoundError:
	    print(f"Error: File '{file_path}' not found.")
	    exit()
	
	for ip in ip_addresses: # Perform ICMP ping precheck for each IP address
	    precheck_ping(ip)

(continued)

#	Task

	When the `else` condition is met, for example, 192.168.127.133 (R1) is unreachable, it will continue to run the program until it reaches the last IP address.
	
	`[jdoe@f38s1 ch5_snmp]$ `**`python precheck_tool.py`**
	`[...omitted for brevity]`
	`-----------------------------------------------------------------`
	`PING 192.168.127.133 (192.168.127.133) 56(84) bytes of data.`
	`From 192.168.127.10 icmp_seq=1 Destination Host Unreachable`
	`From 192.168.127.10 icmp_seq=2 Destination Host Unreachable`
	`From 192.168.127.10 icmp_seq=3 Destination Host Unreachable`
	
	`--- 192.168.127.133 ping statistics ---`
	`3 packets transmitted, 0 received, +3 errors, 100% packet loss, time 2056ms`
	`pipe 3`
	`192.168.127.133 is either offline or ICMP is filtered. Continuing...`
	`-----------------------------------------------------------------`
	`[...omitted for brevity]`
	`PING 192.168.127.244 (192.168.127.244) 56(84) bytes of data.`
	`64 bytes from 192.168.127.244: icmp_seq=1 ttl=255 time=16.7 ms`
	`64 bytes from 192.168.127.244: icmp_seq=2 ttl=255 time=14.2 ms`
	`64 bytes from 192.168.127.244: icmp_seq=3 ttl=255 time=31.9 ms`
	
	`--- 192.168.127.244 ping statistics ---`
	`3 packets transmitted, 3 received, 0% packet loss, time 2003ms`
	`rtt min/avg/max/mdev = 14.154/20.926/31.947/7.860 ms`
	`192.168.127.244 is reachable.`
	`-----------------------------------------------------------------`

(continued)

#	Task
④	You'll now proceed to create a Cisco device SNMP agent configuration script, called `snmp_config.py`. This `netmiko` application will configure the same configuration shown in Step ① on four switches. Following best practices, you'll designate the SNMP engine ID using each device's IP address. The script will remove the dots from the IP addresses and repurpose the resulting whole number as the device's SNMP engine ID.

```
[jdoe@f38s1 ch5_snmp]$ nano snmp_config.py
[jdoe@f38s1 ch5_snmp]$ cat snmp_config.py
#!/usr/bin/python3
import time
from getpass import getpass
from netmiko import ConnectHandler
from precheck_tool import precheck_ping

def configure_snmp(device, eng_id):
    print(f"Now connecting to {device['ip']}")

    try:
        # Establish SSH connection using Netmiko
        with ConnectHandler(**device) as ssh_conn:
            print(f"Successful connection to {device['ip']}\n")

            # Add SNMP configurations
            print(f"Adding SNMP configuration to {device['ip']}")
            output = ssh_conn.send_config_set([
                "show clock",
                "configure terminal",
                "snmp-server engineID local " + eng_id,
                "snmp-server group GROUP1 v3 priv",
                "snmp-server user SNMPUser1 GROUP1 v3 auth sha
                AUTHPass1 priv aes 128 PRIVPass1",
                "do write",
                "exit"
            ])
```

(continued)

#	Task
| | ```python
            time.sleep(2)

            # Print output
            print()
            time.sleep(2)
            print(output)
            print(f"Successfully configured {device['ip']} &
            Disconnecting.")
            print("-" * 80)

    except Exception as e:
        print(f"Error: Unable to establish SSH connection to
        {device['ip']}. Details: {str(e)}")
        print("-" * 80)

# Read IP addresses from the file
file_path = '/home/jdoe/ch5_snmp/ip_addresses.txt'

try:
    with open(file_path, 'r') as file:
        ip_addresses = file.read().splitlines()
except FileNotFoundError:
    print(f"Error: File '{file_path}' not found.")
    exit()

# Perform ICMP ping precheck for each IP address
for ip in ip_addresses:
    precheck_ping(ip)

# Get login credentials
username = input("Enter username: ")
password = getpass("Enter password: ")
``` |

(continued)
```

CHAPTER 5   PYTHON NETWORK AUTOMATION LABS: SNMP DISCOVERY WITH PYTHON

#	Task
| | ```
# Define common device parameters for Netmiko
device_params = {
    'device_type': 'cisco_ios',
    'username': username,
    'password': password,
}

# Perform SNMP configuration for each IP address
for ip in ip_addresses:
    eng_id = ip.replace(".", "")  # Remove the dot in the IP address
                                  # and use it as SNMP Engine ID
    device_params['ip'] = ip
    configure_snmp(device_params, eng_id)
print("All tasks were completed successfully. Bye!")
``` |
| ⑤ | You should see the last line of code as a successful configuration script, which is outside of the for loop in the main script. Run snmp_config.py and configure SNMP group settings on all devices. The script will check the network connection using ICMP initially. If any device is unreachable, the application will exit. If all devices are on the network, the script will run and complete the SNMP configuration on all devices.

[jdoe@f38s1 ch5_snmp]$ **cat ip_addresses.txt**
192.168.127.101
192.168.127.102
192.168.127.153
192.168.127.244

[jdoe@f38s1 ch5_snmp]$ **python snmp_config.py**
PING 192.168.127.101 (192.168.127.101) 56(84) bytes of data.
64 bytes from 192.168.127.101: icmp_seq=1 ttl=255 time=12.0 ms
64 bytes from 192.168.127.101: icmp_seq=2 ttl=255 time=20.1 ms
64 bytes from 192.168.127.101: icmp_seq=3 ttl=255 time=8.64 ms |

(continued)
```

#	Task
| | ```
--- 192.168.127.101 ping statistics ---
3 packets transmitted, 3 received, 0% packet loss, time 2004ms
rtt min/avg/max/mdev = 8.641/13.561/20.059/4.793 ms
192.168.127.101 is reachable.
-------------------------------------------------------------------
[...omitted for brevity]
-------------------------------------------------------------------
Now logging into 192.168.127.244
Successful connection to 192.168.127.244

Now completing the following tasks:

*********************************************************************
* IOSv is strictly limited to use for evaluation, demonstration and
IOS  *
* education. IOSv is provided as-is and is not supported by Cisco's *
* Technical Advisory Center. Any use or disclosure, in whole or in
part, *
* of the IOSv Software or Documentation to any third party for any   *
* purposes is expressly prohibited except as otherwise authorized by *
* Cisco in writing.                                                  *
*********************************************************************
SW4#show clock
*11:11:19.276 UTC Sat Nov 11 2023
SW4#configure terminal
Enter configuration commands, one per line.  End with CNTL/Z.
SW4(config)#snmp-server engineID local 192168127244
SW4(config)#snmp-server group GROUP1 v3 priv
SW4(config)#$User1 GROUP1 v3 auth sha AUTHPass1 priv aes 128 PRIVPass1
SW4(config)#do write
Building configuration...

Successfully configured 192.168.127.244 & Disconnecting.
-------------------------------------------------------------------
All tasks were completed successfully. Bye!
``` |

(continued)
```

#	Task
⑥	At the end of the successful SNMP configuration run, log in to each device to confirm SNMP-related configurations. An example of the SW4 configuration is shown, but all other devices should have the SNMP user and group configurations. SW1#**show snmp user** User name: SNMPUser1 Engine ID: 192168127101 storage-type: nonvolatile active Authentication Protocol: SHA Privacy Protocol: AES128 Group-name: GROUP1 [...omitted for brevity] SW4#**show snmp user** User name: SNMPUser1 Engine ID: 192168183244 storage-type: nonvolatile active Authentication Protocol: SHA Privacy Protocol: AES128 Group-name: GROUP1 SW4#show run | in snmp snmp-server engineID local 192168183244 snmp-server group GROUP1 v3 priv

(*continued*)

#	Task
⑦	Now power on r3 and run the same script so it has the same SNMP-server configuration as the other devices. For R1, due to the age of this device, it does not support the same level of AES (Advanced Encryption Standard), so you do not have to configure SNMPv3 on R1. [jdoe@f38s1 ch5_snmp]$ **python test_ssh3.py** Enter the username: jdoe Enter the password: SSH login to 192.168.127.4 successful. Time on 192.168.127.4: *21:31:05.875 AEST Sat Nov 11 2023 \- [jdoe@f38s1 ch5_snmp]$ **cat ip_addresses.txt** 192.168.127.4 [jdoe@f38s1 ch5_snmp]$ python snmp_config.py PING 192.168.127.4 (192.168.127.4) 56(84) bytes of data. 64 bytes from 192.168.127.4: icmp_seq=1 ttl=255 time=30.8 ms 64 bytes from 192.168.127.4: icmp_seq=2 ttl=255 time=12.3 ms 64 bytes from 192.168.127.4: icmp_seq=3 ttl=255 time=28.4 ms --- 192.168.127.4 ping statistics --- 3 packets transmitted, 3 received, 0% packet loss, time 2004ms rtt min/avg/max/mdev = 12.294/23.837/30.849/8.224 ms 192.168.127.4 is reachable. \- Enter username: jdoe Enter password: Now logging into 192.168.127.4 Successful connection to 192.168.127.4 Now completing the following tasks:

(continued)

#	Task
| | ```
*******************************************************************
* IOSv is strictly limited to use for evaluation, demonstration and IOS *
* education. IOSv is provided as-is and is not supported by Cisco's *
* Technical Advisory Center. Any use or disclosure, in whole or in part, *
* of the IOSv Software or Documentation to any third party for any *
* purposes is expressly prohibited except as otherwise authorized by *
* Cisco in writing.                                                  *
*******************************************************************
R2#show clock
21:27:08.072 AEST Sat Nov 11 2023
R2#configure terminal
Enter configuration commands, one per line.  End with CNTL/Z.
R2(config)#snmp-server engineID local 1921681274
R2(config)#snmp-server group GROUP1 v3 priv
R2(config)#$User1 GROUP1 v3 auth sha AUTHPass1 priv aes 128 PRIVPass1
R2(config)#do write
Building configuration...

Successfully configured 192.168.127.4 & Disconnecting.
-------------------------------------------------------------------
All tasks were completed successfully. Bye!
``` |

(continued)
```

CHAPTER 5   PYTHON NETWORK AUTOMATION LABS: SNMP DISCOVERY WITH PYTHON

#	Task
(8)	When you log in and check r3 for the SNMP user information, you should see the Engine ID and all other authentication information required for secure SNMP communications. r3#**show snmp user** User name: SNMPUser1 Engine ID: 1921681274 storage-type: nonvolatile active Authentication Protocol: SHA Privacy Protocol: AES128 Group-name: GROUP1
(9)	Now quickly update the show clock command in test_ssh3.py to write memory and rename the copied version to save_all.py. Then run save_all.py to save the current running configuration on all devices. [jdoe@f38s1 ch5_snmp]$ **cp test_ssh3.py save_all.py** [jdoe@f38s1 ch5_snmp]$ **nano save_all.py** [jdoe@f38s1 ch5_snmp]$ **python save_all.py** Enter the username: **jdoe** Enter the password: ******** SSH login to 192.168.127.3 successful. Time on 192.168.127.3: Building configuration... [OK] --- *[...omitted for brevity]* --- SSH login to 192.168.127.244 successful. Time on 192.168.127.244: Building configuration... Compressed configuration from 3915 bytes to 1865 bytes[OK] ---

Your Cisco devices are now ready for SNMP communication with your SNMP server, f38s1.

Doing the SNMPwalk on the Linux Server

Having invested significant effort in the preparation leading up to this moment, you are now poised to execute SNMP commands from your designated SNMP server, f38s1. You'll initiate an SNMPwalk entering SNMPv3 commands directly on your SNMP server's console. This exercise holds substantial learning potential, so it is essential to approach it with a fresh perspective. Ensure that you perform these commands in your lab setting to maximize the insights gained. In one glance, an image speaks volumes beyond a thousand words.

#	Task
①	The Fedora server uses the following `snmpwalk` command to retrieve the OID information used by R2's vIOS MIB via SNMP. If you use this command, all OID information used by R2 is loaded, as shown here: ``` [jdoe@f38s1 ch5_snmp]$ snmpwalk -v3 -l authPriv -u SNMPUser1 -a SHA -A "AUTHPass1" -x AES -X "PRIVPass1" 192.168.127.3 SNMPv2-MIB::sysDescr.0 = STRING: Cisco IOS Software, IOSv Software (VIOS-ADVENTERPRISEK9-M), Version 15.7(3)M3, RELEASE SOFTWARE (fc2) Technical Support: http://www.cisco.com/techsupport Copyright (c) 1986-2018 by Cisco Systems, Inc. Compiled Wed 01-Aug-18 16:45 by prod_rel_team SNMPv2-MIB::sysObjectID.0 = OID: SNMPv2-SMI::enterprises.9.1.1041 DISMAN-EVENT-MIB::sysUpTimeInstance = Timeticks: (899951) 2:29:59.51 SNMPv2-MIB::sysContact.0 = STRING: SNMPv2-MIB::sysName.0 = STRING: R2.pynetauto.local SNMPv2-MIB::sysLocation.0 = STRING: SNMPv2-MIB::sysServices.0 = INTEGER: 78 SNMPv2-MIB::sysORLastChange.0 = Timeticks: (0) 0:00:00.00 SNMPv2-MIB::sysORID.1 = OID: SNMPv2-SMI::enterprises.9.7.129 ```

(*continued*)

CHAPTER 5   PYTHON NETWORK AUTOMATION LABS: SNMP DISCOVERY WITH PYTHON

#	Task
	[...omitted for brevity] SNMPv2-SMI::mib-2.197.1.3.2.1.2.39924 = Counter32: 150 SNMPv2-SMI::mib-2.197.1.3.2.1.3.39924 = Counter32: 0 SNMPv2-SMI::mib-2.207.1.2.1.0 = Counter64: 0 SNMPv2-SMI::mib-2.207.1.2.2.0 = Counter64: 0 SNMPv2-SMI::mib-2.207.1.2.3.0 = Counter64: 0 SNMPv2-SMI::mib-2.207.1.2.4.0 = Timeticks: (0) 0:00:00.00 [jdoe@f38s1 ch5_snmp]$
②	If you run the same command on other devices, you should also see all the MIB information for each device. The same device type will use the same MIBs. [jdoe@f38s1 ch5_snmp]$ **snmpwalk -v3 -l authPriv -u SNMPUser1 -a SHA -A "AUTHPass1" -x AES -X "PRIVPass1" 192.168.127.4** [jdoe@f38s1 ch5_snmp]$ **snmpwalk -v3 -l authPriv -u SNMPUser1 -a SHA -A "AUTHPass1" -x AES -X "PRIVPass1" 192.168.127.101** [jdoe@f38s1 ch5_snmp]$ **snmpwalk -v3 -l authPriv -u SNMPUser1 -a SHA -A "AUTHPass1" -x AES -X "PRIVPass1" 192.168.127.102** [jdoe@f38s1 ch5_snmp]$ **snmpwalk -v3 -l authPriv -u SNMPUser1 -a SHA -A "AUTHPass1" -x AES -X "PRIVPass1" 192.168.127.153** [jdoe@f38s1 ch5_snmp]$ **snmpwalk -v3 -l authPriv -u SNMPUser1 -a SHA -A "AUTHPass1" -x AES -X "PRIVPass1" 192.168.127.244**
③	Practice retrieving SW1's system information using the snmpwalk or snmpget command. 1.3.6.1.2.1.1.3.0 (sysUpTime.0) is an OID indicating the time the system was started. You can use the snmpget command to retrieve and check information from the SNMP manager to the SNMP agent. [jdoe@f38s1 ch5_snmp]$ **snmpwalk -v3 -l authPriv -u SNMPUser1 -a SHA -A "AUTHPass1" -x AES -X "PRIVPass1" 192.168.127.101 sysUpTime.0** DISMAN-EVENT-MIB::sysUpTimeInstance = Timeticks: (852361) 2:22:03.61 The SNMP agent SW1 tells you that the uptime since the system restart is 2 hours 22 minutes and 3 seconds.

(continued)

#	Task
④	Check a router or switch's name using 1.3.6.1.2.1.1.5.0 (sysName.0 or sysName). As you can see here, you can use the OID number or the unique names. They all return the same value—in this case, the fully qualified name of R2. `[jdoe@f38s1 ch5_snmp]$ `**`snmpwalk -v3  -l authPriv -u SNMPUser1 -a SHA -A "AUTHPass1"  -x AES -X "PRIVPass1" 192.168.127.3 sysName.0`** `SNMPv2-MIB::sysName.0 = STRING: R2.pynetauto.local` `[jdoe@f38s1 ch5_snmp]$ `**`snmpwalk -v3  -l authPriv -u SNMPUser1 -a SHA -A "AUTHPass1"  -x AES -X "PRIVPass1" 192.168.127.3 sysName`** `SNMPv2-MIB::sysName.0 = STRING: R2.pynetauto.local` `[jdoe@f38s1 ch5_snmp]$ `**`snmpwalk -v3  -l authPriv -u SNMPUser1 -a SHA -A "AUTHPass1"  -x AES -X "PRIVPass1" 192.168.127.3 1.3.6.1.2.1.1.5.0`** `SNMPv2-MIB::sysName.0 = STRING: R2.pynetauto.local`

⑤	If you use 1.3.6.1.2.1.2.2.1.7(ifAdminStatus.1), you can check the interface's setting status. This information confirms that the administrator has activated the interface.
	
	`[jdoe@f38s1 ch5_snmp]$ `**`snmpwalk -v3  -l authPriv -u SNMPUser1 -a SHA -A "AUTHPass1"  -x AES -X "PRIVPass1" 192.168.127.3 1.3.6.1.2.1.2.2.1.7`**
	`IF-MIB::ifAdminStatus.1 = INTEGER: up(1)`
	`IF-MIB::ifAdminStatus.2 = INTEGER: down(2)`
	`IF-MIB::ifAdminStatus.3 = INTEGER: down(2)`
	`IF-MIB::ifAdminStatus.4 = INTEGER: down(2)`
	`IF-MIB::ifAdminStatus.5 = INTEGER: up(1)`
	`IF-MIB::ifAdminStatus.6 = INTEGER: up(1)`
	`IF-MIB::ifAdminStatus.7 = INTEGER: up(1)`
	
	`[jdoe@f38s1 ch5_snmp]$ `**`snmpwalk -v3  -l authPriv -u SNMPUser1 -a SHA -A "AUTHPass1"  -x AES -X "PRIVPass1" 192.168.127.3 ifAdminStatus`**
	`IF-MIB::ifAdminStatus.1 = INTEGER: up(1)`
	`IF-MIB::ifAdminStatus.2 = INTEGER: down(2)`
	`IF-MIB::ifAdminStatus.3 = INTEGER: down(2)`
	`IF-MIB::ifAdminStatus.4 = INTEGER: down(2)`
	`IF-MIB::ifAdminStatus.5 = INTEGER: up(1)`
	`IF-MIB::ifAdminStatus.6 = INTEGER: up(1)`
	`IF-MIB::ifAdminStatus.7 = INTEGER: up(1)`

(continued)

#	Task

6	As you may have noticed previously, the OID of the first interface is the ID with `.1` appended to 1.3.6.1.2.1.2.2.1.7. If `.1` is added and executed, as shown next, the status of the GigabitEthernet0/0 port is displayed as up(1).
	
	`[jdoe@f38s1 ch5_snmp]$ snmpwalk -v3 -l authPriv -u SNMPUser1 -a SHA -A "AUTHPass1" -x AES -X "PRIVPass1" 192.168.127.3 1.3.6.1.2.1.2.2.1.7.1`
	`IF-MIB::ifAdminStatus.1 = INTEGER: up(1)`

7	Likewise, adding `.2` to the end means GigabitEthernet0/1 is down. This port is currently disabled, so it is displayed as down(2). The OID 1.3.6.1.2.1.2.2.1.7.X represents the administrative status (ifAdminStatus) of the second network interface, where the X, can be up (1), down (2), or testing (3).
	
	`[jdoe@f38s1 ch5_snmp]$ snmpwalk -v3 -l authPriv -u SNMPUser1 -a SHA -A "AUTHPass1" -x AES -X "PRIVPass1" 192.168.127.3 1.3.6.1.2.1.2.2.1.7.2`
	`IF-MIB::ifAdminStatus.3 = INTEGER: down(2)`

8	If you use OID 1.3.6.1.2.1.2.2.1.8 (ifOperStatus), you can query all the interfaces' operational statuses. Quickly run an snmpwalk command from your server for confirmation.
	
	`[jdoe@f38s1 ch5_snmp]$ snmpwalk -v3 -l authPriv -u SNMPUser1 -a SHA -A "AUTHPass1" -x AES -X "PRIVPass1" 192.168.127.3 ifOperStatus`
	`IF-MIB::ifOperStatus.1 = INTEGER: up(1)`
	`IF-MIB::ifOperStatus.2 = INTEGER: down(2)`
	`IF-MIB::ifOperStatus.3 = INTEGER: down(2)`
	`IF-MIB::ifOperStatus.4 = INTEGER: down(2)`
	`IF-MIB::ifOperStatus.5 = INTEGER: up(1)`
	`IF-MIB::ifOperStatus.6 = INTEGER: up(1)`
	`IF-MIB::ifOperStatus.7 = INTEGER: up(1)`

9	Use the `snmpget` command to query information about an interface, in this case, GigabitEthernet0/0. IfDescr.1 queries the interface's description, and ifOperStatus.1 queries the interface's operational status. You can replace the number 1 with 2 to query the next GigabitEthernet2 interface. Normally, snmpget is used to fetch a single data point from a device by querying a specific OID, while snmpwalk explores an entire branch of the MIB tree, gathering a set of values related to a starting OID.

(continued)

#	Task

	`[jdoe@f38s1 ch5_snmp]$ `**`snmpget -v3  -l authPriv -u SNMPUser1 -a SHA -A "AUTHPass1"  -x AES -X "PRIVPass1" 192.168.127.3 ifDescr.1  ifOperStatus.1`**
	`IF-MIB::ifDescr.1 = STRING: GigabitEthernet0/0`
	`IF-MIB::ifOperStatus.1 = INTEGER: up(1)`

⑩	The following `snmpget` command queries for the interface status of GigabitEthernet0/3; because it is not in use, it is marked as down(2).
	
	`[jdoe@f38s1 ch5_snmp]$ `**`snmpget -v3  -l authPriv -u SNMPUser1 -a SHA -A "AUTHPass1"  -x AES -X "PRIVPass1" 192.168.127.3 ifDescr.4  ifOperStatus.4`**
	`IF-MIB::ifDescr.4 = STRING: GigabitEthernet0/3`
	`IF-MIB::ifOperStatus.4 = INTEGER: down(2)`

⑪	Many OIDs can reveal each component and configuration's current status. Another fact worth noting is the rmon OID; it can reveal the software version number and the booting information. The example shows the router rmon OID.
	
	`[jdoe@f38s1 ch5_snmp]$ `**`snmpwalk -v3  -l authPriv -u SNMPUser1 -a SHA -A "AUTHPass1"  -x AES -X "PRIVPass1" 192.168.127.3 rmon`**
	`RMON-MIB::rmon.19.1.0 = Hex-STRING: FF C0 00 40`
	`RMON-MIB::rmon.19.2.0 = STRING: "15.7(3)M3"`
	`RMON-MIB::rmon.19.3.0 = ""`
	`RMON-MIB::rmon.19.4.0 = Hex-STRING: 07 E7 0B 0B 16 04 00 00`
	`RMON-MIB::rmon.19.5.0 = INTEGER: 1`
	`RMON-MIB::rmon.19.6.0 = STRING: "flash0:/vios-adventerprisek9-m"`
	`RMON-MIB::rmon.19.7.0 = IpAddress: 0.0.0.0`
	`RMON-MIB::rmon.19.8.0 = INTEGER: 1`
	`RMON-MIB::rmon.19.9.0 = INTEGER: 1`
	`RMON-MIB::rmon.19.12.0 = IpAddress: 0.0.0.0`
	`RMON-MIB::rmon.19.15.0 = Hex-STRING: 00`
	`RMON-MIB::rmon.19.16.0 = Hex-STRING: 7E 00`

The OID information used in vIOS is almost identical to real IOS equipment. There is a lot you can study from this lab and gain useful information from the previous examples. Once again, you must type the commands yourself on the keyboard and learn the previous concept (see Figure 5-2).

Tip

Accessing Cisco Feature Navigator for more MIB and OID information

For Cisco networking devices, you can find the required OID information using the Cisco Feature Navigator.

URL: https://cfnng.cisco.com/mibs

For your information, the following MIBs are among the most useful for Cisco devices:

- IF-MIB: Interface counter
- IP-MIB: Contains IP address
- IP-FORWARD-MIB: Contains a routing table
- ENTITY-MIB: Contains inventory information
- LLDP-MIB: Contains neighbor information

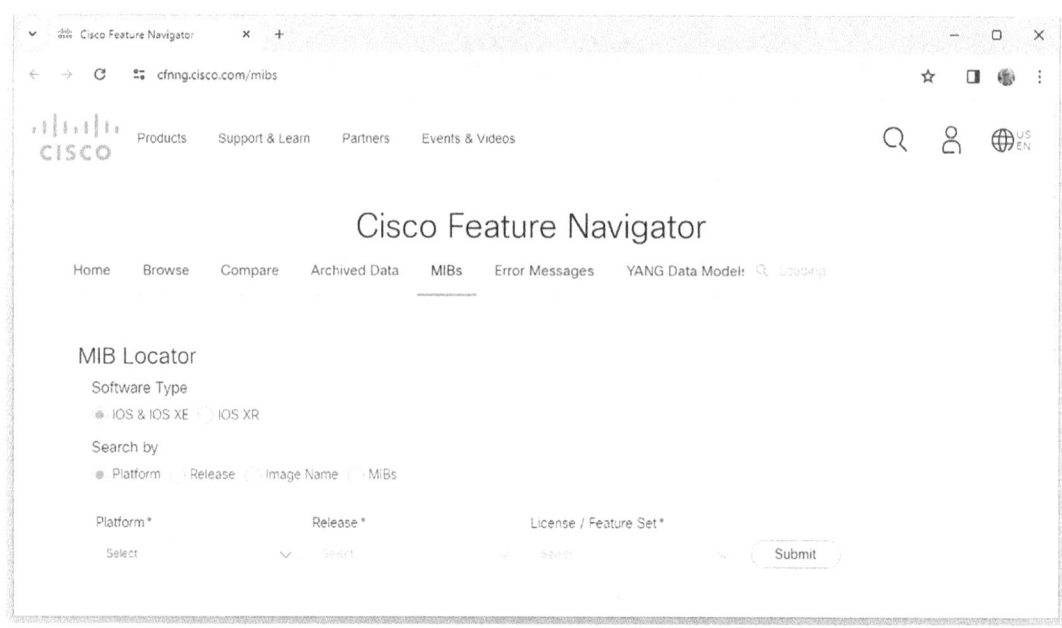

Figure 5-2. Cisco Feature Navigator, MIBs

Borrowing Some Example Python SNMP Code

Throughout this book, I have emphasized that mastering a programming language, be it Python, JavaScript, or Perl, does not automatically transform you into a network engineer capable of developing applications. Building on your solid foundational network knowledge, you must be diligent to expand your proficiency in supporting various technologies. This journey may extend over a considerable period, potentially progressing at a pace that some individuals might find frustratingly slow. Fortunately, a valuable strategy is to leverage existing code available on the Internet for your specific needs. Another strategy is to borrow the power of an AI such as ChatGPT to help you shape, structure, and write the code. However, you will have to work hard to make the code work for your production environment. **Why reinvent the wheel? If you encounter a particular problem, chances are someone else has already tackled it. The key is knowing how and where to find the pertinent information.** (Nevertheless, it is worth noting that certain knowledge is more effectively conveyed through alternative means.)

While exploring Python use cases with SNMP on the web, I discovered an insightful article, which I promptly bookmarked for future reference. When the time came to incorporate SNMP and Python into this book, the request was made to include practical examples. However, I preferred not to invest weeks in writing new code. Instead, I sought out an existing SNMP Python script to exemplify Python's application in SNMP. The SNMP Python code featured in this section is sourced from Alessandro Maggio's home page. I am grateful to Alessandro for granting permission to include his script in this book. For a comprehensive understanding of the code, I encourage you to visit Alex's blog.

Tip

Visit Alessandro Maggio's blog and IT tutorials

Find Alessandro Maggio's blog and code tutorial here:

URL: `https://www.ictshore.com/`

URL: `https://www.ictshore.com/sdn/python-snmp-tutorial/`

CHAPTER 5   PYTHON NETWORK AUTOMATION LABS: SNMP DISCOVERY WITH PYTHON

SNMP Lab Source Code

The provided Python code is designed for SNMPv2c, utilizing read-only community strings that simplify SNMP configuration for both the server and agent sides. Despite its simplicity, SNMPv2c is acknowledged to be less secure than SNMPv3. You will now see how to quickly review the code and make minor adjustments to transform it into SNMPv3 code. Additionally, based on a scenario, you'll create a Python integration script that collaborates seamlessly with this code. Begin by examining the original SNMP Python code, where explanations are provided using comments (#) or docstrings (""" """).

[jdoe@f38s1 ch5_snmp]$ **nano quicksnmp.py**

```
# This code utilizes the High-Level API module included in the pysnmp
library.
from pysnmp import hlapi
"""
The first function, get(), defines the prerequisite information to
initiate and make device information requests. Typically, you need to
provide the target IP (or DNS name), object ID (OID), credentials, SNMP
port number (161), EngineID, or SNMP context. This script forms a handler,
enabling the server to communicate with the agent to retrieve information
about itself.
"""

def get(target, oids, credentials, port=161, engine=hlapi.SnmpEngine(),
context=hlapi.ContextData()):
  handler = hlapi.getCmd(
    engine,
    credentials,
    hlapi.UdpTransportTarget((target, port)),
    context,
    *construct_object_types(oids)
  )
  return fetch(handler, 1)[0]
```

(continued)

```
# The second function, construct_object_types, needs a single argument,
list_of_oids. A blank list is used to collect and return object-type
information.
def construct_object_types(list_of_oids):
  object_types = []
  for oid in list_of_oids:
    object_types.append(hlapi.ObjectType(hlapi.ObjectIdentity(oid)))
  return object_types
```

```
# The third function, fetch, requires two arguments and uses the result
list to handle any errors in the script.
def fetch(handler, count):
  result = []
  for i in range(count):
    try:
      error_indication, error_status, error_index, var_binds =
      next(handler)
      if not error_indication and not error_status:
        items = {}
        for var_bind in var_binds:
          items[str(var_bind[0])] = cast(var_bind[1])
        result.append(items)
      else:
        raise RuntimeError('Got SNMP error: {0}'.format(error_indication))
    except StopIteration:
      break
  return result
```

(continued)

```
# The fourth function, cast(), uses the information received from PySNMP
to pass the values, changing the value into a int, float, or string type.
def cast(value):
  try:
    return int(value)
  except (ValueError, TypeError):
    try:
      return float(value)
    except (ValueError, TypeError):
      try:
        return str(value)
      except (ValueError, TypeError):
        pass
  return value
# This code is written for SNMPv2c and uses a community string named
'ICTSHORE'.
hlapi.CommunityData('ICTSHORE')
# If SNMPv3 is used, a user ID and two authentication keys should be used.
The script uses this information to communicate with the desired SNMP
agent to get device information.
hlapi.UsmUserData('testuser', authKey='authenticationkey',
privKey='encryptionkey',
authProtocol=hlapi.usmHMACSHAAuthProtocol, privProtocol=hlapi.
usmAesCfb128Protocol)
# This specifies the specific OID needed to obtain the specified
information from the SNMP agent. '1.3.6.1.2.1.1.5.0' denotes the dot IOD
for a hostname.
print(get('192.168.47.10', ['1.3.6.1.2.1.1.5.0'], hlapi.
CommunityData('ICTSHORE')))
```

Source code: https://www.ictshore.com/sdn/python-snmp-tutorial/

SNMPv3 Python Code to Query Cisco Device Information

This lab revisits the previous SNMPv2c Python code to support SNMPv3. **By studying someone else's code and testing the script in a lab environment, you can gain insight into how Python interacts with different applications and behaves.** This is also a perfect opportunity to observe and understand the real differences between SNMPv2c and SNMPv3 in action.

#	Task
①	For the server (f38s1) to communicate with network devices over SNMP, you will first install the server's pysnmp library. Complete the installation using the sudo pip3 install pysnmp command. [jdoe@f38s1 ch5_snmp]$ **sudo pip3 install pysnmp** [sudo] password for jdoe: ******** Collecting pysnmp Downloading pysnmp-4.4.12-py2.py3-none-any.whl (296 kB) [...omitted for brevity]
②	After creating the SNMP Python code called quicksnmp_v3.py, open the file with the nano editor and copy and paste all lines of code from quicksnmp.py into quicksnmp_v3.py. [jdoe@f38s1 ch5_snmp]$ **nano quicksnmp_v3.py**
③	The following modifications are necessary to transform quicksnmp.py into SNMP version 3 script. First, upon consulting the documentation on pysnmp.com, it becomes evident that it is essential to import and use UsmUserData from pysnmp.hlapi. For further details, refer to this link. 3-1. Add from pysnmp.hlapi import UsmUserData to the top of your code, as illustrated here: **from pysnmp import hlapi** # Importing the High-Level API module from the pysnmp library **from pysnmp.hlapi import UsmUserData** # Importing UsmUserData for SNMPv3 user configuration

(continued)

#	Task
| | ```python
def get(target, oids, credentials, port=161, engine=hlapi.
SnmpEngine(), context=hlapi.ContextData()):
    handler = hlapi.getCmd(
        engine,
        credentials,
        hlapi.UdpTransportTarget((target, port)),
        context,
        *construct_object_types(oids)  # Constructing object types
                                       for SNMP queries
    )
    return fetch(handler, 1)[0]

def construct_object_types(list_of_oids):
    # Creating object types for requested OIDs
    object_types = [hlapi.ObjectType(hlapi.ObjectIdentity(oid)) for
    oid in list_of_oids]
    return object_types

def fetch(handler, count):
    result = []
    for i in range(count):
        try:
            error_indication, error_status, error_index, var_binds =
            next(handler)
            if not error_indication and not error_status:
                # Extracting SNMP query results
                items = {str(var_bind[0]): cast(var_bind[1]) for
                var_bind in var_binds}
                result.append(items)
            else:
                raise RuntimeError('Got SNMP error: {0}'.
                format(error_indication))
        except StopIteration:
            break
    return result
``` |

(continued)
```

CHAPTER 5   PYTHON NETWORK AUTOMATION LABS: SNMP DISCOVERY WITH PYTHON

#	Task
	``python
def cast(value):
    try:
        # Converting value to an integer
        return int(value)
    except (ValueError, TypeError):
        try:
            # Converting value to a float
            return float(value)
        except (ValueError, TypeError):
            try:
                # Converting value to a string
                return str(value)
            except (ValueError, TypeError):
                pass
    return value
```

3-2. You need to input your user information, authentication key, and privacy key. As demonstrated in the earlier SNMP user configuration, the SNMPv3 user, SNMPUser1, was configured with SHA authentication keys and the AES128 privacy key. Therefore, add the following line exactly as shown:

**hlapi.UsmUserData('SNMPUser1', authKey='AUTHPass1', privKey='PRIVPass1', authProtocol=hlapi.usmHMACSHAAuthProtocol, privProtocol=hlapi.usmAesCfb128Protocol)**

3-3. Lastly, insert the `print` statement so that the `get()` function can convey the correct agent IP information and other credential variables, subsequently printing the information on the screen:

**print(get('192.168.127.3', ['1.3.6.1.2.1.1.5.0'], hlapi.UsmUserData('SNMPUser1', authKey='AUTHPass1', privKey='PRIVPass1', authProtocol=hlapi.usmHMACSHAAuthProtocol, privProtocol=hlapi.usmAesCfb128Protocol)))** |

(*continued*)
```

CHAPTER 5 PYTHON NETWORK AUTOMATION LABS: SNMP DISCOVERY WITH PYTHON

#	Task
	User name: SNMPUser1 Engine ID: 1921681273 storage-type: nonvolatile active Authentication Protocol: SHA Privacy Protocol: AES128 Group-name: GROUP1
4	After the SNMPv2 to SNMPv3 conversion, your file should look like this: [jdoe@f38s1 ch5_snmp]$ **nano quicksnmp_v3.py** [jdoe@f38s1 ch5_snmp]$ **cat quicksnmp_v3.py** **from pysnmp import hlapi** **def get(target, oids, credentials, port=161, engine=hlapi. SnmpEngine(), context=hlapi.ContextData()):** **handler = hlapi.getCmd(** **engine,** **credentials,** **hlapi.UdpTransportTarget((target, port)),** **context,** ***construct_object_types(oids)** **)** **return fetch(handler, 1)[0]** **def construct_object_types(list_of_oids):** **object_types = []** **for oid in list_of_oids:** **object_types.append(hlapi.ObjectType(hlapi. ObjectIdentity(oid)))** **return object_types** **def fetch(handler, count):** **result = []** **for i in range(count):**

(continued)

#	Task

```python
            try:
                error_indication, error_status, error_index, var_binds =
                next(handler)
                if not error_indication and not error_status:
                    items = {}
                    for var_bind in var_binds:
                        items[str(var_bind[0])] = cast(var_bind[1])
                    result.append(items)
            else:
                raise RuntimeError('Got SNMP error: {0}'.format(error_
                indication))
            except StopIteration:
            break
    return result
def cast(value):
    try:
        return int(value)
    except (ValueError, TypeError):
    try:
        return float(value)
    except (ValueError, TypeError):
    try:
        return str(value)
    except (ValueError, TypeError):
        pass
    return value

# Specify SNMP user data
usm_user_data = hlapi.UsmUserData(
    'SNMPUser1',
    authKey='AUTHPass1',
    privKey='PRIVPass1',
```

(continued)

#	Task
| | ```
        authProtocol=hlapi.usmHMACSHAAuthProtocol,
        privProtocol=hlapi.usmAesCfb128Protocol
)
print(get('192.168.127.3', ['1.3.6.1.2.1.1.5.0'], usm_user_data))
``` |
| (5) | You are currently running `pysnmp` version 4.4.12 and this version is not compatible with `pyasn1==0.5.0`. To make this work, you have to downgrade the `pyasn1` to a lower but working version. Use the `pip install pyasn1==0.4.8` command to manually downgrade `pyasn1`.<br><br>```
[jdoe@f38s1 ch5_snmp]$ pip freeze | grep pysnmp # Check installed pysnmp
pysnmp==4.4.12
[jdoe@f38s1 ch5_snmp]$ pip freeze | grep pyasn1 # check installed pyasn1
pyasn1==0.5.0 # Incompatible version
[jdoe@f38s1 ch5_snmp]$ pip install pyasn1==0.4.8 # downgrade to compatible version
Defaulting to user installation because normal site-packages is not writeable
Collecting pyasn1==0.4.8
Downloading pyasn1-0.4.8-py2.py3-none-any.whl (77 kB)
─────────── 77.1/77.1 kB 2.2 MB/s eta 0:00:00
Installing collected packages: pyasn1
Successfully installed pyasn1-0.4.8
[jdoe@f38s1 ch5_snmp]$ pip freeze | grep pyasn1
pyasn1==0.4.8 # Compatible version
```<br><br>You already wrote the dot OID '1.3.6.1.2.1.1.5.0' in the last line of code, so when you execute the SNMPv3 `quicksnmp_v3.py` script, you expect it to return the device name (FQDN) of 192.168.127.3.<br><br>```
[jdoe@f38s1 ch5_snmp]$ python quicksnmp_v3.py
{'1.3.6.1.2.1.1.5.0': 'R2.pynetauto.local'}
``` |

(continued)
```

CHAPTER 5   PYTHON NETWORK AUTOMATION LABS: SNMP DISCOVERY WITH PYTHON

#	Task

(6)	If you want to retrieve the operational status of GigabitEthernet0/0, you can add another dot OID line with an OID value of '1.3.6.1.2.1.2.2.1.7.1' in the last line. When you run the changed script, the interface state will be returned as a number: 1 for on and 2 for off. Add the following line to the end of quicksnmp_v3.py:
	
	``python
	# Specify SNMP user data
	[...omitted for brevity]
	usm_user_data = hlapi.UsmUserData(
	    'SNMPUser1',
	    authKey='AUTHPass1',
	    privKey='PRIVPass1',
	    authProtocol=hlapi.usmHMACSHAAuthProtocol,
	    privProtocol=hlapi.usmAesCfb128Protocol
	)
	
	print(get('192.168.127.3', ['1.3.6.1.2.1.1.5.0'], usm_user_data))
	print(get('192.168.127.3', ['1.3.6.1.2.1.2.2.1.7.1'], usm_user_data))
	``
	
	``
	[jdoe@f38s1 ch5_snmp]$ python quicksnmp_v3.py
	{'1.3.6.1.2.1.1.5.0': 'R2.pynetauto.local'}
	{'1.3.6.1.2.1.2.2.1.7.1': 1} # 1 indicates On, 2 denotes Off
	``

(continued)

#	Task
| ⑦ | When you increase the last digit and run the script for the GigabitEthernet 0/3 interface status, it returns 2. It is not connected to anything. 4 is the interface reference number, and 2 denotes the off status.<br><br>```
# Specify SNMP user data
[...omitted for brevity]
usm_user_data = hlapi.UsmUserData(
    'SNMPUser1',
    authKey='AUTHPass1',
    privKey='PRIVPass1',
    authProtocol=hlapi.usmHMACSHAAuthProtocol,
    privProtocol=hlapi.usmAesCfb128Protocol
)

print(get('192.168.127.3', ['1.3.6.1.2.1.1.5.0'], usm_user_data))
print(get('192.168.127.3', ['1.3.6.1.2.1.2.2.1.7.4'], usm_user_data))
```<br>`[jdoe@f38s1 ch5_snmp]$ python quicksnmp_v3.py`<br>`{'1.3.6.1.2.1.1.5.0': 'R2.pynetauto.local'}`<br>`{'1.3.6.1.2.1.2.2.1.7.4': 2} # 1 indicates On, 2 denotes Off` |

Using Python SNMP Code to Run a Query for an Interface Description

This lab is a continuation of the previous one, and you first must make a copy of the quicksnmp_v3.py script and create another script to complete it. The following script will let you run SNMP queries on interfaces and their descriptions using the SNMPv3 Python code. Because this is an introductory book that tries to introduce as much Python and network automation as possible, I will not go too deep into SNMP knowledge, leaving that task for the Cisco documentation. What is important here is to learn how Python is used to interact with Cisco devices using basic SNMP MIB and OID information.
```

#	Task
①	Using the Linux cp command, copy quicksnmp_v3.py and create another file named quicksnmp_v3_get_bulk_auto.py. [jdoe@f38s1 ch5_snmp]$ **cp quicksnmp_v3.py quicksnmp_v3_get_bulk_auto.py** [jdoe@f38s1 ch5_snmp]$ **nano quicksnmp_v3_get_bulk_auto.py**
| ② | At the end of the line, add the following code and run it. You are appending the next line of code to the bottom. Also, make sure you use the SW3 IP address of 192.168.127.153. This code runs the SNMP query on the interface's name and descriptions.<br><br>[jdoe@f38s1 ch5_snmp]$ **cat quicksnmp_v3_get_bulk_auto77.py**<br>```
from pysnmp import hlapi
from pysnmp.hlapi import UsmUserData

def get(target, oids, credentials, port=161, engine=hlapi.
SnmpEngine(), context=hlapi.ContextData()):
handler = hlapi.getCmd(
    engine,
    credentials,
    hlapi.UdpTransportTarget((target, port)),
    context,
    *construct_object_types(oids)
)
return fetch(handler, 1)[0]

def construct_object_types(list_of_oids):
    object_types = []
    for oid in list_of_oids:
        object_types.append(hlapi.ObjectType(hlapi.
        ObjectIdentity(oid)))
    return object_types
``` |

(continued)
```

#	Task
| | ```
def fetch(handler, count):
    result = []
    for i in range(count):
        try:
            error_indication, error_status, error_index, var_binds =
            next(handler)
            if not error_indication and not error_status:
                items = {}
                for var_bind in var_binds:
                    items[str(var_bind[0])] = cast(var_bind[1])
                result.append(items)
            else:
                raise RuntimeError('Got SNMP error: {0}'.format(error_
                indication))
        except StopIteration:
            break
    return result

def cast(value):
    try:
        return int(value)
    except (ValueError, TypeError):
        try:
            return float(value)
        except (ValueError, TypeError):
            try:
                return str(value)
            except (ValueError, TypeError):
                pass
    return value
``` |

(continued)
```

#	Task
| | ```
def get_bulk(target, oids, credentials, count, start_from=0, port=161,
engine=hlapi.SnmpEngine(), context=hlapi.ContextData()):
handler = hlapi.bulkCmd(
    engine,
    credentials,
    hlapi.UdpTransportTarget((target, port)),
    context,
    start_from, count,
    *construct_object_types(oids)
)
return fetch(handler, count)

def get_bulk_auto(target, oids, credentials, count_oid, start_from=0, port=161,
engine=hlapi.SnmpEngine(), context=hlapi.ContextData()):
count = get(target, [count_oid], credentials, port, engine, context)[count_oid]
return get_bulk(target, oids, credentials, count, start_from, port, engine, context)

its = get_bulk_auto('192.168.127.153', ['1.3.6.1.2.1.2.2.1.2 ',
'1.3.6.1.2.1.31.1.1.1.18'], hlapi.UsmUserData('SNMPUser1',
authKey='AUTHPass1', privKey='PRIVPass1', authProtocol=hlapi.
usmHMACSHAAuthProtocol, privProtocol=hlapi.usmAesCfb128Protocol),
'1.3.6.1.2.1.2.1.0')
for it in its:
for k, v in it.items():
print("{0}={1}".format(k, v))
print('')
``` |

(*continued*)
```

CHAPTER 5   PYTHON NETWORK AUTOMATION LABS: SNMP DISCOVERY WITH PYTHON

#	Task

③	Execute the bulk auto code. Finally, run the `quicksnmp_v3_get_bulk_auto.py` code, and you should get the interface names and their descriptions. Because you only configured the description for `Gi0/0` and `vlan1`, you will see the names of these interfaces.
	
	`[jdoe@f38s1 ch5_snmp]$` **`python quicksnmp_v3_get_bulk_auto77.py`**
	
	`1.3.6.1.2.1.2.2.1.2.1=GigabitEthernet0/0`
	`1.3.6.1.2.1.31.1.1.1.18.1="1GB Main Connection" !For later SNMP lab`
	
	`1.3.6.1.2.1.2.2.1.2.2=GigabitEthernet0/1`
	`1.3.6.1.2.1.31.1.1.1.18.2=`
	
	`1.3.6.1.2.1.2.2.1.2.3=GigabitEthernet0/2`
	`1.3.6.1.2.1.31.1.1.1.18.3=`
	
	`[...omitted for brevity]`
	
	`1.3.6.1.2.1.2.2.1.2.16=GigabitEthernet3/3`
	`1.3.6.1.2.1.31.1.1.1.18.16=`
	
	`1.3.6.1.2.1.2.2.1.2.17=Null0`
	`1.3.6.1.2.1.31.1.1.1.18.17=`
	
	`1.3.6.1.2.1.2.2.1.2.18=Vlan1`
	`1.3.6.1.2.1.31.1.1.1.18.18="Native interface" !For later SNMP lab`

There are hundreds of SNMP OID objects you can query and learn about. If you have some spare time, you can explore more about Cisco Device MIBs, especially the CPU and processes related to MIB information. The following Cisco link shows you how to query CPUs using SNMP:

URL: https://www.cisco.com/c/en/us/support/docs/ip/simple-network-management-protocol-snmp/15215-collect-cpu-util-snmp.html

Note that Cisco may change the URL address during their maintenance or even remove the article, so if this link does not work, search for "Cisco query CPUs using SNMP" to locate the new address.

CHAPTER 5   PYTHON NETWORK AUTOMATION LABS: SNMP DISCOVERY WITH PYTHON

Summary

In this chapter, you embarked on a focused exploration of SNMP, laying the groundwork for its integration into Python applications dedicated to network automation. Beginning with a concise guide to SNMP, I offered valuable insights into its history, types of system monitoring, and a comparative analysis alongside Syslog and APIs. The exploration extended to understanding SNMP standard operations, examining the differences between SNMP polling and the trap (event reporting) methods, and going over various SNMP versions. The chapter provided an in-depth examination of the SNMP protocol within the TCP/IP stack, covering SNMP message types, SMI, MIB, and OID. You delved into SNMP access policies, and synchronous vs. asynchronous communication in SNMP traps. You also learned about the practical use of Python for executing SNMPv3 queries. The practical application took center stage, as I guided you through the installation and configuration of an SNMP server on Fedora. This included configuring Cisco routers and switches to support SNMPv3 using Python scripts, troubleshooting SSH connections, and setting up SNMP agents on Cisco devices. The true essence of this chapter lies in a comprehensive summary, consolidating your understanding and providing a robust foundation for Python code development in SNMPv3 applications. The exploration of SNMP lab source code offered insights into how Python can effectively query Cisco device information and run queries for interface descriptions. You have acquired the skills necessary to seamlessly integrate Python into SNMPv3 application development, enhancing your capabilities in the realm of network automation.

In Chapter 6, you will have a taste of simple Python application development, which can be used for your next project or further development for your work. You will apply the knowledge acquired thus far to explore how Python technologies are effectively used in various contexts, including additional network automation scenarios. This hands-on exploration includes learning to use Python in `virtualenv` mode, a virtual environment for Python testing, as well as Ansible, pyATS, Docker, sendmail, and Twilio SMS in Python Network Automation application development. Anticipate engaging with Python application labs that promise excitement and practical learning experiences.

CHAPTER 6

Python Network Automation Virtual Lab 1: Ansible and pyATS in Python virtualenv

This chapter aims to familiarize you with Python virtual environment concepts and introduce alternative methods for exploring Python's external modules and functionalities. Until now, the focus of this book has been on utilizing Python's fundamental Telnet and SSH remote access techniques to examine, configure, and manage network devices directly on VMware virtual machine servers. In a production environment, dedicated Linux servers for testing or implementation aren't always available. Frequently, you will work within limited resources on a shared tenancy on a single Linux server, sharing resources with other engineers. It's crucial to navigate these limitations effectively while implementing solutions or applications in such production environments. By the end of this chapter, you will have gained the ability to operate within a Python virtual environment for testing purposes without impacting the existing Python setup on your host virtual machine. You'll take initial steps in understanding Ansible and pyATS (Genie) and mastering email alert delivery using Gmail, Yahoo Mail, Python applications, and Sendmail. Additionally, you will explore adapting to various open-source tools and writing custom Python programs as required.

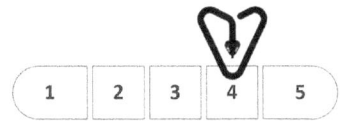

Python Virtual Lab 1: Ansible and pyATS in Python virtualenv

You are approaching the half point of this book, and in this chapter, you explore more engaging Python Network Automation labs. This chapter, followed by the next, introduce various tools, which spark creative thinking and reignite your journey in network automation. Extensive resources and documentation for Ansible and pyATS are widely available online. Consequently, mastering these tools solely through a single chapter in this book has its technical limitations. Nonetheless, installing and familiarizing yourself with these tools holds value, enabling you to evaluate their potential in fulfilling your network automation objectives. While not every network engineer aims to specialize in writing Python code for a living, readers of this book likely have a desire to explore the extensive possibilities of network automation, using various tools. In these instances, out-of-the-box innovative tools cater to prevalent network automation scenarios, reducing the necessity for extensive Python application development. In simpler terms, you often don't need to "reinvent the wheel" because someone else, perhaps a vendor, has already encountered similar challenges and taken extensive measures to create ready-made tools for your use. However, it's vital to recognize that no two client network environments are identical, and just as human fingerprints differ, the IP traffic of different organizations exhibits distinct priorities and characteristics.

If, having read this far, you still believe that network automation merely involves automating routine tasks and the thought process of a network engineer, there's a misconception. A significant aspect of the role involves managing data—specifically, extracting and leveraging that data to benefit your organization, substituting repetitive tasks with lines of code. Consequently, customization and code development are indispensable in every scenario. This was highlighted in the Regular Expression chapter in the Part I book. To elevate your skills, comprehending how data is manipulated, stored, and transmitted can set you apart from peers who have approached network automation with a one-sided perspective. In essence, this chapter serves as an introductory guide to Ansible and pyATS, with minimal use of programming language. Subsequently, the following chapter will provide a quick-start guide to running Python applications on Docker.

CHAPTER 6   PYTHON NETWORK AUTOMATION VIRTUAL LAB 1: ANSIBLE AND PYATS IN PYTHON VIRTUALENV

Navigating the Ansible and pyATS Networking Tools

Red Hat's Ansible has gained significant popularity in network configuration management over the past decade. Initially developed as a configuration management tool for Linux operating systems, it has evolved into a highly favored solution among network engineers, extensively utilized for network device configuration management and engineer task automation tools. The primary advantage of Ansible lies in its ability to avoid direct coding in languages such as Python, JavaScript, Perl, or PowerShell. Instead, it utilizes a pseudocode called YAML (YAML Ain't Markup Language). Writing YAML code offers a more user-friendly experience, sitting between conventional coding and GUI-based code editors. Essentially, Ansible operates as a simplified YAML application, relying on Python's `paramiko`, `netmiko`, and `nornir` libraries.

Ansible implements YAML files for endpoint automation, adopting an agentless push method rather than the pull method seen in its competitors such as Chef, Puppet, SaltStack, and others. **The agentless feature of Ansible holds immense appeal for most network engineers, primarily because the majority of network and security devices operate on their unique operating systems, often restricting the installation of agentful software due to system stability and security concerns.** While many network OSs are based on diverse *nux flavors, network vendors typically provide these devices with appliance versions, limiting the available commands. Therefore, not mandating the installation of an agent and solely managing these devices via SSH connections grants Ansible a definitive advantage over its competitors. This sets Ansible apart and endears it to numerous network engineers. Moreover, Ansible's open-source nature significantly contributes to its appeal. Although the Ansible Automation Platform (AAP) and Controller offer a more user-friendly interface and additional features, the core advantage lies in Ansible's open-source foundation. For many engineers, Ansible can serve as the gateway into the realm of network automation, preceding a deeper dive into Python or appealing to experienced Python programmers who appreciate its user-friendly interface and YAML workflow design. While Ansible is open-source, Red Hat's AAP is the commercial version, primarily derived from its upstream development, AWX.

This tool appears more as an application than a scripted programming language due to its YAML usage, catering to users with limited programming language knowledge, although familiarity with Python proves beneficial. Not every network engineer aims to master Python programming; some prefer a pre-built tool, **making Ansible an attractive** solution. However, while Ansible offers a lower entry barrier than Python, this accessibility could disappoint engineers seeking deeper programming language expertise, potentially leading to premature abandonment of their programming journey.

You will also delve into Cisco pyATS, previously known as Genie. Both pyATS and Nornir are endorsed and developed by the Cisco DevNet Team. In this section, you get a brief introduction to using pyATS. pyATS primarily serves as a network probing and information-gathering tool, distinct from a configuration tool. For handling configurations, Nornir offers an enhanced version of `netmiko` with additional user-friendly features. `netmiko`, an improved iteration of `paramiko`, specifically targets multivendor network automation. On a personal note, I hope that Cisco refrains from exclusively focusing on its proprietary tool development and instead embraces a more open approach toward users. Historically, Cisco hasn't been a software-centric company, and some of its applications may lack in comparison. pyATS, in my view, might be perceived as another unsuccessful attempt, representing misplaced efforts. **I have yet to encounter any network engineer who gets genuinely excited about the opportunity to use pyATS or even knows what pyATS is, except for those die-hard Cisco fanboys.** Unfortunately, I still retain some aspects of being a Cisco fanboy.

These tools are exceptional and merit exploration. You will examine each of these Python-friendly tools in their respective quick-start guides, followed by simple labs. Initially, you will set up Python's virtual environment to familiarize yourself with working in a Python virtual environment. Subsequently, you will be introduced to Ansible and pyATS through `virtualenv` labs.

Lab 1: Quick-Start Guide to Ansible in virtualenv

While developing your Python applications, you might face scenarios requiring testing of new Python modules without impacting a production Python server. As discussed in previous chapters, Python servers typically run on Linux operating systems. When operating seamlessly, Linux systems offer a delightful experience. However, introducing new software, especially different Python libraries, into a functional Python environment can lead to unexpected issues. These situations often result in extensive troubleshooting, involving incorrectly installed programs, software removal, and restoring the operating system to a stable state. `virtualenv` **and Pipenv stand as Python tools tailored to create isolated environments housing independent copies of Python, pip, and**

dedicated storage for testing libraries sourced from PyPI. They empower users to work on multiple projects with distinct dependencies on a single host machine. Think of `virtualenv` and `Pipenv` as quick sandbox environments for experimenting with Python libraries.

In this lab, you learn to set up a Python virtual environment on your Ubuntu Server, create a virtual environment specific to Ansible, and explore Ansible to collect information and efficiently configure Cisco devices in your lab. It's essential to note that Ansible isn't a programming language like Python, Perl, PowerShell, or JavaScript; rather, it's an automation tool utilizing YAML for managing IP device configurations. Although primarily written in Python, Ansible can support agentful device management as well as agentless network device management via SSH connections, relying on Python's SSH module, `paramiko`, as detailed in Chapter 3.

Many individuals contemplate whether they should prioritize learning Ansible or Python first. The suitable choice hinges upon their career aspirations in the network. Considering your progress in the book, it's likely that you already discerned your inclination. I don't expect you to ask me this question, but I started by mastering Python, and then gradually studied Ansible. Ansible from Red Hat (and to some extent, pyATS [Genie] from Cisco) offers robust capabilities without necessitating engineers to delve deeply into coding. However, this lower barrier might sometimes mislead Python learners into assuming that learning Python isn't necessary if you have chosen Ansible first—an erroneous assumption and a significant drawback of prioritizing Ansible over Python in the learning sequence. Ultimately, each person's career choices and educational paths are personal decisions. Drawing parallels from computer programming idioms, there are myriad ways to achieve a goal, and no definitive right or wrong methods—only what aligns with your career aspirations and where you are standing in life. Recommendations, including mine, serve merely as useful pointers. Define your career objectives and tailor your learning according to your current circumstances.

Before diving into the lab exercises, read the following "Expand Your Knowledge" to familiarize yourself with the two distinct Python virtual environment options: `virtualenv` and `Pipenv`. Yes, Python provides not just one, but two ways to create a Python virtual environment! While the primary focus here revolves around `virtualenv` labs, if your aspirations extend toward software development or Python Web Framework backend development, mastering `Pipenv` at your own pace would be a worthwhile investment of your time and resources.

Expand your knowledge

Python development essentials: Don't get confused between virtualenv and Pipenv

virtualenv and Pipenv play crucial roles in the Python ecosystem, each serving distinct yet complementary purposes. virtualenv is primarily designed for creating isolated Python environments, allowing the creation of separate environments for various projects. This segregation ensures that each project maintains a unique set of Python packages and dependencies, preventing conflicts between packages used in different projects. For instance, if there's a need to test a specific version of Python software or modules without affecting the host's Python configuration and module support, virtualenv is the ideal choice. On the other hand, Pipenv takes a more comprehensive approach, aiming to provide a complete package and environment management solution for Python projects. It combines the functionalities of virtualenv, pip (the package manager), and a Pipfile (to manage dependencies) into a unified tool. Pipenv is purposely designed to simplify and streamline the development and deployment processes of Python projects. For example, when developing a web frontend for a Python tool using the Django/Flask library, Pipenv is preferred over virtualenv, as it automatically takes care of various dependencies of required libraries, making it better suited for software development settings.

In terms of isolation, virtualenv focuses on creating isolated Python environments without directly managing project-specific dependencies. In contrast, Pipenv not only generates isolated environments but also effectively handles project-specific dependencies. It automatically updates a Pipfile (and a Pipfile.lock) to track the packages used in the project, facilitating the re-creation of the same environment on different systems. Regarding dependency management, virtualenv relies on the manual use of pip for installing, upgrading, or uninstalling packages within the virtual environment. Pipenv, however, incorporates a built-in dependency management system, utilizing a Pipfile and Pipfile.lock to specify and pin dependencies, contributing to maintaining consistent environments across various machines and deployments.

When working with `Pipenv`, your primary task typically revolves around software development within a project. As a result, you often find yourself relying heavily on feature-rich Python Integrated Development Environments (IDEs) such as PyCharm, Sublime Text, Atom, and Visual Studio Code. IDEs are software tools that streamline coding by offering features like code editing, debugging, and project management in a unified user interface. These robust IDEs offer comprehensive tools and features tailored to streamline the software development process within a project. In contrast, when using virtualenv, your primary focus is aligned toward testing specific features or assessing software compatibility. This scenario involves less emphasis on active software development. Consequently, you can opt for basic text editors, such as Notepad/Notepad++ on Windows, vi/nano/Joe on Linux, or TextEdit on macOS, installed on your preferred operating system. The less feature-intensive nature of these text editors aligns well with the testing and compatibility assessment tasks carried out in a virtualenv environment. This distinction allows you to choose the most suitable tools based on your specific objectives, ensuring efficiency and effectiveness in both software development and testing processes.

In terms of usage, `virtualenv` is primarily for creating virtual environments and can be used in conjunction with pip for package management. In contrast, `Pipenv` is designed as a comprehensive solution, serving as a one-stop-shop for both virtual environment creation and dependency management. It streamlines workflows by combining these aspects into a single, integrated tool. **The key distinction between** `virtualenv` **and** `Pipenv` **lies in their scope and purpose.** `virtualenv` **concentrates on creating isolated environments, while** `Pipenv` **provides a more inclusive solution by merging environment isolation with effective dependency management.** The choice between these tools depends on individual needs and preferences, allowing developers to select one or both for their Python development workflows.

The following content provides a quick-startup guide for installing the `virtualenv` modules on an Ubuntu 22.04 LTS Server environment. Follow these steps to set up your environment for the Ansible startup lab. If you are using Fedora OS (or any Red Hat distribution), the process changes slightly, but the fundamental features remain the same. Figure 6-1 displays the logical connection and devices used in this lab. Now, let's put your knowledge to work.

CHAPTER 6   PYTHON NETWORK AUTOMATION VIRTUAL LAB 1: ANSIBLE AND PYATS IN PYTHON VIRTUALENV

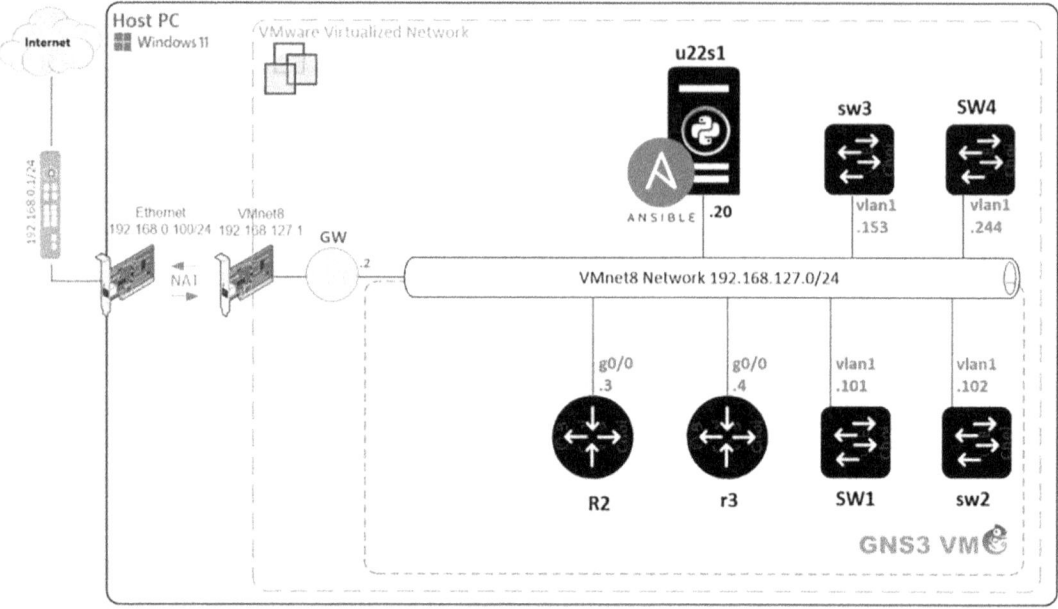

Figure 6-1. *Ansible virtualenv lab used devices*

Installing virtualenv and Getting Started with Ansible

You'll start by installing the virtualenv module on your Ubuntu 22 LTS Server and setting up the first lab in this chapter. Follow each step to complete the virtualenv installation.

#	Task
①	SSH into your Ubuntu Server and perform the following steps to get up and running for the next lab. Update and upgrade the Ubuntu packages to the latest version using Ubuntu's Advanced Packaging Tool (apt). `jdoe@u22s1:~$ `**`sudo apt update`** `jdoe@u22s1:~$ `**`sudo apt -y upgrade`**

(*continued*)

#	Task
| ② | To make your Python environment more useful, install some extra packages and development tools to ensure that the Linux server is set up for a programming environment. It is best to install these packages as a root user.<br><br>```
jdoe@u22s1:~$ su -
Password: ********
root@u22s1:~$ apt update
root@u22s1:~$ apt install -y
root@u22s1:~$ apt install -y build-essential python3-dev libssl-dev libffi-dev
root@u22s1:~$ apt install -y python3-pip # if you have not installed pip already
``` |
| ③ | You will be using the venv Python module to configure the python3 virtual environment, which is part of the standard Python 3 library.<br><br>```
root@u22s1:~$ apt install -y python3-venv
root@u22s1:~$ apt install sshpass # For ansible ssh login
root@u22s1:~$ su - jdoe
``` |
| ④ | Use mkdir to create a directory (environment) where your virtual environments will live. Use the cd command to move into the new directory.<br><br>```
jdoe@u22s1:~$ mkdir ch6_virtualenv && cd ch6_virtualenv
jdoe@u22s1:~/ch6_virtualenv $ mkdir venv1
jdoe@u22s1:~/ch6_virtualenv $ cd venv1
jdoe@u22s1:~/ch6_virtualenv/venv1$
``` |
| ⑤ | Once you are in the directory, create your testing environment. Use the ls Linux command to view items in your environment directory, in this case, ansible_venv.<br><br>```
jdoe@u22s1:~/ch6_virtualenv/venv1$ python -m venv ansible_venv
jdoe@u22s1:~/ch6_virtualenv/venv1$ ls ansible_venv
bin   include   lib   lib64   pyvenv.cfg
``` |

(*continued*)

CHAPTER 6   PYTHON NETWORK AUTOMATION VIRTUAL LAB 1: ANSIBLE AND PYATS IN PYTHON VIRTUALENV

#	Task
⑥	Activate and start your new virtual environment. ./ means the currently working directory or the directory you are working within. Once you have activated your virtual environment, you will see (ansible_venv) at the beginning of the CLI. jdoe@u22s1:~/ch6_virtualenv/venv1$ **source ./ansible_venv/bin/activate** (ansible_venv) jdoe@u22s1:~/ch6_virtualenv/venv1$
⑦	If you have (ansible_venv) in front of your Linux command line, it means you are in the virtual environment. Run Python -V or python3 -V to confirm the Python version. If you only have Python version 3, you can use either Python or Python3 to run your script in this environment. (ansible_venv) jdoe@u22s1:~/ch6_virtualenv/venv1$ **python -V** Python 3.10.12
⑧	To exit or deactivate the environment, use the deactivate command to exit your virtualenv. (ansible_venv) jdoe@u22s1:~/ch6_virtualenv/venv1$ **deactivate** jdoe@u22s1:~/ch6_virtualenv/venv1$
⑨	To reactivate the same environment, if you are already in the ansible_venv directory, use the same ./ansible_venv/bin/activate command. jdoe@u22s1:~/ch6_virtualenv/venv1$ **source ./ansible_venv/bin/activate** (ansible_venv) jdoe@u22s1:~/ch6_virtualenv/venv1$
⑩	To start your Ansible lab, install paramiko first and then Ansible using the pip3 install commands. Ansible uses paramiko for its SSH connection to network devices, so it is a prerequisite. (ansible_venv) jdoe@u22s1:~/ch6_virtualenv/venv1$ **pip3 install paramiko** (ansible_venv) jdoe@u22s1:~/ch6_virtualenv/venv1$ **pip3 install ansible** (ansible_venv) jdoe@u22s1:~/ch6_virtualenv/venv1$ **ansible --version** ansible [core 2.15.6] config file = None

(*continued*)

#	Task
	configured module search path = ['/home/jdoe/.ansible/plugins/modules', '/usr/share/ansible/plugins/modules'] ansible python module location = /home/jdoe/ch6_virtualenv/venv1/ansible_venv/lib/python3.10/site-packages/ansible ansible collection location = /home/jdoe/.ansible/collections:/usr/share/ansible/collections executable location = /home/jdoe/ch6_virtualenv/venv1/ansible_venv/bin/ansible python version = 3.10.12 (main, Jun 11 2023, 05:26:28) [GCC 11.4.0] (/home/jdoe/ch6_virtualenv/venv1/ansible_venv/bin/python3) jinja version = 3.1.2 libyaml = True
⑪	Generate the SSH key on the u22s1 server again after the Ansible installation. (ansible_venv) jdoe@u22s1:~/ch6_virtualenv/venv1$ **ssh-keygen** Generating public/private rsa key pair. Enter file in which to save the key (/home/jdoe/.ssh/id_rsa): /home/jdoe/.ssh/id_rsa already exists. Overwrite (y/n)? **y** Enter passphrase (empty for no passphrase): ******** Enter same passphrase again: ******** Your identification has been saved in /home/jdoe/.ssh/id_rsa Your public key has been saved in /home/jdoe/.ssh/id_rsa.pub The key fingerprint is: SHA256:x7vPXl3hHplbvLOmmVoJDyr8S5SZ95ptVV+7dcXYENo jdoe@u22s1.pynetauto.com

(*continued*)

#	Task
| | The key's randomart image is:<br>```
+---[RSA 3072]----+
|              .. |
|              o. |
|            . E* |
|         .+   o.X|
|        S=o+   =X|
|      . ..o.= o+@|
|       o o.  ==+*|
|       + o=* o.  |
|        oo**.o.  |
+----[SHA256]-----+
``` |
| ⑫ | Use ls ansible_venv/bin/ansible* to check the available commands from Ansible.

(ansible_venv) jdoe@u22s1:~/ch6_virtualenv/venv1$ **ls ansible_venv/bin/ansible***
ansible_venv/bin/ansible
ansible_venv/bin/ansible-config
ansible_venv/bin/ansible-console
ansible_venv/bin/ansible-galaxy
ansible_venv/bin/ansible-playbook
ansible_venv/bin/ansible-test
ansible_venv/bin/ansible-community
ansible_venv/bin/ansible-connection
ansible_venv/bin/ansible-doc
ansible_venv/bin/ansible-inventory
ansible_venv/bin/ansible-pull
ansible_venv/bin/ansible-vault |

(continued)
```

#	Task
⑬	Next, use the sudo /home/jdoe/.ssh/config command to open the SSH configuration file and check/add the security key exchange supported by Cisco devices. After the check, confirm that your host's ~/.ssh/config file is for the correct security key exchange, as shown here. (ansible_venv) jdoe@u22s1:~/ch6_virtualenv/venv1$ **nano /home/jdoe/.ssh/config** Host 192.168.127.133 KexAlgorithms diffie-hellman-group1-sha1 HostkeyAlgorithms ssh-rsa Ciphers aes128-cbc,3des-cbc,aes192-cbc,aes256-cbc Host * KexAlgorithms +diffie-hellman-group-exchange-sha1,diffie-hellman-group14-sha1,diffie-hellman-group1-sha1 HostkeyAlgorithms +ssh-rsa
⑭	You will run into an issue with the next step, so you can prevent the login issue by removing the known_hosts entries first. (ansible_venv) jdoe@u22s1:~/ch6_virtualenv/venv1$ **sudo nano /home/jdoe/.ssh/known_hosts** # Remove all entries so, the 'known_hosts' file is empty.
⑮	Now, try to SSH into each device using the ssh username@host_ip_address commands. Make sure you get the correct prompt and can successfully log in using your network device administrator credentials. Note that R1 (the old IOS router) uses an older AES-type key exchange method and is not compatible with the modern Ubuntu Server, so it was excluded from this lab. For CML or newer Cisco devices, you can also force the server to use diffie-hellman-group1-sha1 for the key exchange. Complete the following manual task before test driving Ansible.

(continued)

#	Task
	It is important that you SSH into each router and switch. On the first prompt, accept the new RSA key fingerprints to allow the SSH connection into each device from your Python server.

```
(ansible_venv) jdoe@u22s1:~/ch6_virtualenv/venv1$ ssh
jdoe@192.168.127.3 # SSH command
The authenticity of host '192.168.127.3 (192.168.127.3)' can't be
established.
RSA key fingerprint is SHA256:LMfxTWTD4Etwmit2kF7Fji/8ORrOdLBGRy72zMk
PXV0.
This key is not known by any other names
Are you sure you want to continue connecting (yes/no/[fingerprint])?
Yes # Accept the fingerprint
Warning: Permanently added '192.168.127.3' (RSA) to the list of known
hosts.

**************************************************************
* IOSv is strictly limited to use for evaluation, demonstration and
IOS  *
* education. IOSv is provided as-is and is not supported by Cisco's *
* Technical Advisory Center. Any use or disclosure, in whole or in
part, *
* of the IOSv Software or Documentation to any third party for any *
* purposes is expressly prohibited except as otherwise authorized by *
* Cisco in writing.                                                 *
(jdoe@192.168.127.3) Password: **************** # Enter your
password

**************************************************************
* IOSv is strictly limited to use for evaluation, demonstration and IOS *
* education. IOSv is provided as-is and is not supported by Cisco's *
* Technical Advisory Center. Any use or disclosure, in whole or in
part, *
* of the IOSv Software or Documentation to any third party for any *
```
|

(continued)

#	Task
| | ```
 * purposes is expressly prohibited except as otherwise authorized by *
 * Cisco in writing.                                                  *
 **********************************************************************
R2#exit # logout
Connection to 192.168.127.3 closed.
(ansible_venv) jdoe@u22s1:~/ch6_virtualenv/venv1$ ssh
jdoe@192.168.127.4 # Log into the next device
[...omitted for brevity]
r3#exit
Connection to 192.168.127.4 closed.
(ansible_venv) jdoe@u22s1:~/ch6_virtualenv/venv1$ ssh
jdoe@192.168.127.101 # Repeat the process
[...omitted for brevity]
SW1#exit
Connection to 192.168.127.101 closed.
(ansible_venv) jdoe@u22s1:~/ch6_virtualenv/venv1$ ssh
jdoe@192.168.127.102 # Repeat the process
``` |
| | ```
[...omitted for brevity]
sw2#exit
Connection to 192.168.127.102 closed.
(ansible_venv) jdoe@u22s1:~/ch6_virtualenv/venv1$ ssh
jdoe@192.168.127.153 # Repeat the process
[...omitted for brevity]
sw3#exit
Connection to 192.168.127.153 closed.
(ansible_venv) jdoe@u22s1:~/ch6_virtualenv/venv1$ ssh
jdoe@192.168.127.244 # Repeat the process
The authenticity of host '192.168.127.244 (192.168.127.244)' can't be
established.
[...omitted for brevity]
SW4#exit
Connection to 192.168.127.244 closed.
``` |

(*continued*)

#	Task
⑯	Do not ask what the command does; simply run the command shown here. The following command is an Ansible ad hoc command to SSH into R2 (192.168.127.3) and check Cisco IOS facts. I will explain this command in full later, so, just run the command for now.

```
(ansible_venv) jdoe@u22s1:~/ch6_virtualenv/venv1$ ansible all -i
192.168.127.3, -c network_cli -u jdoe -k -m ios_facts -e ansible_
network_os=ios
SSH password: ********
[WARNING]: ansible-pylibssh not installed, falling back to paramiko
192.168.127.3 | FAILED! => {
  "changed": false,
  "msg": "Failed to authenticate: Authentication failed: transport
  shut down or saw EOF"
}
```

As you can see, it spits a dummy and fails to run the command correctly. You need to install the Python module using the `pip` command. Type the following `pip3` command and install the required module.

```
(ansible_venv) jdoe@u22s1:~/ch6_virtualenv/venv1$ pip3 install
ansible-pylibssh
Collecting ansible-pylibssh
Downloading ansible_pylibssh-1.1.0-cp310-cp310-manylinux_2_24_x86_64.
whl (2.3 MB)
    ----------------------------------------------------------------
    2.3/2.3 MB 6.1 MB/s eta 0:00:00
Installing collected packages: ansible-pylibssh
Successfully installed ansible-pylibssh-1.1.0
``` |

(continued)

CHAPTER 6   PYTHON NETWORK AUTOMATION VIRTUAL LAB 1: ANSIBLE AND PYATS IN PYTHON VIRTUALENV

#	Task
⑰	After installing the required Python module for Ansible, run the same command for the second time. `(ansible_venv) jdoe@u22s1:~/ch6_virtualenv/venv1$` **`ansible all -i 192.168.127.3, -c network_cli -u jdoe -k -m ios_facts -e ansible_network_os=ios`** `SSH password:` `192.168.127.3	SUCCESS => {` ` "ansible_facts": {` ` "ansible_net_api": "cliconf",` ` "ansible_net_gather_network_resources": [],` ` "ansible_net_gather_subset": [` ` "default"` `],` ` "ansible_net_hostname": "R2",` ` "ansible_net_image": "flash0:/vios-adventerprisek9-m",` ` "ansible_net_iostype": "IOS",` ` "ansible_net_model": "IOSv",` ` "ansible_net_operatingmode": "autonomous",` ` "ansible_net_python_version": "3.10.12",` ` "ansible_net_serialnum": "91N1192IYKJFPLDPNMJ8Q",` ` "ansible_net_system": "ios",` ` "ansible_net_version": "15.7(3)M3",` ` "ansible_network_resources": {}` ` },` ` "changed": false` `}`

(continued)

#	Task
	Just like that, one simple command using Ansible will fetch you valuable IOS device information. Here is the breakdown of the ad hoc command used. At this stage, you do not need to understand this command but read the explanation for future reference only. Just follow along to get familiar with Ansible. Emulate someone, and you will attain mastery; therefore, follow the tasks closely to catch the rhythm and gain familiarity. This is how babies learn to speak from their parents. `ansible all -i 192.168.127.3, -c network_cli -u jdoe -k -m ios_facts -e ansible_network_os=ios` This Ansible ad hoc command targets all hosts in the specified host (inventory), connecting to the Cisco device at 192.168.127.3 using the `network_cli` connection plugin. It uses the remote user `jdoe` with a password prompt (`-k`). The command executes the `ios_facts` Ansible module to gather information about the Cisco device, like the hostname, serial number, and OS version (`ansible_net_version`). Additionally, the network operating system is defined as `ios` through the `ansible_network_os` variable. Ensure accurate credentials, SSH access, and host reachability for successful execution. The iostype, `ios`, works for both the older Cisco IOS and newer Cisco IOS-XE devices.
18	You can visit Ansible's official site for helpful information and technical documentation. Ansible is open-source software shared by Red Hat, and Red Hat provides very detailed documentation on how to get started with Ansible, along with easily understandable network module documents. URL1: https://docs.ansible.com/ansible/latest/network/getting_started/first_playbook.html URL2: https://docs.ansible.com/ansible/latest/modules/ios_facts_module.html URL3: https://docs.ansible.com/ansible/2.9/modules/list_of_network_modules.html

(continued)

#	Task

	Now you'll explore the first playbook and quickly test what Ansible can do for you. Once you're on the website, navigate to first_playbook.yml, and within your virtualenv, create a new YAML file with the same name. Modify it by replacing Vyatta with Cisco, and the first playbook will look similar to this. This playbook will fetch the hostname and OS of your device.
	
	(ansible_venv) jdoe@u22s1:~/ch6_virtualenv/venv1$ **pwd**
	/home/jdoe/ch6_virtualenv/venv1
	(ansible_venv) jdoe@u22s1:~/ch6_virtualenv/venv1$ **nano first_playbook.yml**
	(ansible_venv) jdoe@u22s1:~/ch6_virtualenv/venv1$ **cat first_playbook.yml**
	--- # indicates that this is a YAML file
	- name: Network Getting Started First Playbook
	connection: network_cli
	gather_facts: false
	hosts: all
	tasks:
	  - name: Get config for Cisco devices
	  ios_facts:
	    gather_subset: all
	  - name: Display the config
	  debug:
	  msg: "The hostname is {{ ansible_net_hostname }} and the OS is {{ ansible_net_version }}"
	
	Source: https://docs.ansible.com/ansible/latest/network/getting_started/first_playbook.html
	Direct: https://docs.ansible.com/ansible/latest/_downloads/588d4b6e9316c8eb903fbe2485b14d64/first_playbook.yml

(continued)

#	Task

⑲	Follow the documentation to run `first_playbook.yml`. The following example uses R2 (192.168.127.3), but you can use another device's IP address to run the `ansible-playbook` command. Notice the comma after the IP address—it must be appended, so be careful with any typos. Go ahead and run your script.
	
	(ansible_venv) jdoe@u22s1:~/ch6_virtualenv/venv1$ **ansible-playbook -i 192.168.127.3, -u jdoe -k -e ansible_network_os=ios first_playbook.yml** # Run your first playbook
	SSH password: ******** # Enter your router/switch password
	PLAY [Network Getting Started First Playbook]
	**
	TASK [Get config for Cisco devices]
	**
	ok: [192.168.127.3]
	TASK [Display the config]
	**
	ok: [192.168.127.3] => {
	"msg": "The hostname is R2 and the OS is 15.7(3)M3"
	}
	PLAY RECAP
	**
	192.168.127.3 : ok=2 changed=0 unreachable=0 failed=0 skipped=0 rescued=0 ignored=0
	Change the IP address and try it on other devices.
	(ansible_venv) jdoe@u22s1:~/ch6_virtualenv/venv1$ **ansible-playbook -i 192.168.127.244, -u jdoe -k -e ansible_network_os=ios first_playbook.yml**
	[...omitted for brevity]

(*continued*)

#	Task

⑳	20-1. First, create an inventory file in the present working directory. Add the variable information for all devices and specify the network devices' OS type. Then, group the devices into device types, so the routers are under the same group and the switches under the same group, as shown here:
	
	(ansible_venv) jdoe@u22s1:~/ch6_virtualenv/venv1$ **nano inventory**
	(ansible_venv) jdoe@u22s1:~/ch6_virtualenv/venv1$ **cat inventory**
	[all:vars]
	ansible_network_os = ios
	
	[routers]
	192.168.127.3
	192.168.127.4
	
	[switches]
	192.168.127.101
	192.168.127.102
	192.168.127.153
	192.168.127.244
	
	20-2. Now create a simple show_command.yml file to get the clock and uptime on routers. Because you do not need to reinvent the wheel, you will be borrowing an Ansible playbook from my blog to perform the show clock command. The blog page referenced for this exercise is https://wordpress.com/post/italchemy.wordpress.com/2649. Write the YAML playbook as shown here. If you want to see the process of writing the YAML playbook, visit the blog.
	
	(ansible_venv) jdoe@u22s1:~/ch6_virtualenv/venv1$ **nano show_command.yml**
	(ansible_venv) jdoe@u22s1:~/ch6_virtualenv/venv1$ **cat show_command.yml**
	--- 	
	- name: Running show commands on Cisco IOS
	  hosts: routers # target group, routers
	  gather_facts: false # disable fact gathering to speed up
	  connection: network_cli # use network_cli network module

(continued)

#	Task
	``yaml
tasks:
  - name: Run multiple commands on Cisco IOS XE nodes
    ios_command:
      commands:
        - show clock   # first command
        - show version | include uptime   # second command

    register: print_output

  - debug: var=print_output.stdout_lines
```
...

20-3. Once you are completely happy with your YAML playbook, run the following command to probe the clock times and uptimes of the routers only.

```
(ansible_venv) jdoe@u22s1:~/ch6_virtualenv/venv1$ ansible-playbook -i
./inventory show_command.yaml -u jdoe -k
SSH password: ********
PLAY [Running show commands on Cisco IOS]
****

TASK [Run multiple commands on Cisco IOS XE nodes]
****
ok: [192.168.127.3]
ok: [192.168.127.4]
TASK [debug]
****
ok: [192.168.127.3] => {
  "print_output.stdout_lines": [
    [
      "*00:35:39.608 AEST Thu Nov 16 2023"
    ],
```
|

*(continued)*
```

#	Task
	``
    [
      "R2 uptime is 2 hours, 30 minutes"
    ]
  ]
}

ok: [192.168.127.4] => {
  "print_output.stdout_lines": [
    [
      "*00:36:34.563 AEST Thu Nov 16 2023"
    ],
    [
      "r3 uptime is 2 hours, 31 minutes"
    ]
  ]
}
PLAY RECAP
****************888888**
192.168.127.3   : ok=2   changed=0   unreachable=0
failed=0   skipped=0   rescued=0   ignored=0
192.168.127.4   : ok=2   changed=0   unreachable=0
failed=0   skipped=0   rescued=0   ignored=0
```

20-4. Now open the show_command.yml file and update the hosts to switches.

(ansible_venv) jdoe@u22s1:~/ch6_virtualenv/venv1$ **nano show_command.yml**
**Before:**
hosts: **routers**
**After:**
hosts: **switches** |

(*continued*)
```

#	Task

	20-5. Now execute the playbook one more time and you will be able to get the clock and uptime information from all four switches instantly. If you run into any errors, go over your code and check for typos. Also, ensure that your `ansible-playbook` execution command is correct.
	
	(ansible_venv) jdoe@u22s1:~/ch6_virtualenv/venv1$ **ansible-playbook -i ./inventory show_command.yaml -u jdoe -k**
	SSH password: ********
	PLAY [Running show commands on Cisco IOS]
	**
	TASK [Run multiple commands on Cisco IOS XE nodes]
	**
	ok: [192.168.127.244]
	ok: [192.168.127.153]
	ok: [192.168.127.102]
	ok: [192.168.127.101]
	TASK [debug]
	**
	ok: [192.168.127.101] => {
	  "print_output.stdout_lines": [
	    [
	      "*00:20:02.046 AEST Thu Nov 16 2023"
	    ],
	    [
	      "SW1 uptime is 2 hours, 14 minutes"
	    ]
	  ]
	}
	
	(continued)

#	Task
| | ```
[...omitted for brevity]
ok: [192.168.127.244] => {
  "print_output.stdout_lines": [
    [
      "*14:22:40.039 UTC Wed Nov 15 2023"
    ],
    [
      "SW4 uptime is 2 hours, 16 minutes"
    ]
  ]
}
PLAY RECAP
*********************************************************************
192.168.127.101            : ok=2    changed=0    unreachable=0
failed=0    skipped=0    rescued=0    ignored=0
192.168.127.102            : ok=2    changed=0    unreachable=0
failed=0    skipped=0    rescued=0    ignored=0
192.168.127.153            : ok=2    changed=0    unreachable=0
failed=0    skipped=0    rescued=0    ignored=0
192.168.127.244            : ok=2    changed=0    unreachable=0
failed=0    skipped=0    rescued=0    ignored=0
``` |

(*continued*)
```

CHAPTER 6   PYTHON NETWORK AUTOMATION VIRTUAL LAB 1: ANSIBLE AND PYATS IN PYTHON VIRTUALENV

#	Task

21	Now, deactivate the testing Python virtual environment and check if you have Ansible installed on your local server. Your `venv1` virtualization Ansible installation did not affect the Python configurations on your host OS. In other words, the Ansible was only installed in your Python `virtualenv`.
	
	`(ansible_venv) jdoe@u22s1:~/ch6_virtualenv/venv1$ `**`deactivate`**
	`jdoe@u22s1:~/ch6_virtualenv/venv1$ `**`python`**
	`Python 3.10.12 (main, Jun 11 2023, 05:26:28) [GCC 11.4.0] on linux`
	`Type "help", "copyright", "credits" or "license" for more information.`
	`>>> `**`import ansible`**
	`Traceback (most recent call last):`
	`File "<stdin>", line 1, in <module>`
	`ModuleNotFoundError: No module named 'ansible'`
	`>>>`

You've explored `virtualenv` and Ansible in this section, broadening your knowledge of testing packages within the Python virtual environment. Integrating Ansible into your Python setup was a sizable step, revealing its power in managing enterprise devices. Imagine what Ansible can do for you in network automation. Although Ansible was born out of the necessity of managing Linux servers, it can be used for network automation, and it does it well. Ansible helps modern network engineers automate various manual tasks using YAML, a user-friendly language that packs a serious punch. Its capabilities go beyond the norm, making it essential for anyone stepping into the world of Network Automation Engineering. For those keen on this path, diving deeper into Ansible and its advanced version, the Ansible Automation Platform (AAP, formerly Ansible Tower), is highly recommended. If obtaining licensing for Red Hat AAP poses challenges, AWX serves as a viable alternative. The domain of Ansible and network automation is vast and involved, deserving a whole book dedicated solely to its intricacies. This is not a book about Ansible network automation, so this is where the Ansible discussion ends. If your thirst for Ansible knowledge persists, Red Hat's official Ansible site or dedicated study materials can fulfill that curiosity. I cover these depths in my book titled *Introduction to Ansible Network Automation: A Practical Primer*, 2023, by Brendan Choi and Erwin Medina.

Let's now shift focus to integrating pyATS in another Python virtual environment. Exploring how pyATS sets itself apart from other tools will uncover its unique strengths in the network automation space.

Lab 2: Quick-Start Guide to pyATS (Genie) in virtualenv

Around 2016, pyATS was introduced by Cisco atop the Genie framework, redefining network automation. pyATS indeed stands as the core framework supported within Cisco Systems, originally developed for Cisco's internal engineering team's needs. It utilizes the Genie library as the standard library for pyATS and XPRESSO serves as its web dashboard. It swiftly evolved into an automation solution for Cisco DevNet. pyATS, rooted in Genie's capabilities, became influential in streamlining network testing and operations across Cisco platforms and Cisco DevNet functions. Currently, pyATS plays an essential role in Cisco's DevNet automation environment. Visit the "Getting Started with pyATS" page to get more information about this useful tool. However, within the realm of networking automation, pyATS, although relatively new, primarily serves as a tool for information gathering and network system monitoring, rather than operating as a comprehensive configuration management tool like Red Hat's open-source tool, Ansible.

It's important to note that pyATS isn't positioned as the primary automation tool but serves as a supplementary network tool. Some key supported features include:

- Being Cisco's default test framework for internal engineers, utilized across various Cisco platforms and functions for tasks such as CI/CD, sanity, regression, scale, HA, and solution testing.

- It has a decent-sized user base among Cisco engineers and developers.

- pyATS operates as an open-source, platform-agnostic library system (Genie) and features a web dashboard (XPRESSO) for managing test suites, testbeds, results, and insights.

- It offers integration support for other tools via libraries, extensions, plugins, and APIs.

- It allows the creation of custom business logic and solutions atop the infrastructure stack.

- It primarily focuses on Cisco devices and is not as popular as other tools.

CHAPTER 6   PYTHON NETWORK AUTOMATION VIRTUAL LAB 1: ANSIBLE AND PYATS IN PYTHON VIRTUALENV

While Cisco's pyATS documentation is extensive, it was not designed for network configuration management tasks. **If your network automation ambitions stretch beyond the Cisco ecosystem, exploring vendor-neutral tools aligns better with diverse use cases.** Staying within Cisco's realm? Investing time in pyATS and related Cisco tools proves beneficial. However, for device-focused automation, augmenting pyATS with Nornir becomes crucial. Nornir, predominantly Python-based, shines in handling network devices, particularly Cisco's. It complements pyATS by focusing on device orchestration and streamlined network management. For even broader and multivendor network automation, embracing NAPALM alongside pyATS and Nornir creates a robust trio. NAPALM, a cross-vendor library, augments both pyATS and Nornir, offering versatile support for network device interactions and configurations across various platforms. Choosing these tools strategically enables a comprehensive approach to network automation, tailored to your specific requirements and network environment. Yet, the requirement for Nornir underscores the significance of acquiring proficiency in Python.

URL: https://developer.cisco.com/docs/pyats/ (pyATS introduction)

URL: https://developer.cisco.com/docs/pyats-getting-started/ (pyATS getting started)

Review Figure 6-2 before starting this lab. In line with the previous `virtualenv` lab, you will create a fresh directory and initiate `virtualenv` within it. Follow these steps to begin your new `virtualenv` setup for pyATS.

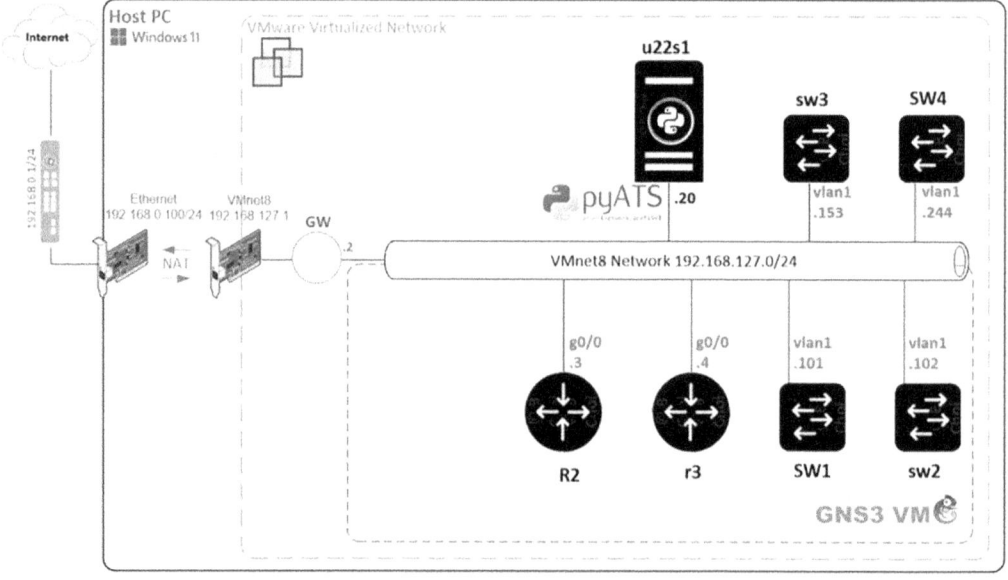

Figure 6-2. *pyATS virtualenv lab used devices*

#	Task
①	Create a new directory called mygenie and cd into the new directory. Once you are inside the directory, create a new virtual environment to test pyATS. To create and activate the pyATS test environment, type the following commands: jdoe@u22s1:~/ch6_virtualenv/venv1$ **mkdir mygenie && cd mygenie** jdoe@u22s1:~/ch6_virtualenv/venv1/mygenie$ **python -m venv genie_venv** jdoe@u22s1:~/ch6_virtualenv/venv1/mygenie$ **ls genie_venv** bin include lib lib64 pyvenv.cfg jdoe@u22s1:~/ch6_virtualenv/venv1/mygenie$ **source ./genie_venv/bin/activate** (genie_venv) jdoe@u22s1:~/ch6_virtualenv/venv1/mygenie$
②	Install pyATS with the library. If you get some bdist_wheel-related messages, you can ignore the messages and continue. (genie_venv) jdoe@u22s1:~/ch6_virtualenv/venv1/mygenie$ **pip install pyATS[library]** Collecting pyATS[library] Downloading pyats-23.10-cp310-cp310-manylinux2014_x86_64.whl (4.4 MB) ─────────────────────────4.4/4.4 MB 6.2 MB/s eta 0:00:00 Collecting pyats.log<23.11.0,>=23.10.0 [...omitted for brevity] l.yaml.clib-0.2.8 six-1.16.0 smmap-5.0.1 tftpy-0.8.0 tqdm-4.66.1 typing-extensions-4.8.0 unicon-23.10 unicon.plugins-23.10 urllib3-2.1.0 wcwidth-0.2.10 wheel-0.41.3 xmltodict-0.13.0 yamllint-1.33.0 yang.connector-23.10 yarl-1.9.2

(*continued*)

#	Task
③	Run the `git clone` command to download the Cisco Test Automation examples from GitHub.com. (genie_venv) jdoe@u22s1:~/ch6_virtualenv/venv1/mygenie$ **git clone https://github.com/CiscoTestAutomation/examples** Cloning into 'examples'... remote: Enumerating objects: 1414, done. remote: Counting objects: 100% (262/262), done. remote: Compressing objects: 100% (156/156), done. remote: Total 1414 (delta 137), reused 150 (delta 87), pack-reused 1152 Receiving objects: 100% (1414/1414), 1.17 MiB | 3.83 MiB/s, done. Resolving deltas: 100% (709/709), done.

④	Then run the `pyats run job` command to check the operational status of pyATS. If `basic_example_job.py` runs successfully, your installation is good and you are ready to go.
	
	(genie_venv) jdoe@u22s1:~/ch6_virtualenv/venv1/mygenie$ **cd examples**
	(genie_venv) jdoe@u22s1:~/ch6_virtualenv/venv1/mygenie/examples$ **pyats run job basic/basic_example_job.py**
	2023-11-17T19:28:33: %EASYPY-INFO: Starting job run: basic_example_job
	2023-11-17T19:28:33: %EASYPY-INFO: Runinfo directory: /home/jdoe/.pyats/runinfo/basic_example_job.2023Nov17_19:28:32.478051
	[...omitted for brevity]
	Pro Tip
	------- 	
	  Try the following command to view your logs:
	    pyats logs view

(continued)

#	Task
⑤	Before proceeding, it's essential to install additional libraries for Excel, which are prerequisites for proper functionality with pyATS later. (genie_venv) jdoe@u22s1:~/ch6_virtualenv/venv1/mygenie/examples$ **cd ..** (genie_venv) jdoe@u22s1:~/ch6_virtualenv/venv1/mygenie$ **pip install xlrd xlwt xlsxwriter** Collecting xlrd Downloading xlrd-2.0.1-py2.py3-none-any.whl (96 kB) [...omitted for brevity] Successfully installed xlrd-2.0.1 xlsxwriter-3.1.9 xlwt-1.3.0
⑥	Genie uses a YAML testbed JSON file for device connection and authentication. Install pyats.contrib, which is a requirement. (genie_venv) jdoe@u22s1:~/ch6_virtualenv/venv1/mygenie$ **pip install pyats.contrib** [...omitted for brevity] Successfully installed pyats.contrib-23.10 requests-toolbelt-1.0.0 xlrd-1.2.0
⑦	Next, create a testbed.yml file to be used for authentication. Use the same pyats create command to create a testbed.yml file. The following example creates a testbed.yml file for two switches, sw3 and SW4: (genie_venv) jdoe@u22s1:~/ch6_virtualenv/venv1/mygenie$ **genie create testbed interactive --output testbed.yml --encode-password** Start creating Testbed yaml file ... Do all of the devices have the same username? [y/n] **y** Common Username: **jdoe** Do all of the devices have the same default password? [y/n] **y** Common Default Password (leave blank if you want to enter on demand): ************

(continued)

#	Task

	Do all of the devices have the same enable password? [y/n] **y**
	Common Enable Password (leave blank if you want to enter on demand):
	********
	
	Device hostname: sw3
	  IP (ip, or ip:port): **192.168.127.153**
	  Protocol (ssh, telnet, ...): **ssh**
	  OS (iosxr, iosxe, ios, nxos, linux, ...): **iosxe**
	More devices to add ? [y/n] **y**
	
	Device hostname: **SW4**
	  IP (ip, or ip:port): **192.168.127.244**
	  Protocol (ssh, telnet, ...): **ssh**
	  OS (iosxr, iosxe, ios, nxos, linux, ...): **iosxe**
	More devices to add ? [y/n] **n**
	Testbed file generated:
	testbed.yml
	
	Note: Depending on the installed pyATS version, the `genie create testbed interactive --output testbed.yml --encode-password` command should be replaced with the `pyats create testbed interactive --output testbed.yml --encode-password` command.

⑧	Quickly review the newly created `testbed.yml` file. In this case, the format represents device configurations in YAML format, including details such as connections, credentials, OS, and device types. (genie_venv) jdoe@u22s1:~/ch6_virtualenv/venv1/mygenie$ **ls** examples genie_venv testbed.yml (genie_venv) jdoe@u22s1:~/ch6_virtualenv/venv1/mygenie$ **cat testbed.yml** *(continued)*

#	Task
| | ```
credentials:
  default:
    password: '%ENC{w5PDosOUw5fDosKQwpbCmA==}'
    username: jdoe
  enable:
    password: '%ENC{w5PDosOUw5fDosKQwpbCmA==}'
  os: iosxe
  type: iosxe
sw3:
  connections:
    cli:
      ip: 192.168.127.153
      protocol: ssh
  credentials:
    default:
      password: '%ENC{w5PDosOUw5fDosKQwpbCmA==}'
      username: jdoe
    enable:
      password: '%ENC{w5PDosOUw5fDosKQwpbCmA==}'
  os: iosxe
  type: iosxe
devices:
  SW4:
    connections:
      cli:
        ip: 192.168.127.244
        protocol: ssh
``` |

(*continued*)
```

Task

⑨ First, use the `genie parse` command, followed by the `show version` command on both switches. You can use the device name to run the command as follows. The following example displays the result from sw3 and SW4.

For sw3:
(genie_venv) jdoe@u22s1:~/ch6_virtualenv/venv1/mygenie$ **genie parse "show version" --testbed-file testbed.yml --devices sw3**
Using the default YAML encoding key since no key was specified in configuration.
THIS IS A SHARED KEY AND IS NOT SECURE, PLEASE RUN 'pyats secret keygen' AND ADD TO YOUR pyats.conf FILE BEFORE ENCODING ANY VALUES.
0%| | 0/1 [00:00<?, ?it/s]/home/jdoe/ch6_virtualenv/venv1/mygenie/genie_venv/lib/python3.10/site-packages/genie/libs/parser/iosxe/c9300/show_idprom.py:7: DeprecationWarning: The asyncore module is deprecated and will be removed in Python 3.12. The recommended replacement is asyncio
  from asyncore import poll3
{
  "version": {
  "chassis": "IOSv",
  "chassis_sn": "92Y1L232FYM",
  "compiled_by": "sasyamal",
  "compiled_date": "Tue 28-Jul-15 18:52",
  "copyright_years": "1986-2015",
  "curr_config_register": "0x0",
  "hostname": "sw3",
  "image_id": "vios_l2-ADVENTERPRISEK9-M",
  "image_type": "production image",
  "label": "TEST ENGINEERING ESTG_WEEKLY BUILD, synced to END_OF_FLO_ISP",

(continued)

#	Task

```
      "last_reload_reason": "Unknown reason",
      "main_mem": "574709",
      "mem_size": {
      "non-volatile configuration": "256"
    },
      "number_of_intfs": {
      "Gigabit Ethernet": "16",
      "Virtual Ethernet": "1"
    },
      "os": "IOS",
      "platform": "vios_l2",
      "processor_board_flash": "OK",
      "processor_type": "",
      "returned_to_rom_by": "reload",
      "rom": "Bootstrap program is IOSv",
      "rtr_type": "IOSv",
      "system_image": "flash0:/vios_l2-adventerprisek9-m",
      "uptime": "11 minutes",
      "version": "15.2(4.0.55)E",
      "version_short": "15.2"
  }
}
100%|████████████████████████████████████████████████
████████████████████████| 1/1 [00:02<00:00, 2.02s/it]

For SW4:
(genie_venv) jdoe@u22s1:~/ch6_virtualenv/venv1/mygenie$ genie parse
"show version" --testbed-file testbed.yml --devices SW4
[...omitted for brevity]
```

(continued)

#	Task

⑩	Use the following `genie parse` command to retrieve your device's interface information:
	
	(genie_venv) jdoe@u22s1:~/ch6_virtualenv/venv1/mygenie$ **genie parse "show ip int brief" --testbed-file testbed.yml --devices sw3**
	
	Using the default YAML encoding key since no key was specified in the configuration.
	
	THIS IS A SHARED KEY AND IS NOT SECURE, PLEASE RUN 'pyats secret keygen' AND ADD TO YOUR pyats.conf FILE BEFORE ENCODING ANY VALUES. 0%| | 0/1 [00:00<?, ?it/s]/home/jdoe/ch6_virtualenv/venv1/mygenie/genie_venv/lib/python3.10/site-packages/genie/libs/parser/iosxe/c9300/show_idprom.py:7: DeprecationWarning: The asyncore module is deprecated and will be removed in Python 3.12. The recommended replacement is asyncio
	from asyncore import poll3
	{
	  "interface": {
	    "GigabitEthernet0/0": {
	    "interface_is_ok": "YES",
	    "ip_address": "unassigned",
	    "method": "unset",
	    "protocol": "up",
	    "status": "up"
	    },
	  [...omitted for brevity]
	    "Vlan1": {
	    "interface_is_ok": "YES",
	    "ip_address": "192.168.127.153",
	    "method": "NVRAM",
	    "protocol": "up",

(continued)

#	Task

      "status": "up"
     }
    }
  }
}
100%|■■| 1/1 [00:01<00:00, 1.57s/it]

(11) Revise the `testbed.yml` file in the nano text editor to incorporate all CML routers and switches. Adjust settings for sw3 and SW4 to utilize Telnet while configuring the rest for SSH for testing.

```
(genie_venv) jdoe@u22s1:~/ch6_virtualenv/venv1/mygenie$
(genie_venv) jdoe@u22s1:~/ch6_virtualenv/venv1/mygenie$ cat testbed.yml
devices:
  SW4:
    connections:
    cli:
      ip: 192.168.127.244
      protocol: telnet
    credentials:
      default:
        password: '%ENC{w5PDosOUw5fDosKQwpbCmA==}'
        username: jdoe
      enable:
        password: '%ENC{w5PDosOUw5fDosKQwpbCmA==}'
    os: iosxe
    type: iosxe
  sw3:
    connections:
      cli:
        ip: 192.168.127.153
        protocol: telnet
```

(continued)

#	Task

```
            credentials:
              default:
                password: '%ENC{w5PDosOUw5fDosKQwpbCmA==}'
                username: jdoe
              enable:
                password: '%ENC{w5PDosOUw5fDosKQwpbCmA==}'
            os: iosxe
            type: iosxe
          sw2:
            connections:
              cli:
                ip: 192.168.127.102
                protocol: ssh
            credentials:
              default:
                password: '%ENC{w5PDosOUw5fDosKQwpbCmA==}'
                username: jdoe
              enable:
                password: '%ENC{w5PDosOUw5fDosKQwpbCmA==}'
            os: iosxe
            type: iosxe
          SW1:
            connections:
              cli:
                ip: 192.168.127.101
                protocol: ssh
            credentials:
              default:
                password: '%ENC{w5PDosOUw5fDosKQwpbCmA==}'
                username: jdoe
```

(*continued*)

#	Task

```
          enable:
            password: '%ENC{w5PDosOUw5fDosKQwpbCmA==}'
        os: iosxe
        type: iosxe
      r3:
        connections:
          cli:
            ip: 192.168.127.4
            protocol: ssh
        credentials:
          default:
            password: '%ENC{w5PDosOUw5fDosKQwpbCmA==}'
            username: jdoe
          enable:
            password: '%ENC{w5PDosOUw5fDosKQwpbCmA==}'
        os: iosxe
        type: iosxe
      R2:
        connections:
          cli:
            ip: 192.168.127.3
            protocol: ssh
        credentials:
          default:
            password: '%ENC{w5PDosOUw5fDosKQwpbCmA==}'
            username: jdoe
          enable:
            password: '%ENC{w5PDosOUw5fDosKQwpbCmA==}'
        os: iosxe
        type: iosxe
```

(continued)

CHAPTER 6   PYTHON NETWORK AUTOMATION VIRTUAL LAB 1: ANSIBLE AND PYATS IN PYTHON VIRTUALENV

#	Task
	Rerun the `pyats create testbed interactive --output testbed.yml --encode-password` command to regenerate this file. Retrieve the `testbed.yml` file from the chapter's download page. Due to page limitations, certain content has been omitted. `(genie_venv) jdoe@u22s1:~/ch6_virtualenv/venv1/mygenie$` **`pyats create testbed interactive --output testbed.yml --encode-password`**

⑫	Use the `show clock` command to test-drive Genie. Genie will connect to R2, r3, SW1, and sw2 via SSH and run the command, and for sw3 and SW4, Genie will Telnet in and run the command. Ensure that on sw3 and SW4, the `transport input all` command has been configured under `line vty 0 15`. You should receive the current time from all six devices.
	
	`(genie_venv) jdoe@u22s1:~/ch6_virtualenv/venv1/mygenie$` **`genie parse "show clock" --testbed-file testbed.yml --devices[hostname]`**
	`[...omitted for brevity]`
	`0%|                         | 0/1 [00:00<?, ?it/s]{`
	`  "day": "17",`
	`  "day_of_week": "Fri",`
	`  "month": "Nov",`
	`  "time": "09:46:47.219",`
	`  "timezone": "UTC",`
	`  "year": "2023"`
	`}`
	
	`100%|████████████████████████████████████████████| 1/1`
	`[00:01<00:00,  1.02s/it]`

(continued)

#	Task
⑬	This time, change the command to show cdp neighbor and execute the command one more time.

```
(genie_venv) jdoe@u22s1:~/ch6_virtualenv/venv1/mygenie$ genie parse
"show cdp neighbor" --testbed-file testbed.yml --devices[hostname]
[...omitted for brevity]
{
  "cdp": {
    "index": {
      "1": {
        "capability": "R S I",
        "device_id": "SW1.pynetauto.local",
        "hold_time": 129,
        "local_interface": "GigabitEthernet0/0",
        "platform": "Gig",
        "port_id": "0/0"
      },
      "2": {
        "capability": "S I",
        "device_id": "SW4.jdoe.local",
        "hold_time": 170,
        "local_interface": "GigabitEthernet0/0",
        "platform": "",
        "port_id": "GigabitEthernet0/0"
      }
    },
    "total_entries": 2
  }
}
100%|████████████████████████████████████████████████████████████████████████| 1/1
[00:01<00:00,  1.80s/it]
``` |

(*continued*)

#	Task
	If the connected interface has CDP disabled, the result will come back as an error: Parsed command 'show cdp neigh' or if it returned empty. You may have to enable the cdp enable command on the affected interface. If you have all the CDP information, then you have completed this task successfully.
⑭	Let's quickly install the pandas library to be used in this lab. Let's install the pandas module for data analytics and store the data in Excel. pandas is a Python library that facilitates data manipulation and analysis, offering structures like Series and DataFrame for efficient data handling. (genie_venv) jdoe@u22s1:~/ch6_virtualenv/venv1/mygenie$ **pip install pandas** *[...omitted for brevity]* `Successfully installed numpy-1.26.2 pandas-2.1.3 python-dateutil-2.8.2 pytz-2023.3.post1 tzdata-2023.3`

⑮	pyATS, backed by Cisco, excels in data collection. Combining its prowess with Python's data handling enables effortless data collection and storage. The example showcases pyATS with Python's re module in an interactive session, allowing variable extraction and Excel file storage. You can follow this interactive example or save it as a .py script file.
	
	(genie_venv) jdoe@u22s1:~/ch6_virtualenv/venv1/mygenie$ **python**
	`Python 3.10.12 (main, Jun 11 2023, 05:26:28) [GCC 11.4.0] on linux`
	`Type "help", "copyright", "credits" or "license" for more information.`
	`>>> `**`import os`**
	`>>> `**`show_ver = os.popen('genie parse "show version" --testbed-file testbed.yml --devices[hostname]')`**
	`>>> Using the default YAML encoding key since no key was specified in the configuration.`
	`THIS IS A SHARED KEY AND IS NOT SECURE, PLEASE RUN 'pyats secret keygen' AND ADD TO YOUR pyats.conf FILE BEFORE ENCODING ANY VALUES.`

(continued)

CHAPTER 6   PYTHON NETWORK AUTOMATION VIRTUAL LAB 1: ANSIBLE AND PYATS IN PYTHON VIRTUALENV

#	Task

	0%| | 0/1 [00:00<?, ?it/s]/home/jdoe/ch6_virtualenv/venv1/mygenie/genie_venv/lib/python3.10/site-packages/genie/libs/parser/iosxe/c9300/show_idprom.py:7: DeprecationWarning: The asyncore module is deprecated and will be removed in Python 3.12. The recommended replacement is asyncio
	from asyncore import poll3
	100%|██| 1/1 [00:01<00:00, 1.68s/it]
	[...omitted for brevity]
	100%|██| 1/1 [00:01<00:00, 1.31s/it]
	# Press the \<Enter\> key once
	\>\>\> **output = show_ver.read()**
	\>\>\> **print(output)**
	{
	"version": {
	"chassis": "IOSv",
	"chassis_sn": "91N1192IYKJFPLDPNMJ8Q",
	"compiled_by": "prod_rel_team",
	"compiled_date": "Wed 01-Aug-18 16:45",
	"copyright_years": "1986-2018",
	"curr_config_register": "0x0",
	"hostname": "R2",
	"image_id": "VIOS-ADVENTERPRISEK9-M",
	"image_type": "production image",
	"label": "RELEASE SOFTWARE (fc2)",
	"last_reload_reason": "Unknown reason",
	"main_mem": "460009",

(continued)

#	Task

```
        "mem_size": {
        "non-volatile configuration": "256"
        },
        "number_of_intfs": {
        "Gigabit Ethernet": "4"
        },
        "os": "IOS",
        "platform": "IOSv",
        "processor_board_flash": "OK",
        "processor_type": "revision 1.0",
        "returned_to_rom_by": "reload",
        "rom": "Bootstrap program is IOSv",
        "rtr_type": "IOSv",
        "system_image": "flash0:/vios-adventerprisek9-m",
        "uptime": "52 minutes",
        "version": "15.7(3)M3",
        "version_short": "15.7"
        }
        }
        [...omitted for brevity]
        {
        "version": {
        "chassis": "IOSv",
        "chassis_sn": "92Y1L232FYM",
        "compiled_by": "sasyamal",
        "compiled_date": "Tue 28-Jul-15 18:52",
        "copyright_years": "1986-2015",
        "curr_config_register": "0x0",
        "hostname": "sw3",
        "image_id": "vios_l2-ADVENTERPRISEK9-M",
```

(continued)

#	Task
| | ```
"image_type": "production image",
"label": "TEST ENGINEERING ESTG_WEEKLY BUILD, synced to  END_OF_FLO_
ISP",
"last_reload_reason": "Unknown reason",
"main_mem": "574709",
"mem_size": {
"non-volatile configuration": "256"
},
"number_of_intfs": {
"Gigabit Ethernet": "16",
"Virtual Ethernet": "1"
},
"os": "IOS",
"platform": "vios_l2",
"processor_board_flash": "OK",
"processor_type": "",
"returned_to_rom_by": "reload",
"rom": "Bootstrap program is IOSv",
"rtr_type": "IOSv",
"system_image": "flash0:/vios_l2-adventerprisek9-m",
"uptime": "1 hour, 25 minutes",
"version": "15.2(4.0.55)E",
"version_short": "15.2"
}
}
``` |

(*continued*)
```

#	Task
⑯	As confirmed in the interactive session, the output data type is a string type that is saved to the variable output.

```
>>> type(output)
<class 'str'>
```

Import the re module and use one of the cool Regular Expressions you learned in Chapter 9 of the Part I book to capture the specific information you are after. Here, the example uses the lookahead (?=) and lookbehind (?<=) Regular Expression examples.

In the following code, a positive lookbehind and positive lookahead are used to get the hostnames of each device from the output in the previous step:

```
>>> import re
>>> p1 = re.compile(r'(?<=\"hostname\": \").+(?=\")')
>>> m1 = p1.findall(output)
>>> m1
['R2', 'SW1', 'SW4', 'r3', 'sw2', 'sw3']
```

In the following lines, once again, lookbehind and lookahead are used to get each device's uptime from the output:

```
>>> p2 = re.compile(r'(?<=\"uptime\": \").+(?=\")')
>>> m2 = p2.findall(output)
>>> m2
['6 hours, 52 minutes', '6 hours, 52 minutes', '11 hours, 14 minutes', '5 hours, 13 minutes', '6 hours, 12 minutes', '11 hours, 39 minutes']
```

If you want to convert the uptime information into a bar graph using Python, you must first convert the hours and minutes into the correct decimal format. Look at the following conversion to convert the time into two decimal places. The following Python code will work only if all of your devices have been up and running for more than an hour (60 minutes), so the output of the uptime is "x hours y minutes." |

(continued)

#	Task
	``
>>> uptime = [] # create empty list called uptime
>>> for x in m2: # for loop to call out each item in list m2
...   y = [int(s) for s in x.split() if s.isdigit()] # If the string is a digit, then save it as y
...   z = (y[1]/60) # y[1] or the second number from y is the minute value, divide it by 60 minutes and convert it into a decimal points
...   a = round(y[0] + z, 2) # y[0] or the first number from y is the hour value, combine it with z, then round it to 2 decimal places
...   uptime.append(a) # append the value a to the uptime list
...
>>> print(uptime) # print the result of the uptime list
[6.87, 6.87, 11.23, 5.22, 6.2, 11.65]
```
If your device uptimes were more than an hour and you have completed this step successfully, you can skip Step ⑰ and go straight to Step ⑱. |
| ⑰ | Perform this task only if your devices' uptime were less than 60 minutes. Use the following Python code:

```
>>> uptime = []
>>> for x in m2:
...   y = [int(s) for s in x.split() if s.isdigit()]
...   z = (y[0]/60)
...   a = round(z, 2)
...   uptime.append(a)
...
>>> print(uptime)
[0.25, 0.22, 0.24, 0.19, 0.18, 0.21]
``` |

(continued)
```

#	Task
| 18 | Now, use the dictionary zip feature to turn the two lists into a Python dictionary. You need to combine the device name list and uptime list into decimal using the dict(zip(m1, uptime)) function. The outcome should look as follows:<br><br>```
>>> device_uptime = dict(zip(m1,uptime)) # Combine m1 and uptime lists and make them into a dictionary
>>> print(device_uptime) # display device_uptime dictionary
{'R2': 6.87, 'SW1': 6.87, 'SW4': 11.23, 'r3': 5.22, 'sw2': 6.2, 'sw3': 11.65}
```<br><br>Now use the pandas module to turn the dictionary into a pandas data frame and save it as an Excel spreadsheet for reporting purposes. While converting the dictionary into a pandas data frame, you will add the headers: host for device names and uptime for running time. The key success factor in Python Network Automation is how you can process and handle this crucial data to suit your company's needs.<br><br>```
>>> type(device_uptime) # Check type
<class 'dict'>
>>> import pandas as pd # import pandas
>>> df = pd.DataFrame(list(device_uptime.items()),columns = ['host','uptime']) # convert to dataframe
>>> df # Check dataframe
  host  uptime
0  R2    6.87
1  SW1   6.87
2  SW4  11.23
3  r3    5.22
4  sw2   6.20
5  sw3  11.65
>>> df.to_excel('device_uptime.xlsx') # Write data frame to excel using panda's to_excel feature.
>>> exit() # exit Python
``` |

(continued)

CHAPTER 6 PYTHON NETWORK AUTOMATION VIRTUAL LAB 1: ANSIBLE AND PYATS IN PYTHON VIRTUALENV

#	Task

⑲	19-1. Check if the file exists in the `mygenie` directory.
	
	`(genie_venv) jdoe@u22s1:~/ch6_virtualenv/venv1/mygenie$` **`ls -lh`**
	`total 20K`
	`-rw-rw-r--  1 jdoe jdoe 5.5K Nov 17 21:45 device_uptime.xlsx`
	`drwxrwxr-x 24 jdoe jdoe 4.0K Nov 17 19:26 examples`
	`drwxrwxr-x  5 jdoe jdoe 4.0K Nov 17 19:19 genie_venv`
	`-rw-rw-r--  1 jdoe jdoe 1.7K Nov 17 20:50 testbed.yml`
	
	19-2. Once you confirm that the file is under your directory, issue a `deactivate` command and stop the virtual environment.
	
	`(genie_venv) jdoe@u22s1:~/ch6_virtualenv/venv1/mygenie$` **`deactivate`**
	`jdoe@u22s1:~/ch6_virtualenv/venv1/mygenie$`
	
	19-3. Use WinSCP to log in to the u22s1 (192.168.127.20) server via SCP, download a copy of the `device_uptime.xlsx` file to your Windows host PC, and open it in Excel to confirm the data has been correctly saved in Excel format. See Figures 6-3 and 6-4.
	
	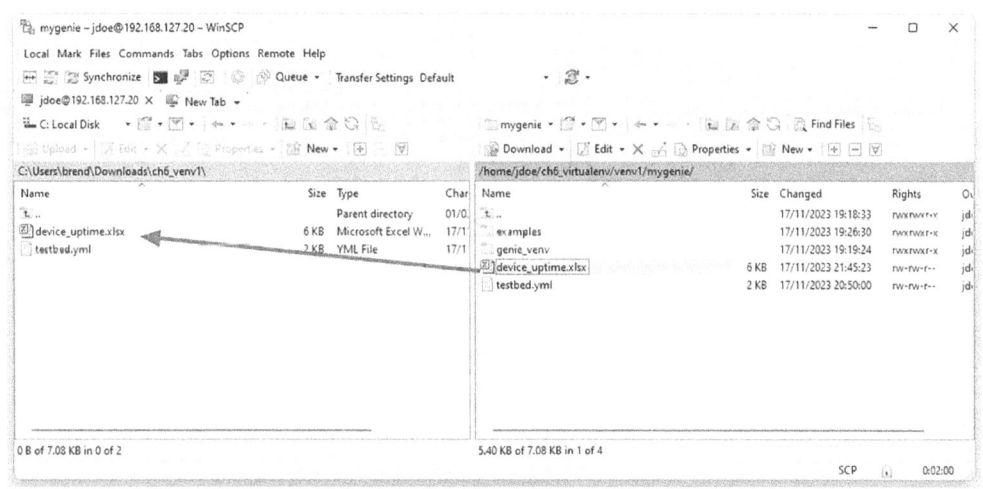
	
	Figure 6-3. *WinSCP, retrieving device_uptime.xlsx from u22s1 server*

(*continued*)

#	Task

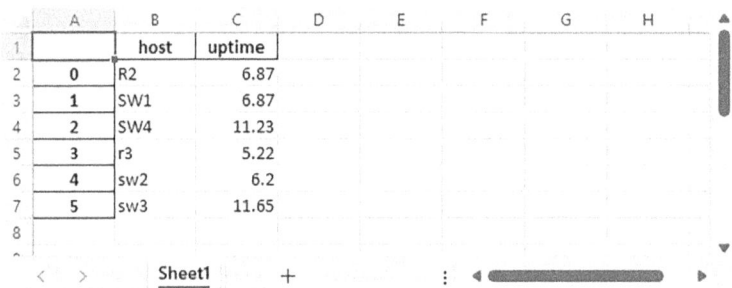

Figure 6-4. Host PC, opening device_uptime.xlsx in Excel

20 Now, on your Windows host PC (or desktop of your Ubuntu Server), you can write a simple Python code to convert the dictionary from Step 18 into a graph. You also can read the Excel file using pandas and convert it into a data frame to achieve the same result. (This will not work well with the Linux command line, as it does not have direct graphical interface support.)

Before writing the script from your Windows host PC's Command Prompt, install pandas and matplotlib using the pip3 commands shown here:

```
C:\Users\brendan>cd Desktop # Update the user name to your user name
C:\Users\brendan\Desktop>pip3 install pandas
[...omitted for brevity]
Successfully installed numpy-1.26.2 pandas-2.1.3 python-
dateutil-2.8.2 pytz-2023.3.post1 six-1.16.0 tzdata-2023.3
C:\Users\brendan\Desktop>pip3 install matplotlib
[...omitted for brevity]
Successfully installed contourpy-1.2.0 cycler-0.12.1
fonttools-4.44.3 kiwisolver-1.4.5 matplotlib-3.8.1 packaging-23.2
pillow-10.1.0 pyparsing-3.1.1
```

(continued)

#	Task
	Write the following code and save the script in your Windows Documents folder. Create a Python file called device_uptime_graph.py and save it. Watch out for any typos.

```python
import matplotlib.pyplot as plt # import matplotlib library
import pandas as pd # import pandas library
device_uptime = {'R2': 6.87, 'SW1': 6.87, 'SW4': 11.23, 'r3': 5.22,
'sw2': 6.2, 'sw3': 11.65}
# Resulting dictionary from Step ⑱
df = pd.DataFrame(list(device_uptime.items()),columns =
['host','uptime']) # Convert dictionary into pandas dataframe with
column titles 'host' & 'uptime'
#print(df)
df.plot(kind='bar',x='host',y='uptime') # Plot a bar graph with host
as x-axis and uptime(float) as y-axis
plt.show() # Display graph
```

Now, run the script from PowerShell or Windows Command Prompt or double-click the script. If everything works correctly, your data will come out as a bar graph, as shown in Figure 6-5. Alternatively, you can turn the data into different types of graphs and apply the same method to convert data into data frames and create graphs as you want. You can create the same graph from Excel, but you now know another method to create a graph using Python pandas and matplotlib libraries for some reporting.

C:\Users\brend\Documents>**python device_uptime_graph.py** |

(continued)

CHAPTER 6   PYTHON NETWORK AUTOMATION VIRTUAL LAB 1: ANSIBLE AND PYATS IN PYTHON VIRTUALENV

#	Task

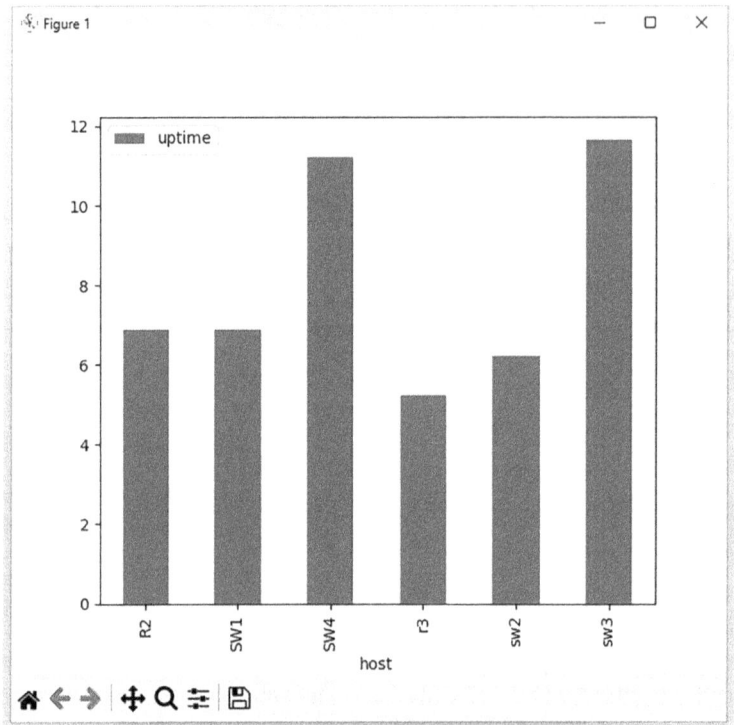

Figure 6-5. matplotlib graph of device uptime example

You can save the file to your folder. Now you have completed all the tasks required in this lab.

In the second `virtualenv` lab, Cisco's pyATS was introduced as one of the essential tools for Cisco device network automation. Gaining enough coding practice involves typing the presented Python code independently. Skills from prior chapters combined with new modules like `pandas` and `matplotlib` were utilized for data manipulation and visualization. Employing Python's `re` and `pandas`, device uptime data was collected and converted into a `pandas` data frame and saved to an Excel file. This data was then translated into a simple bar graph using `pandas` and `matplotlib`. This comprehensive lab encompassed various Python aspects, using network device data to manipulate, convert to Excel, and create insightful visualizations. While intricate, if you have managed to complete this lab, that indicates significant progress in Python mastery.

Summary

Upon completing all the tasks in this initial Python Virtual Lab chapter, you've engaged in exercises designed to encourage innovative thinking, connecting Python concepts to practical network infrastructure management. At this point, consider the automation challenges present in your team or organization, particularly when working within limited IT budget constraints. Throughout this chapter, you've implemented `virtualenv` in Python and briefly experimented with Ansible and pyATS. Now, it's time to explore various ways Python applications can be applied in your professional environment. Take on challenges that are often avoided, utilizing both open-source and proprietary tools available. Looking ahead to Chapter 7, the focus remains on Python within a virtual environment, specifically exploring Python on Docker. Docker, a software containerization platform, offers an intriguing solution for scaling down computing power, effectively reducing both operational expenditure (OPEX) and capital expenditure (CAPEX), commonly managed within a Kubernetes environment.

Storytime 2: Doubling Down!

Authoring a book is a monumental task. It's mentally and physically taxing, and filled with multifaceted challenges. Often, I advise against it—while anyone can write, not everyone can craft a thousand-page tome on a topic. Some strive to become authorities in their field; others pursue the title of author. For many, like myself, sharing knowledge takes precedence over a title or financial rewards. Success is not solely defined by monetary gains; it manifests diversely, encompassing career advancements, personal achievements, and community contributions through avenues like books, blogs, or software tools. In 2023, no one writes a book for financial gains as their primary reason, especially in technical domains where the return on investment hardly values each hour at $1. The current publishing industry landscape often requires you to author a thousand-page book like this based on passion. Additionally, my view of the world resembled Figure 6-6 for over ten months. I have used up two weeks of annual leave to complete this book on the promised date. There are a lot of personal sacrifices you must make, so, the last ten months have been all about work and authoring books.

Figure 6-6. *My view of the world during the entire year of 2023*

Delving deeper into Ansible Network Automation surpasses a mere exercise; my initial attempt at Ansible only scratched the surface, failing to grasp its full potential for network automation. Its true power lies in automating a myriad of mundane manual tasks. It had so much potential, yet I had to shift out of the Python programmer's mindset. Teaching Python for network automation to colleagues proved challenging, spurring my comprehensive mastery of Ansible. Approximately three years ago, I aspired to join Red Hat as an ISP network automation engineer, but my proficiency in Ansible fell short, closing that door. Determined to excel, I intensified my studies, aiming to become a subject matter expert in Ansible for network automation. However, mastering Linux automation proved crucial before delving deep into Ansible for network automation. Implementing Ansible extensively in my workplace, I documented experiences, shared insights through my blog, and conducted training sessions for colleagues. Yet, the aspiration to author a book on Ansible Network Automation lingered for nearly two years amid job transitions and adapting to new organizational cultures.

Figure 6-7. My previously published books

On the second anniversary of my second book's release, Apress's chief acquisition manager approached me to write the second edition of this book. The proposal was inviting: the first edition, *Introduction to Python Network Automation: The First Journey*, 2021, not only broke even but sold well. Eagerly, I agreed under one condition—to author the sequel, a book on Ansible Network Automation, before revisiting the second edition of my initial book. Accepting my terms, I completed my third book centered on Ansible Network Automation (see Figure 6-7) in six months. This marks my fourth and fifth books and the third book this year! Though hours in front of a computer are not my favorite activity, translating thoughts into words and publishing them brings immense satisfaction. If you plan to author a technical book, follow your passion, not just your rationale. Doubling down and embarking on the audacious task of penning two tomes in just ten months might seem a bit outlandish. This year has been an electrifying roller coaster ride, testing my limits and pushing the boundaries of what I thought possible in my writing journey. If you have been contemplating writing what you've experienced and sharing it with others has crossed your mind, consider penning a tome—you'll never know until you take that leap.

CHAPTER 7

Python Network Automation Virtual Lab 2: Sendmail Email Notification and Twilio SMS Notification on Docker

Welcome to Chapter 7, where you continue focusing on cost-efficient IT operations by harnessing the power of Docker for Python networks. This chapter is a comprehensive exploration of Docker's impact on expense optimization in Python networking. This chapter delves into various labs and exercises designed to enable efficient utilization of Docker for network automation. Beginning with an introduction to Docker components, account registration, and installation procedures, this chapter lays the groundwork for firsthand learning. You will dive into practical labs, starting with Sendmail tasks utilizing Docker images, followed by the development of email notification scripts within the Docker environment. Additionally, you explore CPU utilization monitoring labs, integrating API-driven Twilio for SMS notifications, highlighting the creation of Twilio accounts, Twilio Python module installation, and SMS setup. At the end of the chapter, you will have gained valuable insights into leveraging Docker's capabilities for efficient network automation, culminating in expense optimization strategies. This quick journey equips you with the firsthand skills to drive Docker effectively for Python network operations, emphasizing cost efficiency and lowering resource utilization.

Cost-Efficient IT Operations: Leveraging Docker for Python Networks

This chapter serves as the continuation of Python in a virtual environment, aiming to acquaint you with Docker concepts and alternative methods for implementing and utilizing Python in your network automation endeavors. In the previous chapter, you received a quick introduction to Python's `virtualenv` with Ansible. In this chapter, you start with a quick introduction to Docker, and the approach is hands-on. By remaining open to innovative work methods, you might discover flexible approaches that can often save time and resources by deviating from conventional practices. All organizations operate with OPEX (Operational Expenditure), CAPEX (Capital Expenditure), or both. OPEX involves the ongoing day-to-day expenses to sustain operations and generate revenue, while CAPEX represents funds used for acquiring or enhancing long-term assets vital for future revenue or operational efficiency. **Avoiding the waste of computing and human resources translates into cost savings, thereby yielding greater profits for IT companies**. In this chapter, you will discover how to establish a simple Python development environment within Docker. Additionally, you will learn to monitor your network devices' CPU utilization using a specially crafted Python application that integrates with a free API-driven Twilio SMS account, enabling you to send SMS messages to your smartphone through your Python API application. You will explore adapting to paid service tools on a free trial and writing custom Python programs accordingly. Enjoy the learning process as you delve into new ways of integrating different systems. **Fundamentally, saving cost is the main driving force behind the strong inclination of numerous organizations toward adopting automation: to efficiently manage their resource expenditures and actualize substantial cost savings, which leads to higher profitability.**

CHAPTER 7   PYTHON NETWORK AUTOMATION VIRTUAL LAB 2: SENDMAIL EMAIL NOTIFICATION AND TWILIO SMS NOTIFICATION ON DOCKER

Docker is a platform that allows you to package and run applications in containers. These containers are isolated environments that encapsulate the application along with its dependencies, ensuring consistency and portability across different environments. Virtual Machines (VMs) rely on a hypervisor to operate multiple OS instances. In contrast, Dockers utilize the Host OS directly, eliminating the need for a hypervisor. The Docker Hub acts as a repository, enabling, sharing, and accessing pre-built Docker images, thus streamlining application deployment and scalability. In the context of Python Network Automation, Docker offers several advantages. You don't need a dedicated VM to run your Python application's tasks—you only need a portion of the OSs' resources to run the packaged application on your Docker image. Also, the same image can be spun up as multiple Docker instances. Imagine Docker as a beneficial parasite on a host machine. It operates separately from the host but must run on top of a host OS, enabling multiple applications to run on the same server while efficiently sharing system resources. This sharing aspect is advantageous because it optimizes resource utilization. When applied to organizations aiming to curtail IT operational costs, Docker's ability to co-share resources becomes crucial. If organizations effectively plan for capacity and accurately predict future usage, they can leverage Docker to achieve more with less. For Python Network Automation, Docker provides a sandboxed environment where Python scripts and applications can be packaged into containers. These containers ensure that the Python environment and dependencies remain consistent, regardless of where they are deployed. This consistency streamlines the deployment process, making it easier to move applications between development, testing, and production environments. Moreover, Docker's lightweight (portable) nature allows for rapid deployment and scaling of Python-based network automation solutions. It facilitates the creation of modular, reusable components that can be orchestrated to automate various networking tasks. This flexibility aligns well with the dynamic and evolving nature of network infrastructures, enabling teams to adapt quickly to changing requirements.

In short, Docker is a platform for creating and managing containers, while Kubernetes is a platform for orchestrating and managing containerized applications at scale. Docker sparked the adoption of containers, and Kubernetes emerged to streamline the management of containerized applications. They work together, with Docker providing the containers and Kubernetes orchestrating them efficiently. Red Hat OpenShift is an enterprise container platform built on Kubernetes. It extends Kubernetes with additional features, tools, and support to simplify the deployment, management, and scaling of containerized applications, offering a streamlined developer experience and enterprise-grade support.

Expense Optimization in Python Networking: Docker, Kubernetes Impact

Understanding the financial impact of Docker as the container and Kubernetes as the orchestration tool in Python-based network automation application development is essential for strategic application development and planning. In layman's terms, Docker and Kubernetes are like a team that helps make computer programs work better and easier. Docker keeps everything needed for programs to run neatly packed, while Kubernetes helps manage and organize these packages efficiently. They save time and money by making sure programs run smoothly. Although there are popular Docker alternatives in the form of Podman, OpenShift, LXD, Docker Swarm, BuildKit, and Apache Mesos in the IT market today. Docker and Kubernetes still play instrumental roles in streamlining the development, deployment, and management of automation solutions. Their contributions to operational efficiency primarily align with operational expenditure (OPEX). Docker's containerization of Python scripts and dependencies ensures consistency across environments, optimizing resource utilization and enhancing development agility. Similarly, Kubernetes orchestrates these containers efficiently, supporting ongoing operations. These tools, considered part of day-to-day operations, contribute significantly to network automation capabilities, falling under OPEX due to their role in ensuring consistent performance and reducing deployment complexities.

However, as Python-based network automation applications evolve, substantial internal development efforts aimed at creating proprietary tools or enhancing automation frameworks might lead to considerations of capital expenditure (CAPEX). Instances where Docker or Kubernetes are significantly customized for long-term asset creation or contribute to enduring assets for network automation could shift expenses into the CAPEX realm. Strategic decisions around the allocation of resources for enhancing network automation capabilities must consider the potential long-term benefits against immediate operational expenses. Balancing ongoing operational costs with potential CAPEX implications involves careful financial planning, and aligning technological advancements with financial strategies for optimized expenses.

CHAPTER 7   PYTHON NETWORK AUTOMATION VIRTUAL LAB 2: SENDMAIL EMAIL NOTIFICATION AND TWILIO SMS NOTIFICATION ON DOCKER

Navigating the financial categorization of Docker and Kubernetes expenses requires collaboration with your finance team or accountants within your organization. Their expertise ensures accurate classification based on the organization's specific circumstances and accounting guidelines. This collaboration facilitates a balanced approach, leveraging the operational efficiency gains from Docker and Kubernetes for Python-based network automation applications while aligning with the organization's financial objectives. Continuously adapting financial strategies to optimize expenses and maximize technology potential remains an ongoing process, integrating financial planning with the dynamic advancements in Python-based network automation applications and associated technologies.

You've explored the recent trends, observing a shift toward multi-cloud infrastructures and the rising prevalence of software-defined environments. **With this evolution, many organizations are considering serverless cloud services for application development, even eliminating the need for a dedicated Docker instance.** The landscape for automation engineers and developers is undoubtedly rich with an array of solutions, providing an abundance of options to deploy their applications. There are more possibilities than before, from traditional on-premises servers offering familiarity and control to the flexibility of containers, allowing for efficient packaging and deployment. Additionally, the advent of cloud services has opened avenues for code as a service, empowering them to harness the scalability and convenience of cloud environments. This wealth of choices grants them the freedom to tailor their deployment strategies, ensuring optimal performance, scalability, and adaptability for their applications across diverse technological landscapes.

Python Network Automation Virtual Lab 2: Docker and Twilio SMS

This chapter touches on Docker as a resource-saving tool in production. Docker is a containerized solution that runs on Linux or Windows systems, finding many use cases in enterprise network automation. However, Docker is primarily used as an instance on a Linux server to run and execute tasks or applications, thanks to its small footprint that can operate where a full virtual machine might be excessive for a single task, such as running a nightly system check on a dedicated VM. Docker and Kubernetes present an interesting proposition and serve as stepping stones before transitioning into full serverless computing in the public cloud. Serverless computing, as seen in AWS Lambda, Azure Functions, and Google Cloud Functions, eliminates the need for a dedicated server and operates on-demand as the application code executes and runs. So, if designed and used properly, it can offer a lot of cost savings for various custom-developed enterprise applications. However, these topics are beyond the scope of this book. Here, you will be guided through setting up the Docker lab, installing Docker, and downloading my purpose-built Python Network Automation Docker container from the official Docker Hub. You will learn how to use a Docker container as a substitute for an actual virtual machine. In certain scenarios, Docker might be the optimal alternative environment to run standalone Python applications, efficiently utilizing computing power without unnecessary resource waste.

Lab 1: Sendmail Lab Using an Imported Docker Image

Docker, an open-source platform for developers and system administrators, champions the "Build, ship, run" approach. It offers scalable container management services, optimizing server resources for deploying Linux network automation servers. Introduced in March 2013, Docker is a containerization tool based on platform-as-a-service (PAAS) software, using virtualization to isolate software into containers with well-defined communication channels. These act as self-contained units, like virtual machines but without their kernels or operating systems. They rely on the host server's kernel. Containerization, vital in Agile and DevOps-based projects, maximize server efficiency. Understanding Docker is pivotal to grasping Kubernetes (often abbreviated as "K8s"), which extends Docker's containerization concept. Proficiency in Docker's principles, container creation, and image management is essential to understanding Kubernetes' orchestration of containers within larger systems. Mastering Docker aids comprehension of Kubernetes' automated deployment, scaling, and management for diverse applications, saving companies substantial operational costs by optimizing IT resources.

Docker's slogan, "develop, ship, and run anywhere," underscores its role as a developer's tool, facilitating swift application development and deployment into network-deployable containers. Docker's minimal footprint utilizes host server resources, distinct from a full virtual machine, functioning through runnable instances of snapshot images. Containers operate with their file system, networking, and isolated process tree, simplifying collaboration across development, Quality Assurance (QA), and operations teams. Docker containers deploy seamlessly across physical, virtual, or cloud environments.

It's not possible to master Docker in a single chapter, but a hands-on lab with real Docker images can speed up learning. Self-paced learning beyond this book is encouraged. To accelerate your learning journey, I've created a Docker image available for download from Docker Hub, specifically tailored for this Python Network Automation lab. Utilize this image from an Ubuntu or Fedora Server. An Internet connection is required, and the topology and overview of what's ahead for use are illustrated in Figure 7-1.

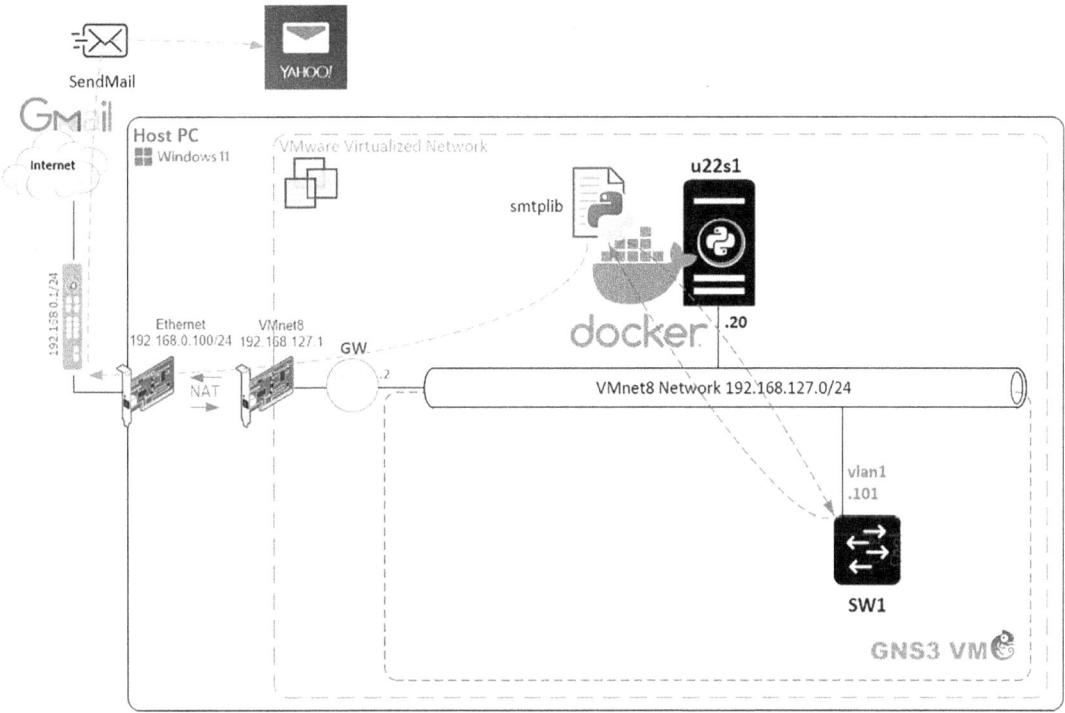

Figure 7-1. Sendmail Docker lab used devices

CHAPTER 7   PYTHON NETWORK AUTOMATION VIRTUAL LAB 2: SENDMAIL EMAIL NOTIFICATION
AND TWILIO SMS NOTIFICATION ON DOCKER

Docker Components, Account Registration, and Installation

Docker operates like a purpose-built virtual machine, leveraging the host's Linux kernels to function with minimal resource consumption. It facilitates running multiple Docker containers on a single platform, alleviating system resource contention issues. Let's explore the essential components composing Docker, focusing on practical insights rather than exhaustive details.

Key Docker components include:

- Docker boasts cross-platform support, seamlessly installing on Linux, macOS, and Windows as a service.

- The Docker Engine serves to build Docker images and create containers efficiently.

- Docker Hub, a registry, hosts a myriad of Docker images, streamlining access to diverse resources.

- Docker Compose defines applications through multiple Docker containers, enhancing application structuring.

Registering a Docker account is essential for leveraging Docker's full suite online. Post-registration, users gain access to a wealth of purpose-built Docker images, simplifying application deployment. You need a Docker Hub account to log in and download the Docker images. To prepare for this lab, visit the following URL and register your Docker Hub account: `https://hub.docker.com/`. This step is crucial for following this section. Once you've registered, take some time to acquaint yourself with Docker's functionality by visiting the Docker Hub page: `https://www.docker.com/get-started/`.

Docker Installation Procedure

In this lab, all tasks will be executed from the Ubuntu Server. Follow these straightforward steps to install Docker on your `u22s1` server. Fortunately, Docker's installation on Ubuntu is hassle-free, and within 15 minutes, you will be up and running.

CHAPTER 7   PYTHON NETWORK AUTOMATION VIRTUAL LAB 2: SENDMAIL EMAIL NOTIFICATION
            AND TWILIO SMS NOTIFICATION ON DOCKER

#	Task

①	As always, begin with an `apt update` command and upgrade your Ubuntu 22 LTS Server. This will take a few minutes if you have not updated your server package for some time.
	
	`jdoe@u22s1:~$ `**`sudo apt update`**
	`[sudo] password for jdoe: ***********`
	`[...omitted for brevity]`
	`jdoe@u22s1:~$ `**`sudo apt upgrade -y`**
	`[...omitted for brevity]`

②	Enter the following command to download and install the Docker package: `jdoe@u22s1:~$ `**`sudo apt install docker.io -y`** `[...omitted for brevity]`

③	To enable Docker, enter the `enable` command and use the `docker -version` command to check which version is installed.
	
	`jdoe@u22s1:~$ `**`sudo systemctl enable --now docker.service`**
	`jdoe@u22s1:~$ `**`sudo systemctl status docker`**
	`● docker.service - Docker Application Container Engine`
	  `Loaded: loaded (/lib/systemd/system/docker.service; enabled; vendor preset: enabled)`
	  `Active: active (running) since Sat 2023-11-18 12:39:04 AEDT; 2min 6s ago`
	`TriggeredBy: ● docker.socket`
	  `Docs: https://docs.docker.com`
	`Main PID: 36048 (dockerd)`
	`Tasks: 8`
	`Memory: 28.4M`
	  `CPU: 390ms`
	  `CGroup: /system.slice/docker.service`
	     `└─36048 /usr/bin/dockerd -H fd:// --containerd=/run/containerd/containerd.sock`
	`[...omitted for brevity]`
	`jdoe@u22s1:~$ `**`sudo docker --version`**
	`Docker version 24.0.5, build 24.0.5-0ubuntu1~22.04.1`

(continued)

CHAPTER 7   PYTHON NETWORK AUTOMATION VIRTUAL LAB 2: SENDMAIL EMAIL NOTIFICATION
            AND TWILIO SMS NOTIFICATION ON DOCKER

#	Task
④	Add yourself to the Docker group to run the sudo commands. Replace jdoe with your user ID. Run the last command for the change to take effect. jdoe@u22s1:~$ **sudo usermod -aG docker jdoe** jdoe@u22s1:~$ **sudo gpasswd -a $USER docker** Adding user jdoe to group docker jdoe@u22s1:~$ **newgrp docker**
⑤	Test Docker by running the hello-world command, which will open a container to run the hello-world command. jdoe@u22s1:~$ **docker run hello-world** Unable to find image 'hello-world:latest' locally latest: Pulling from library/hello-world 719385e32844: Pull complete Digest: sha256:c79d06dfdfd3d3eb04cafd0dc2bacab0992ebc243e083cabe208bac4dd7759e0 Status: Downloaded newer image for hello-world:latest Hello from Docker! [...omitted for brevity]
⑥	Run docker ps -a to check if the hello-world command ran successfully. Also, run the docker images command to check the downloaded Docker images. jdoe@u22s1:~$ **docker ps -a** CONTAINER ID IMAGE COMMAND CREATED STATUS PORTS NAMES 10744d60f2e1 hello-world "/hello" 2 minutes ago Exited (0) 2 minutes ago boring_bardeen jdoe@u22s1:~$ **docker images** REPOSITORY TAG IMAGE ID CREATED SIZE hello-world latest 9c7a54a9a43c 6 months ago 13.3kB

414

CHAPTER 7 PYTHON NETWORK AUTOMATION VIRTUAL LAB 2: SENDMAIL EMAIL NOTIFICATION AND TWILIO SMS NOTIFICATION ON DOCKER

Test-Driving Docker

A Python Network Automation Docker image has been pre-built for readers of this book, using the Dockerfile method to save time. The image you are about to download onto your u22s1 server is an Ubuntu 20.04 LTS Server, equipped with numerous pre-installed IP services and network automation libraries. Creating a template Docker image from a Dockerfile involves multiple steps, a topic beyond the scope of this book, potentially warranting its dedicated chapter. Consider exploring the Internet, YouTube, and ChatGPT for basic Docker training. In this lab, you download my pre-installed Python Network Automation Docker image and execute it on your Ubuntu Server.

Table 7-1 lists the top 20 Docker commands for your review. The commands that will be used in this book have been marked in bold. For other commands, you can study them from Docker's documentation page at https://www.docker.com/get-started/.

Table 7-1. *Docker Top 20 Commands*

Command	Description
docker run	Launches a new container from an image
docker ps	Lists running containers
docker stop	Halts a running container gracefully
docker start	Starts a stopped container
docker pull	Fetches an image from a registry
docker rm	Removes one or more containers
docker rmi	Deletes one or more images
docker exec	Runs a command inside a running container
docker logs	Displays logs of a container
docker-compose	Manages multi-container applications with a YAML configuration
docker images	Lists available images
docker network ls	Lists available networks
docker volume ls	Lists available volumes

(*continued*)

CHAPTER 7   PYTHON NETWORK AUTOMATION VIRTUAL LAB 2: SENDMAIL EMAIL NOTIFICATION
            AND TWILIO SMS NOTIFICATION ON DOCKER

Table 7-1. (*continued*)

Command	Description
`docker inspect`	Displays detailed information about a container, image, network, or volume
`docker build`	Builds an image from a Dockerfile
`docker-compose up`	Builds, starts, and attaches to containers for a service
`docker-compose down`	Stops and removes containers, networks, volumes, and images
`docker-compose logs`	Displays logs for services in a Docker Compose file
`docker stats`	Displays live system resource usage statistics for containers
`docker login`	Logs into a Docker registry

Now it's time to dive in and get going.

#	Task
①	After you have created a Docker Hub ID, go to the following Docker Hub site to review the image you will download. URL: https://hub.docker.com/r/pynetauto/pynetauto_ubuntu20
②	From your Linux command line, use `docker login` to log in to the Docker Hub. Replace the username with yours and enter your password to log in. You need to get a "Login Succeeded" message at the bottom. When you create an account and log in to Docker Hub, you can push the Docker image to Docker Hub, save the image as a template, and share it with others in the community. jdoe@u22s1:~$ **docker login** Log in with your Docker ID to push and pull images from the Docker Hub. If you don't have a Docker ID, head over to https://hub.docker.com to create one. Username: **pynetauto** Password: *********

(*continued*)

#	Task
	WARNING! Your password will be stored unencrypted in /home/jdoe/.docker/config.json. Configure a credential helper to remove this warning. See https://docs.docker.com/engine/reference/commandline/login/#credentials-store Login Succeeded
③	Use the following docker pull command to download this image onto your virtual machine's Docker image repository. The Docker image is about 1.69GB, so it is recommended that you connect to the Internet through your home network, not on a 4G or 5G network, as it will use up your mobile data. jdoe@u22s1:~$ **docker pull pynetauto/pynetauto_ubuntu20:latest** latest: Pulling from pynetauto/pynetauto_ubuntu20 d51af753c3d3: Pull complete *[... omitted for brevity]* f92a9a96eae3: Pull complete Digest: sha256:86f2178825cf09a1b7f7c370a460b34b109f62ce4471d944ef108d0a29162ed4 Status: Downloaded newer image for pynetauto/pynetauto_ubuntu20:latest docker.io/pynetauto/pynetauto_ubuntu20:latest
④	Using the docker images command, check the new image with the correct details, as shown here: jdoe@u22s1:~$ **docker images** REPOSITORY TAG IMAGE ID CREATED SIZE hello-world latest 9c7a54a9a43c 6 months ago 13.3kB pynetauto/pynetauto_ubuntu20 latest 39ea52cc1e39 3 years ago 1.69GB

(continued)

CHAPTER 7   PYTHON NETWORK AUTOMATION VIRTUAL LAB 2: SENDMAIL EMAIL NOTIFICATION
            AND TWILIO SMS NOTIFICATION ON DOCKER

#	Task
⑤	To initiate a Docker container instance in a bash shell with a local server directory mounting point with the container, use the exact command shown here. Be sure to replace the local username with your own. jdoe@u22s1:~$ **docker run -it --entrypoint=/bin/bash -v /home/jdoe/mnt/:/home --name pynetauto_ubuntu20 pynetauto/pynetauto_ubuntu20** root@05121eddd81b:/# Here is the breakdown and explanation of this command: docker run -it --entrypoint=/bin/bash: Run Docker with the -i and -t options where the entry point is the bash shell. -i denotes interactive mode, whereas -t denotes allocation of a pseudo-tty. -v /home/jdoe/mnt/:/home: -v is related to a volume (shared file system); in this case, u22s1's /home/jdoe/mnt directory is linked to the /home directory of the new Docker container. Replace your user ID accordingly. --name pynetauto_ubuntu20: The --name option allows you to provide a more meaningful name to this instance of the container. If –name is not used, then Docker will assign a random name automatically. pynetauto/pynetauto_ubuntu20: This is the actual Docker image name in the local image pool. For full Docker run references, visit the following site. URL: https://docs.docker.com/engine/reference/run/

(*continued*)

#	Task
⑥	Check the Docker image version, Python version, and Python modules installed on the pynetauto_ubuntu20 Docker container. root@05121eddd81b:/# **cat /etc/*release** DISTRIB_ID=Ubuntu DISTRIB_RELEASE=20.04 *[...omitted for brevity]* UBUNTU_CODENAME=focal root@05121eddd81b:/# **python3 --version** Python 3.8.2 root@05121eddd81b:/# **pip freeze** aiohttp==3.6.2 *[...omitted for brevity]* yarl==1.5.1 Most of the commands are the same as the Linux standard commands and should behave like another Linux machine. However, the basic services, systemctl, or Linux standard software are not available from the Docker container instance; in other words, you must precisely specify which Linux software you want to install during the Docker image build process. If you want to install more software on an existing Docker image, you can make the modifications. However, the Docker base image modification topic is outside of this book's scope.
⑦	Create a file called testfile999.txt within the Docker bash shell. You are going to examine its connection to your host Ubuntu 22 Server, allowing file access between your Docker Ubuntu 20 Server and the host Ubuntu 22 Server. This resembles running a virtual Linux server within your existing Linux server like nested VMs in virtualization terms. root@05121eddd81b:/# **pwd** / root@05121eddd81b:/# **touch home/testfile999.txt** root@05121eddd81b:/# **ls /home** testfile999.txt

(continued)

CHAPTER 7   PYTHON NETWORK AUTOMATION VIRTUAL LAB 2: SENDMAIL EMAIL NOTIFICATION
            AND TWILIO SMS NOTIFICATION ON DOCKER

#	Task

⑧	Detach from the container and inspect the newly created file on the host PC. Use Ctrl+P followed by Ctrl+Q to detach from the Docker container, returning to your Linux host OS. Verify the presence of the `testfile999.txt` file in the `/home/jdoe/mnt` folder, located in the Linux server's directory.
	
	`root@05121eddd81b:/# jdoe@u22s1:~$ # Press 'Ctrl+P', followed by 'Ctrl+Q'`
	`jdoe@u22s1:~$ `**`pwd`**
	`/home/jdoe # Host Ubuntu 22`
	`jdoe@u22s1:~$ `**`ls /home/jdoe/mnt`**
	`testfile999.txt # file is available from the mount point`

⑨	From your Linux host, create a new file under the shared directory. Then, reattach to the Docker instance to check the file from the Docker container instance. When you reattach to the running Docker instance, you can use either the container ID or names.
	
	`jdoe@u22s1:~$ `**`sudo touch /home/jdoe/mnt/testfile123.txt`**
	`[sudo] password for jdoe: **********`
	`jdoe@u22s1:~$ `**`docker ps`**
	`CONTAINER ID    IMAGE                            COMMAND         CREATED`
	`STATUS            PORTS                          NAMES`
	`05121eddd81b    pynetauto/pynetauto_ubuntu20    "/bin/bash"     14`
	`minutes ago    Up 14 minutes    20-22/tcp, 25/tcp, 12020-12025/tcp`
	`pynetauto_ubuntu20`
	`jdoe@u22s1:~$ `**`docker attach pynetauto_ubuntu20`**
	`root@05121eddd81b:/# `**`ls /home`**
	`testfile123.txt   testfile999.txt`

(*continued*)

#	Task
⑩	To stop and exit Docker, press Ctrl+D or type `exit`. This will stop and exit the currently logged-in Docker container. root@05121eddd81b:/# **exit** exit jdoe@u22s1:~$ **docker ps -a** CONTAINER ID IMAGE COMMAND CREATED STATUS PORTS NAMES 05121eddd81b pynetauto/pynetauto_ubuntu20 "/bin/bash" 16 minutes ago Exited (130) 6 seconds ago pynetauto_ubuntu20 10744d60f2e1 hello-world "/hello" 35 minutes ago Exited (0) 35 minutes ago boring_bardeen
⑪	To restart a stopped Docker instance, use the `docker start instance_name` command. Check the status and reattach it to the Docker instance. jdoe@u22s1:~$ **docker start pynetauto_ubuntu20** pynetauto_ubuntu20 jdoe@u22s1:~$ **docker ps -a** CONTAINER ID IMAGE COMMAND CREATED STATUS PORTS NAMES 05121eddd81b pynetauto/pynetauto_ubuntu20 "/bin/bash" 18 minutes ago Up 6 seconds 20-22/tcp, 25/tcp, 12020-12025/tcp pynetauto_ubuntu20 10744d60f2e1 hello-world "/hello" 37 minutes ago Exited (0) 37 minutes ago boring_bardeen jdoe@u22s1:~$ **docker attach pynetauto_ubuntu20** root@05121eddd81b:/#

(continued)

#	Task
⑫	To prune non-running Docker instances and keep your environment clean, run the `docker system prune` command. You are pruning the `hello-world` image you created earlier. This action does not delete the image—you will do this in the next step.

```
root@05121eddd81b:/# read escape sequence # 'Ctrl+P', then 'Ctrl+Q'
jdoe@u22s1:~$ docker system prune
WARNING! This will remove:
- all stopped containers
- all networks not used by at least one container
- all dangling images
- all dangling build cache
Are you sure you want to continue? [y/N] y
Deleted Containers:
10744d60f2e149645e3f50c5a7b852b1b2ffde01ee65dc1a20283fbbe0e03b9b
Total reclaimed space: 0B
jdoe@u22s1:~$ docker ps -a
CONTAINER ID    IMAGE                              COMMAND          CREATED
STATUS          PORTS                              NAMES
05121eddd81b    pynetauto/pynetauto_ubuntu20      "/bin/bash"      20
minutes ago     Up 2 minutes    20-22/tcp, 25/tcp, 12020-12025/tcp
pynetauto_ubuntu20
``` |

(continued)

CHAPTER 7　PYTHON NETWORK AUTOMATION VIRTUAL LAB 2: SENDMAIL EMAIL NOTIFICATION AND TWILIO SMS NOTIFICATION ON DOCKER

#	Task
| ⑬ | To delete Docker images, use `docker rmi image_name:version`. At the end of this task, you should only have the `pynetauto_ubuntu20` container image available for your next lab.<br><br>```
jdoe@u22s1:~$ docker images
REPOSITORY                    TAG      IMAGE ID       CREATED       SIZE
hello-world                   latest   9c7a54a9a43c   6 months ago  13.3kB
pynetauto/pynetauto_ubuntu20  latest   39ea52cc1e39   3 years ago   1.69GB
jdoe@u22s1:~$ docker rmi hello-world:latest
Untagged: hello-world:latest
Untagged: hello-world@sha256:c79d06dfdfd3d3eb04cafd0dc2bacab0992ebc243e083cabe208bac4dd7759e0
Deleted: sha256:9c7a54a9a43cca047013b82af109fe963fde787f63f9e016fdc3384500c2823d
Deleted: sha256:01bb4fce3eb1b56b05adf99504dafd31907a5aadac736e36b27595c8b92f07f1
jdoe@u22s1:~$ docker images
REPOSITORY                    TAG      IMAGE ID       CREATED      SIZE
pynetauto/pynetauto_ubuntu20  latest   39ea52cc1e39   3 years ago  1.69GB
``` |

Great job! You've tackled the fundamental Docker procedures. You are set to advance to the following lab, where you will delve into sending emails using your Docker image server.
```

CHAPTER 7   PYTHON NETWORK AUTOMATION VIRTUAL LAB 2: SENDMAIL EMAIL NOTIFICATION AND TWILIO SMS NOTIFICATION ON DOCKER

Lab 2: Sendmail Python Lab Using Docker

Now that you are getting familiar with Docker, you'll learn how to benefit from Docker and run your Python scripts. In this lab, you use the pre-installed Sendmail on your `pynetauto/pynetauto_ubuntu20` latest Docker container and send a test email from your Python script. Then you will be guided through writing a Python script to monitor the CPU utilization of one of the devices in the topology. When the utilization goes above the yellow watermark, your Python script will trigger an email alert to your email inbox. First, let's get you set up for Sendmail on your Docker image. For Sendmail to work on your Docker container, Sendmail needs to be pre-installed and configured, and port 25 must be opened up for SMTP traffic. Also, to receive the test email, your email account security level must be lowered on your SMTP server side. In this example, a Gmail account is used for demonstration purposes, and you can follow the steps to achieve the same result. If you already have an exchange server up and running, you may replace the public email accounts with your internal PoC lab email accounts.

Lab Prerequisites

To make the Sendmail labs work, you are going to need two test email accounts. One is to send the email using Python's `smtplib` library from the Docker image's Sendmail service and another email to receive the email. This can be any free email service provider. I could use two testing email accounts from Gmail, but to make things a little more interesting, I am using a Gmail account as the sender account with an app password enabled and a Yahoo account to receive the sent email. Replace the testing accounts with your accounts and follow these instructions carefully to get your Gmail app password. Note, to get your app password, you must first enable multifactor authentication on your Gmail account. Follow the instructions.

Warning

This lab requires two test email accounts

I've set up two test accounts on different public email services: Gmail and Yahoo Mail. To follow along, you can create an account on each email service provider with your details. Alternatively, you can create two test accounts solely on the Gmail service.

Here are my test emails:

Gmail account: pynetauto1@gmail.com # Replace this sender email

Yahoo Mail account: pynetauto@yahoo.com # Replace this recipient email

Substitute these emails with your test email account aliases.

In the first edition of this book, you used the "Less Secure Apps" feature to perform the Docker Sendmail Python email testing, but since May 30, 2022, Google no longer supports the use of third-party apps or devices that ask you to sign in to your Google account using only your username and password. To work around this problem, you must enable the multifactor authentication (MFA) on your Gmail account to enhance its security and then use the app password method to generate the key to be able to send emails from your account.

Here are the steps to enable MFA and create an app password (passkey) for your Python application testing on Docker's Sendmail service.

To enable MFA on your Gmail account, follow these steps:

#	Task
①	Sign in to your Google Account: https://myaccount.google.com/
②	Find two-step verification. Look for the Signing In To Google section. Under this, locate and click 2-Step Verification.
③	Start setup. You may need to re-enter your Gmail password to confirm. Click Get Started to begin setting up 2-Step Verification.
④	Enter phone number. At the phone number prompt for verification purposes, enter your cell/mobile phone number and choose whether you want to receive a verification code via text message or phone call.
⑤	Verify your phone number. Google will send a verification code to the phone number you provided.
⑥	Enter this code to verify your phone.

(*continued*)

CHAPTER 7   PYTHON NETWORK AUTOMATION VIRTUAL LAB 2: SENDMAIL EMAIL NOTIFICATION
              AND TWILIO SMS NOTIFICATION ON DOCKER

#	Task
⑦	Enable 2-step verification. Once your phone is verified, you can turn on 2-Step Verification. Follow the on-screen instructions to enable this feature for your account.
⑧	Access security settings. Once you're logged in, go to Manage My Google Account, then select Security from the left sidebar. You may need to re-enter your password for security verification.

Reference URL: https://support.google.com/accounts/answer/185839?hl=en&co=GENIE.Platform%3DDesktop

After enabling 2-Step Verification, you must create an app password (also known as an app-specific password) for your Docker Sendmail Python applications that do not support MFA directly. Follow these steps to do so.

#	Task
①	Go to Manage Your Google Account, use the search bar to search for "app passwords," and select it when it appears on the search result, as shown in Figure 7-2.

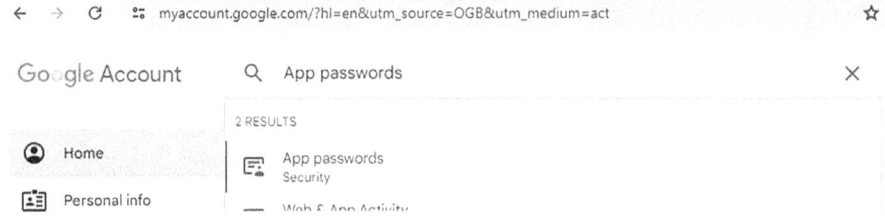

Figure 7-2. *Search for App passwords in your Google account*

(*continued*)

#	Task
②	When the App Passwords window opens, type a name for your app password. I named mine Docker_Sendmail_Python_App, but you can give any meaningful name you prefer (see Figure 7-3).

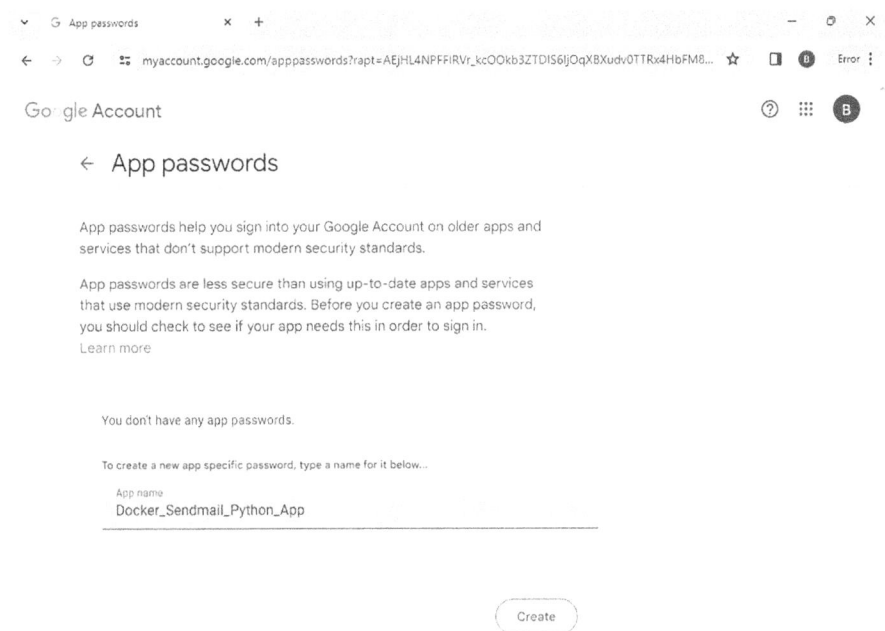

Figure 7-3. Name the app passwords

(continued)

#	Task
③	After you have created the app name, you can click the Create Password button to generate your app password. This only appears once, so make sure you document this information for later use. This is the required information for your Sendmail email script to trigger SMTP email triggers from your sender account (see Figure 7-4).

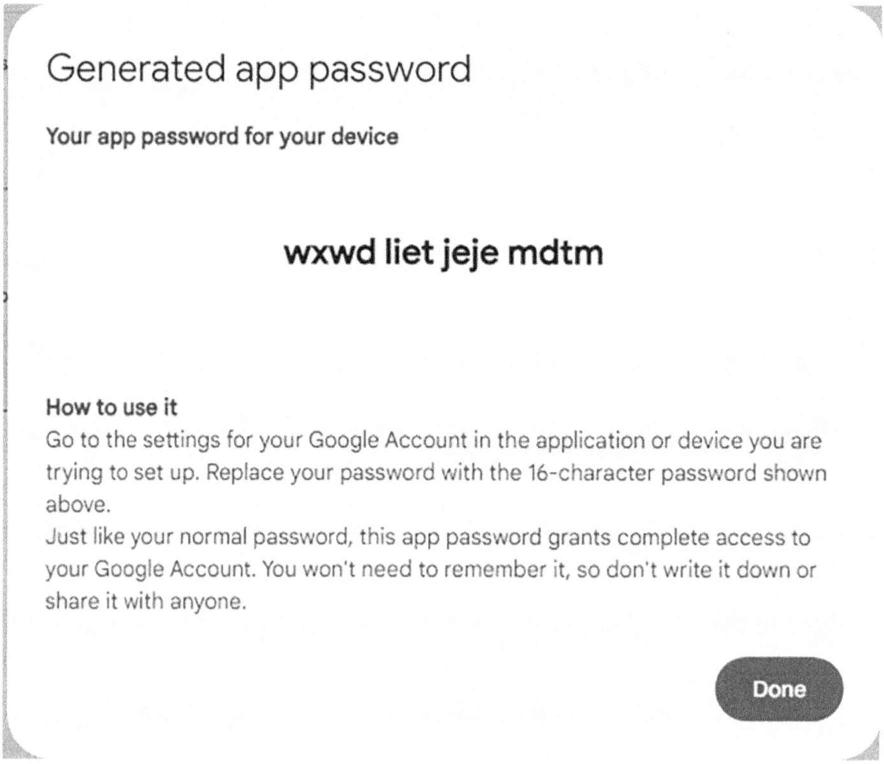

Generated app password

Your app password for your device

wxwd liet jeje mdtm

How to use it
Go to the settings for your Google Account in the application or device you are trying to set up. Replace your password with the 16-character password shown above.
Just like your normal password, this app password grants complete access to your Google Account. You won't need to remember it, so don't write it down or share it with anyone.

Done

Figure 7-4. Generate an app password

(*continued*)

#	Task
④	Now you can go back to your email home page and continue with the lab tasks (see Figure 7-5).

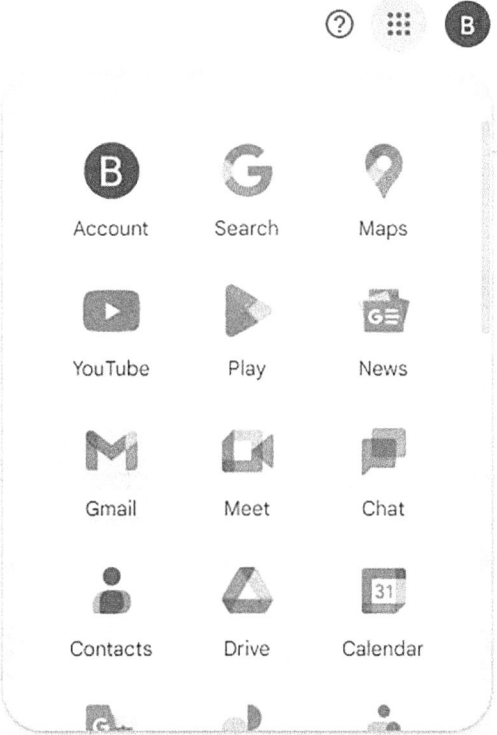

Figure 7-5. *Go back to your Gmail home page*

Now it's time to kick off the practical part of this lab.

CHAPTER 7   PYTHON NETWORK AUTOMATION VIRTUAL LAB 2: SENDMAIL EMAIL NOTIFICATION
            AND TWILIO SMS NOTIFICATION ON DOCKER

#	Task
①	If you followed the Docker installation processes in the previous steps, you should have the Docker image named pynetauto_ubuntu20. Also, the previous Docker container is currently running in this environment as shown. jdoe@u22s1:~$ **docker images** REPOSITORY TAG IMAGE ID CREATED SIZE pynetauto/pynetauto_ubuntu20 latest 39ea52cc1e39 3 years ago 1.69GB jdoe@u22s1:~$ **docker ps -a** CONTAINER ID IMAGE COMMAND CREATED STATUS PORTS NAMES 05121eddd81b pynetauto/pynetauto_ubuntu20 "/bin/bash" 44 minutes ago Up 26 minutes 20-22/tcp, 25/tcp, 12020-12025/tcp pynetauto_ubuntu20
②	Start a new Docker instance by rerunning the following command. You will let the previous Docker instance with the ID, 05121eddd81b run. Create a new Docker instance here. When your new Docker instance starts, you will be logged in to the bash shell of your new Docker instance with the new container ID and should be ready to start the real lab. jdoe@u22s1:~$ **docker run -it --entrypoint=/bin/bash -v /home/jdoe/mnt/:/home --name pynetauto_ubuntu20sendmail pynetauto/pynetauto_ubuntu20** root@22ac05fa791a:/#

(continued)

#	Task
③	Check if port 25 is in listening mode. Check for open ports using the netstat -tuna command. No result was returned, so it looks like you will have to configure Sendmail and allow port 25 on this Docker instance or machine. root@22ac05fa791a:/# **netstat -tuna** Active Internet connections (servers and established) Proto Recv-Q Send-Q Local Address Foreign Address State Also, check the Sendmail installation status by quickly running apt install Sendmail; it should be installed already. Unlike the real Linux host, the Docker container does not allow you to check the Sendmail using the standard systemctl command, but it does allow you to use the other commands. Use one of these commands to check the installation of Sendmail. root@22ac05fa791a:/# **apt install sendmail** Reading package lists... Done Building dependency tree Reading state information... Done sendmail is already the newest version (8.15.2-18). 0 upgraded, 0 newly installed, 0 to remove, and 2 not upgraded. root@22ac05fa791a:/# **ps aux | grep sendmail** root 52 0.0 0.0 3312 732 pts/0 S+ 14:00 0:00 grep --color=auto sendmail root@22ac05fa791a:/# **service sendmail status** MSP: is run via cron (20m) MTA: is not running QUE: Same as MTA

(continued)

CHAPTER 7   PYTHON NETWORK AUTOMATION VIRTUAL LAB 2: SENDMAIL EMAIL NOTIFICATION
AND TWILIO SMS NOTIFICATION ON DOCKER

#	Task
④	Now you'll configure Sendmail on your Docker container instance. Execute the sendmailconfig command. When prompted for confirmation (Y), press Y and then the Enter key three times. Following the second Y, it might take a moment for the files to update, so exercise patience. Once it's done, proceed to reload the Sendmail service for the changes to take effect. root@22ac05fa791a:/# **sendmailconfig** Configure sendmail with the existing /etc/mail/sendmail.conf? [Y] **Y** Reading configuration from /etc/mail/sendmail.conf. *[... omitted for brevity]* Configure sendmail with the existing /etc/mail/sendmail.mc? [Y] **Y** # may take a few minutes Updating sendmail environment ... *[... omitted for brevity]* Reload the running sendmail now with the new configuration? [Y] **Y** Reloading sendmail ...
⑤	Use the following command to check that Sendmail-related directories have been created successfully on Docker: root@22ac05fa791a:/# **ls /usr/sbin/send*** /usr/sbin/sendmail /usr/sbin/sendmail-msp /usr/sbin/sendmail-mta /usr/sbin/sendmailconfig
⑥	Run the netstat -tuna command again to confirm that port 25 is in listening mode. root@22ac05fa791a:/# **netstat -tuna** Active Internet connections (servers and established) Proto Recv-Q Send-Q Local Address Foreign Address State tcp 0 0 127.0.0.1:587 0.0.0.0:* LISTEN tcp 0 0 127.0.0.1:25 0.0.0.0:* LISTEN

(continued)

#	Task
⑦	Go back to your Docker instance; you will send two test emails to yourself using two different email-sending methods, using simple Python scripts. First, create a Python file and copy the contents of the following script, which uses email.mime.text and the Linux subprocess. Update the testing email addresses to your email addresses. ```\nroot@22ac05fa791a:/# cd /home\nroot@22ac05fa791a:/# nano docker_sendmail_01.py\nroot@22ac05fa791a:/# cat docker_sendmail_01.py\n# This email app uses Sendmail with Gmail App password\n# to send Python test emails to recipients. Use with care.\nimport smtplib\nfrom email.mime.text import MIMEText\n\nsender_email = "pynetauto1@gmail.com" # Gmail with App password\nrecipient_email = "pynetauto@yahoo.com" # Recipient email for testing\nsubject = "Docker Sendmail Python Email App Test #01"\nbody = (\n"This is a test email sent using Docker Sendmail Python Email Application utilizing. "\n"Gmail's SMTP server with app passwords. You can safely discard this test email. Thank!"\n)\n\n# Create the email message\nmsg = MIMEText(body)\nmsg["From"] = sender_email\nmsg["To"] = recipient_email\nmsg["Subject"] = subject\n\n# Gmail SMTP server configuration\nsmtp_server = "smtp.gmail.com"\nsmtp_port = 587 # Gmail SMTP port\n```

(*continued*)

#	Task
| | ```python
# Your Gmail account credentials
gmail_user = sender_email
gmail_password = "wxwdlietjejemdtm"  # Replace with your app password

try:
    # Connect to Gmail's SMTP server
    server = smtplib.SMTP(smtp_server, smtp_port)
    server.starttls()  # Enable TLS encryption
    server.login(gmail_user, gmail_password)  # Log in to your Gmail account
    # Send email
    server.sendmail(gmail_user, recipient_email, msg.as_string())
    print("Email sent using Gmail's SMTP server with app passwords!")
    # Quit the server
    server.quit()

except Exception as e:
    print(f"Something went wrong: {str(e)}")

print("Your requested task has been completed!")
``` |
| ⑧ | After you complete the script, run it from your Docker instance. There will be a slight delay in sending the email. Wait about two to five minutes for the email to be sent (see Figure 7-6).

root@22ac05fa791a:/home# **python3 docker_sendmail_01.py**
Email sent using Gmail's SMTP server with app passwords!
Your requested task has been completed!

After the script runs, go to your sender email account and check the Sent folder. |

(continued)
```

CHAPTER 7   PYTHON NETWORK AUTOMATION VIRTUAL LAB 2: SENDMAIL EMAIL NOTIFICATION AND TWILIO SMS NOTIFICATION ON DOCKER

#	Task

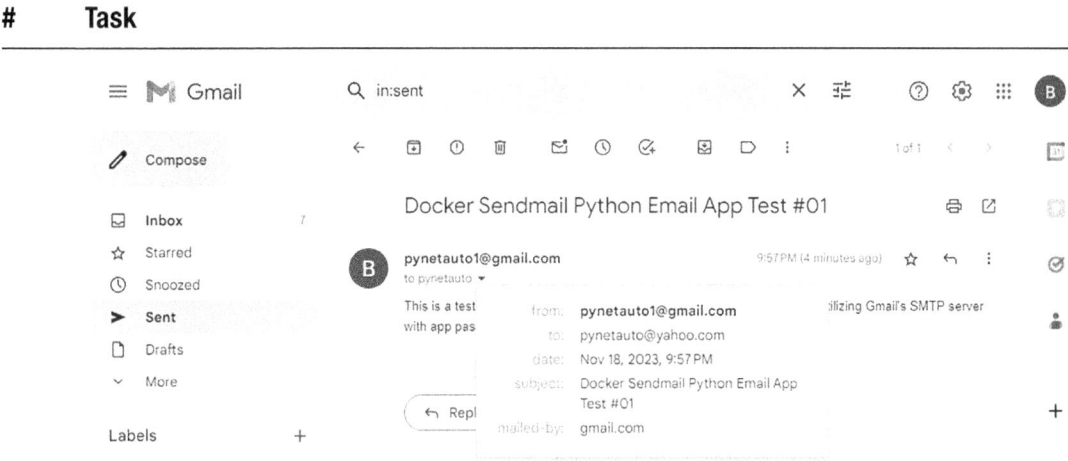

Figure 7-6. Check the sender's Gmail Sent folder

⑨	Go to your recipient email account's inbox or spam. I have used a Yahoo test account and it received the email and saved it to the Spam folder (see Figure 7-7).

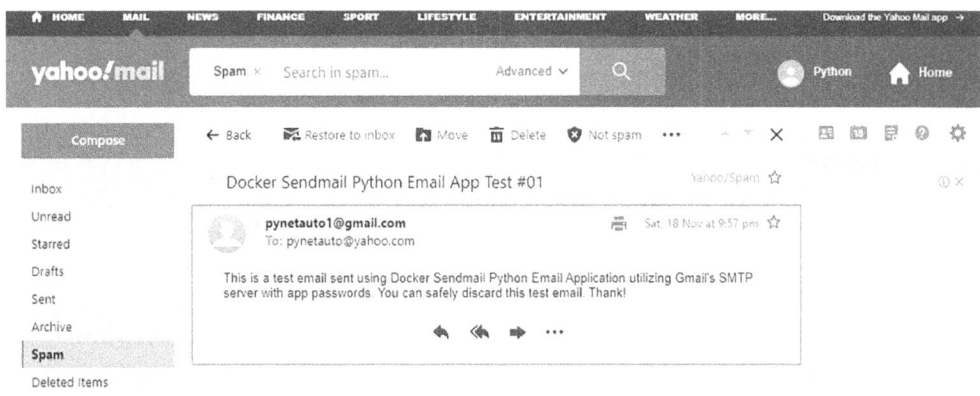

Figure 7-7. Check the recipient's Yahoo Spam folder

⑩	You will continue to use the same Docker instance, so exit the Docker instance using Ctrl+P followed by Ctrl+Q to keep it running in the background.

```
root@22ac05fa791a:/home# read escape sequence # Press 'Ctrl+P'
followed by 'Ctrl+Q'
jdoe@u22s1:~ $
```

435

CHAPTER 7 PYTHON NETWORK AUTOMATION VIRTUAL LAB 2: SENDMAIL EMAIL NOTIFICATION
           AND TWILIO SMS NOTIFICATION ON DOCKER

If you've successfully sent a test email from your Gmail to another account using the Python `smtplib` module used in your Docker's Sendmail Python application, you are now equipped to leverage this feature within your GNS3 network lab and apply it to networking devices. Next, you'll step into more engaging territory and have some fun—it is time to explore and experiment further!

Lab 3: Sendmail Email Notification Script Development in Docker

In this lab, your objective is to develop a Python monitoring script that actively monitors the accessibility of `SW1`'s open socket, using port 22 for this demonstration, and triggers an email notification if the switch becomes unreachable, effectively simulating a link outage. In practical scenarios, these email alerts often route to the 24/7 service desk team via SNMP monitoring servers. Moreover, in sophisticated IT environments, these notifications can initiate automated case logging within IT enterprise ticketing or ITSM systems. Integration with these systems relies on API calls to facilitate seamless ticket creation. Noteworthy ITSM tools that facilitate such integrations encompass ServiceNow, Jira Service Management, Zendesk, BMC Helix ITSM, and Autotask, showcasing their compatibility with automation through APIs for efficient incident management. Furthermore, you delved into the realm of enterprise network and system monitoring, highlighting key players like SolarWinds, PRTG Network Monitor, Nagios, Zabbix, ManageEngine OpManager, ScienceLogic, and LogicMonitor. These industry leaders offer comprehensive suites of monitoring tools based on both cloud and on-prem solutions, each renowned for specific strengths in network and system monitoring. SolarWinds, for instance, excels in network performance monitoring, while LogicMonitor specializes in cloud-based monitoring solutions. Understanding and leveraging the capabilities of monitoring tools becomes pivotal in optimizing enterprise network automation. Mastering enterprise network automation involves a nuanced understanding of both ITSM and network monitoring tools due to their synergistic relationship. Proficiency in these tools empowers professionals to streamline operations effectively. Hence, investing time to study and comprehend the intricacies of these tools at your own pace lays a strong foundation for excelling in enterprise network automation.

CHAPTER 7   PYTHON NETWORK AUTOMATION VIRTUAL LAB 2: SENDMAIL EMAIL NOTIFICATION
AND TWILIO SMS NOTIFICATION ON DOCKER

All tasks in this lab are performed within the Docker environment, providing you with exposure to Docker's capabilities in a practical context. Building on previous labs, you'll write a functional socket monitoring tool that checks SSH service availability every six seconds. If the socket remains unavailable for ten consecutive checks, signifying it is offline for 60 seconds, the script will trigger an email notification to your designated test email inbox via Sendmail. This script will continuously monitor the socket until you halt the application. Optionally, this type of application can run as the background job on Linux systems or be scheduled to run at specified times using cron jobs. The beauty of setting up a lab is the freedom it offers to explore and test a wide array of concepts. What's showcased in this book merely scratches the surface of what you can automate in your network using Python and other Python-based automation tools. **Everyone brings their unique background and perspective, influencing how they think, research, and approach certain work challenges. This diversity fuels creativity when writing code. The area where you excel might differ from those of your peers.** In coding, there's rarely a clear right or wrong answer; it is more about recommendations and preferences. Keeping an open mind and nurturing your imagination is key. Consider what you want to achieve with Python, be it in your work or studies. While there are sophisticated tools specifically tailored for SNMP monitoring, API integration, email notifications, and automated ticket logging, creating these fundamental tools through your code provides deeper insights into their functionality within IT ecosystems.

Imagine dealing with a device exhibiting a link flap that requires periodic monitoring. "Link flap" in networking describes when a network connection, like an Ethernet link or interface, repeatedly switches between being active (up) and inactive (down) quickly. This happens for various reasons like physical faults, setup errors, hardware issues, ISP maintenance work, or problems with network protocols. You don't want to spend your day or night staring at console messages. Instead, create a simple script, log the events, and set up the script to alert the relevant personnel via email notifications. Let's explore how this can be achieved with ease.

CHAPTER 7 PYTHON NETWORK AUTOMATION VIRTUAL LAB 2: SENDMAIL EMAIL NOTIFICATION
          AND TWILIO SMS NOTIFICATION ON DOCKER

#	Task
①	You will continue working within the Docker instance from the previous section. If you have stopped or detached from the Docker instance, reattach it to the Docker instance. The following example shows how to check and reattach to the running Docker instance: ``` jdoe@u22s1:~ $ docker ps -a CONTAINER ID IMAGE COMMAND CREATED STATUS PORTS NAMES 22ac05fa791a pynetauto/pynetauto_ubuntu20 "/bin/bash" 9 hours ago Up About an hour 20-22/tcp, 25/tcp, 12020-12025/tcp pynetauto_ubuntu20sendmail 05121eddd81b pynetauto/pynetauto_ubuntu20 "/bin/bash" 10 hours ago Up 10 hours 20-22/tcp, 25/tcp, 12020-12025/tcp pynetauto_ubuntu20 jdoe@u22s1:~ $ docker attach pynetauto_ubuntu20sendmail root@22ac05fa791a:/home# ```
②	For simplicity purposes, create a new directory named monitoring under the /home directory, which is mapped to your host Linux. ``` root@22ac05fa791a:/home# mkdir /home/monitoring && cd /home/monitoring root@22ac05fa791a:/home/monitoring# pwd /home/monitoring ```
③	Now, create two empty Python files, one for the script and another for the email message. ``` root@22ac05fa791a:/home/monitoring# touch monitor_sw1.py send_email.py root@22ac05fa791a:/home/monitoring# ls monitor_sw1.py send_email.py ```

(*continued*)

#	Task

④	The following is a socket-checking script that I often use to check the SSH port, port 22. If you have another port opened on your devices, such as port 23 (Telnet), 69 (TFTP), or 80 (HTTP), they can also be replaced in this example. However, you already know that port 22 is an open port on all of your Cisco devices in your topology; hence, this script will check port 22. Using this basic port-checking script and what you have learned from the previous exercises, you will be rewriting the code and applying it to solving your challenges.
	
	``
	root@22ac05fa791a:/home/monitoring# nano check_port_22.py
	root@22ac05fa791a:/home/monitoring# cat check_port_22.py
	import socket
	
	IP_ADDRESS = '192.168.127.101' # Device IP address
	PORT_TO_CHECK = 22 # Port number to check
	TIMEOUT = 3 # Connection timeout in seconds
	
	def check_port(ip, port):
	try:
	with socket.socket(socket.AF_INET, socket.SOCK_STREAM) as s:
	s.settimeout(TIMEOUT) # Set socket timeout
	s.connect((ip, port)) # Connect to IP and port
	print(f"Port {port} is open on {ip}. This device is on the network.")
	# Success message
	except socket.error:
	print(f"Port {port} is closed on {ip}. FAILED to reach the device. Check connectivity.") # Error message
	exit() # Exit the application in case of failure
	
	def check_port_22():
	check_port(IP_ADDRESS, PORT_TO_CHECK) # Check port 22 on the specified IP
	
	if __name__ == "__main__":
	check_port_22() # Run the port check function
	``

(continued)

#	Task
⑤	Now, open monitor_sw1.py in the nano text editor to begin creating your Python application. While the script is available for download, cutting and pasting the code might seem to save time. However, in the real world, it is rarely that easy. Typically, you will need to manually type all the code and test it yourself, unless your company has hired a contractor to write a complete and functional application. Therefore, aim to follow the script and input each line of code. Inline explanations accompany the code to guide you through this main script. ```\nroot@22ac05fa791a:/home/monitoring# nano monitor_sw1.py\nroot@22ac05fa791a:/home/monitoring# cat monitor_sw1.py\nimport socket\nimport time\nfrom datetime import datetime\nfrom send_email import send_email # Importing the send_email function\n\nIP_ADDRESS = "192.168.127.101" # Device IP address\nPORT = 22 # Port number to check\nTIMEOUT = 3 # Timeout for socket connection\nMAX_CHECKS = 10 # Maximum number of checks before sending email\n\ndef check_sw1():\n checks_passed = 0 # Counter for successful checks\n while checks_passed < MAX_CHECKS:\n with open('./monitoring_logs.txt', 'a') as f: # Open log file for appending\n try:\n with socket.socket(socket.AF_INET, socket.SOCK_STREAM) as s:\n s.settimeout(TIMEOUT)\n s.connect((IP_ADDRESS, PORT))\n checks_passed = 0 # Reset checks count on success\n f.write(f"{datetime.now().strftime('%Y-%m-%d %H:%M:%S')} Port {PORT} is open\n")\n print(f"Port {PORT} is open") # Log message for an open port\n time.sleep(3) # Wait before the next check\n```

(continued)

#	Task
|   | ```
except socket.error:
    checks_passed += 1  # Increment checks count on failure
    f.write(f"{datetime.now().strftime('%Y-%m-%d %H:%M:%S')} Port {PORT} is closed\n")
    print(f"Port {PORT} is closed")  # Log message for a closed port
time.sleep(3)  # Wait before next check
if checks_passed == MAX_CHECKS:  # Check if maximum checks reached
    print("Sending failed email notification")  # Notification for sending email
    send_email()  # Call function to send email
    break  # Break the inner loop to continue monitoring, this is important!

if __name__ == "__main__":
    while True:  # Runs the script forever until interrupted, this is important!
        check_sw1()  # Continuous rerun of the check_sw1 function
``` |
| ⑥ | Just for good measures and courtesy, check the connectivity between your Docker instance (same as host 192.168.127.20) and SW1 (192.168.127.101).

On the Docker instance:
root@22ac05fa791a:/home/monitoring# **ping -c 4 192.168.127.101**
PING 192.168.127.101 (192.168.127.101) 56(84) bytes of data.
64 bytes from 192.168.127.101: icmp_seq=1 ttl=254 time=13.7 ms
64 bytes from 192.168.127.101: icmp_seq=2 ttl=254 time=11.6 ms
64 bytes from 192.168.127.101: icmp_seq=3 ttl=254 time=30.7 ms
64 bytes from 192.168.127.101: icmp_seq=4 ttl=254 time=9.55 ms
[...omitted for brevity]

On SW1:
SW1#**ping 192.168.127.20**
Type escape sequence to abort.
Sending 5, 100-byte ICMP Echos to 192.168.127.20, timeout is 2 seconds:
.!!!!
Success rate is 80 percent (4/5), round-trip min/avg/max = 5/6/8 ms |

(*continued*)
```

#	Task
| ⑦ | Now you'll write an smtplib Python module file, so the main script can import this custom module and send a failure notification email using Sendmail. This is the email part of the script. By breaking up scripts into multiple functional parts, you can keep your code clean and more manageable.<br><br>```
root@22ac05fa791a:/home/monitoring# nano send_email.py
root@22ac05fa791a:/home/monitoring# cat send_email.py
import smtplib
from email.mime.text import MIMEText

def send_email():
# Email details
sender_email = "pynetauto1@gmail.com"
recipient_email = "pynetauto@yahoo.com"
subject = " SW1 not reachable for more than 60 seconds !"
body = ("SW1 is not reachable. Please investigate." )

# Construct the email message
msg = MIMEText(body)
msg["From"] = sender_email
msg["To"] = recipient_email
msg["Subject"] = subject

# SMTP server configuration
smtp_server = "smtp.gmail.com"
smtp_port = 587
gmail_user = sender_email
gmail_password = "wxwdlietjejemdtm" # Replace with your app password

try:
# Establish connection to SMTP server
with smtplib.SMTP(smtp_server, smtp_port) as server:
server.starttls()  # Enable TLS encryption
server.login(gmail_user, gmail_password) # Log in to Gmail account
``` |

(*continued*)
```

#	Task
	server.send_message(msg) # Send the email message **print("Failure notification email sent by Docker Sendmail!")** **except Exception as e:** **print(f"Something went wrong: {str(e)}")**
⑧	Once both the main monitoring and supplementary email scripts have been created, run the main application from your Docker instance using the following command. Once you see the message on the screen, you can let the script run continuously. root@22ac05fa791a:/home/monitoring# **python3 monitor_sw1.py** Port 22 is open Port 22 is open Port 22 is open [...Running continuously]
⑨	Next, log in to SW1's console (192.168.127.101) and quickly shut down the uplink port Gi0/0 by executing the shut command. Then wait for approximately 120 seconds. SW1(config)# **interface Gi0/0** SW1(config-if)#**shut** # wait for approximately 2 minutes After about two minutes, open port 22 on SW1 again by executing the no shut command so the SSH port goes into open status again. SW1(config-if)#**no shut**

(continued)

CHAPTER 7   PYTHON NETWORK AUTOMATION VIRTUAL LAB 2: SENDMAIL EMAIL NOTIFICATION
            AND TWILIO SMS NOTIFICATION ON DOCKER

#	Task
⑩	Return to the Docker instance console screen; you will notice outputs like the following. When you encounter the Port 22 is open message, use the Ctrl+Z key combination to halt the application, as demonstrated. You have set the port checking interval to three seconds, incorporating a three-second pause within the loop. Additionally, the script has a maximum wait counter of 10, signifying that if the port remains down for 60 seconds without recovery, the main script monitor_sw1.py will trigger send_email.py. Consequently, this operation will persist, sending an email every 60 seconds until you restore the GigabitEthernet 0/0 port on SW1. After the port becomes operational, the script will resume periodic checks on the status of port 22 on SW1. As long as port 22 remains open (up) and recovers at short intervals, the email trigger won't be activated. ```\nroot@22ac05fa791a:/home/monitoring# python3 monitor_sw1.py\nPort 22 is open\nPort 22 is open\nPort 22 is open\nPort 22 is closed\nPort 22 is closed\nPort 22 is closed\nPort 22 is closed\nPort 22 is closed\nPort 22 is closed\nPort 22 is closed\nPort 22 is closed\nPort 22 is closed\nPort 22 is closed\nPort 22 is closed\nSending failed email notification # Output by monitor_sw1.py\nFailure notification email sent by Docker Sendmail! # Output by send_email.py\nPort 22 is closed\nPort 22 is closed\nPort 22 is closed\nPort 22 is closed\nPort 22 is closed\n```

(continued)

#	Task
| | ```
Port 22 is closed
Port 22 is closed
Port 22 is closed
Port 22 is closed
Port 22 is closed
Sending failed email notification # Output by monitor_sw1.py
Failure notification email sent by Docker Sendmail! # Output by send_
email.py
Port 22 is closed
Port 22 is closed
Port 22 is closed
Port 22 is open  # Open port 22 detected and now performing a
continuous check
Port 22 is open
Port 22 is open
Port 22 is open
Port 22 is open
...
^Z # Press 'Ctrl+Z' to stop the application.
[2]+  Stopped                 python3 monitor_sw1.py
root@22ac05fa791a:/home/monitoring#
``` |
| (11) | Check the Spam folder of your recipient email account. If you kept the interface shut down for two minutes, you should receive two email notifications, as shown in Figure 7-8. |

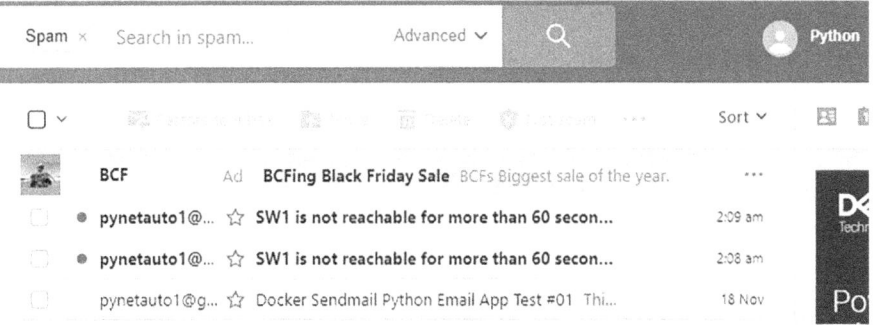

Figure 7-8. *Checking the failure email received in the Spam folder*

(*continued*)
```

CHAPTER 7   PYTHON NETWORK AUTOMATION VIRTUAL LAB 2: SENDMAIL EMAIL NOTIFICATION
            AND TWILIO SMS NOTIFICATION ON DOCKER

#	Task
⑫	Also, check your sender email account. You should see two email notifications sent at one-minute intervals (see Figure 7-9).

Figure 7-9. *Checking the failure email in the sender's Sent folder*

#	Task
⑬	The monitoring activities have been stored in the monitoring_logs.txt file for your convenience. If executed as a background job in Linux or a scheduled task, this script enables diverse network device monitoring. It is easily adaptable to various monitoring tasks, utilizing only a fraction of your system's resources, owing to Docker's lightweight nature, operating as a resource-efficient entity. **This characteristic aligns with the essence of Docker, which functions like a parasite, not demanding the entire OS or significant resources.** Such efficiency contributes to the rising popularity of Kubernetes in recent years and indicates its sustained success in the evolving landscape of multi-cloud environments.

```
root@22ac05fa791a:/home/monitoring# ls -lh monitoring*
-rw-r--r-- 1 root root 7.4K Nov 19 02:10 monitoring_logs.txt
root@22ac05fa791a:/home/monitoring# cat monitoring_logs.txt
2023-11-18 23:58:38 Port 22 is closed
2023-11-18 23:58:44 Port 22 is closed
[...omitted for brevity]
2023-11-19 01:27:06 Port 22 is open
2023-11-19 01:27:09 Port 22 is open
```

#	Task
⑭	Press Ctrl+P and then the Ctrl+Q keys to detach from the Docker instance. This will keep the script running continuously. To stop the script, press Ctrl+Z and, to exit the Docker instance, type `exit` and press Enter or use the Ctrl+D keys to exit. You can also use `docker stop <container>` to shut down the container instance gracefully. Now you should be back in your host Linux machine for another exciting Python exercise next. `jdoe@u22s1:~ $`

Imagine it is 5p.m. on a Friday, and your boss tasks you with monitoring an unreliable network device like a router, switch, or firewall. If this device loses connection, it is critical to your organization's operations, as your company is an online shopping company. Your job is to monitor and promptly inform all stakeholders if there is another link flap or a total link failure. As demonstrated here, your script operating within a Docker instance can efficiently handle network communication checks on your behalf. Upon detecting an incident, it triggers email notifications. With this setup, you can conveniently monitor emails from your smartphone at any time, even while commuting home or after dinner. As you leave the office at 5p.m., use the cron to schedule the script or run the script as a background job, enabling your Python application to autonomously handle the monitoring tasks for you. Next, you'll check if you can write a Python monitoring script to send SMS messages using an API service provider.

Lab 4: CPU Utilization Monitoring Lab: Send an SMS Message Using Twilio

In this lab, you will develop a REST API SMS message tool. Then, you will write the Python code to monitor your router's CPU utilization. When a certain threshold is reached, the script will trigger an SMS message to your smartphone using REST API calls. As the newer network device platforms and most of the virtual and cloud networking platforms start to support the REST API, understanding the REST API and applying the theory to practical use cases will be one of the must-have skillsets in network automation. This book does not discuss how to start your REST API studies or attempt to cover the topic extensively; this is one simple example to give you a taste. However, you can start learning about the REST API from YouTube or LinkedIn video tutorials. Most of

CHAPTER 7 PYTHON NETWORK AUTOMATION VIRTUAL LAB 2: SENDMAIL EMAIL NOTIFICATION
AND TWILIO SMS NOTIFICATION ON DOCKER

the tutorials will have you install one of the REST clients such as POSTMAN or REST for Visual Studio. If you have some time, go to the following sites and do some preliminary reading before completing this lab. Review Figure 7-10 before moving on to the lab task.

URL: https://www.postman.com/ (POSTMAN REST client)

URL: https://marketplace.visualstudio.com/items?itemName=humao.rest-client (REST client for Microsoft Visual Studio)

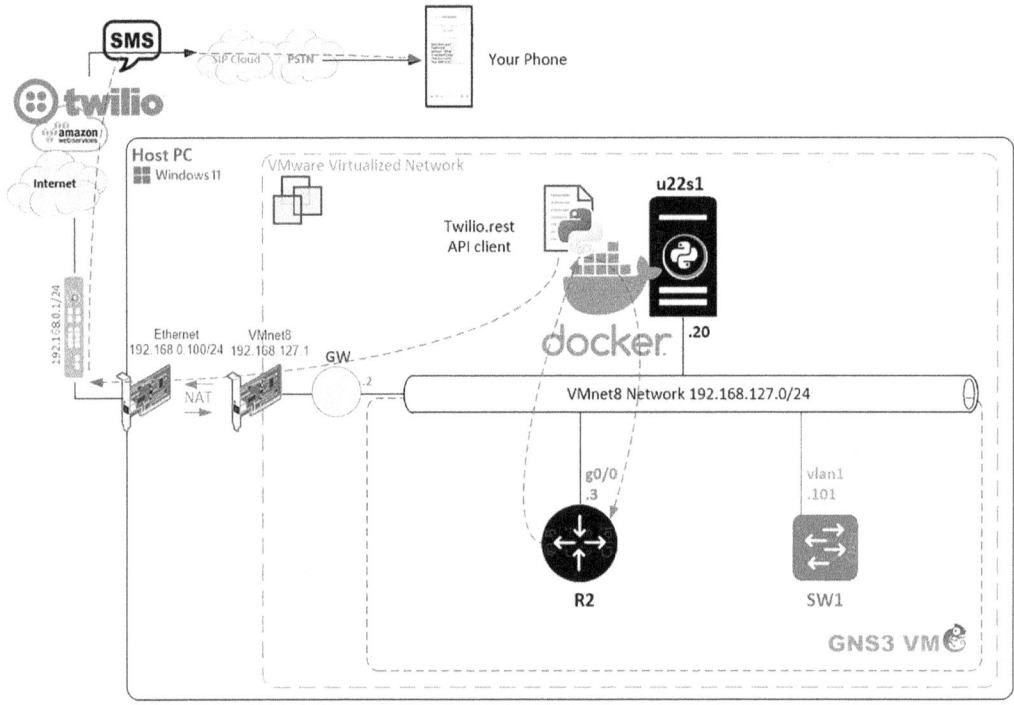

Figure 7-10. R2 CPU utilization monitoring lab used devices

Lab 5: TWILIO Account Creation: Install Twilio Python Module and Set Up an SMS Message

In this section, you will set up a Twilio testing account to send SMS messages for free. According to Wikipedia, Twilio is a cloud communications platform as a service company, allowing software developers to programmatically make and receive phone calls, send, and receive text messages, and perform other communication functions using Twilio's web service APIs. For the use here, you need to send SMS notification messages when a particular condition is met during your system monitoring of the

CHAPTER 7   PYTHON NETWORK AUTOMATION VIRTUAL LAB 2: SENDMAIL EMAIL NOTIFICATION
AND TWILIO SMS NOTIFICATION ON DOCKER

network devices in the topology. First, create an account to receive a U.S. testing number, account SID, and authorization code; then install the Twilio module on Python and write a simple SMS message to send to your smartphone number. After this, you will be writing a CPU utilization monitoring script, emulating a high CPU utilization scenario under a security attack, and triggering an SMS notification. First, set up the account and get the first testing message out to your smartphone number.

#	Task
①	Go to the Twilio trial page and create a trial account to start (see Figure 7-11). URL: https://www.twilio.com/try-twilio

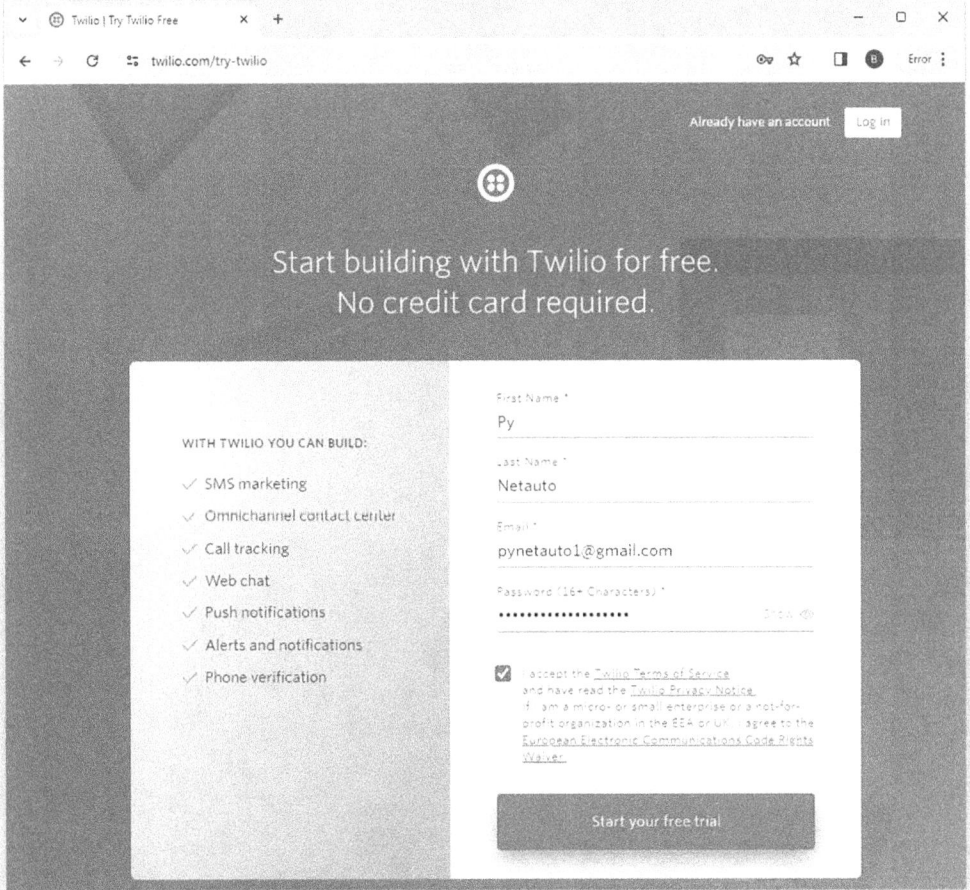

Figure 7-11. Twilio account registration

(*continued*)

CHAPTER 7   PYTHON NETWORK AUTOMATION VIRTUAL LAB 2: SENDMAIL EMAIL NOTIFICATION
            AND TWILIO SMS NOTIFICATION ON DOCKER

#	Task
	Go to your email inbox to locate and open the Twilio trial account verification email. Click the Verify Your Email button from the email to activate the account. Next, enter your smartphone number to receive an SMS verification code. Once you receive the code, enter the code and verify your phone. When you get to the welcome to Twilio page, scroll down and fill in the product survey form to complete the account registration (see Figure 7-12). 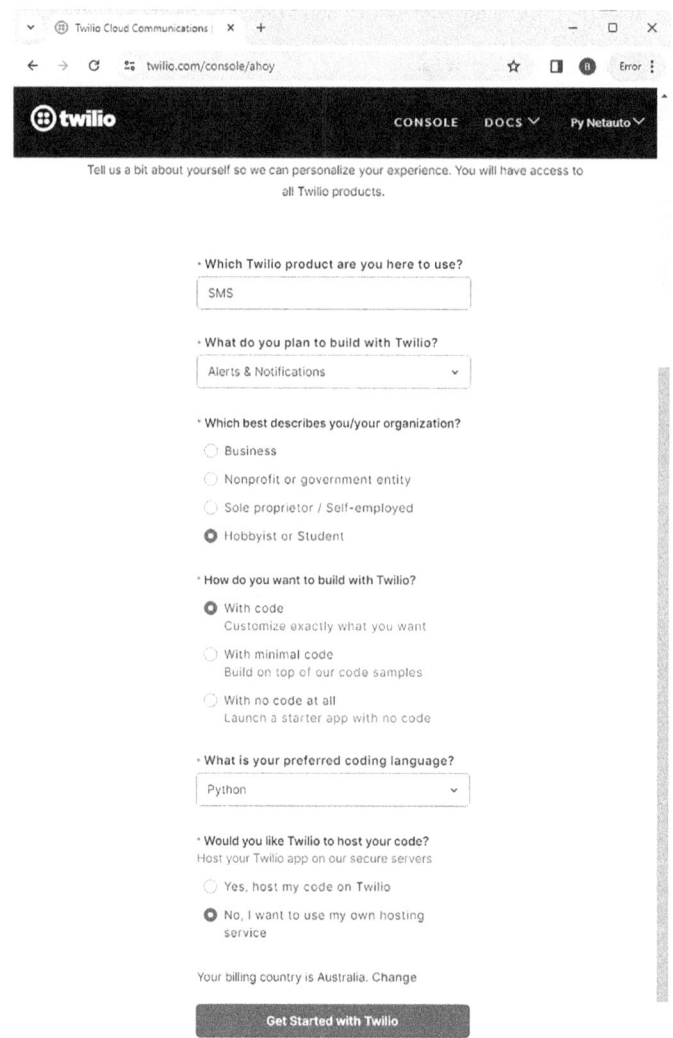 *Figure 7-12. Twilio – Fill in the product survey to complete account setup*

(continued)

450

#	Task
	Follow Steps 1 to 5 to complete your registration. Step 1-1. Twilio will provide you with a U.S. test number to send an SMS. Note the unique phone number assigned to your account (see Figure 7-13). 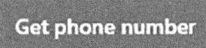 ***Figure 7-13.*** *Twilio – Get Phone Number button* Step 1-2. Write a quick message and send a test SMS message from your Twilio phone number to your number. Send an SMS message to check the functionality. If your registration worked, you should receive an SMS from your U.S. Twilio number (see Figure 7-14). 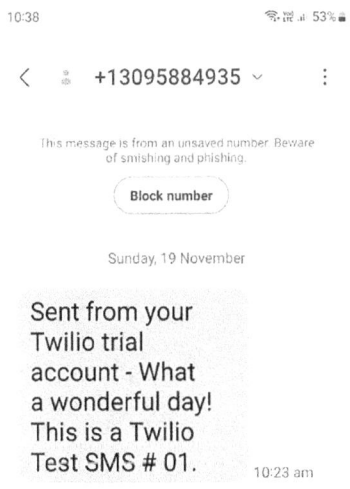 ***Figure 7-14.*** *First SMS sent from the Twilio trial account*

(continued)

CHAPTER 7 PYTHON NETWORK AUTOMATION VIRTUAL LAB 2: SENDMAIL EMAIL NOTIFICATION AND TWILIO SMS NOTIFICATION ON DOCKER

#	Task
	In the Request section, you will find the range of programming languages supported by Twilio, such as curl, Java, Ruby, PHP, Python, C#, and Node.js. **Scroll down the same page to access your account information details. It is essential to note that all three pieces of information are correct.** You have set up a Twilio account to receive this trial account SID and token (see Figure 7-15). *Figure 7-15. Twilio account info* Account SID: **ACd4a53a28d711420bd6ee19c67ea477e7** Auth Token: **ef861c2130126cb336292230741cb52a** My Twilio phone number: **+13095884935** Step 1-3. To start building an app, visit the Launch demo app (https://www.twilio.com/code-exchange/browser-based-sms-notifications) to find out more about how to use Twilio. Optionally, visit Get Sample Code to learn more about Twilio: https://www.twilio.com/code-exchange/sms-notifications Step 1-4. Read the tutorial at https://www.twilio.com/docs/sms/tutorials/how-to-send-sms-messages. Step 1-5. Launch and scale your application. You can close the window, as you have your account information for your Python application development and testing.
②	To read more about how to get started with Twilio, go to the following website and read the official documentation. URL: https://www.twilio.com/docs/sms/quickstart/python

(continued)

#	Task
③	In this lab, you will be working within the Docker instance to isolate the instance, so it does not directly impact your lab setup. With Docker, you can delete the instance and re-create as many instances as required on demand. Go ahead and create another Docker instance to run your Twilio lab. `jdoe@u22s1:~$ `**`docker run -it --entrypoint=/bin/bash -v /home/jdoe/mnt/:/home --name pynetauto_ubuntu20twilio pynetauto/pynetauto_ubuntu20`**

④	To get ready for Twilio SMS messaging, use the `pip3` command to install Twilio on your Docker instance. Because you are evaluating this in a Docker environment, you do not have to worry about breaking the software compatibility issues with the real Linux host.
	
	`root@76bd265d0146:/# `**`pip3 install Twilio`**
	`[...omitted for brevity]`
	`Successfully installed PyJWT-2.8.0 Twilio-8.10.2 aiohttp-3.9.0`
	`aiohttp-retry-2.8.3 aiosignal-1.3.1 async-timeout-4.0.3`
	`frozenlist-1.4.0`
	
	If you are getting a `bash: snmpwalk: command not found` error message, detach from the Docker session to go back to your Ubuntu host and install `snmp` first, then try to install the Twilio installation again.
	
	`jdoe@u22s1:~$ `**`apt-get install snmp -y`**

⑤	You are going to create two files to send the testing SMS message from the Python script. Create a credential feeder script called `credentials.py` and then the SMS script to send an SMS called `twilio_sms.py`. Update and replace the credentials and numbers with yours. `root@76bd265d0146:/# `**`mkdir /home/monitor_cpu && cd /home/monitor_cpu`** `root@76bd265d0146:/home/monitor_cpu# `**`nano credentials.py`** `root@76bd265d0146:/home/monitor_cpu# `**`cat credentials.py`** **`account_sid = "ACd4a53a28d711420bd6ee19c67ea477e7"`** **`auth_token = "ef861c2130126cb336292230741cb52a "`** **`my_smartphone = "+61414123456"`** `# Replace with your Cell Phone number` **`twilio_trial = "+13095884935"`** `# Replace with your Twilio trial number`

(continued)

#	Task
6	This is the SMS script that will send the message to the SMS-receiving smartphone. root@76bd265d0146:/home/monitor_cpu# **nano twilio_sms.py** root@76bd265d0146:/home/monitor_cpu# **cat twilio_sms.py** **from twilio.rest import Client** **from credentials import account_sid, auth_token, my_smartphone, twilio_trial** **client = Client(account_sid, auth_token)** **my_message = f"High CPU utilization Alert! R2 has reached 99% CPU utilization!"** **message = client.messages.create(body=my_message, from_=twilio_trial, to=my_smartphone)** **print(message.sid)**
7	From the Docker container command line, run the python3 command to run the SMS script to send your first Twilio SMS. If everything has been set up correctly, it will send an SMS to your smartphone. root@76bd265d0146:/home/monitor_cpu# **python3 twilio_sms.py** SM946cccd303bca5cee7cdcd3e26d2344f

(continued)

CHAPTER 7 PYTHON NETWORK AUTOMATION VIRTUAL LAB 2: SENDMAIL EMAIL NOTIFICATION AND TWILIO SMS NOTIFICATION ON DOCKER

#	Task
(8)	On your smartphone, check the SMS message. You should receive an SMS message similar to Figure 7-16.

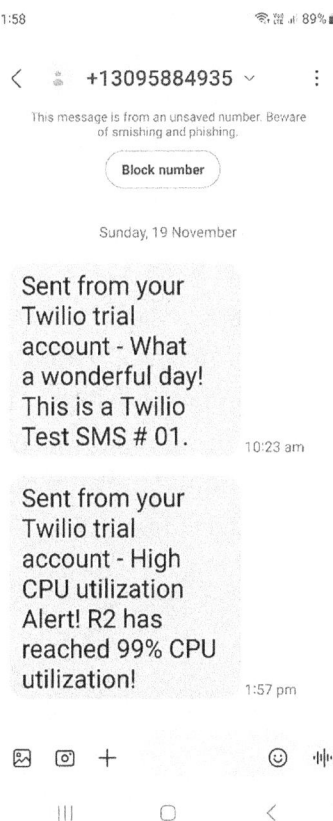

Figure 7-16. *Test SMS message received example*

If you have received an SMS message successfully on your smartphone, you are ready for the next lab.

CHAPTER 7   PYTHON NETWORK AUTOMATION VIRTUAL LAB 2: SENDMAIL EMAIL NOTIFICATION
            AND TWILIO SMS NOTIFICATION ON DOCKER

Lab 6: CPU Utilization Monitoring Lab with an SMS Message

Now that you know how to send an SMS message to your smartphone using a simple Python script, it is time to write some code and integrate the previous code snippet for use in a real-world simulation. Let's write a simple CPU monitoring application for network devices using SNMP v3, and this application can monitor the five-minute interval CPU utilization levels of R2. You could use a third-party traffic generator to make your router very busy and push the CPU utilization up above 90 percent, but that means you must learn another tool to complete this lab. To keep it simple, you will use the `debug all` command and multiple ping packets to push up the CPU utilization of R2. Yes, you will enable the debugging of all commands on R2, which will increase the CPU utilization to around 50 percent. You will be sending large ping packets to push the CPU utilization above 90 percent from other routers and switches. Be careful using the `debug all` command in the production environment and leaving it running on your Cisco devices in the production environment; you may bring down your network and cause an outage.

Review the devices that will be used for this lab, as depicted in Figure 7-17.

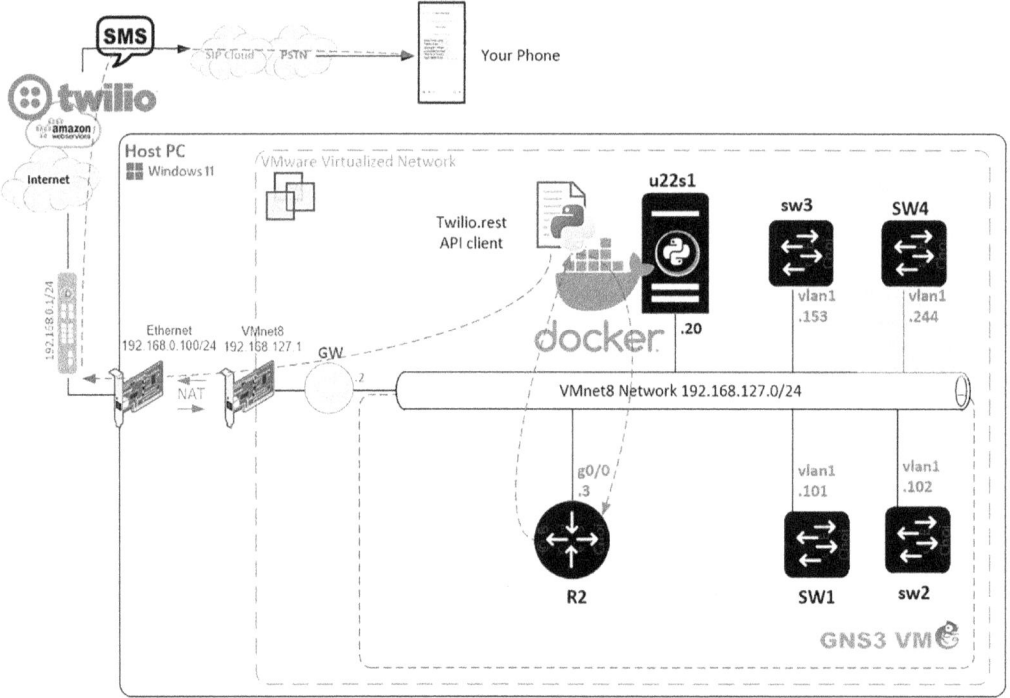

Figure 7-17. *Twilio – CPU utilization overloading lab topology*

First, you'll write a simple CPU monitoring script and integrate the SMS script into it, so when the CPU utilization is more than 90 percent for more than five minutes, your script will send out an SMS alert message to your smartphone stating that the CPU utilization has exceeded 90 percent for more than five minutes.

For this, you need to find out the exact OID reference number, as shown in Table 7-2. Go to https://oidref.com/1.3.6.1.4.1.9.9.109.1.1.1.1 and check out the various Cisco IOS device CPU-related OID values.

Table 7-2. Cisco Device CPU Utilization OID IDs

OID ID	Description
1.3.6.1.4.1.9.9.109.1.1.1.1.3	The overall CPU busy percentage in the last five-second period
1.3.6.1.4.1.9.9.109.1.1.1.1.4	The overall CPU busy percentage in the last one-minute period
1.3.6.1.4.1.9.9.109.1.1.1.1.5	The overall CPU busy percentage in the last five-minute period

Source: https://oidref.com/

You are going to be using an OID ID of 1.3.6.1.4.1.9.9.109.1.1.1.1.5 for CPU utilization and check every five minutes. You can use crontab on the Ubuntu Server to run the script every five minutes. When the CPU utilization for the last five minutes is more than 90 percent, the script will send an SMS alert using the Twilio API account from the previous lab.

#	Task
①	This lab continues to work in the `pynetauto_ubuntu20twilio` Docker container.

You'll first write a simple `snmpwalk` script that sends a `snmpwalk` command to R2 and retrieves the CPU utilization information. The following command should return the CPU utilization of the router in the last five minutes. Under regular operation, the CPU utilization of a router should be less than 80 percent, but while some change or update work is performed, it is normal to see high CPU utilization during the normal change window. This is the original script. Create the following file and check if it still works. If you have not yet read the SNMP chapter, do so for further reading.

```
root@76bd265d0146:/home/monitor_cpu# apt-get update
[...omitted for brevity]
root@76bd265d0146:/home/monitor_cpu# apt-get install snmp
[...omitted for brevity]
root@76bd265d0146:/home/monitor_cpu# nano cpu_util_5min.py
root@76bd265d0146:/home/monitor_cpu# cat cpu_util_5min.py
# cpu_util_5min.py
import os
stream = os.popen('snmpwalk -v3  -l authPriv -u SNMPUser1 -a SHA -A "AUTHPass1"  -x AES -X "PRIVPass1" 192.168.127.3 1.3.6.1.4.1.9.9.109.1.1.1.1.5')
output = stream.read()
print(output)

root@76bd265d0146:/home/monitor_cpu# python3 cpu_util_5min.py
iso.3.6.1.4.1.9.9.109.1.1.1.1.5.1 = Gauge32: 15 # R2 CPU is at 15%
``` |

(continued)

#	Task
| ② | Now use the Twilio SMS scripts from the previous exercises to update the information.<br><br>```
root@76bd265d0146:/home/monitor_cpu# cat credentials.py
account_sid = "ACd4a53a28d711420bd6ee19c67ea477e7"
auth_token = "ef861c2130126cb336292230741cb52a "
my_smartphone = "+61414123456" # Replace with your Cell Phone number
twilio_trial = "+13095884935" # Replace with your Twilio trial number
```<br><br>Copy the previous Python file, called twilio_sms.py, and create another file called twilio_sms1.py. Modify the message only, as shown here:<br><br>```
root@76bd265d0146:/home/monitor_cpu# cp twilio_sms.py twilio_sms1.py
root@76bd265d0146:/home/monitor_cpu# nano twilio_sms1.py
root@76bd265d0146:/home/monitor_cpu# cat twilio_sms1.py
from twilio.rest import Client
from credentials import account_sid, auth_token, my_smartphone, twilio_trial

client = Client(account_sid, auth_token)
my_message = f"R2 has reached 99% CPU utilization! Investigate the root cause immediately!"
message = client.messages.create(body=my_message, from_=twilio_trial, to=my_smartphone)
print(message.sid)
``` |

(continued)

CHAPTER 7 PYTHON NETWORK AUTOMATION VIRTUAL LAB 2: SENDMAIL EMAIL NOTIFICATION AND TWILIO SMS NOTIFICATION ON DOCKER

#	Task
| ③ | Now you'll get the exact value you want from the output and extract the CPU utilization value using the power of good old Regular Expressions again. Create the new code and run it against R2.<br><br>```
root@76bd265d0146:/home/monitor_cpu# cp cpu_util_5min.py cpu_util_5min1.py
root@76bd265d0146:/home/monitor_cpu# nano cpu_util_5min1.py
root@76bd265d0146:/home/monitor_cpu# cat cpu_util_5min1.py
# cpu_util_5min.py
import os
import re

stream = os.popen('snmpwalk -v3  -l authPriv -u SNMPUser1 -a SHA -A "AUTHPass1"  -x AES -X "PRIVPass1" 192.168.127.3 1.3.6.1.4.1.9.9.109.1.1.1.1.5')
output = stream.read()
print(output)
p1 = re.compile(r"(?:Gauge32: )(\d+)") # Use re positive lookbehind and match the digit
m1 = p1.findall(output) # Match the CPU utilization value (digit) in output
cpu_util_value = int(m1[0]) # Index item 0 in the list and convert to an integer
print(cpu_util_value) # print the integer
``` |
| ④ | When you run the code from your server, you should get a CPU utilization value between 1 and 100.<br><br>In the following example, 9 percent utilization is returned.<br><br>```
root@76bd265d0146:/home/monitor_cpu# python3 cpu_util_5min1.py
iso.3.6.1.4.1.9.9.109.1.1.1.1.5.1 = Gauge32: 9
9
``` |

(continued)

CHAPTER 7   PYTHON NETWORK AUTOMATION VIRTUAL LAB 2: SENDMAIL EMAIL NOTIFICATION
            AND TWILIO SMS NOTIFICATION ON DOCKER

#	Task
⑤	Combining the scripts shown earlier, it is time to complete the script. For the simplicity of the lab, all three scripts—credentials.py, twilio_sms1.py, and cpu_util_5min1.py—will be rewritten into cpu_util_5min_monitor.py. Reiterations and integration are the keys to developing a working application like this, and this approach will be extended into the final IOS upgrade lab. Upon combining all three scripts, your working script should look like this: root@76bd265d0146:/home/monitor_cpu# **nano cpu_util_5min_refactor1.py** root@76bd265d0146:/home/monitor_cpu# **cat cpu_util_5min_refactor1.py** **from twilio.rest import Client** # Importing Twilio Client module **import os** # Importing OS module **import re** # Importing Regular Expression module **import time** # Importing Time module **from time import strftime** # Importing strftime function from Time module **current_time = strftime("%a, %d %b %Y %H:%M:%S")** # Getting current time **account_sid = "ACd4a53a28d711420bd6ee19c67ea477e7"** # Twilio account SID **auth_token = "ef861c2130126cb336292230741cb52a"** # Twilio authentication token **my_smartphone = "+61414123456"** # Replace with your Cell Phone number **twilio_trial = "+13095884935"** # Replace with your Twilio trial number **def send_sms():** # Function to send SMS via Twilio **client = Client(account_sid, auth_token)** **my_message = f"High CPU utilization notice, R2 has reached 90% CPU utilization."**

(*continued*)

#	Task

	`message = client.messages.create(body=my_message, from_=twilio_trial, to=my_smartphone)`
	`print(message.sid)` # Printing SMS SID for confirmation
	
	`stream = os.popen('snmpwalk -v3 -l authPriv -u SNMPUser1 -a SHA -A "AUTHPass1"  -x AES -X "PRIVPass1" 192.168.127.3 1.3.6.1.4.1.9.9.109.1.1.1.1.5')`
	`output = stream.read()` # Reading SNMP output
	`time.sleep(3)` # Pausing execution for 3 seconds
	`print("-"*79)` # Printing separator for clarity
	`print(current_time, output)` # Printing timestamp and SNMP output
	
	`with open('./cpu_oid_log.txt', 'a+') as f:` # Opening log file in append mode
	`if "Gauge32:" in output:` # Checking for 'Gauge32' in output
	`p1 = re.compile(r"(?:Gauge32: )(\d+)")` # Compiling regex to find CPU utilization
	`m1 = p1.findall(output)` # Finding CPU utilization in output
	`cpu_util_value = m1[0]` # Extracting CPU utilization value
	
	`if int(cpu_util_value) < 90:` # Checking if CPU utilization is less than 90%
	`f.write(f"{current_time} {cpu_util_value}%, OK\n")` # Writing to log file
	`print("OK")` # Printing status to console
	`elif int(cpu_util_value) >= 90:` # If CPU utilization is 90% or more
	`f.write(f"{current_time} {cpu_util_value}%, High CPU\n")` # Writing to log file
	`print("High CPU")` # Printing status to console
	`send_sms()` # Sending SMS notification
	
	`elif "Timeout:" in output:` # Handling Timeout scenario
	`f.write(f"{current_time} High CPU, Timeout: No Response\n")` # Writing to log file

(continued)

#	Task
| | ```python
print("Timeout: No Response") # Printing status to console
send_sms() # Sending SMS notification

elif "snmpwalk:" in output: # Handling 'snmpwalk' scenario
f.write(f"{current_time} High CPU, snmpwalk: Timeout\n") # Writing to log file
print("No Response") # Printing status to console
send_sms() # Sending SMS notification

else: # Handling other scenarios
f.write(f"{current_time} High CPU utilization, IndexError\n") # Writing to log file
print("IndexError occurred due to High CPU Utilization") # Printing status to console
send_sms() # Sending SMS notification

print("Finished") # Indicating end of script
``` |
| ⑥ | Refactor the code once again. It should look like this:<br><br>root@76bd265d0146:/home/monitor_cpu# **cp cpu_util_5min_refactor1.py cpu_util_5min_refactor2.py**<br>root@76bd265d0146:/home/monitor_cpu# **nano cpu_util_5min_refactor2.py**<br>root@76bd265d0146:/home/monitor_cpu# **cat cpu_util_5min_refactor2.py**<br>```python
from twilio.rest import Client
import subprocess
import re
from time import strftime, sleep

current_time = strftime("%a, %d %b %Y %H:%M:%S")
account_sid = "ACd4a53a28d711420bd6ee19c67ea477e7"
auth_token = "ef861c2130126cb336292230741cb52a "
my_smartphone = "+61414123456" # Replace with your Cell Phone number
``` |

(continued)

#	Task

	`twilio_trial = "+13095884935"` # Replace with your Twilio trial number
	
	`def send_sms(message):`
	`client = Client(account_sid, auth_token)`
	`client.messages.create(body=message, from_=twilio_trial, to=my_smartphone)`
	
	`def get_cpu_util():`
	`cmd = 'snmpwalk -v3 -l authPriv -u SNMPUser1 -a SHA -A "AUTHPass1" -x AES -X "PRIVPass1" 192.168.127.3 1.3.6.1.4.1.9.9.109.1.1.1.1.5'`
	`output = subprocess.getoutput(cmd)`
	`return output`
	
	`def log_and_notify(message):`
	`with open('./cpu_oid_log.txt', 'a+') as f:`
	`f.write(f"{current_time} {message}\n")`
	`send_sms(message)`
	
	`def monitor_cpu():`
	`output = get_cpu_util()`
	`print("-" * 80)`
	`print(current_time, output)`
	
	`if "Gauge32:" in output:`
	`cpu_util_match = re.search(r"(?:Gauge32: )(\d+)", output)`
	`if cpu_util_match:`
	`cpu_util_value = cpu_util_match.group(1)`
	`if int(cpu_util_value) >= 90:`
	`log_and_notify(f"{cpu_util_value}%, High CPU")`
	`print("High CPU")`
	`else:`
	`print("OK")`

(continued)

#	Task
| | ```
else:
    log_and_notify("Error: Unable to parse CPU utilization value")
    print("Error: Unable to parse CPU utilization value")
elif "Timeout:" in output:
    log_and_notify("High CPU, Timeout: No Response")
    print("Timeout: No Response")
elif "snmpwalk:" in output:
    log_and_notify("High CPU, snmpwalk: Timeout")
    print("No Response")
else:
    log_and_notify("High CPU utilization, IndexError")
    print("IndexError occurred due to High CPU Utilization")
    print("Finished")
monitor_cpu()
``` |
| (7) | To induce high CPU utilization on R2, do the following. You want to increase the CPU utilization on this router so that your script can go to work and send an SMS when a certain condition is met.<br><br>7-1. Here is how to turn on all debugging on R2:<br><br>```
R2#debug all # Enable all debugging on R2. Do not do this on a
production device!
This may severely impact network performance. Continue? (yes/[no]):
yes # Lab environment, a big yes
Persistent variable debugs All enabled
```<br><br>7-2. From other switches, send a large-size datagram continuously. What you want is a CPU utilization of over 90 percent or more on R2, so the router's CPU is overworked and becomes unresponsive.<br><br>```
SW1#ping # initialize ping configuration
Protocol [ip]:
Target IP address: 192.168.127.3 # R2 IP address
Repeat count [5]: 100000 # number of continuous pings
``` |

(continued)
```

#	Task
	Datagram size [100]: **1000** # datagram, also try 2-3000 to overload the router's CPU more Timeout in seconds [2]: Extended commands [n]: Sweep range of sizes [n]: Repeat and enable the same `ping` commands from `sw2` and `sw3`. Wait about three to five minutes for R2 to reach 90 percent or above CPU utilization. Optionally, if R2's CPU utilization does not reach more than 90 percent, repeat the `ping` commands from `sw4`. Note: To stop the large-size datagram continuous pings, you must press Ctrl+Alt+6 to stop the operation. As soon as you turn off the ping, CPU utilization will drop and R2's performance will be normalized.
⑧	While you are alternating between `devices` to overload and halt R2's CPU, use the `show CPU process` and `show CPU process history` commands on an SSH-connected console of R2 to monitor the CPU performance (see Figure 7-18).

(continued)

CHAPTER 7 PYTHON NETWORK AUTOMATION VIRTUAL LAB 2: SENDMAIL EMAIL NOTIFICATION AND TWILIO SMS NOTIFICATION ON DOCKER

#	Task

```
R2#show processes cpu history

R2    09:51:33 PM Sunday Nov 19 2023 AEST




      9977777444445555555554444466666555544449999999999999999
      2222222999992222211111999991111100000999922222222222222000
  100
   90 **                                   ********************
   80 **                                   ********************
   70 ******                               ********************
   60 ******                     *****     ********************
   50 ********************************************************
   40 ********************************************************
   30 ********************************************************
   20 ********************************************************
   10 ********************************************************
      0....5....1....1....2....2....3....3....4....4....5....5....6
           0    5    0    5    0    5    0    5    0    5    0
               CPU% per second (last 60 seconds)




      99999999999965776555555565556555555555555656565556655655556
      322234353443306440767577165659978656777794882777788572572772
  100         *
   90 #############*
   80 #############
   70 #############   **           *                    *       *
   60 #############**##****************************************
   50 ##########################################################
   40 ##########################################################
   30 ##########################################################
   20 ##########################################################
   10 ##########################################################
      0....5....1....1....2....2....3....3....4....4....5....5....6
           0    5    0    5    0    5    0    5    0    5    0
               CPU% per minute (last 60 minutes)
             * = maximum CPU%    # = average CPU%
```

Figure 7-18. *R2 - show process CPU history output*

(*continued*)

CHAPTER 7 PYTHON NETWORK AUTOMATION VIRTUAL LAB 2: SENDMAIL EMAIL NOTIFICATION AND TWILIO SMS NOTIFICATION ON DOCKER

#	Task
(9)	Under normal load, the CPU utilization will remain relatively low and steady. When debugging significant changes or odd problems, the CPU utilization can spike above 90 percent. If the CPU utilization continues to operate at or above 90 percent for a prolonged period, this can have a significant impact on the performance of the services running on this device, and the issue becomes critical if the device is a core or edge router. Use your Python script to get the Gauge32 digit. Whenever the CPU utilization goes above 90 percent or the router becomes unresponsive with performance issues, the application will trigger the SMS message and send it to your phone. Under normal conditions: root@76bd265d0146:/home/monitor_cpu# **python3 cpu_util_5min_refactor2.py** -- Sun, 19 Nov 2023 17:08:13 iso.3.6.1.4.1.9.9.109.1.1.1.1.5.1 = Gauge32: 9 OK Finished After five minutes, debug all is enabled: root@76bd265d0146:/home/monitor_cpu# **python3 cpu_util_5min_refactor2.py** -- Sun, 19 Nov 2023 22:54:26 iso.3.6.1.4.1.9.9.109.1.1.1.1.5.1 = Gauge32: 54 OK Finished After an extended ping has started on SW1: root@76bd265d0146:/home/monitor_cpu# **python3 cpu_util_5min_refactor2.py**

(continued)

#	Task
	--
	Sun, 19 Nov 2023 22:56:42 iso.3.6.1.4.1.9.9.109.1.1.1.1.5.1 = Gauge32: 83 OK Finished
	After an extended ping from multiple switches, over 90 percent CPU utilization, the router becomes unresponsive and will trigger an SMS message.
	root@76bd265d0146:/home/monitor_cpu# **python3 cpu_util_5min_refactor2.py** -- Sun, 19 Nov 2023 22:56:42 iso.3.6.1.4.1.9.9.109.1.1.1.1.5.1 = Gauge32: 91 91 SM9bc812d18dd74a6d95cb8c4cc4c8d758 Finished
	If the snmpwalk times out, this will trigger an SMS.
	root@76bd265d0146:/home/monitor_cpu# **python3 cpu_util_5min_refactor2.py** -- Sun, 19 Nov 2023 23:00:08 snmpwalk: Timeout No Response Finished
	If there is no response, this will also trigger an SMS.
	root@76bd265d0146:/home/monitor_cpu# **python3 cpu_util_5min_refactor2.py** -- Sun, 19 Nov 2023 23:04:36 Timeout: No Response from 192.168.127.3 Timeout: No Response Finished

(continued)

CHAPTER 7   PYTHON NETWORK AUTOMATION VIRTUAL LAB 2: SENDMAIL EMAIL NOTIFICATION
            AND TWILIO SMS NOTIFICATION ON DOCKER

#	Task
⑩	It's a little bit tricky to get things right, as you simulate the CPU overloading. But if you followed the steps correctly, when the router meets the high CPU utilization conditions you set, it should send an SMS message, as shown in Figure 7-19. Remember, if this were in a real production environment, the ultimate goal would be to make you spend less time in front of your computer physically monitoring the high CPU utilization in use.

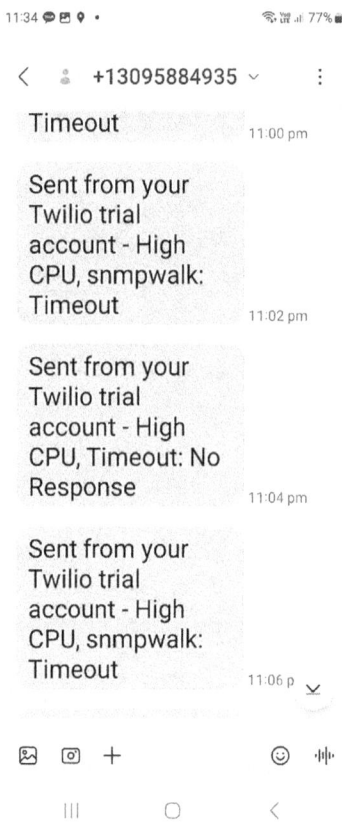

Figure 7-19. *High CPU/unresponsive device trigged SMS message example*

Well done reaching this point! If you are still eager to explore more, you can optionally monitor CPU utilization levels using your custom script and schedule a task on Linux cron. It is your turn to apply your skills to the next level by creating a CPU monitoring Python application and scheduling it on Linux cron. Your application can monitor the CPU utilization of your network devices—routers, switches, or firewalls—at scheduled intervals: every minute, every five minutes, or every hour. Additionally, it will notify you via API-enabled SMS if the CPU utilization surpasses a specified threshold. By combining your knowledge of cron scheduler with Python, you should be able to write an efficient monitoring system that keeps your team informed about potential issues in real time. Moreover, mastering this level means you are primed to delve into more thrilling territories, like developing Python applications to upgrade Cisco routers. Embrace this opportunity to further enhance your networking prowess and delve into the dynamic world of network automation and optimization.

Summary

Congratulations on completing this chapter, a crucial segment in your exploration of innovative Python use cases for network infrastructure. This chapter, alongside its predecessor, aimed to foster innovative thinking and explore diverse Python applications in effectively managing the network infrastructure. A key takeaway is the necessity of documenting urgent automation tasks and establishing a starting point for your automation initiatives. Throughout this chapter, you delved into Docker concepts and extensively experimented with Python application development within a Docker environment. Notably, you successfully implemented an email-sending application on Docker and established a free Twilio account. Additionally, you engineered a Python application capable of monitoring a router's CPU utilization and triggering SMS alerts to your smartphone under specific conditions.

This experience should encourage you to consider other potential Python application use cases at work, prompting exploration and problem-solving by leveraging available open-source and proprietary tools. Looking ahead to Chapter 8, you will dive deeply into Cisco IOS XE upgrades, marking the initial phase of your application development journey. Stressing the importance of documentation, meticulous task outlining remains pivotal for successful automation through programming languages. It is crucial not only to discuss but also to act on ideas. Effective implementation hinges on thoughtful planning and execution. The transformation of good ideas into impactful outcomes necessitates diligent follow-through with well-planned actions. This chapter serves as a stepping stone toward realizing the potential of Python applications in network automation and the efficient management of network infrastructures.

CHAPTER 8

Upgrading Multiple Cisco IOS Routers

This chapter offers an exploration of upgrading multiple Cisco IOS routers and delves into Python Network Automation through the lens of real-world scenarios. It begins with an examination of object-oriented programming (OOP) applied in networking, followed by proactive patch management strategies and discussions on Zero Trust approaches to strengthen network security. This chapter guides readers through a hands-on experience in creating virtual routers for IOS upgrade tasks in subsequent chapters, utilizing Cisco Catalyst 8000v IOS XE software. It covers the software's download, installation on VMware Workstation, and a detailed discussion of the intricate process of upgrading IOS (XE/XR) on Cisco devices. Additionally, it outlines the tasks integral to a Cisco IOS upgrade process. This chapter culminates with a comprehensive summary that encapsulates the key learnings and insights from the entire chapter.

Empowering Python Network Automation: Unveiling OOP and Real-world Scenarios

In this chapter, there are a few concepts and topics you must understand to prepare for the next three chapters. I hope your initial enthusiasm for your Python Network Automation journey has not diminished too much along the way. If you found the preceding chapters enjoyable, that's fantastic. However, it is likely you faced a few obstacles and areas needing further attention. Every chapter is designed to empower you with new IT skills. Get ready for the most exciting four chapters in this book! Throughout this book, you have gained various IT system management skills, and now you are here. Python, like many other popular programming languages, falls under the category of an object-oriented programming language. I have purposely saved the discussion of object-oriented programming (OOP) until now, understanding that delving into it might seem daunting for novice Python learners. In this chapter, I present a practical example demonstrating how OOP can be applied in a real work scenario, allowing you to immediately relate OOP concepts to your work. Following that, you will explore how to implement basic flow controls using user ID and password inputs in your scripts. In many scenarios, secure password vaults might not be available, and you will often find yourself writing interactive code, requiring users to input their user IDs and passwords into the applications you've developed. This chapter includes a meaningful discussion about why IOS upgrade is critical to protect your enterprise network. As the final task, you will install and configure two Cisco Catalyst 8000v routers on VMware Workstation, preparing you for the IOS upgrade tools development in Chapters 9 and 10. Last, but vital to the next three chapters, toward the end of this chapter, you will delve into a detailed discussion of the Cisco Internetwork Operating System (IOS) upgrade manual processes and strategies to write them into lines of Python code. It is worth mentioning that Cisco has established itself as the go-to Enterprise networking standard since the early 90s. In our course, we use Cisco IOS upgrades as a case study to teach upgrade processes, but the underlying concepts are universally applicable to other networking vendors.

Applying OOP Concepts to Your Network

Until this point, I have not discussed object-oriented programming (OOP) in Python. Even without knowing a lot about OOP, your Python scripts are working fine. However, intermediate to advanced Python users often emphasize that Python is an OOP language when optimizing scripts. So, you must take full advantage of the OOP concepts. **Just like with another popular OOP language, Java, you need to know four fundamental concepts: encapsulation, abstraction, inheritance, and polymorphism.** OOP aims to bind the data better so a lot of duplication is removed, and the same code can be reused throughout the same program or application.

Let's quickly cover the basics, and then you will look at OOP in action by writing an OOP example using routers and switches. I keep the theory to a minimum and get into the practical example to help you better understand the OOP concepts.

- **Object-oriented programming**: From an OOP perspective, everything is considered an object. The C language is called a procedure-oriented programming language because it runs based on the processes running in a functional order. On the other hand, in OOP, the objects are related to each other and connected to run the program. OOP treats anything and everything as an object, and it is viewed that they are linked to each other by some relationship.

- **Object**: As the word suggests, an object in OOP is a thing. For example, a human or a robot can be considered an object, and also a book or a router is an object. Because the same router (or switch) models look the same, every router can be called an object. If you drop a screwdriver and make a dent on one of the routers, it does not make dents in the other routers. Objects with similar characteristics or features are in the same group of objects, meaning that they can be grouped. Certain characteristics are the same or similar between the same objects in a common group. Although every human is unique, even humans can be considered objects based on the Python OOP concept.

- **Class**: People generally have almost identical attributes such as two eyes, a nose, a mouth, hands, feet, and other physical body parts. Books also have the same attributes, such as book title, author, publisher, and publication dates. Switches from different vendors will still have similar attributes, such as switch model, serial number, rack unit size, and Ethernet ports. Classes are defined by a collection of common properties that objects (such as people, books, and switches) have in common.

- **Abstraction**: This concept refers to showing only essential attributes and hiding unnecessary objects from the user. For example, a complex function in a class hides (abstracts) detailed information from the user, so they can implement logic on top of abstraction without understanding the actual function or even thinking about all the hidden complexities of the class.

- **Encapsulation**: This concept refers to the idea of binding the data and methods that manipulate the data into a unit. Again, a class is an excellent example of encapsulation as the objects inside a class keep their state private and not directly accessible. Instead, the state of the objects is accessed through methods by calling a list of public functions.

- **Polymorphism**: This process involves utilizing a single operator in multiple ways, maintaining a singular form while serving diverse use cases. A fitting analogy is found in pens: despite being categorized under the same label, pens vary significantly based on their intended purposes. For instance, some pens are meant for note-taking, while others excel at drawing and artistic creation. Similar principles apply to network devices. There exists a wide array of network devices, each designed to fulfill distinct functions. Routers handle routing, switches manage switching, firewalls bolster security measures, fabric switches specialize in storage area network (SAN) connections, and this list continues. Despite their varied roles, they are all classified as network devices.

- **Inheritance**: This is the process of making a class from an existing class. Consider parents to children and children to parents; if you are born in a family and your parents are your birth parents, you and your siblings as children take or inherit both parents' characteristics. If you create a new class that derives from and is associated with an existing class, the new class inherits the old class's characteristics.

Now that the basic OOP theory is out of the way, you can use router and switch examples to write classes and to understand the OOP style of coding better. As your applications become more complex, the power of OOP truly shines as it binds the data. Let's look at a real example. To complete this task, open the Python Interpreter and read through, following along within the Interpreter.

#	Task
①	Open your Python Interpreter and follow along by typing the commands. You'll construct classes for routers and switches, where they share similar attributes but differ in functionality or features. Using inheritance in this example, you can consolidate or relocate the def init method to the parent class.

In the Router and Switch classes, you will notice attributes such as serial number, hostname, management IP address, model number, and device type. init serves as a reserved method, known as a constructor in OOP, initializing the class's attributes. The term self within the class denotes the instance of the class, enabling access to its attributes and methods. It binds attributes with the provided arguments similar to the @ syntax in Java, adding a distinctive attribute to the class, method, and variable.

```
class Router:
def __init__(self, serialnumber, hostname, ipaddress, modelnumber, devicetype):
self.serialnumber = serialnumber  # 12 Hexadecimal numbers
self.hostname = hostname
self.ipaddress = ipaddress  # format 1.0.0.X - 254.254.254.XXX
self.modelnumber = modelnumber
self.devicetype = devicetype

def process(self):
print("Packet routing")

class Switch:
def __init__(self, serialnumber, hostname, ipaddress, modelnumber, devicetype):
self.serialnumber = serialnumber  # 12 Hexadecimal numbers
self.hostname = hostname
self.ipaddress = ipaddress  # format 1.0.0.X - 254.254.254.XXX
self.modelnumber = modelnumber
self.devicetype = devicetype

def process(self):
print("Packet switching")
```
 |

(continued)

CHAPTER 8   UPGRADING MULTIPLE CISCO IOS ROUTERS

#	Task
	The Router and Switch classes handle packet processing; one performs packet routing while the other conducts packet switching. The process function encapsulates their respective functionalities. Congratulations, you've successfully established your initial classes for network devices!
②	Follow along by typing the commands. Start by creating a parent class named Devices, similar to the Router and Switch classes introduced in Step 1. The Devices class mirrors the attributes of the Router and Switch classes, so replicate the Devices class accordingly. Additionally, incorporate a new function named show to display information when invoked. **class Devices:** **def __init__(self, serialnumber, hostname, ipaddress, modelnumber, devicetype):** **self.serialnumber = serialnumber** # 12 Hexadecimal numbers **self.hostname = hostname** **self.ipaddress = ipaddress** # format 1.0.0.X - 254.254.254.XXX **self.modelnumber = modelnumber** **self.devicetype = devicetype** **def show(self):** # Display all attributes **print(f"{self.serialnumber},{self.hostname},{self.ipaddress},{self.modelnumber},{self.devicetype}")**
③	Now proceed to create three classes: Router, Switch, and Firewall. Ensure each device type class inherits from the Devices class. By inheriting from the parent class, Devices, all three classes acquire their shared attributes. Notice how this approach streamlines your code; you've established a single set of attributes in the parent class, which is applied universally across the child classes. **class Router(Devices):** **def process(self):** **print("Packet routing")**

(continued)

#	Task
| | ```
class Switch(Devices):
    def process(self):
        print("Packet switching")

class Firewall(Devices):
    def process(self):
        print("Packet filtering")
``` |
| ④ | Next, create three variables—one for each child class—as demonstrated here:<br><br>```
rt1 = Switch("001122AABBCC", "RT01", "1.1.1.1", "C4431-K9", "RT")
sw1 = Router("001122DDEEFF", "SW02", "2.2.2.2", "WS-3850-48T", "SW")
fw1 = Firewall("003344ABCDEF", "FW02", "3.3.3.3", "PA-5280", "FW")
``` |
| ⑤ | From your Python Interpreter, execute the following commands to retrieve device information. The show() method inherited from the parent class, Devices, will automatically display the details. While this example is showcased in the interpreter, you can also integrate it into your code or download oop1.py and run it as a script. Congratulations, you've completed a study on the inheritance of OOP classes.<br><br>```
>>> rt1.show()
001122AABBCC,RT01,1.1.1.1,C4431-K9,RT
>>> sw1.show()
001122DDEEFF,SW02,2.2.2.2,WS-3850-48T,SW
>>> fw1.show()
003344ABCDEF,FW02,3.3.3.3,PA-5280,FW
```<br><br>These commands will showcase the detailed information for each device type—Router, Switch, and Firewall—reflecting the attributes inherited from the Devices class. |

(continued)
```

CHAPTER 8 UPGRADING MULTIPLE CISCO IOS ROUTERS

#	Task

⑥	Now you'll reinforce your understanding of OOP by constructing hierarchical classes, applicable to real-life scenarios. The following example demonstrates a hierarchical structure starting with the top-tier class, Pynetauto_co, representing your company. Subsequently, there's a child class named Cisco, denoting the vendor of your company's networking devices. Further down the hierarchy, you have the Devices class, and nested within Devices are two child classes—Router and Switch. This inheritance structure resembles a waterfall, organized top-down.
	
	In a text editor (such as Notepad or Nano), create or modify the oop_task1.py script as shown here:
	
	``python
	class Pynetauto_co:
	"Parent class of Cisco, HP, Juniper, Arista"
	def __init__(self, companyname):
	self.companyname = companyname
	
	def PrintInfo_P1(self):
	print("Pynetauto Company")
	
	class Cisco(Pynetauto_co):
	"Parent class of Switch, Router, and Firewall"
	def __init__(self, vendor):
	self.Vendor = vendor
	
	def PrintInfo_P2(self):
	print("Cisco")
	
	class Devices(Cisco):
	"Parent class of Switch, Router, and Firewall"
	def __init__(self, serialnumber, hostname, ipaddress, modelnumber, devicetype):
	self.serialnumber = serialnumber # 12 Hexadecimal numbers
	self.hostname = hostname
	self.ipaddress = ipaddress # format 1.0.0.X - 254.254.254.XXX
	self.modelnumber = modelnumber
	self.devicetype = devicetype
	``

(continued)

#	Task
| | ```python
def show(self):
    print(f"{self.serialnumber},{self.hostname},{self.ipaddress},\
{self.modelnumber},{self.devicetype}")

class Router(Devices):
    def process(self):
        print("Packet switching")

class Switch(Devices):
    def process(self):
        print("Packet routing")

rt3 = Router("000111222222", "RT03", "2.2.2.2", "C4351/K9", "RT")
sw3 = Switch("000111222111", "SW03", "1.1.1.1", "WS-3850-48T", "SW")

rt3.PrintInfo_P1()
rt3.PrintInfo_P2()
rt3.show()

sw3.PrintInfo_P1()
sw3.PrintInfo_P2()
sw3.show()
``` |
| ⑦ | Running this script in Python will produce the following output:<br><br>```
Pynctauto Company
Cisco
000111222222,RT03,2.2.2.2,C4351/K9,RT
Pynetauto Company
Cisco
000111222111,SW03,1.1.1.1,WS-3850-48T,SW
```<br><br>This output demonstrates the structure of the hierarchical classes and their attributes. Save the script as oop_task1.py and execute it in Python to observe the inheritance structure and output. |

OOP provides the framework to consolidate data and streamline code, offering a clean and organized approach. When faced with numerous objects sharing common attributes, OOP becomes a valuable tool, sparing you from individually defining each object's properties. Moreover, in scenarios involving data manipulation within data frames or SQL structures, OOP's structure proves instrumental in efficiently managing and processing data within Python scripts. Now, let's swiftly transition to the next topic.

Managing Flow Control and User Input: UID, PWD, and Information Collection

This section dives into a practical and engaging aspect. When crafting Python applications—be they interactive or non-interactive—controlling user input becomes crucial. Ensuring that user input aligns precisely with your script's expectations empowers you to dictate the script's flow. For instance, prompting users with yes or no input leads to four potential scenarios: typing yes or y, typing no or n, entering something other than the specified options, or even providing no input. Introducing Regular Expressions into user input allows you to narrow down accepted information and reject any unmatched variable values. While the upcoming example demonstrates this technique, let's begin with an entertaining exploration of the yes/no user input scenario. Later, you will refine the script, exploring how it can significantly alter Python script flow and how enhancements can elevate a simple script. The initial script focuses on creating a username and password input application using a yes/no script.

CHAPTER 8　UPGRADING MULTIPLE CISCO IOS ROUTERS

#	Task
①	The following is a simple yes/no user input example using the `if ~ else` statement. Create a new Python script called yes_or_no1.py and run it. There are a few flaws with this script, but you will enhance this first input script to make it useful in your real Python applications. Create a separate directory on your server and create your first script, as shown here: jdoe@u22s1:~$ **mkdir ch8_user_input && cd ch8_user_input** jdoe@u22s1:~/ch8_user_input$ **nano yes_or_no1.py** jdoe@u22s1:~/ch8_user_input$ **cat yes_or_no1.py** **def yes_or_no():** # Create function called yes_or_no **yes_or_no = input("Enter yes or no : ")** # Ask for user input **yes_or_no = yes_or_no.lower()** # change all input casing to lower casing **if yes_or_no == "yes":** # if answer is 'yes' take the following action **print("Oh Yes!")** **else:** # If the answer is 'no' or other input, take the following action **print("Oh No!")** **yes_or_no()** # Run the function When you run the previous script and provide input, the output should look like this for each response. This Python code is pretty basic. Think about where you can improve on this script. jdoe@u22s1:~/ch8_user_input$ **python yes_or_no1.py** Enter yes or no : **Yes** Oh Yes! jdoe@u22s1:~/ch8_user_input$ python yes_or_no1.py Enter yes or no : **no** Oh No! jdoe@u22s1:~/ch8_user_input$ python yes_or_no1.py Enter yes or no : **maybe** Oh No!

(*continued*)

#	Task
| ② | The first example has the following flaws:<br><br>a. Does not consider an abbreviated response from the user, y for yes and n for no.<br><br>b. No response and an incorrect response are treated as the same response.<br><br>Now you'll correct the previous and see if you can reiterate this simple code and make it better. Name your code file yes_or_no2.py.<br><br>```
jdoe@u22s1:~/ch8_user_input$ nano yes_or_no2.py
def yes_or_no():
yes_or_no = input("Enter yes or no : ")
yes_or_no = yes_or_no.lower()
if yes_or_no == "yes" or yes_or_no == "y": # Takes abbreviated response
print("Oh Yes!")
elif yes_or_no == "no" or yes_or_no == "n": # Takes abbreviated response
print("Oh No!")
else: # For Any other responses, print the following statement
print("You have not entered the correct response.")
yes_or_no()
```<br><br>The expected output examples are as shown here, but you can still improve this further:<br><br>```
jdoe@u22s1:~/ch8_user_input$ python yes_or_no2.py
Enter yes or no : y
Oh Yes!
jdoe@u22s1:~/ch8_user_input$ python yes_or_no2.py
Enter yes or no : no
Oh No!
jdoe@u22s1:~/ch8_user_input$ python yes_or_no2.py
Enter yes or no : maybe
You have not entered the correct response.
``` |

(continued)

CHAPTER 8 UPGRADING MULTIPLE CISCO IOS ROUTERS

#	Task
| ③ | To restrict responses to yes or no and prompt the user until obtaining one of these valid responses, you must take control of the script's flow. How can this be achieved? Which part of the script requires modification to ensure a yes or no response? Various approaches can accomplish this, and in Python coding, there are no absolute right or wrong answers—just optimal recommendations. The next example helps fulfill this objective.<br><br>In the upcoming example, a straightforward list of the expected answers is used. The script proceeds only upon receiving one of these predefined values as a response. Continuously prompting the user until the correct response is received conveys the expectation of a yes or no response.<br><br>jdoe@u22s1:~/ch8_user_input$ **nano yes_or_no3.py**<br><br>```python
def yes_or_no():
yes_or_no = input("Enter yes or no : ")
yes_or_no = yes_or_no.lower()
expected_response = ['yes', 'y', 'no', 'n'] # Expected responses
while yes_or_no not in expected_response: # Prompt until 'yes' or 'no' response is given
    yes_or_no = input("Expecting yes or no : ")
if yes_or_no == "yes" or yes_or_no == "y": # 'yes' or 'y' action
    print("Oh Yes!")
else: # 'no' or 'n' action
    print("Oh No!")
yes_or_no()
```<br><br>The expected output should resemble the following, and the response needs to fulfill the criteria to exit the while loop. Otherwise, the user is repeatedly prompted for the correct response. This demonstrates a promising approach, but where could you apply this code? The subsequent example explores a real-life scenario and its practical implementation. |

(continued)
```

CHAPTER 8   UPGRADING MULTIPLE CISCO IOS ROUTERS

#	Task
	jdoe@u22s1:~/ch8_user_input$ **python yes_or_no3.py** Enter yes or no : **Y** Oh Yes! jdoe@u22s1:~/ch8_user_input$ **python yes_or_no3.py** Enter yes or no : **NO** Oh No! jdoe@u22s1:~/ch8_user_input$ **python yes_or_no3.py** Enter yes or no : **maybe** Expecting yes or no : **maybe** Expecting yes or no : **yes** Oh Yes!
④	The following script collects two sets of network administrator ID and password. To start, get the first administrator's credentials, and then ask if the second administrator's credentials are the same. If the response is yes or y, the user ID and passwords are the same. If the response is no or n, then collect the second set of user ID and password. jdoe@u22s1:~/ch8_user_input$ **nano yes_or_no4.py** **from getpass import getpass** **def get_credentials():** #Prompts for, and returns a username1 and password1 **username1 = input("Enter Network Admin1 ID: ")** # Request for username 1 **password1 = getpass("Enter Network Admin1 PWD: ")** # Request for password 1 **print("Username1 :", username1, "Password1 :", password1)** # Print username and password #Prompts for username2 and password2 **yes_or_no = input("Network Admin2 credentials same as Network Admin1 credentials? (Yes/No): ").lower()** # Ask if Network Admin 2 has the same credentials as Admin 1

(*continued*)

#	Task
	``
expected_response = ['yes', 'y', 'no', 'n'] # Expect any of these
four responses
while yes_or_no not in expected_response: # Prompt until 'yes' or
'no' response is given
    yes_or_no = input("Expecting yes or no : ")
if yes_or_no == "yes" or yes_or_no == "y": # If 'yes' or 'y',
credentials ate the same as Admin1
    username2 = username1
    password2 = password1
    print("Username2 :", username2, "Password2: ", password2) # Print
username and password
else: # If 'no' or 'n', request for Admin2 username and password
    username2 = input("Enter Network Admin2 ID : ")
    password2 = getpass("Enter Network Admin2 Password : ")
    print("Username2 :", username2, "Password2 :", password2) # Print
username and password
get_credentials()
```
Run the script to test the user ID and password collector script. The expected output should be similar to the following output. Do you think you can reiterate this script and improve it? Let's see the final example.

```
jdoe@u22s1:~/ch8_user_input$ python yes_or_no4.py
Enter Network Admin1 ID: hugh
Enter Network Admin1 PWD:*********
Username1 : john Password1 : password1
Network Admin2 credentials same as Network Admin1 credentials? (Yes/
No): yes
Username2 : john Password2: password1
jdoe@u22s1:~/ch8_user_input$ python yes_or_no4.py
``` |

(*continued*)

487
```

CHAPTER 8   UPGRADING MULTIPLE CISCO IOS ROUTERS

#	Task
	Enter Network Admin1 ID: **hugh** Enter Network Admin1 PWD: ************* Username1 : hugh Password1 : **password1** Network Admin2 credentials same as Network Admin1 credentials? (Yes/No): **no** Enter Network Admin2 ID : **john** Enter Network Admin2 Password : *************** Username2 : john Password2 : **password777**
⑤	In this final iteration of the yes-or-no example, a password verification feature is included to ensure the user's first password matches the second one. Utilizing the getpass module from the getpass library conceals the password during input. This addition of the password verification feature minimizes the probability of an incorrect password entry. jdoe@u22s1:~/ch8_user_input$ **nano yes_or_no5.py** **from getpass import getpass** **def get_credentials():** #Prompts for, and returns a username1 and password1 **username1 = input("Enter Network Admin1 ID: ")** # Request for username 1 **password1 = None** # Set password1 to None (initial value to None) **while not password1:** # Until password1 is given **password1 = getpass("Enter Network Admin1 PWD : ")** # Get password1 **password1_verify = getpass("Confirm Network Admin1 PWD : ")** # Request for validation **if password1 != password1_verify:** # If the password1 and verification password does not match **print("Passwords do not match. Please try again.")** # Print this information **password1 = None** # Set the password to None and ask for password1 again **print("Username1 :", username1, "Password1 :", password1)** # Print username and password

(*continued*)

#	Task

```python
#Prompts for  username2 and password2
yes_or_no = input("Network Admin2 credentials same as Network Admin1 
credentials? (Yes/No): ").lower() # Ask if Network Admin 2 has the 
same credentials as Admin 1
expected_response = ['yes', 'y', 'no', 'n'] # Expect any of these 
four responses
while yes_or_no not in expected_response: # Prompt until 'yes' or 
'no' response is given
yes_or_no = input("Expecting yes or no : ")
if yes_or_no == "yes" or yes_or_no == "y": # If 'yes' or 'y', 
credentials ate the same as Admin1
username2 = username1
password2 = password1
print("Username2 :", username2, "Password2 :", password2) # Print 
username and password
else: # If 'no' or 'n', request for Admin2 username and password
username2 = input("Enter Network Admin2 ID: ") # Request for 
username 2
password2 = None # Explanation same as above
while not password2: # Explanation same as above
password2 = getpass("Enter Network Admin2 PWD : ") # Explanation same 
as above
password2_verify = getpass("Confirm Network Admin2 PWD : ")
if password2 != password2_verify: # Explanation same as above
print("Passwords do not match. Please try again.")
password2 = None # Explanation same as above
print("Username2 :", username2, "Password2 :", password2) # Print 
username and password
get_credentials()
```

(continued)

CHAPTER 8   UPGRADING MULTIPLE CISCO IOS ROUTERS

#	Task
	You can grasp the iterative process to enhance the original code. Once you're satisfied with the quality of your code, you have the option to save it to your code repository and document it. Alternatively, you can share it with your team or the online community. In reality, Python coding isn't always about having fun; precision is key in your coding endeavors. Often, you will need to refine and iterate your code multiple times. Begin with a basic script and progressively refine it through iterations. Here is the expected output: ``` jdoe@u22s1:~/ch8_user_input$ python yes_or_no5.py Enter Network Admin1 ID: hugh Enter Network Admin1 PWD :********** Confirm Network Admin1 PWD : ********* # incorrect password typed Passwords do not match. Please try again. Enter Network Admin1 PWD : ********** Confirm Network Admin1 PWD : ********** # correct password typed Username1 : hugh Password1 : password123 Network Admin2 credentials same as Network Admin1 credentials? (Yes/No): yes Username2 : hugh Password2 : password123 jdoe@u22s1:~/ch8_user_input$ python yes_or_no5.py Enter Network Admin1 ID: john Enter Network Admin1 PWD : ********** Confirm Network Admin1 PWD : ********** Username1 : john Password1 : password777 Network Admin2 credentials same as Network Admin1 credentials? (Yes/No): no Enter Network Admin2 ID: bill Enter Network Admin2 PWD : ********** Confirm Network Admin2 PWD : ********** Username2 : bill Password2 : password888 ```

You've finished an exercise involving writing a basic Python application and subsequently enhancing the code by introducing new features and ideas through iterations. The initial code is seldom flawless; refining even the most straightforward Python programs demands time and imagination in your work. Now you'll get ready for the final lab and strategize for the development of an IOS upgrade application by exploring the actual IOS upgrade tasks and processes.

Proactive Patch Management and Zero Trust: Shaping Network Security

Security has undergone significant evolution over the past decade, transitioning from a pressing issue to an escalating concern projected to intensify in the future. The complexity of connecting IP services, driven by advancements such as cloud computing, SD-WAN, edge computing, Zero Trust security, serverless frameworks, and microservices (service-oriented architecture) has elevated the importance of enterprise security to unprecedented levels. To safeguard enterprise networks against both external and internal threats, the Zero Trust security model emerged, advocating for a meticulous, untrusting approach by network and security engineers. In the context of medium to larger organizations utilizing enterprise-grade technologies, the network infrastructure heavily relies on edge devices working in tandem with robust firewalls from industry-leading vendors like Palo Alto PAN-OS, Fortinet FortiGate, and Cisco Firepower. These firewalls seamlessly facilitate user connections across different geographic locations and contexts to various IP services. Instances of hackers breaching enterprise and government networks have become more frequent, leading network and security engineers to operate under the assumption of incessant security breaches. This realization has propelled the widespread adoption of Zero Trust security protocols across the board.

Since the inception of Zero Trust security, proactive OS patch management has taken center stage, becoming an undeniable priority for numerous CTOs, network operations managers, and business owners. This heightened focus emphasizes the pivotal role of proactive OS upgrading in today's dynamic IT environment. Its primary aim is to proactively diminish the risks associated with breaches and cyberattacks, ensuring a proactive stance and steer clear of unwanted media attention, reputational damage, and red faces. A decade ago, network engineers primarily addressed vulnerabilities in platforms reactively, triggered by the announcement of Common

Vulnerabilities and Exposures (CVEs). Organizations today find themselves at varying stages of proactive patch management, using different vendor products. Some may grapple with end-of-life systems lacking support and updates, while others boast cutting-edge technologies, maintaining meticulous OS alignment with their patch management schedule. These disparities often stem from the diverse approaches and capabilities of the teams supporting each organization's IT infrastructure.

The CVE database, collated by the CVE Program, aggregates information from diverse sources, ensuring comprehensive coverage of vulnerabilities across systems and applications. Each CVE entry is a unique marker for identified vulnerabilities affecting services, software, or platforms, heavily relied on by both network vendors and corporations for publicly available security information. In recent years, corporations have shifted toward proactive measures, embracing Zero Trust security and proactive patch management strategies. Upgrading the operating systems of critical enterprise network devices, whether in hardware or virtual form, now stands as a foundational responsibility for network engineers.

Router IOS Upgrade Lab Preparation

One issue with IOS upgrading using old IOS or IOSvL3 images on GNS3 is its limited support for reloading, a shortcoming compared to real IOS software on hardware or supported virtual environment. To overcome IOS upgrade scenarios, two options exist: acquiring real hardware or finding a suitable virtual IOS XE image for upgrades in a virtual environment. While I've promised emulation and practice without costly vendor hardware in this book, upgrading a Cisco IOS necessitates real hardware or supported virtual IOS XE images in a production-like setting.

Here it is worth noting a difference between the old Cisco IOS and newer Cisco IOS XE software. Cisco IOS stands as a monolithic system, while IOS XE is a modular version based on a Linux kernel, both managing Cisco network devices. IOS XE offers enhanced functionalities—modularity, improved resilience, and support for SDN features.

Virtualized Cisco routers and switches lack physical components such as motherboards, ports, power supply, and cooling fans, operating as logical parts. Although IOSvL2 switches emulate some switching aspects, they lack complete ASCII support for LAN switching. While the IOSvL3 router provides many features for testing routing concepts, it doesn't support real-life IOS XE upgrades.

To emulate IOS/IOS XE upgrades using Python scripts without actual hardware, utilizing the virtual edge router Cisco Catalyst 8000v is the closest thing you can get to the life-like scenario, as these are used in the production edge network. In the first edition, two Cisco CSR 1000v Cloud routers were used. This device has now reached the end of life (EoL), so, in this edition, Cisco Catalyst 8000v is used for IOS upgrading application development. To practice IOS XE upgrades, you need two different versions with two different file extension types (.ova and .bin), which will be explained soon.

Unfortunately, this software is Cisco proprietary, and the author or the publisher cannot share them publicly. You must source the software through recommended channels for practice. Alternatively, without access to Catalyst 8000v IOSXE images, you can still use the older CSR 1000v IOS-XE or even consider purchasing two outdated Cisco 2621XM/2651XM routers for $20 each on eBay to emulate the router upgrade. I also recommend that starting network engineers purchase four older switches along with the router purchases for learning.

Even the times have changed, working with actual hardware teaches various aspects of infrastructure management not otherwise taught in today's training courses. While most network engineers need exposure to OSI model L5 to L7 technologies, the traditional network engineers' strength indeed lies in their expertise in OSI model L1 to L4. Even in the age of cloud computing, physical connections remain vital. If you're a network student or engineer in 2024, you must cover all OSI layers' technologies and dedicate time to L7, as you are doing with this book. Let's move forward with current trends, and become more competitive with the developers, without "buts" or "ifs."

Cisco Catalyst 8000v IOS XE Software and Download

To prepare for this lab, you need to download a set of Cisco Catalyst 8000v IOS XE files. Pay close attention and make notes on the different file extensions included in this set for your final lab. Initially, to create the virtual machines (routers), you need the file with the .ova extension, specifically the VMware ESXi7.0/8.0 virtual machine template for router installation. To practice IOS upgrading, you need the latest IOS XE file, denoted by the .bin file extension. Detailed information about the example files used in this book can be found in Table 8-1. Your software version doesn't need to precisely match what's recommended here, provided one file is in .ova format and the second file is a .bin file with a higher IOS XE version number.

CHAPTER 8   UPGRADING MULTIPLE CISCO IOS ROUTERS

Table 8-1. *Catalyst 8000v IOS XE Files Used in This Chapter*

Item	From IOS XE Detail	To IOS XE Detail
Release (Train)	Bengaluru-17.6.3a	Bengaluru-17.6.5a
File Name	c8000v-universalk9.17.06.03a.ova	c8000v-universalk9.17.06.05a.bin
Release Date	18-Apr-2022	28-Oct-2023
Min Memory	DRAM 4096 Flash 8192	DRAM 4096 Flash 8192
File Size	844.25 MB	770.79 MB
MD5 Value	09b8f40158f79b793690bad8a534ad0e	13f0161a50210f2f21618fc59c5f5343

As you've noticed, the demanding minimum memory and flash size requirements make it challenging to run Cisco 8000v Edge routers in the GNS3 environment. Consequently, you will install these routers on VMware Workstation 17 Pro as virtual machines, setting up two 8000v VMs. Network automation's strength lies in its ability to operate across multiple devices or applications. Hence, you will create two virtual Catalyst 8000v router VMs on VMware Workstation, collectively consuming 8GB of memory. Unlike with the GNS3 setup with limited rebooting emulation, you will import the .ova file and create two Cisco IOS XE routers on VMware Workstation 17 Pro. This allows you to emulate the IOS upgrade by transitioning to the latest .bin file within the same IOS XE train. Although this lab setup won't extensively test routing concepts, it adequately facilitates the emulation of IOS upgrades and aids in developing a functional Python application for end-to-end IOS upgrade scenarios.

If you have an active Cisco service contract and access to the correct Cisco software, log in to download the files needed for your lab practice (see Figure 8-1). These files need not match exactly. If you prefer an alternate IOS XE train for Catalyst 8000v, download the necessary files for this lab.

CHAPTER 8   UPGRADING MULTIPLE CISCO IOS ROUTERS

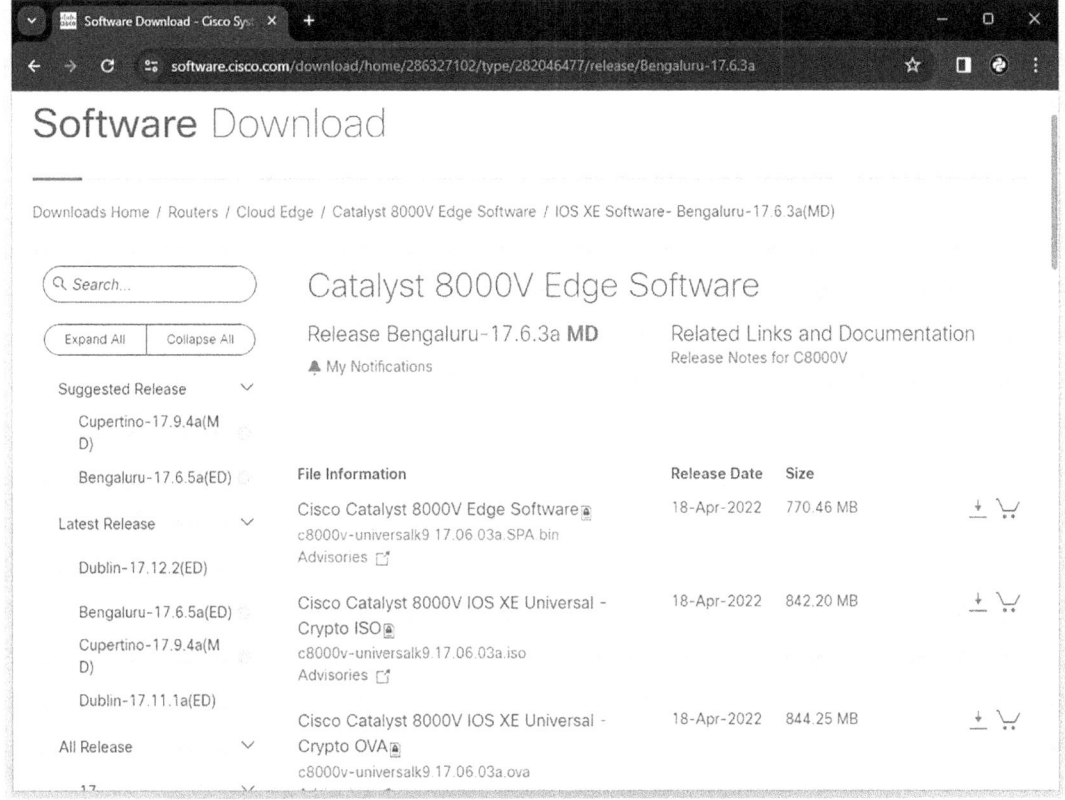

Figure 8-1. *Catalyst 8000v IOS XE downloads from Cisco software download site*

Cisco Catalyst 8000v Installation on VMware Workstation

Installing Catalyst 8000v on VMware Workstation 17 is no different from installing the GNS3 VM .ova file you previously installed. But this time, you are going to create two virtual machines instead of one. Also, for this lab, you can keep the GNS3 completely powered off, but you will still need the Python server, u22s1, powered on for login and connectivity testing. As soon as you build the virtual routers, you will configure both routers with minimal configuration and get on with the IOS upgrade application development (see Figure 8-2).

CHAPTER 8   UPGRADING MULTIPLE CISCO IOS ROUTERS

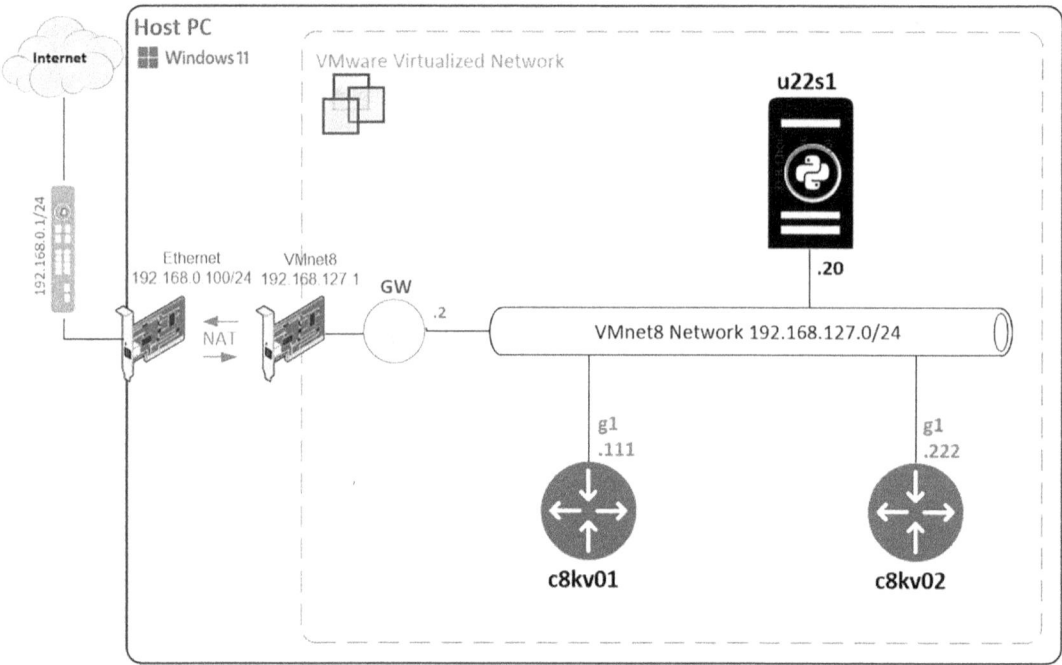

Figure 8-2. Chapter 8 lab equipment

Now follow these tasks to install the c8kv01 and c8kv02 virtual routers. The installation of Catalyst 8000v is straightforward.

CHAPTER 8 UPGRADING MULTIPLE CISCO IOS ROUTERS

#	Task
1	To create the virtual machine, first open the VMware Workstation's main window and choose File ➤ Open. Then go to the Downloads folder and select the c8000v-universalk9.17.06.03a.ova file (see Figure 8-3).

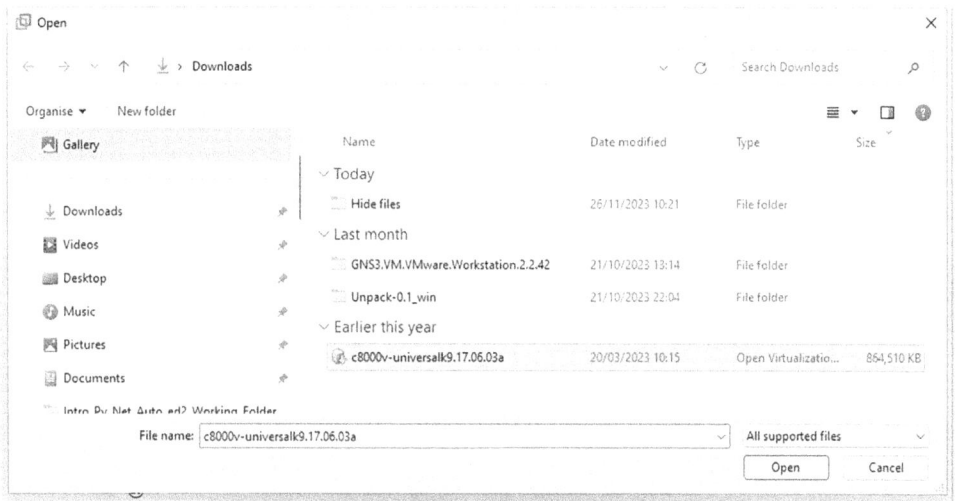

Figure 8-3. *VMware Workstation, selecting the Catalyst 8000v .ova file*

(*continued*)

CHAPTER 8 UPGRADING MULTIPLE CISCO IOS ROUTERS

#	Task
②	For the first virtual router, provide c8kv01 as the name of this device. When you create the second virtual router, you just need to change the last digit of the router name to 2, so it will be c8kv02. See Figure 8-4. 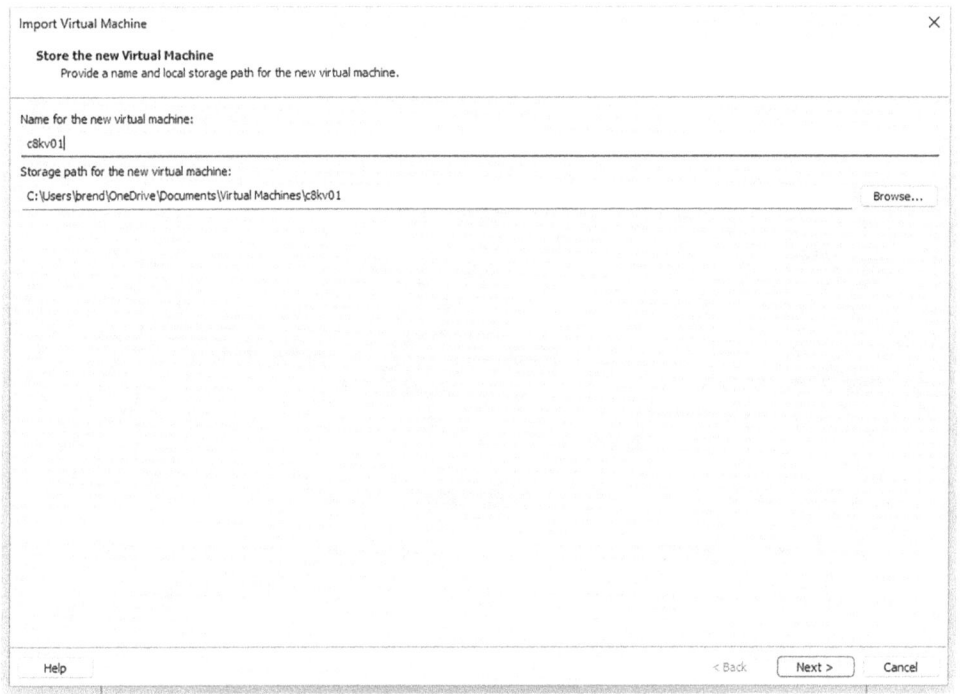 *Figure 8-4. c8kv01, giving the new router name*

(continued)

CHAPTER 8 UPGRADING MULTIPLE CISCO IOS ROUTERS

#	Task
③	Leave the Deployment Options setting as Small and click the Next button. See Figure 8-5.

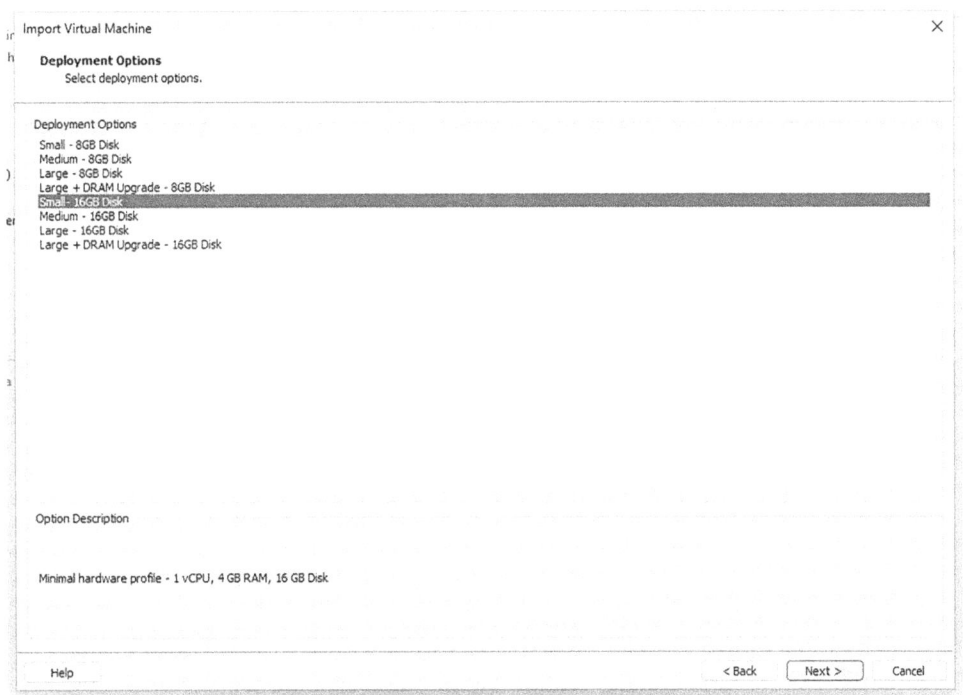

Figure 8-5. *c8kv01, deployment options*

(*continued*)

CHAPTER 8   UPGRADING MULTIPLE CISCO IOS ROUTERS

#	Task
④	To import the virtual router, you are going to fill out the following properties. Configuring the properties on the GUI is the same as configuring these settings through the CLI. Just click the Import button to create your first virtual router (see Figure 8-6).

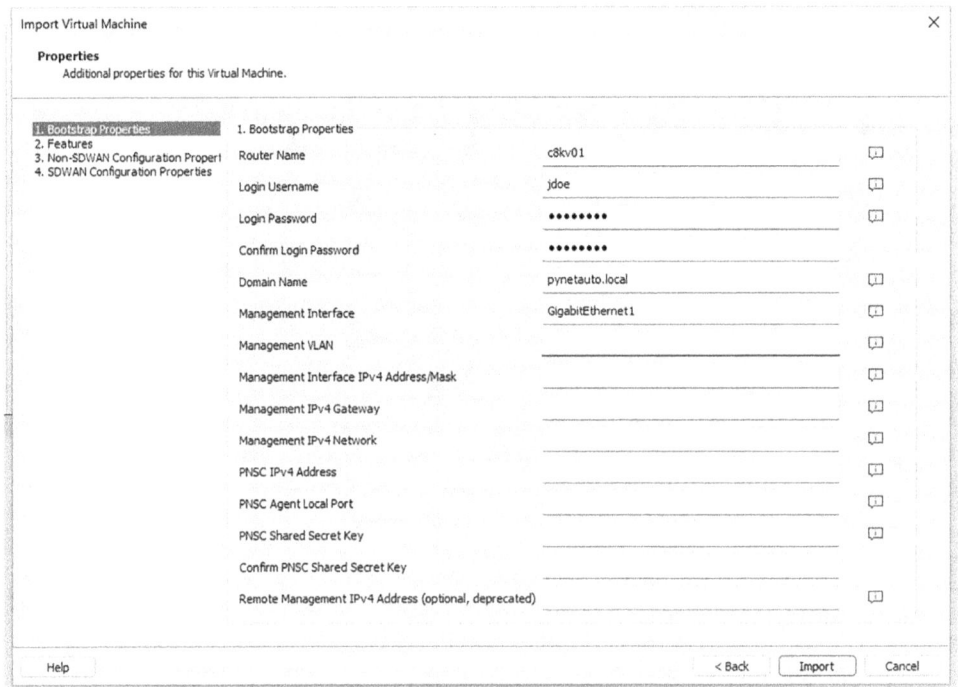

Figure 8-6. c8kv01, Import Virtual Machine Properties

(*continued*)

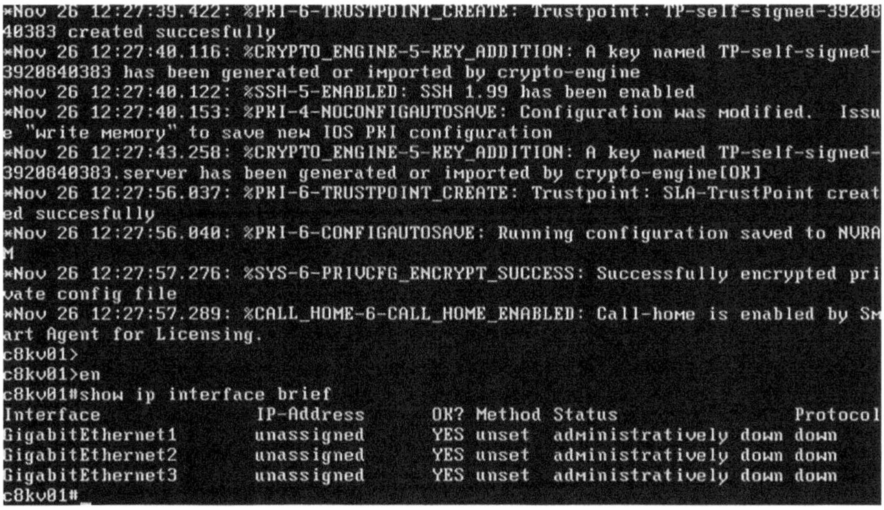

#	Task
| ⑤ | Leave the router to complete its installation and wait approximately five to seven minutes, until it goes through the full installation process.<br><br>You may have to wait for three to five minutes for the virtual machine to be imported to the workstation. Given enough time, the router import will be completed, as shown next. As soon as the importing is done, log in and check the interface status (see Figure 8-7).<br><br>```
*Nov 26 12:27:39.422: %PKI-6-TRUSTPOINT_CREATE: Trustpoint: TP-self-signed-39208
40383 created succesfully
*Nov 26 12:27:40.116: %CRYPTO_ENGINE-5-KEY_ADDITION: A key named TP-self-signed-
3920840383 has been generated or imported by crypto-engine
*Nov 26 12:27:40.122: %SSH-5-ENABLED: SSH 1.99 has been enabled
*Nov 26 12:27:40.153: %PKI-4-NOCONFIGAUTOSAVE: Configuration was modified.  Issu
e "write memory" to save new IOS PKI configuration
*Nov 26 12:27:43.258: %CRYPTO_ENGINE-5-KEY_ADDITION: A key named TP-self-signed-
3920840383.server has been generated or imported by crypto-engine[OK]
*Nov 26 12:27:56.037: %PKI-6-TRUSTPOINT_CREATE: Trustpoint: SLA-TrustPoint creat
ed succesfully
*Nov 26 12:27:56.040: %PKI-6-CONFIGAUTOSAVE: Running configuration saved to NVRA
M
*Nov 26 12:27:57.276: %SYS-6-PRIVCFG_ENCRYPT_SUCCESS: Successfully encrypted pri
vate config file
*Nov 26 12:27:57.289: %CALL_HOME-6-CALL_HOME_ENABLED: Call-home is enabled by Sm
art Agent for Licensing.
c8kv01>
c8kv01>en
c8kv01#show ip interface brief
Interface              IP-Address      OK? Method Status                Protocol
GigabitEthernet1       unassigned      YES unset  administratively down down
GigabitEthernet2       unassigned      YES unset  administratively down down
GigabitEthernet3       unassigned      YES unset  administratively down down
c8kv01#
``` |

Figure 8-7. *c8kv01, installation completed, run show ip interface brief command*

(*continued*)
```

CHAPTER 8   UPGRADING MULTIPLE CISCO IOS ROUTERS

#	Task
⑥	While the VM c8kv01 is selected, go to the VM menu from the main menu and select the Settings… option (see Figure 8-8). Alternatively, you can use right-click the c8kv01 VM and get to the same Settings window.

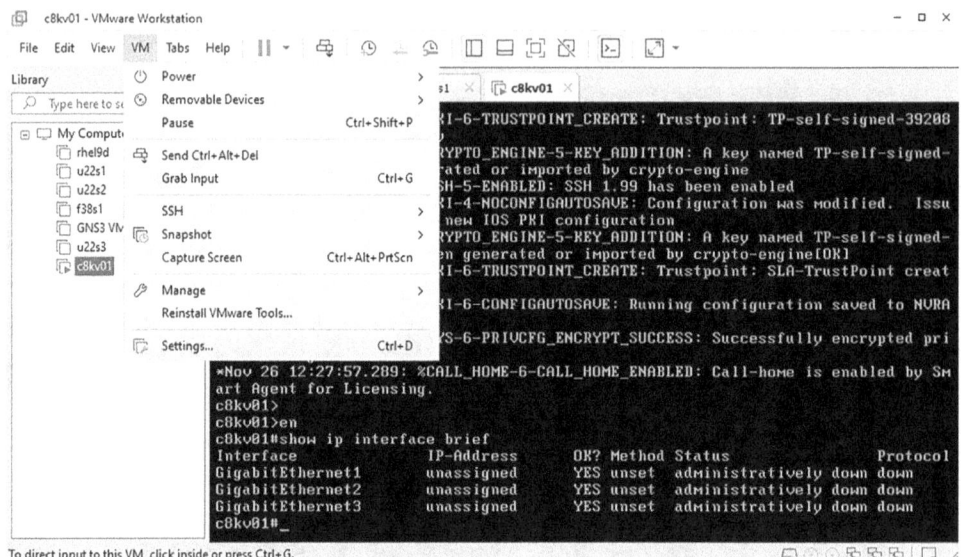

Figure 8-8.   c8kv01, opening Settings…

(*continued*)

CHAPTER 8 UPGRADING MULTIPLE CISCO IOS ROUTERS

#	Task
⑦	Under Hardware, change the Network Adapter configuration from Bridged (Automatic) to NAT: Used to Share the Host's IP Address (see Figure 8-9).

Figure 8-9. *c8kv01, changing network adapter (GigabitEthernet1) to NAT*

(*continued*)

#	Task
⑧	The interfaces are in an administrative down/down state, like all other Cisco routers. Before bringing up the interface, you'll quickly assign the correct IP address for the GigabitEthernet1 interface with 192.168.127.111/24. Because you need only one interface for your IOS upgrade tool development, you do not have to configure the other interfaces. c8kv01> **en** c8kv01#**config terminal** c8kv01(config)#**interface GigabitEthernet1** c8kv01(config-if)#**ip address 192.168.127.111 255.255.255.0** c8kv01(config-if)#**no shut** c8kv01(config)#**enable secret cisco123** # important for IOS file uploading c8kv01(config-if)#**end**
⑨	Now repeat the Steps 1-7 to create c8kv02. You can simply create another VM by copying c8kv01, but you will just create a clean VM, as it only takes a few minutes to create a new virtual router.
⑩	Note that once you create the second virtual router, c8kv02, you should configure the GigabitEthernet1 interface with 192.168.127.222/24. c8kv02> **en** c8kv02#**config terminal** c8kv02(config)#**interface GigabitEthernet1** c8kv02(config-if)#**ip address 192.168.127.222 255.255.255.0** c8kv02(config-if)#**no shut** c8kv02(config)#**enable secret cisco123** # important for IOS file uploading c8kv02(config-if)#**end**

(continued)

CHAPTER 8   UPGRADING MULTIPLE CISCO IOS ROUTERS

#	Task
⑪	On your Windows host, open the PuTTY SSH client and log in to c8kv01 (192.168.127.111) and c8kv02 (192.168.127.222). When you are prompted with a PuTTY security alert, make sure you click the Yes button to accept the device's host key (see Figure 8-10).

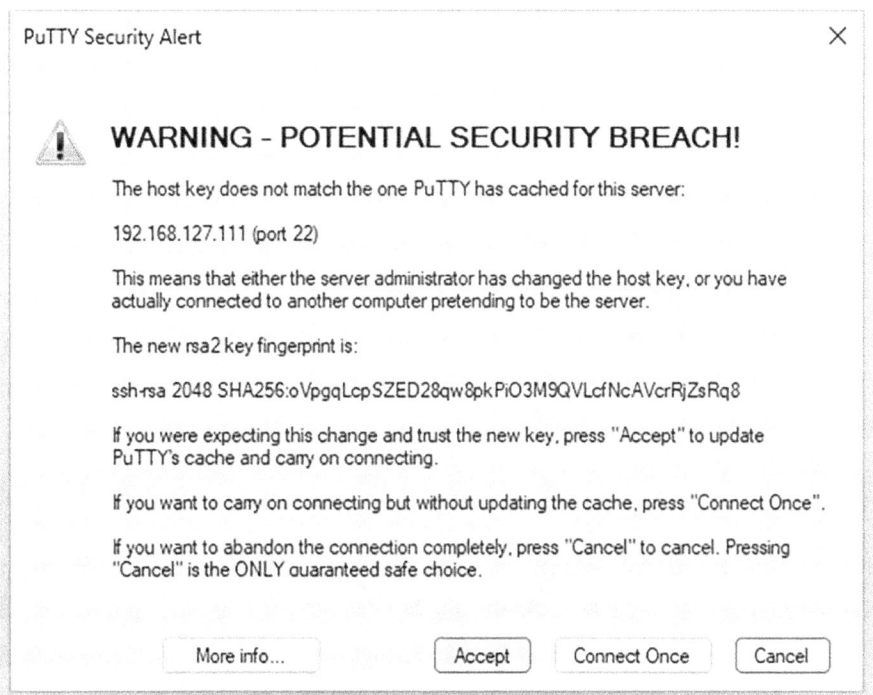

Figure 8-10. Accept the PuTTY security alert

Also, don't forget to send some ICMP commands to the u22s11 server and between routers to test the connectivity (see Figure 8-11).

(*continued*)

CHAPTER 8   UPGRADING MULTIPLE CISCO IOS ROUTERS

#	Task
	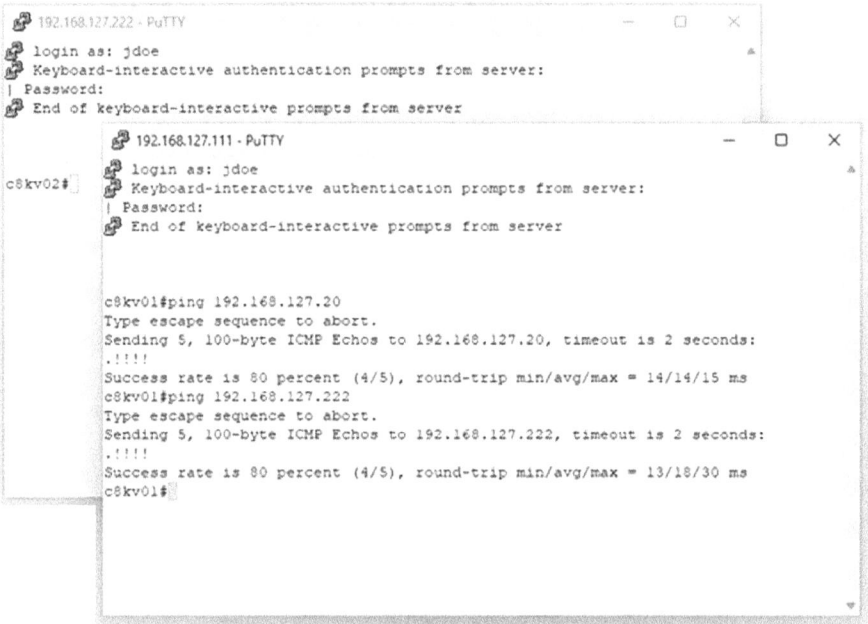 *Figure 8-11. c8kv01 and c8kv02, PuTTY login for the first time*
⑫	To complete the basic configuration, you'll now check the domain and create the RSA key for secure key exchange. Also, disable the timeout for VTY lines to stop the SSH session timeouts; only use this feature for the lab environment. Save the configuration by running write memory. c8kv01#**show run | in ip domain** ip domain name pynetauto.local c8kv01#**conf t** Enter configuration commands, one per line. End with CNTL/Z. c8kv01(config)#**crypto key generate rsa** The name for the keys will be: c8kv01.pynetauto.local Choose the size of the key modulus in the range of 512 to 4096 for your General Purpose Keys. Choosing a key modulus greater than 512 may take a few minutes. *(continued)*

#	Task
	How many bits in the modulus [2048]: **2048** % Generating 2048 bit RSA keys, keys will be non-exportable... [OK] (elapsed time was 1 seconds) c8kv01(config)#**line vty 0 15** c8kv01(config-line)#**exec-timeout 0** c8kv01(config-line)#**end** c8kv01#**write memory** Building configuration... [OK]
⑬	Repeat the same task on c8kv02. c8kv02#**show run | in ip domain** ip domain name pynetauto.local c8kv02#**conf t** Enter configuration commands, one per line. End with CNTL/Z. c8kv02(config)#**crypto key generate rsa** The name for the keys will be: c8kv02.pynetauto.local Choose the size of the key modulus in the range of 512 to 4096 for your General Purpose Keys. Choosing a key modulus greater than 512 may take a few minutes. How many bits in the modulus [2048]: **2048** % Generating 2048 bit RSA keys, keys will be non-exportable... [OK] (elapsed time was 1 seconds) c8kv02(config)#**line vty 0 15** c8kv02(config-line)#**exec-timeout 0** c8kv02(config-line)#**end** c8kv02#**write memory** Building configuration... [OK]

(*continued*)

CHAPTER 8   UPGRADING MULTIPLE CISCO IOS ROUTERS

#	Task
⑭	Once you are happy with the basic configuration, take a snapshot of the first virtual router, as shown in Figure 8-12. 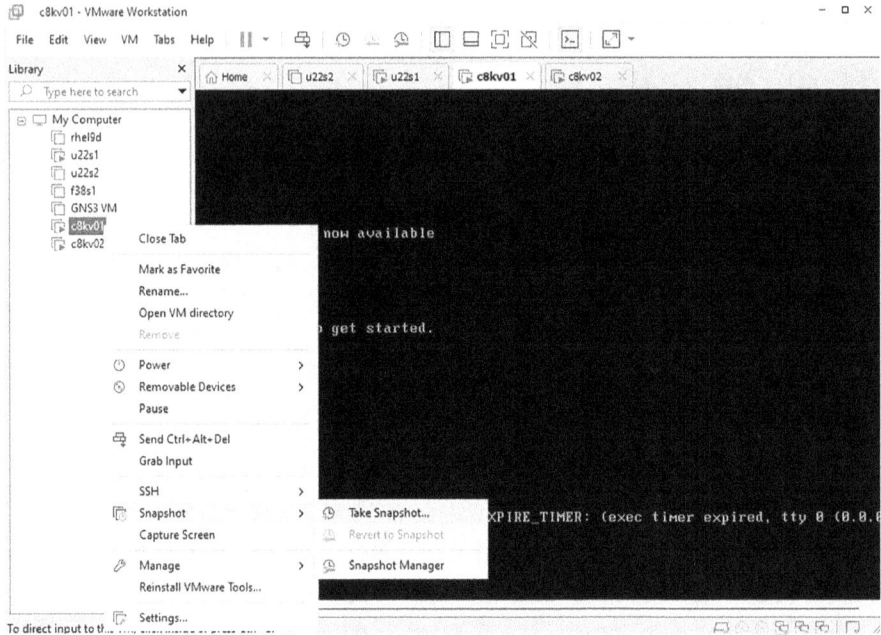 ***Figure 8-12.*** *c8kv01, taking a snapshot*
⑮	Once the first router's snapshot-taking completes, repeat this process on the second virtual router, c8kv02. These snapshots are required and will come in handy during IOS upgrade application development. Having snapshots of the last known good configuration adds flexibility to your lab.

(continued)

CHAPTER 8   UPGRADING MULTIPLE CISCO IOS ROUTERS

#	Task
⑯	Once you have the two Catalyst 8000v routers (c8kv01 and c8kv02) installed, you are ready to move on to the development phase (see Figure 8-13).

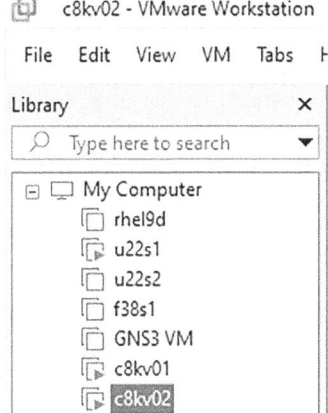

Figure 8-13. *c8kv01 and c8kv02 are up and running*

You have now installed two Cisco Catalyst 8000v routers for IOS upgrade application development. The next section covers the process of upgrading the Cisco IOS devices.

How an IOS (IOS XE/XR) Is Upgraded on Cisco Devices

Upgrading Cisco IOS (IOS/IOS XE/IOS-XR) on Cisco routers and switches isn't a task for the faint-hearted, especially when numerous business-critical applications are traversing the network—both during and after business hours. Even when system and DevOps engineers are off-duty, enterprise routers and switches must ensure seamless IP connectivity for end-users and facilitate uninterrupted IP services for system backups, especially in an online business where secure transactions need to proceed without service disruptions. Enterprise networking devices operate almost non-stop, 24x7, 365 days a year. This is why enterprises and reputable companies prioritize purchasing products from established vendors like Cisco for their networking needs. Any service interruptions often lead to the finger being pointed at IP gateway devices (routers), even for issues related to company emails, cloud services, or system backups.

In ITIL-driven organizations, executing an IOS upgrade involves navigating through significant red tape via change control processes. Reasons for upgrading IOS devices vary, including dealing with outdated IOS versions, bug fixes, adherence to Cisco TAC recommendations, or fulfilling obligations outlined in third-party agreements. IOS upgrade commitments should typically be part of support contracts (internally or externally) and are integral to network patch management practices. The general rule is that the latest IOS versions often come with built-in protection against well-known security threats, reducing attackers' opportunities. However, it is essential to note that complete protection isn't guaranteed if devices are connected to the network; isolated labs or unpowered devices offer only foolproof safety. This is the basic essence of Zero Trust security, and it is nothing new to network engineers. For some organizations, OS patching is a resource-intensive and critical task. Due to resource constraints, many opt to outsource OS patch management to larger managed service providers (MSPs) specializing in such tasks and equipped with the right tools.

Tasks Involved in a Cisco IOS Upgrade

This section delves into how a seasoned network engineer undertakes a Cisco IOS upgrade from an outdated version to the latest one. The IOS upgrade process encompasses a predefined set of tasks, but variations exist in the pre-checks and post-checks across different engineers. Typically, the process is relatively standardized, demanding meticulous attention to detail and scheduling during non-business hours to prevent significant outages. In some environments, Cisco IOS upgrades are restricted to after midnight, with a narrow change window in extreme cases. These changes often happen during odd hours, leading to sleep disturbances among veteran network engineers and potentially straining long-term relationships. As described in Chapter 1 of the Part I book, the constraint on the relationship is also related to the way engineers think—**they have a tenancy to think that they are always logical and are in the right, usually not interested in listening to others' opinions or arguments.** Anyway, let's explore a generic IOS upgrade workflow that engineers must adhere to when performing an IOS upgrade on Cisco routers or switches. Figure 8-14 outlines a typical Cisco IOS upgrade workflow across many common platforms.

CHAPTER 8   UPGRADING MULTIPLE CISCO IOS ROUTERS

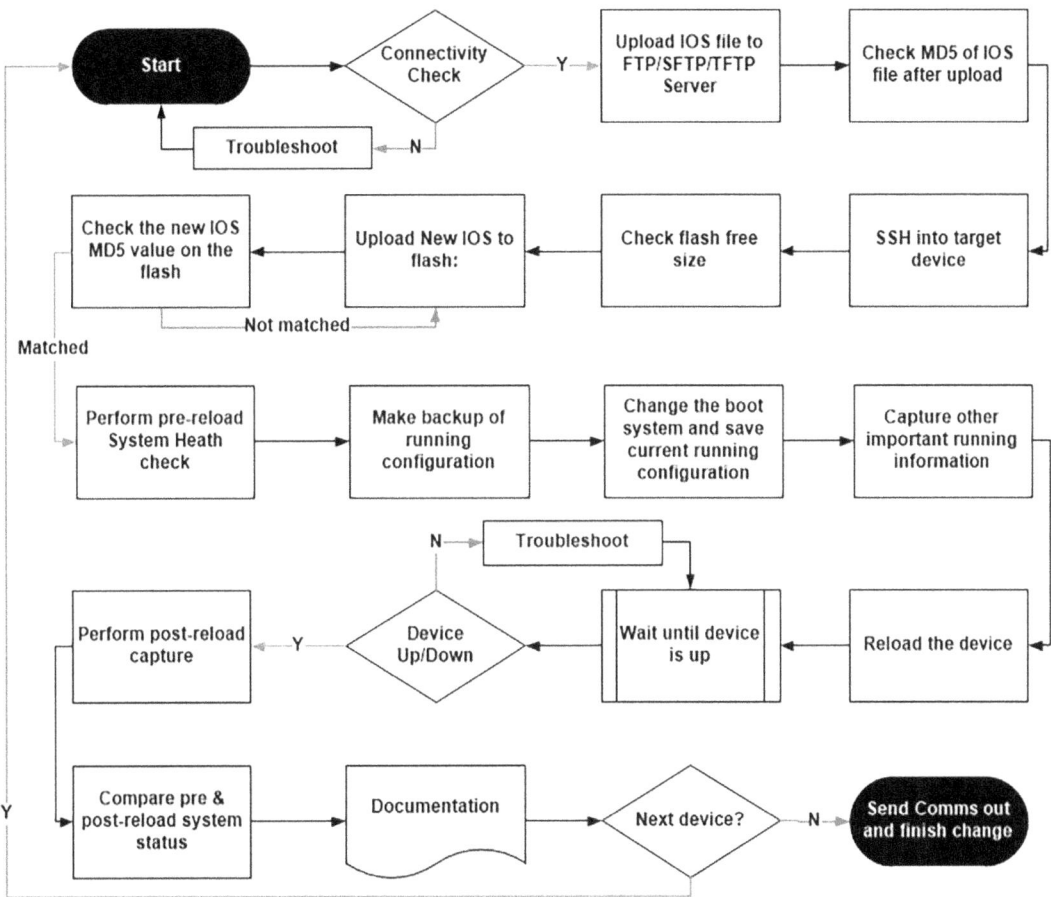

Figure 8-14. *General Cisco IOS upgrade workflow*

As showcased in Figure 8-14, the Cisco IOS upgrade process is intricate and labor intensive. When dealing with multiple devices, this process quickly becomes repetitive. The solution? Why not automate these tasks for engineers? In the final lab across the next two chapters, you will be guided to write small Python modules, replacing most of the tasks illustrated in Figure 8-14.

Let's dissect Figure 8-14 into three core stages and several linear steps. This breakdown will provide a clearer view of the tasks to be translated into Python code. Here, you must wear both an application developer's and a network engineer's hat simultaneously. Table 8-2 offers a breakdown of the manual task-driven Cisco IOS upgrade process, albeit variations may occur across generations and platforms. Generally, the underlying principles remain almost identical.

CHAPTER 8   UPGRADING MULTIPLE CISCO IOS ROUTERS

Table 8-2. *IOS Upgrade Task Breakdown*

Phase 1: Pre-check

1. Check the network connectivity between the server and network devices.

2. Gather the user's login credentials and store the user input as variables.

3. Verify the MD5 value of the new IOS as a variable. Compare the user's MD5 value with the server's checked MD5 value.

4. Examine the flash size on the Cisco device as a variable. Offer the user options to delete an old IOS file on `flash:/` or to explore a directory on a Cisco switch and remove an outdated IOS file.

5. Create backups of the running configuration.

Phase 2: IOS uploading and pre-uploading check

6. Upload the new IOS file from a file server to the router's flash.

7. Check the new IOS MD5 value on the Cisco device's flash. Compare the server's MD5 value with the switch MD5 value.

8. Provide the user with a review of the results. Offer the user the option to defer the device reload if necessary.

9. Pre-reload tasks: Change the boot system and save the current running-config to start-up config. Back up the running configuration on local/TFTP/FTP storage, capturing any essential information as needed.

Phase 3: Reloading and post-upgrade check

10. During device reload: Verify that the device is on the network.

11. After the reload (the device is back on the network): Log in to the device and conduct a post-upgrade verification using before and after configurations. Optionally, send an email notification to the engineer upon upgrade completion.

CHAPTER 8   UPGRADING MULTIPLE CISCO IOS ROUTERS

You've explored the manual IOS upgrade processes outlined earlier, but can you convert each step into Python code? Where do you start, and is it feasible to develop such advanced IOS upgrading production applications? In comparison to the offerings of other tools in today's market, numerous tasks (and an engineer's logical thinking) need to be translated into lines of code (a program). Infrastructure as Code (IaC) automates setup and management through configuration scripts, ensuring consistent, repeatable tasks by converting an engineer's vision into code or template. Next, examine Figure 8-15, which is my rendition of Table 8-2 and Figure 8-14, breaking them down into tasks that can be automated.

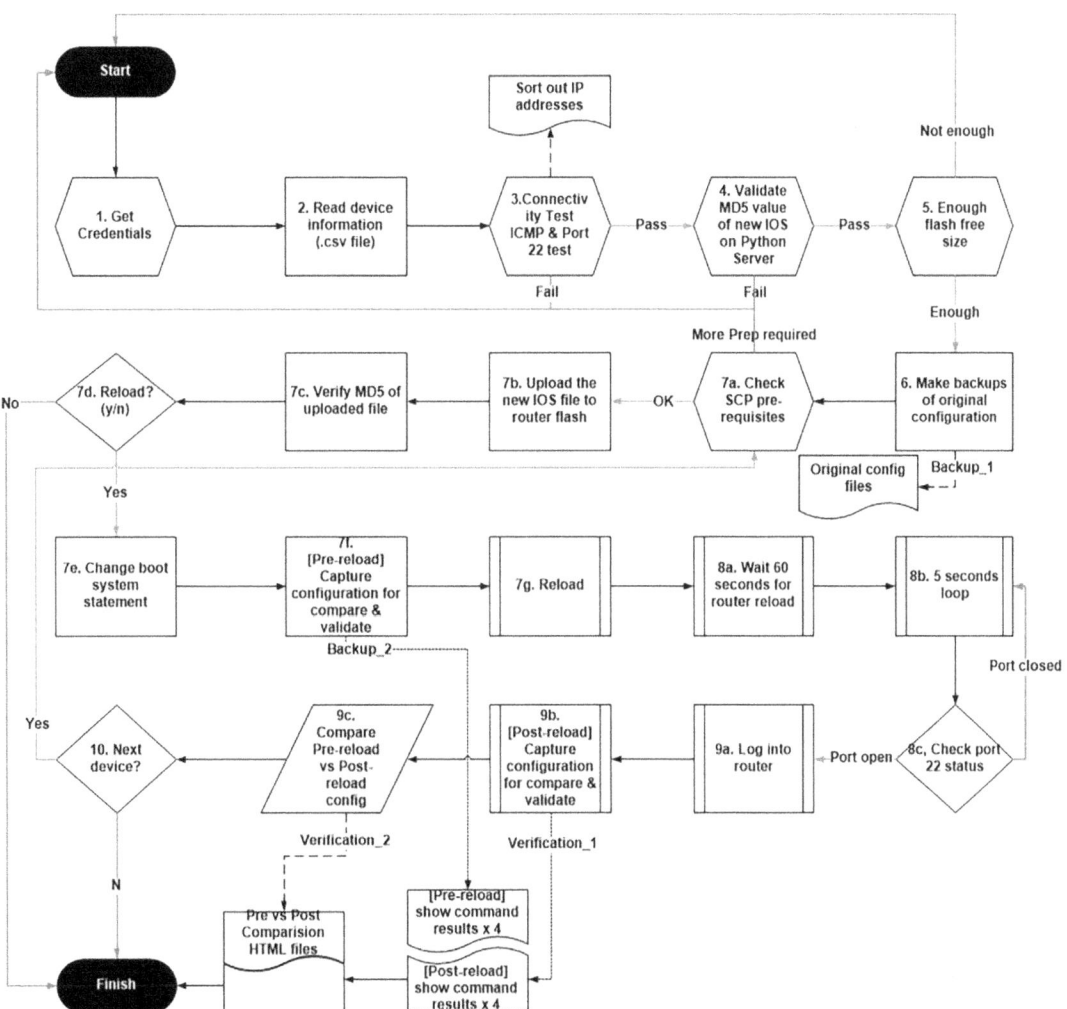

Figure 8-15. *Cisco IOS/IOS XE automated upgrade suggested workflow example*

Figure 8-15 directly mirrors the tasks outlined in Table 8-2, with corresponding numbers. It is essential to thoroughly examine it before proceeding to the next chapter. Chapters 9 and 10 break down each upgrade step into smaller Python code snippets or modules. These snippets will evolve into individual tools themselves within the next two chapters. Subsequently, in Chapter 11, you will amalgamate them into a unified tool (application) to seamlessly replace the Cisco IOS upgrade tasks from start to finish, encompassing the pre-check, IOS upgrade, and post-check processes.

Summary

In this chapter, you embarked on a comprehensive journey, illuminating the intricacies of upgrading multiple Cisco IOS routers while immersing yourself in the realm of Python Network Automation within tangible real-world scenarios. The chapter commenced by demystifying object-oriented programming (OOP) concepts and their alignment within networking, establishing a strong foundational understanding of their pivotal role in network architecture and operations. It delved into proactive patch management strategies and underscored the significance of Zero Trust methodologies, fundamental for fortifying network security in enterprise infrastructure. Throughout the chapter, you explored manual IOS upgrade processes and looked at a development path for Python applications to replace manual tasks with functioning code, aligning with the concept of Infrastructure as Code (IaC). Each task within this process was methodically dissected, facilitating a comprehensive understanding of its intricate workings. The chapter emphasized the critical nature of IOS upgrades in bolstering reliability and fortifying security across critical network devices, emphasizing their integral role as a key performance indicator for network teams, indispensable to a business's network infrastructure success. Furthermore, it highlighted the necessity for network engineers with coding prowess, as they are uniquely positioned to convert intricate network tasks into actionable lines of code. While AI may evolve to potentially replace certain tasks, the indispensability of experienced and knowledgeable network engineers in initiating and developing AI platforms remains irrefutable. In the next two chapters, you will commence developing applications tailored for your IOS upgrading tasks, translating the concepts discussed in this chapter into lines of code for automation.

CHAPTER 9

Cisco IOS-XE Upgrade Tools Development Part 1

This chapter focuses on crafting pivotal tools crucial for Cisco IOS-XE upgrading, catering to engineers' logical and operational requirements. Starting with Tools Development 1, you'll establish dependable connections through network connectivity and socket validation. Tools Development 2 emphasizes secure access via login credentials and user input collection. Extracting vital data, Tools Development 3, retrieves the new IOS file name and MD5 values, while Tools Development 4 validates the new IOS's integrity. Addressing space limitations, Tools Development 5 assesses flash size on Cisco routers. Finally, Tools Development 6 facilitates backups for running configurations and routing tables. These tools are fundamental, forming the backbone of the primary IOS upgrading tool. They aid engineers in transitioning from manual tasks to code writing, adhering to Infrastructure as Code (IaC) principles and ensuring speed, accuracy, and seamless network operations.

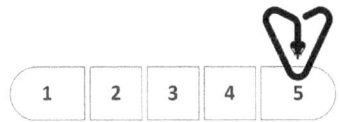

Mastering Cisco IOS Upgrades using Python: A Comprehensive Guide for Network Engineers

The terminology and processes for OS upgrading across enterprise network vendors may vary, yet the fundamental steps of preparation, upgrading, and post-upgrade verification remain consistent industry wide. In this exploration spanning Chapters 8 to 11, you delve into these OS upgrading processes, specifically focusing on Cisco routers. Cisco has sustained its prominence in enterprise networking since the 1990s, serving as the linchpin of the Internet and cloud infrastructure. The emphasis here revolves around Cisco's IOS-XE operating system, utilizing the Cisco Catalyst 8000v image to illustrate and practice the OS upgrading process for network devices. At the end of Chapter 11, you will be able to upgrade Cisco routers with Python scripts, with the insights gained extendable to various network vendor OS upgrading and patching processes, aiming to automate current manual engineering processes into lines of code.

Transitioning to the development phase, you'll aim to create a series of Python tools, akin to building blocks, culminating in a complete network automation application in the forthcoming chapter. To initiate this process, you'll start by discussing methods to input information into a script, converting it into Python variables and arguments. You have three pathways to input user and device information into the Cisco IOS XE upgrade application. First, you can use an interactive information collector tool, gathering both user and device details in real-time. Second, while gathering user login information interactively, you can concurrently read device data from files like text, Excel, CSV, or a database. Third, one option entails retrieving all details from files or a database. Concerning security, the potential threat of retrieving sensitive credentials from insecure text files in a production environment is significant. It's crucial to consider the implementation of an encrypted password vault to secure credentials on the network, albeit a topic beyond this book's scope. Instead, I encourage you to seek more secure user interaction tools. To synthesize this knowledge, you'll opt for the second method of information input for this application. Here, you'll obtain user credentials from an interactive session and fetch device information from a CSV file using the pandas module.

Expand your knowledge

Coding efficiency and ownership: balancing efficiency and accountability

The introduction of conversational AI tools like ChatGPT to the general public in November 2022 marked a significant milestone in the coding landscape. Traditionally, coding and application development required an in-depth understanding of programming languages and extensive experience, where developers wrote code from scratch or modified existing code from multiple sources. However, the accessibility of these AI-powered bots, including subsequent releases such as Bard (March 2023), Claude (March 2023), and Bing AI

(May 2023), has democratized coding. They enable non-programmers to engage provided they possess foundational understanding and clear objectives. These AI tools exhibit impressive efficiency, often generating 95 percent of functional code and refining the structure of less experienced code, streamlining the coding process. However, this efficiency poses a challenge for developers lacking in-depth coding knowledge or insight into AI-generated code. Some may mistakenly assume proficiency without establishing a solid foundation in coding, potentially leading to misconceptions about their skill level.

Nonetheless, this accelerated coding process emphasizes the importance of understanding how code impacts applications and devices within enterprise networks. **In the context of enterprise solutions automation, while AI expedites development, a strong grasp of foundational coding knowledge remains essential.** In practical enterprise scenarios, developers and engineers hold responsibility for their code's performance and outcomes. Having access to such a powerful tool, whether in its free or paid form, underscores the need for finding a delicate balance. Even though AI accelerates coding, it cannot replace the necessity of a deep grasp of coding principles. It's essential to remember that pointing fingers at an AI bot for a major outage caused by code it generated won't hold water. Developers and engineers must both uphold a strong understanding of coding fundamentals, even while utilizing AI-driven coding. Ultimately, they bear the responsibility for a code's performance and impact on enterprise systems. Thus, foundational knowledge and AI-driven coding both play pivotal roles in today's coding landscape.

To explore further details about the magic quadrant for enterprise conversational AI platforms, you can visit the Gartner website or simply search for "magic quadrant for enterprise conversational AI" on Google.

https://www.gartner.com/doc/reprints?id=1-2CSOJHFG&ct=230306&st=sb

---

In this chapter, you'll be utilizing the devices listed in Figure 9-1. Study this figure before you delving into the initial tools' development. Ensure that the u22s1 Python server and the c8KV01 and c8kv02 IOS-XE routers are powered on via VMware

CHAPTER 9   CISCO IOS-XE UPGRADE TOOLS DEVELOPMENT PART 1

Workstation 17 Pro. Additionally, activate the R1 (IOS) router on the GNS3 VM for the development of the first tools. Following the initial tools development, you can power off R1 to conserve processor and memory resources on your host PC.

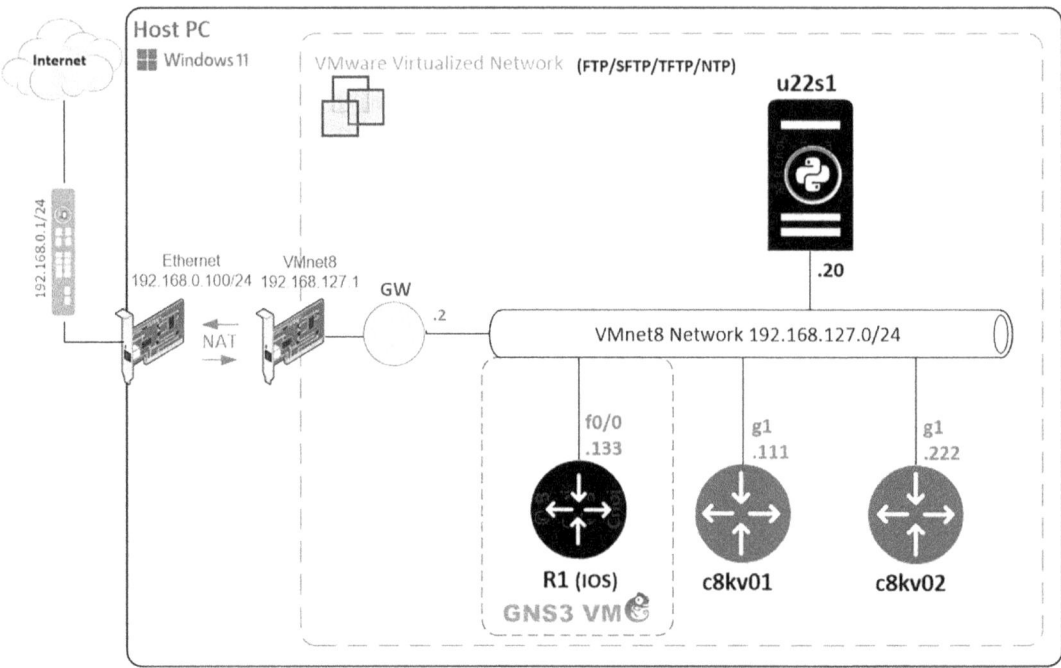

Figure 9-1.   Devices in use in Chapter 9

Tools Development 1: Network Connectivity and Socket Validation Tool

Network engineers constantly endeavor to maintain connectivity among various IP devices, applications, and users within their network, aiming to minimize or eliminate any network interruptions. Long before the buzz around network programmability, infrastructure as code (IaC), or network automation gained momentum, every resilient networked organization prioritized the security and stability of their managed networks. When writing a scripted network automation application, the foremost concern remains network stability and security, obliging developers to adhere to the best networking and security practices. Competence across both networking and programming best practices is imperative for creating robust network automation applications. However, because you are still on the first leg of your journey, this lab and the next explore the different ways to develop snippets of network tools to be combined and used for Cisco router upgrades.

The initial focus revolves around developing a network connectivity tool. While a similar tool was previously introduced, you'll aim to refactor the code, enhancing its functionality. This tool enables the testing of communication between the server (u22s1) and the two routers (c8kv01 and c8kv02). ICMP (ping) and socket (port) scanning are fundamental tasks for every network engineer, and your application will replace these manual tasks in the terminal console with Python code. Throughout this process, you'll refine the ICMP application by iteratively revisiting and improving the same code. The applications developed in this chapter will seamlessly integrate into the final Cisco IOS upgrade application in the concluding chapter. Most of the tools developed here can be used as simple tools on their own.

#	Task
①	SSH into your Python automation server (u22s1) using PuTTY and write the following code. You will write one ICMP script and then one socket script to test the network connectivity between the server and the routers. Then run and validate the scripts and modify them to change the flow of the scripts. Here you will only use the native os and socket modules; although the scapy module is an excellent external tool, the native tools can achieve what you want, so you will use what comes out of the box.

```
jdoe@u22s1:~$ mkdir ch9_tools_dev1 && cd ch9_tools_dev1
jdoe@u22s1:~/ch9_tools_dev1$ mkdir tool1_ping && cd tool1_ping
jdoe@u22s1:~/ch9_tools_dev1/tool1_ping$ nano ping_1.py
jdoe@u22s1:~/ch9_tools_dev1/tool1_ping$ cat ping_1.py
import os

device_list = ['192.168.127.111', '192.168.127.222']

for ip in device_list: # for loop for IP address
    if len(ip) != 0: # Only run this part of the script if the list is
                     not empty
        print(f'Sending icmp packets to {ip}') # Informational
        resp = os.system(f'ping -c 3 {ip}') # send ICMP packets 4 times
        if resp == 0: # If on the network (pingable), run this script
```

(continued)

#	Task
| | ```
print(f'{ip} is on the network.') # Informational
print('-'*79) # Line divider
else:
print(f'{ip} is unreachable.') # Informational
print('-'*79) # Line divider
else:
exit() # If not on the network, exit application
``` |
| ② | If your server can communicate to both routers and run the previous script, you should see a similar result on your SSH console. Run python ping_1.py. The result will look like this:<br>```
jdoe@u22s1:~/ch9_tools_dev1/tool1_ping$ python ping_1.py
Sending ICMP packets to 192.168.127.111
PING 192.168.127.111 (192.168.127.111) 56(84) bytes of data.
64 bytes from 192.168.127.111: icmp_seq=2 ttl=255 time=0.667 ms
[...omitted for brevity]
-----------------------------------------------------------------
Sending ICMP packets to 192.168.127.222
PING 192.168.127.222 (192.168.127.222) 56(84) bytes of data.
64 bytes from 192.168.127.222: icmp_seq=2 ttl=255 time=1.14 ms
[...omitted for brevity]
``` |
| ③ | In this step, you'll update the device_list from ['192.168.127.111', '192.168.127.222'] to ['10.10.10.1', '192.168.127.111', '192.172.1.33', '192.168.127.222']. The additional IPs, 10.10.10.1 and 192.172.1.33, are dummy addresses and don't exist on the network. You'll utilize this revised list to iterate through and refine our script. With this enhancement, if the devices are reachable (pingable), the script will add the IP address to a new list named reachable_ips. Conversely, if the IP addresses are unreachable, they'll be added to a new list called unreachable_ips. In a production network environment, various issues might arise. Working on a single device means focusing on that specific device; hence, if it's not remotely reachable, any upgrade or troubleshooting is on hold until connectivity is restored. However, **dealing with multiple devices simultaneously and encountering a couple of offline devices should not halt the entire process. It's crucial to distinguish between reachable and unreachable IPs using the ICMP tool, allowing you to segregate and manage IP addresses accordingly into separate lists.** |

(continued)

#	Task

```
jdoe@u22s1:~/ch9_tools_dev1/tool1_ping$ cp ping_1.py ping_2.py
jdoe@u22s1:~/ch9_tools_dev1/tool1_ping$ nano ping_2.py
jdoe@u22s1:~/ch9_tools_dev1/tool1_ping$ cat ping_2.py
import os

device_list = ['10.10.10.1', '192.168.127.111', '192.172.1.33',
'192.168.127.222']

reachable_ips = [] # Define a blank list for reachable IPs
unreachable_ips = [] # Define a blank list for unreachable IPs

for ip in device_list:
    if len(ip) != 0:
        print(f'Sending ICMP packets to {ip}') # Display information
                                                 about the IP being
                                                 pinged
        resp = os.system(f'ping -c 3 {ip}') # Ping the IP address
        if resp == 0:
            reachable_ips.append(ip) # Append the IP to reachable_
                                       ips list if reachable
            print('-' * 79) # Display a divider for readability
        else:
            unreachable_ips.append(ip) # Append the IP to unreachable_
                                         ips list if unreachable
            print('-' * 79) # Display a divider for readability
    else:
        exit()

print("Reachable IPs: ", reachable_ips) # Print list of reachable
                                          IPs
print("Unreachable IPs: ", unreachable_ips) # Print list of
                                              unreachable IPs
```

(*continued*)

#	Task
④	Now run the `python3 ping_2.py` command. You should get a similar result to this: jdoe@u22s1:~/ch9_tools_dev1/tool1_ping$ **python ping_2.py** Sending icmp packets to 10.10.10.1 PING 10.10.10.1 (10.10.10.1) 56(84) bytes of data. --- 10.10.10.1 ping statistics --- 3 packets transmitted, 0 received, 100% packet loss, time 2037ms *[...omitted for brevity]* --- Reachable IPs: ['192.168.127.111', '192.168.127.222'] Unreachable IPs: ['10.10.10.1', '192.172.1.33'] As anticipated, the reachable IPs are 192.168.127.111 and 192.168.127.222, enabling you to proceed with the script for these devices. On the other hand, the unreachable IPs are 10.10.10.1 and 192.172.1.33, which aligns with expectations. In a real production scenario, if these devices are intended to be on the network but are unreachable, resolving connectivity issues is crucial before implementing any changes or upgrades.
⑤	The lists as variables are useful, but this information is only displayed while the script is running and you are sitting in front of the console to watch this information. Note that the server's random access memory is only temporary, and Python's `print` statement is only used for the user's convenience. The actual Python script does not use what's printed on the screen. If you want to access this information when the script runs successfully by a job scheduler such as cron, you must write and save the information to a file. Let's write two lists into two separate files for later use. jdoe@u22s1:~/ch9_tools_dev1/tool1_ping$ cp ping_2.py ping_3.py jdoe@u22s1:~/ch9_tools_dev1/tool1_ping$ nano ping_3.py jdoe@u22s1:~/ch9_tools_dev1/tool1_ping$ cat ping_3.py **import os** # Define the list of devices **device_list = ['10.10.10.1', 192.168.127.111', '192.172.1.33', '192.168.127.222']**

(continued)

#	Task

```python
# Open or create a file to store reachable IPs
# reachable_ips = [] # Add '#' to make this line inactive
f1 = open('reachable_ips.txt', 'w+')

# Open or create a file to store unreachable IPs
# unreachable_ips = []# Add '#' to make this line inactive
f2 = open('unreachable_ips.txt', 'w+')

# Loop through each IP in the device list
for ip in device_list:
    # Check if the IP is not empty
    if len(ip) != 0:
        # Display the attempt to send ICMP packets to the IP
        print(f'Sending ICMP packets to {ip}')

        # Send ICMP packets to the IP
        resp = os.system('ping -c 3 ' + ip)

        # Check the response
        if resp == 0:
            # Write reachable IPs to the reachable_ips.txt file
            f1.write(f'{ip}\n')
            # Display separator for readability
            print('-' * 79)
        else:
            # Write unreachable IPs to the unreachable_ips.txt file
            f2.write(f'{ip}\n')
            # Display separator for readability
            print('-' * 79)
    else:
        # Exit the script if the IP is empty
        exit()

# Close the files after writing IPs
f1.close()
f2.close()
```

(continued)

CHAPTER 9   CISCO IOS-XE UPGRADE TOOLS DEVELOPMENT PART 1

#	Task
⑥	Run the third iteration of your original code, and it should produce two text files, one containing the reachable IP addresses and one containing the unreachable IP addresses. Note that the output has been omitted to save space. ``` jdoe@u22s1:~/ch9_tools_dev1/tool1_ping$ ls ping_1.py ping_2.py ping_3.py [... output omitted for brevity] jdoe@u22s1:~/ch9_tools_dev1/tool1_ping$ ls ping_1.py ping_2.py ping_3.py reachable_ips.txt unreachable_ips.txt jdoe@u22s1:~/ch9_tools_dev1/tool1_ping$ more reachable_ips.txt 192.168.127.111 192.168.127.222 jdoe@u22s1:~/ch9_tools_dev1/tool1_ping$ more unreachable_ips.txt 10.10.10.1 192.172.1.33 ``` So, your tool can test network connectivity and sort the IP addresses between online and offline devices at this stage. Also, the IP addresses can be saved (documented) in their respective files.
⑦	You will now transform the IP address list, previously used in the scripts, into a file named ip_addresses.txt. This process enables more efficient management of a larger set of IP addresses, eliminating the need to manually enter each address into a list. While you're currently extracting IP addresses from a text file, you can also directly retrieve data from Excel or CSV files. Importing information from these sources allows you to convert the data into a two-dimensional format, granting greater depth and versatility to the information. Also, with the CI/CD pipeline, you could use the script on the cloud using the correct credentials; however, this is out of this book's scope. It's also worth noting the final IP address, 192.168.127.133, which corresponds to GNS3 R1's router IP address. Therefore, ensure to initiate R1 within GNS3's pynetauto-devops project. Power on GNS3 and start the IOS router, R1 (Cisco 3745). Create an access list to stop any inbound SSH traffic to this device. Create an access list and apply it to FastEthernet 0/0 of R1. Also, you can enable both Telnet and SSH on line vty 0 15.

(*continued*)

#	Task
	R1#**show ver** Cisco IOS Software, 3700 Software (C3745-ADVENTERPRISEK9-M), Version 12.4(25d), RELEASE SOFTWARE (fc1) *[...omitted for brevity]* R1#**configure terminal** R1(config)#**access-list 100 deny tcp any any eq 22** R1(config)#**access-list 100 permit ip any any** R1(config)#**interface f0/0** R1(config-if)#**ip access-group 100 in** R1(config-if)#**line vty 0 15** R1(config-line)#**transport input telnet ssh** R1(config-line)#**do write memory** Building configuration... [OK]

⑧	Now go back to your Python sever, u22s1, and create and write a socket tool, as shown here.
	
	jdoe@u22s1:~/ch9_tools_dev1/tool1_ping$ **nano socket_1.py**
	import socket
	
	device_list = ['10.10.10.1', '192.168.127.111', '192.172.1.33', '192.168.127.222', '192.168.127.133']
	
	for ip in device_list: # Loop through each IP in the device list
	    print("-"*79) # Print a line to separate IP results
	    for port in range (22, 24): # Loop through SSH ports 22 to 23
	        destination = (ip, port) # Create a tuple for IP and port
	        try:
	            with socket.socket(socket.AF_INET, socket.SOCK_STREAM)
	            as s: # Create a socket object

(continued)

#	Task

	`s.settimeout(3)` # Set a timeout for the connection
	`connection = s.connect(destination)`
	# Connect to the IP and port
	`print(f"On {ip}, SSH port {port} is open!")`
	# Print if the port is open
	`except:`
	    `print(f"On {ip}, SSH port {port} is closed.")` # Print if the port is closed or not reachable

(9)	Run this script. You should expect similar responses to what's shown here. If you do get such results, it is time to integrate this script into the `ping_3.py` script.
	
	`jdoe@u22s1:~/ch9_tools_dev1/tool1_ping$ python socket_1.py`
	`-----------------------------------------------------------------`
	`On 10.10.10.1, SSH port 22 is closed.`
	`On 10.10.10.1, SSH port 23 is closed.`
	`-----------------------------------------------------------------`
	`On 192.168.127.111, SSH port 22 is open!`
	`On 192.168.127.111, SSH port 23 is closed.`
	`-----------------------------------------------------------------`
	`On 192.172.1.33, SSH port 22 is closed.`
	`On 192.172.1.33, SSH port 23 is closed.`
	`-----------------------------------------------------------------`
	`On 192.168.127.222, SSH port 22 is open!`
	`On 192.168.127.222, SSH port 23 is closed.`
	`-----------------------------------------------------------------`
	`On 192.168.127.133, SSH port 22 is closed.`
	`On 192.168.127.133, SSH port 23 is open!`

(continued)

CHAPTER 9   CISCO IOS-XE UPGRADE TOOLS DEVELOPMENT PART 1

#	Task
⑩	Now let's combine the two tools and enhance the ping tool's capabilities to send ICMP messages and then check the open ports for Telnet or SSH connections. After you complete this tool, you can use it as a stand-alone tool to check the connectivity and port status. Our preferred connection method is an SSH connection. When the script runs, it will create three files, the first containing the IP addresses with an SSH connection, the second containing the IP addresses with a Telnet connection, and the third containing the unreachable IP addresses and logs of closed ports. You will replace device_list with the information read from the ip_addresses.txt file, so go ahead and create the file with the IP addresses. jdoe@u22s1:~/ch9_tools_dev1/tool1_ping$ **nano ip_addresses.txt** jdoe@u22s1:~/ch9_tools_dev1/tool1_ping$ **cat ip_addresses.txt** **10.10.10.1** **192.168.127.111** **192.172.1.33** **192.168.127.222** **192.168.127.133**
⑪	Copy ping_3.py, save it as ping_4_socket.py, and then merge socket_1.py from the previous step into ping_4_socket.py. Once you have combined the two scripts into one file, it should look like the completed ping_4_socket.py script, which follows. Here, you are breaking the rules of writing more Pythonic code because the code uses multiple loops, and the indentation consists of eight blocks—one block consists of four spaces. For code readability, the recommended number of indentation blocks is four. You'll look at this soon. Note that if port 22 is opened in this script, it will not check port 23 because the interest is only whether port 22 is opened. Also note that the time module has been added to check the time to run the application; the time module was covered in earlier chapters. jdoe@u22s1:~/ch9_tools_dev1/tool1_ping$ **cp ping_3.py ping_4_socket.py** jdoe@u22s1:~/ch9_tools_dev1/tool1_ping$ **nano ping_4_socket.py** jdoe@u22s1:~/ch9_tools_dev1/tool1_ping$ **cat ping_4_socket.py**

(*continued*)

#	Task
| | ```
import os
import socket
import time

t1 = time.mktime(time.localtime()) # Timer start

ip_add_file = './ip_addresses.txt' # IP address file
f1 = open('reachable_ips_ssh.txt', 'w+') # Open f1 file
f2 = open('reachable_ips_telnet.txt', 'w+') # Open f2 file
f3 = open('unreachable_ips.txt', 'w+') # Open f3 file

# device_list = ['10.10.10.1', '192.168.127.111', '192.172.1.33',
'192.168.127.222'] # Hashed out line

with open(ip_add_file, 'r') as ip_addresses: # Read IP addresses
    for ip in ip_addresses: # Loop IPs
        ip = ip.strip() # Remove spaces
        resp = os.system('ping -c 3 ' + ip) # Ping device
        if resp == 0: # Check if reachable
            for port in range(22, 23): # Check SSH port
                destination = (ip, port)
                try:
                    with socket.socket(socket.AF_INET, socket.SOCK_
                    STREAM) as s:
                        s.settimeout(3)
                        connection = s.connect(destination)
                        print(f"{ip} {port} opened") # Print open SSH
                        f1.write(f"{ip}\n") # Write IP to file
                except:
                    print(f"{ip} {port} closed")
                    f3.write(f"{ip} {port} closed\n") # Write
                    IP to file
            for port in range(23, 24): # Check Telnet port
                destination = (ip, port)
                try:
``` |

(continued)
```

#	Task
| | ```
                with socket.socket(socket.AF_INET, socket.SOCK_
                STREAM) as s:
                    s.settimeout(3)
                    connection = s.connect(destination)
                    print(f"{ip} {port} opened") # Print open Telnet
                    f2.write(f"{ip}\n") # Write IP to file
                except:
                    print(f"{ip} {port} closed")
                    f3.write(f"{ip} {port} closed\n")
                    # Write IP to file
        else:
            print(f"{ip} unreachable")
            f3.write(f"{ip} unreachable\n") # Write IP to file

    f1.close() # Close f1
    f2.close() # Close f2
    f3.close() # Close f3

    tt = time.mktime(time.localtime()) - t1 # Timer finish
    print("Total time : {0} sec".format(tt)) # Total time
``` |
| (12) | Run the combined script, ping_4_socket.py. It should create three files with sorted IP addresses based on the ICMP and port verification. **Imagine running such connectivity checks on hundreds or thousands of devices for troubleshooting or making changes; this type of tool can save you a lot of time, and the files created here can be read from another application to carry out further programming tasks.**<br><br>```
jdoe@u22s1:~/ch9_tools_dev1/tool1_ping$ python ping_4_socket.py
PING 10.10.10.1 (10.10.10.1) 56(84) bytes of data.
[... output omitted for brevity]
--- 192.168.127.133 ping statistics ---
3 packets transmitted, 3 received, 0% packet loss, time 2003ms
rtt min/avg/max/mdev = 4.180/7.153/11.242/2.988 ms
Total time : 30.0 sec
jdoe@u22s1:~/ch9_tools_dev1/tool1_ping$ ll
``` |

(continued)

#	Task

	[...omitted for brevity]
	`-rw-rw-r-- 1 jdoe jdoe   32 Nov 25 20:26 reachable_ips_ssh.txt`
	`# Just created file 1`
	`-rw-rw-r-- 1 jdoe jdoe   16 Nov 25 20:26 reachable_ips_telnet.txt`
	`# Just created file 2`
	`-rw-rw-r-- 1 jdoe jdoe  528 Nov 25 20:03 socket_1.py`
	`-rw-rw-r-- 1 jdoe jdoe  126 Nov 25 20:26 unreachable_ips.txt`
	`# Just created file 3`
	`jdoe@u22s1:~/ch9_tools_dev1/tool1_ping$ `**`more reachable_ips_ssh.txt`**
	`192.168.127.111`
	`192.168.127.222`
	`jdoe@u22s1:~/ch9_tools_dev1/tool1_ping$ `**`more reachable_ips_telnet.txt`**
	`192.168.127.133`
	`jdoe@u22s1:~/ch9_tools_dev1/tool1_ping$ `**`more unreachable_ips.txt`**
	`10.10.10.1 unreachable`
	`192.168.127.111 23 closed`
	`192.172.1.33 unreachable`
	`192.168.127.222 23 closed`
	`192.168.127.133 22 closed`

⑬	Now create a `check_port()` function and make this a separate function. When you run the following code, it will produce the same result in a more Pythonic way. The best practice is to separate the function, save it under another file name as a module, and then import it as a tool so the main script is less crowded and easier to digest. But for now, this should be acceptable. Rewrite the code in `ping_4_socket.py` so that it looks like `ping_5_socket_pythonic.py`. `jdoe@u22s1:~/ch9_tools_dev1/tool1_ping$ `**`cp ping_4.py ping_5_socket_pythonic.py`** `jdoe@u22s1:~/ch9_tools_dev1/tool1_ping$ `**`nano ping_5_socket_pythonic.py`** `jdoe@u22s1:~/ch9_tools_dev1/tool1_ping$ `**`cat ping_5_socket_pythonic.py`** **`import os`** **`import socket`** **`import time`**

(continued)

#	Task
| | ```python
t1 = time.mktime(time.localtime())

def check_port(ip):
    for port in range(22, 23):
        destination = (ip, port)
        try:
            with socket.socket(socket.AF_INET, socket.SOCK_STREAM)
            as s:
                s.settimeout(3)
                connection = s.connect(destination)
                print(f"{ip} {port} opened")
                f1.write(f"{ip}\n")
        except:
            print(f"{ip} {port} closed")
            f3.write(f"{ip} {port} closed\n")
    for port in range(23, 24):
        destination = (ip, port)
        try:
            with socket.socket(socket.AF_INET, socket.SOCK_STREAM)
            as s:
                s.settimeout(3)
                connection = s.connect(destination)
                print(f"{ip} {port} opened")
                f2.write(f"{ip}\n")
        except:
            print(f"{ip} {port} closed")
            f3.write(f"{ip} {port} closed\n")
ip_add_file = './ip_addresses.txt'
f1 = open('reachable_ips_ssh.txt', 'w+')
f2 = open('reachable_ips_telnet.txt', 'w+')
f3 = open('unreachable_ips.txt', 'w+')
``` |

(continued)
```

#	Task
| | ```python
with open(ip_add_file, 'r') as ip_addresses:
    for ip in ip_addresses:
        ip = ip.strip()
        resp = os.system('ping -c 3 ' + ip) # Be careful, there is
                                            # a space after the 3.
        if resp == 0:
            check_port(ip)
        else:
            print(f"{ip} unreachable")
            f3.write(f"{ip} unreachable\n")
f1.close()
f2.close()
f3.close()
tt = time.mktime(time.localtime()) - t1
print("Total wait time: {0} seconds".format(tt))
``` |
| (14) | Now run the ping_5_socket_pythonic.py application and check the result. You should expect to see the same result as in ping_4.py. The file will be overwritten, and new files will be created with the different timestamps.

```
jdoe@u22s1:~/ch9_tools_dev1/tool1_ping$ python ping_5_socket_
pythonic.py
PING 10.10.10.1 (10.10.10.1) 56(84) bytes of data.
[... output omitted for brevity]
--- 192.168.127.133 ping statistics ---
3 packets transmitted, 3 received, 0% packet loss, time 2004ms
rtt min/avg/max/mdev = 3.422/10.106/16.943/5.520 ms
192.168.127.133 22 closed
192.168.127.133 23 opened
Total wait time: 31.0 seconds
jdoe@u22s1:~/ch9_tools_dev1/tool1_ping$ ll
[...omitted for brevity]
``` |

(*continued*)
```

#	Task

```
-rw-rw-r-- 1 jdoe jdoe    32 Nov 25 20:39 reachable_ips_ssh.txt
-rw-rw-r-- 1 jdoe jdoe    16 Nov 25 20:39 reachable_ips_telnet.txt
-rw-rw-r-- 1 jdoe jdoe   528 Nov 25 20:03 socket_1.py
-rw-rw-r-- 1 jdoe jdoe   126 Nov 25 20:39 unreachable_ips.txt
```

(15) Let's simplify the main script by separating the port-checking tool into a separate module (a separate file). When you run ping_6_modular.py, you should get the same result, but you can say that this code is a little more Pythonic. It's smart to separate the check_port function as a separate tool, so create and save the module as ping_6_module.py.

```
jdoe@u22s1:~/ch9_tools_dev1/tool1_ping$ nano ping_6_module.py
jdoe@u22s1:~/ch9_tools_dev1/tool1_ping$ cat ping_6_module.py
import socket

def check_port(ip, f1, f2, f3):
    for port in range(22, 23):  # Check ports for SSH (22)
        destination = (ip, port)
        try:
            with socket.socket(socket.AF_INET, socket.SOCK_STREAM)
            as s:
                s.settimeout(3)  # Set timeout
                connection = s.connect(destination)  # Connect to IP
                                                        and port
                print(f"{ip} {port} open")  # Print open status
                f1.write(f"{ip}\n")  # Write IP to file
        except:
            print(f"{ip} {port} closed")  # Print closed status
            f3.write(f"{ip} {port} closed\n")  # Write IP and port
                                                        to file
    for port in range(23, 24):  # Check ports for Telnet (23)
        destination = (ip, port)
        try:
            with socket.socket(socket.AF_INET, socket.SOCK_STREAM)
            as s:
                s.settimeout(3)  # Set timeout
```

(continued)

#	Task

```
                connection = s.connect(destination) # Connect to
                                                      IP and port
                print(f"{ip} {port} open") # Print open status
                f2.write(f"{ip}\n") # Write IP to file
            except:
                print(f"{ip} {port} closed") # Print closed status
                f3.write(f"{ip} {port} closed\n") # Write IP and port
                                                   to file
```

⑯ Copy ping_5_pythonic.py and create the main script, ping_6_modular.py. Make sure you remove the check_port() function and modify the file, as shown here:

```
jdoe@u22s1:~/ch9_tools_dev1/tool1_ping$ cp ping_5.py ping_6_modular.py
jdoe@u22s1:~/ch9_tools_dev1/tool1_ping$ nano ping_6_modular.py
jdoe@u22s1:~/ch9_tools_dev1/tool1_ping$ cat ping_6_modular.py
import os
import time
from ping_6_module import check_port # Importing check_port tool

t1 = time.mktime(time.localtime()) # Current time

# File paths
ip_add_file = './ip_addresses.txt'
f1 = open('reachable_ips_ssh.txt',    'w+') # Open file for reachable
                                              SSH IPs
f2 = open('reachable_ips_telnet.txt', 'w+') # Open file for
                                              reachable Telnet IPs
f3 = open('unreachable_ips.txt', 'w+') # Open file for unreachable IPs

with open(ip_add_file, 'r') as ip_addresses: # Read IPs from file
    for ip in ip_addresses: # Loop through IPs
        ip = ip.strip() # Remove whitespace
        resp = os.system('ping -c 3 ' + ip) # Check if IP is reachable
        if resp == 0: # If reachable, perform port check
            check_port(ip, f1, f2, f3) # Use check_port tool for
                                         SSH and Telnet ports
```

(continued)

#	Task

```
        else:  # If unreachable, write to file
            print(f"{ip} unreachable")
            f3.write(f"{ip} unreachable\n")

# Close files
f1.close()
f2.close()
f3.close()

tt = time.mktime(time.localtime()) - t1  # Total execution time
print("Total wait time : {0} seconds".format(tt))  # Print
                                                    execution time

jdoe@u22s1:~/ch9_tools_dev1/tool1_ping$ ll ping_6*
-rw-rw-r-- 1 jdoe jdoe 771 Nov 25 20:57 ping_6_modular.py # Main
script
-rw-rw-r-- 1 jdoe jdoe 880 Nov 25 20:52 ping_6_module.py # Module
file with check_port()
```

⑰	Now execute the python ping_6_modular.py command, and you should expect to see the same result as the ping_5.py and ping_4.py reiterations. Check the files; they should contain the same information as before, but your code is a lot cleaner, and you now know how to write a tool into its separate module.

```
jdoe@u22s1:~/ch9_tools_dev1/tool1_ping$ python ping_6_modular.py
PING 10.10.10.1 (10.10.10.1) 56(84) bytes of data.
[...output omitted for brevity]
--- 192.168.127.133 ping statistics ---
Total wait time : 30.0 seconds
jdoe@u22s1:~/ch9_tools_dev1/tool1_ping$ ll
-rw-rw-r-- 1 jdoe jdoe   32 Nov 25 21:05 reachable_ips_ssh.txt
-rw-rw-r-- 1 jdoe jdoe   16 Nov 25 21:05 reachable_ips_telnet.txt
-rw-rw-r-- 1 jdoe jdoe  528 Nov 25 20:03 socket_1.py
-rw-rw-r-- 1 jdoe jdoe  126 Nov 25 21:05 unreachable_ips.txt
```

This ping tool has reached its final iteration, primed for immediate use in production. The code quality meets the Pythonic standard for integration into the Cisco IOS upgrade application. As you conclude the development of the first tool, I hope you're becoming familiar with enhancing code through refactoring. Let's now delve into the second tool.

Tools Development 2: Login Credentials and User Input Collector

When a network engineer performs an IOS upgrade on Cisco routers and switches, the engineer must type in their network administrator credentials with the correct user privileges: the level 15 administrator user ID and password. The device access level and security could be using locally configured user credentials, or they could be on the TACACS server.

In this book, you are using the local logins to keep things simple. After successful authentication and login, the network administrator must manually run some commands to check the device's upgradability. That is, the device needs to have enough flash memory (storage) to hold the new IOS image. Then, the network administrator must run a command with the new IOS image file name along with the FTP/SFTP/TFTP server information. In the case of TFTP, no administrator username or password is required. Still, in most cases, you must provide another username and password combination for further authentication for the FTP server. Some information is fed through one piece at a time, line by line. But with scripting, you can simplify and collect the administrator's data right at the beginning of the script run using one of the data collection methods. The easiest way is to use a single file to collect all the required information at once. The most cumbersome way would be to make the user manually enter the information one piece at a time. For demonstration purposes, you can write some code that uses both to get the user ID, password, and secret password. You'll use an interactive data collection tool for the IOS name and MD5 value. The information will be read from a CSV file.

#	Task
| ① | To develop a "get credentials" tool, begin with the input function and the getpass module; the getpass module hides the password entered into the console. As you are developing the tool, you will print the password using the print statement for convenience.<br><br>To collect the user ID and network password and enable the secret password, the most basic tool will look like this:<br><br>```
jdoe@u22s1:~/ch9_tools_dev1$ mkdir tool2_login && cd tool2_login
jdoe@u22s1:~/ch9_tools_dev1/tool2_login$ nano get_cred1.py
from getpass import getpass
uid = input("Enter Network Admin ID : ")
pwd = getpass("Enter Network Admin PWD : ")
secret = getpass("Enter secret password : ")
print(uid, pwd, secret)
```<br><br>Run the initial script for a quick test.<br>```
jdoe@u22s1:~/ch9_tools_dev1/tool2_login$ python get_cred1.py
Enter Network Admin ID : jdoe
Enter Network Admin PWD : ******** # admin login password
Enter secret password : ********* # enable secret password
jdoe cisco123 secret321
```<br><br>With a simple user ID and password collection tool as shown earlier, there are two apparent problems. First, you can enter any information or nothing and still move to the next step, so the script at least needs to get a valid entry from the user before proceeding to the next step. Second, the getpass module does not display the password entered, so you cannot tell if you have entered the correct password until the script runs and errors out during the SSH or Telnet authentication process. So, you must improve this script by adding a validation step. |

(continued)

#	Task
| ② | Let's make some improvements to this tool to bring it up to acceptable standards. In this iteration, you will take care of the second problem, the getpass module, by adding the validation script for the password and secret. You must enter the password and secret twice to make sure that the entered passwords match and are correct. You will also prompt the user if the password and secret are the same as in the production environment. Both the password and secrets are the same, and hence by answering y or yes, the user does not have to type the secret again. After the first reiteration, your code will look like the code written next. Notice that I have moved the secret collection into a separate function to make this script more concise, and, if the user answers other than y, yes, n, or no, they will be prompted to provide a correct response. This function makes the user provide only the expected responses: any one of the y, yes, n, or no responses.<br><br>Referencing get_cred1.py, let's start putting it all together.<br><br>```
jdoe@u22s1:~/ch9_tools_dev1/tool2_login$ nano get_cred2.py
from getpass import getpass

    def get_secret():
        global secret
        resp = input("Is secret the same as password? (y/n) : ")
        resp = resp.lower()
        if resp == "yes" or resp == "y":   # Secret same as password?
            secret = pwd
        elif resp == "no" or resp == "n":  # Secret different from
                                                      password?
            secret = None
        while not secret:
            secret = getpass("Enter the secret : ")   # Input secret
            secret_verify = getpass("Confirm the secret : ")
             # Confirm secret
            if secret != secret_verify:   # Check if secrets match
                print("! Secrets do not match. Please try again.")
                secret = None
            else:
                get_secret()   # Recursively call to get_secret()
```<br><br>*(continued)* |
```

#	Task
| | ```python
def get_credentials():
    global uid
    uid = input("Enter Network Admin ID : ") # Input Admin ID
    global pwd
    pwd = None
    while not pwd:
        pwd = getpass("Enter Network Admin PWD : ") # Input Admin
                                                    Password
        pwd_verify = getpass("Confirm Network Admin PWD : ")
        # Confirm Admin Password
        if pwd != pwd_verify:  # Check if passwords match
            print("! Network Admin Passwords do not match. Please try
            again.")
            pwd = None
    get_secret() # Get secret for password encryption
    return uid, pwd, secret # Return credentials

get_credentials() # Call function to get credentials
print(uid, pwd, secret) # Print retrieved credentials
``` |
| ③ | Once you have written the previous script, test your application.<br><br>```
jdoe@u22s1:~/ch9_tools_dev1/tool2_login$ python get_cred2.py
Enter Network Admin ID : jdoe
Enter Network Admin PWD : ********
Confirm Network Admin PWD : ******* # Enter mismatched password
! Network Admin Passwords do not match. Please try again.
Enter Network Admin PWD : ********
Confirm Network Admin PWD : ********
Is secret the same as password? (y/n) : n
Enter the secret : ********
Confirm the secret : ******** # Enter mismatched password
! secret do not match. Please try again.
Enter the secret : ********
```<br><br>*(continued)* |

CHAPTER 9   CISCO IOS-XE UPGRADE TOOLS DEVELOPMENT PART 1

#	Task
	Confirm the secret : ******** jdoe cisco123 secret123 The previous user ID, password, and secret collection tool look a lot better than the original in Task 1. Still, as mentioned, it has another flaw. The user can input a username, password, or secret of any length. You must address this issue to make the tool more realistic. Look at the following example to see what this means: jdoe@u22s1:~/ch9_tools_dev1/tool2_login$ **python get_cred_b_test.py** Enter Network Admin ID : **a** Enter Network Admin PWD : * Confirm Network Admin PWD : * Is secret the same as password? (y/n) : **n** Enter the secret : * Confirm the secret : *
4	You can use a couple of Regular Expressions to control the user inputs, which will fix the previous problem. In a well-managed IT environment, administrators always enforce conventions for usernames and passwords. Hence, for the script, you will follow the same practice and only allow acceptable inputs to your standards. For the username, the convention is that it needs to be 5 to 30 characters long, must begin with a letter, and use no special characters anywhere in the username except for _ and -. For the password conventions, the password must begin with a lowercase or uppercase letter, and the password must be longer than 8 characters but equal to or less than 50 characters. The characters between the second and 50th can include special characters. Let's see how you can enforce these conventions on usernames and passwords. While writing Python code, you must be flexible and creative with your Regular Expressions to make your code do what you want, and this is one of those instances. Referencing get_cred2.py, start putting the code together. jdoe@u22s1:~/ch9_tools_dev1/tool2_login$ **nano get_cred3.py** **import re** **from getpass import getpass** # Regular Expressions to validate input patterns **p1 = re.compile(r'^[a-zA-Z0-9][a-zA-Z0-9_-]{3,28}[a-zA-Z0-9]$')** # Pattern 1 for username

(continued)

#	Task

```python
    p2 = re.compile(r'^[a-zA-Z].{7,49}') # Pattern 2 for password

    def get_secret():
        global secret
        resp = input("Is secret the same as password? (y/n) : ").lower()
        # Ask user for confirmation
        if resp == "yes" or resp == "y":
            secret = pwd # Assign password to secret if confirmed
        elif resp == "no" or resp == "n":
            secret = None # If not same, initialize secret as None
        while not secret:
            secret = getpass("Enter the secret : ") # Request the secret
                                                    from user
            while not p2.match(secret): # Validate secret with pattern 2
                secret = getpass(r"*Enter the secret : ")
                # Re-enter secret if pattern mismatch
            secret_verify = getpass("Confirm the secret : ")
            # Ask user to confirm the secret
            if secret != secret_verify:
                print("!!! secret do not match. Please try again.")
                secret = None # If mismatch, reset secret to None for re-
                                entry
            else:
                get_secret() # Recursively call to check if secret matches
    def get_credentials():
        global uid
        uid = input("Enter Network Admin ID : ") # Request Network Admin
                                                  ID from user
        while not p1.match(uid): # Validate ID with pattern 1
            uid = input(r"*Enter Network Admin ID : ") # Re-enter ID if
                                                        pattern mismatch
        global pwd
```

(continued)

| Task

```
            pwd = None
            while not pwd:
                pwd = getpass("Enter Network Admin PWD : ")
                # Request password from user
                while not p2.match(pwd): # Validate password with pattern 2
                    pwd = getpass(r"*Enter Network Admin PWD : ")
                    # Re-enter password if pattern mismatch
                pwd_verify = getpass("Confirm Network Admin PWD : ")
                # Ask user to confirm password
                if pwd != pwd_verify:
                    print("!!! Network Admin Passwords do not match. Please
                    try again.")
                    pwd = None # If mismatch, reset password to None for re-
                    entry
            get_secret() # Get secret for the password
        return uid, pwd, secret # Return validated credentials

    get_credentials() # Call the function to get credentials
    print(uid, pwd, secret) # Display credentials
```

(5) Execute the final application, get_cred3.py, and validate the functions. Your test run should look like the following result.

```
jdoe@u22s1:~/ch9_tools_dev1/tool2_login$ python get_cred3.py
Enter Network Admin ID : jdoe # Entered only four characters. A
minimum of five characters required
*Enter Network Admin ID : jdoe
Enter Network Admin PWD : ******* # Entered only seven characters.
Minimum eight characters long
*Enter Network Admin PWD : ********
Confirm Network Admin PWD : ********
Is secret the same as password? (y/n) : n
Enter the secret : ******* # Entered only seven characters, a minimum
of eight characters long
```

(*continued*)

#	Task
	``
*Enter the secret : *********
Confirm the secret : *********
jdoe cisco123 secret321
```
After three reiterations of the original code, it looks almost complete and ready for use in your router IOS upgrade script. Next, you'll see how to read the IOS name and MD5 values from a CSV file and convert them into Python variables. |

Now that you have collected the user ID and password from the application user, you'll see how to collect the new IOS file name and MD5 values by reading a file. If you have multiple devices and values, entering this information through a command line would be too cumbersome. It is also prone to mistakes, so ideally these values are entered into a two-dimensional array form such as Excel or CSV files, letting your Python script read the information. You can save time and reduce human errors during the cut-and-paste operation in the command-line console.

# Tools Development 3: Collect a New IOS File Name and MD5 Value from a CSV File

After the user ID and password are collected, you want to collect even more information required for the Cisco IOS upgrade, but this time by reading the contents from a CSV file. The extra information includes new IOS names and their respective MD5 values. You can also include the hostname, device type, IP address, and other information to feed your script with two-dimensional data.

IOS can be downloaded from the Cisco download site to upgrade to the latest IOS version if you have an active service contract or work for Cisco Partners. The engineer who will perform the IOS upgrade usually downloads the file and obtains the MD5 value from the vendor's download site. After the IOS has been downloaded to the engineer's computer, the engineer confirms the MD5 values of that copy of the IOS, so they know that all the software is intact and not corrupted during the download process. This is one of the first verification steps during an IOS upgrade preparation process. Imagine not checking the MD5 values and using this file to upgrade the Cisco devices' IOS software! It could turn into a nasty situation instead of a straightforward IOS upgrade.
```

CHAPTER 9   CISCO IOS-XE UPGRADE TOOLS DEVELOPMENT PART 1

When you upgrade IOS or any operating system on any vendor product, you must make sure that the vendor's downloaded software version is correct. You need to make sure the download process has not corrupted the IOS due to a unreliable Internet connection or other issues. There are two places in the IOS upgrade tool where new IOS MD5 values will be validated. First, the user-provided MD5 value of the new IOS is validated against the server-checked MD5 value. Second, validate the server checked MD5 value against the MD5 value checked by the Cisco router. Before the router can check the MD5 of a new IOS file, the new IOS must be transferred to the router's flash memory using the TFTP/FTP/SFTP/SCP protocol. To be sure, you want to check all MD5 tests every time to avoid the unthinkable. Before you can compare the MD5 value found on Cisco's website against the server-side MD5 value, you must feed the script the correct MD5 value. An excellent way to feed this information is through an Excel or CSV file. Unlike when using a text file, you can use two-dimensional values (with a header) in Python scripts, and they can be handled by the pandas module.

In the first edition of this book, you created a file named device_info.csv on Microsoft Excel or Windows Notepad. This method still works, but you can create this file on the Linux console directly. Because you have only two devices with similar attributes except the hostname and IP address, you will create the .csv file on u22s1 directly. If you must work with many devices, it would be easier to work on Microsoft Excel or Google Sheets first and save the file as .csv then upload the file to the Linux server. Either way, if you can get the .csv file on the server, you can continue developing and learn how to use pandas module.

#	Task
①	On u22s1, create a file named devices_info.csv and enter the comma-separated information shown here. jdoe@u22s1:~/ch9_tools_dev1$ **mkdir tool3_read_csv && cd tool3_read_csv** jdoe@u22s1:~/ch9_tools_dev1/ tool3_read_csv$ **cat devices_info.csv** devicename,device,devicetype,host,newios,newiosmd5 c8kv01,RT,cisco_xe,192.168.127.111,c8000v-universalk9.17.06.05a.SPA.bin,13f0161a50210f2f21618fc59c5f5343 c8kv02,RT,cisco_xe,192.168.127.222,c8000v-universalk9.17.06.05a.SPA.bin,13f0161a50210f2f21618fc59c5f5343

(continued)

#	Task
②	If you have chosen to create this file using Microsoft Excel or another method, after creating the file, use WinSCP to upload the file to the /home/jdoe/ch9_tools_dev1/tool3_read_csv directory on u22s1. Your file should contain three lines, as shown in Figure 9-2. ``` devicename,device,devicetype,host,newios,newiosmd5 c8kv01,RT,cisco_xe,192.168.127.111,c8000v-universalk9.17.06.05a.SPA.bin,13f0161a50210f2f21618fc59c5f5343 c8kv02,RT,cisco_xe,192.168.127.222,c8000v-universalk9.17.06.05a.SPA.bin,13f0161a50210f2f21618fc59c5f5343 ``` ***Figure 9-2.*** *u23s1, devices_info.csv output*
③	Install pandas module if you have not done so already. pip will install pandas along with the dependent libraries. ``` jdoe@u22s1:~/ch9_tools_dev1/tool3_read_csv$ pip install pandas [...omitted for brevity] Successfully installed numpy-1.26.2 pandas-2.1.3 python-dateutil-2.8.2 tzdata-2023.3 ```
④	Write the base script to read your CSV file. If you are using a .xlsx file, replace the file extension type with .xlsx, and pandas will read the file in the same way as a .csv file. ``` jdoe@u22s1:~/ch9_tools_dev1/tool3_read_csv$ nano read_info1.py jdoe@u22s1:~/ch9_tools_dev1/tool3_read_csv$ cat read_info1.py import pandas as pd df = pd.read_csv(r'./devices_info.csv') print(df) ```
⑤	Execute the first Python application to check its functionality. ``` jdoe@u22s1:~/ch9_tools_dev1/tool3_read_csv$ python read_info1.py devicename device ... newios newiosmd5 0 c8kv01 RT ... c8000v-universalk9.17.06.05a.SPA.bin 13f0161a50210f2f21618fc59c5f5343 1 c8kv02 RT ... c8000v-universalk9.17.06.05a.SPA.bin 13f0161a50210f2f21618fc59c5f5343 [2 rows x 6 columns] ```

(continued)

#	Task
⑥	Now reiterate the script and add two lines of code to read the number of rows. The number of rows can also be used as a variable to control the flow of your application.

```
jdoe@u22s1:~/ch9_tools_dev1/tool3_read_csv$ cp read_info1.py read_info2.py
jdoe@u22s1:~/ch9_tools_dev1/tool3_read_csv$ nano read_info2.py
jdoe@u22s1:~/ch9_tools_dev1/tool3_read_csv$ cat read_info2.py
import pandas as pd

df = pd.read_csv(r'./devices_info.csv')
print(df)
number_of_rows = len(df.index)
print(number_of_rows)
``` |
| ⑦ | Run the second script to get the number of rows; you are expecting 2, as pandas by default reads the first row as the header row.

```
jdoe@u22s1:~/ch9_tools_dev1/tool3_read_csv$ python read_info2.py
  devicename device   ...                    newios                          newiosmd5
0     c8kv01     RT   ...  c8000v-universalk9.17.06.05a.SPA.bin
                                              13f0161a50210f2f21618fc59c5f5343
1     c8kv02     RT   ...  c8000v-universalk9.17.06.05a.SPA.bin
                                              13f0161a50210f2f21618fc59c5f5343

[2 rows x 6 columns]
2
``` |
| ⑧ | You are trying to read each value and use them as variables, so you use the specific information in the script. You want to read each row and convert it into a type of array such as a list or tuple.

```
jdoe@u22s1:~/ch9_tools_dev1/tool3_read_csv$ nano read_info3.py
jdoe@u22s1:~/ch9_tools_dev1/tool3_read_csv$ cat read_info3.py
import pandas as pd
``` |

(continued)

#	Task
| | ```python
df = pd.read_csv(r'./devices_info.csv')
number_of_rows = len(df.index)

# Read the values and save as a list, read the column as df and save it as a list
devicename = list(df['devicename'])
device = list(df['device'])
devicetype = list(df['devicetype'])
ip = list(df['host'])
newios = list(df['newios'])
newiosmd5 = list(df['newiosmd5'])
print(devicename)
print(device)
print(devicetype)
print(ip)
print(newios)
print(newiosmd5)
``` |
| 9 | When you run the previous script, the data is easily accessible now, and it is retrieved as Python lists.<br><br>```
jdoe@u22s1:~/ch9_tools_dev1/tool3_read_csv$ python read_info3.py
['c8kv01', 'c8kv02']
['RT', 'RT']
['cisco_xe', 'cisco_xe']
['192.168.127.111', '192.168.127.222']
['c8000v-universalk9.17.06.05a.SPA.bin', 'c8000v-universalk9.17.06.05a.SPA.bin']
['13f0161a50210f2f21618fc59c5f5343', '13f0161a50210f2f21618fc59c5f5343']
``` |

(continued)

#	Task

⑩	Let's massage the data a little bit and put it into a single list containing list items.
	
	jdoe@u22s1:~/ch9_tools_dev1/tool3_read_csv$ **nano read_info4.py**
	jdoe@u22s1:~/ch9_tools_dev1/tool3_read_csv$ **cat read_info4.py**
	import pandas as pd
	
	df = pd.read_csv(r'./devices_info.csv')
	number_of_rows = len(df.index)
	# Read the values and save as a list, read the column as df and save it as a list
	devicename = list(df['devicename'])
	device = list(df['device'])
	devicetype = list(df['devicetype'])
	ip = list(df['host'])
	newios = list(df['newios'])
	newiosmd5 = list(df['newiosmd5'])
	
	# Convert the list into a device_list
	device_list = []
	for index, rows in df.iterrows():
	    device_append = [rows.devicename, rows.device, rows.devicetype, rows.host, rows.newios, rows.newiosmd5]
	    device_list.append(device_append)
	print(device_list)

⑪	Run the script. You will notice that the device information has now turned into a list of lists. jdoe@u22s1:~/ch9_tools_dev1/tool3_read_csv$ **python read_info4.py** [['c8kv01', 'RT', 'cisco_xe', '192.168.127.111', 'c8000v-universalk9.17.06.05a.SPA.bin', '13f0161a50210f2f21618fc59c5f5343'], ['c8kv02', 'RT', 'cisco_xe', '192.168.127.222', 'c8000v-universalk9.17.06.05a.SPA.bin', '13f0161a50210f2f21618fc59c5f5343']]

(*continued*)

#	Task

⑫	If you need only specific information from the device_list created by reading the CSV file, such as newios and newiosmd5, you can use a simple loop to recall these numbers.
	
	jdoe@u22s1:~/ch9_tools_dev1/tool3_read_csv$ **cp read_info4.py read_info5.py**
	jdoe@u22s1:~/ch9_tools_dev1/tool3_read_csv$ **nano read_info5.py**
	jdoe@u22s1:~/ch9_tools_dev1/tool3_read_csv$ **cat read_info5.py**
	``python
	import pandas as pd
	
	df = pd.read_csv(r'./devices_info.csv')
	number_of_rows = len(df.index)
	
	# Read the values and save as a list, read the column as df and save it as a list
	devicename = list(df['devicename'])
	device = list(df['device'])
	devicetype = list(df['devicetype'])
	ip = list(df['host'])
	newios = list(df['newios'])
	newiosmd5 = list(df['newiosmd5'])
	
	# Convert the list into a device_list
	device_list = []
	for index, rows in df.iterrows():
	    device_append = [rows.devicename, rows.device, rows.devicetype, rows.host, rows.newios, rows.newiosmd5]
	    device_list.append(device_append)
	
	for x in device_list:
	    newios, newiosmd5 = x[4], x[5].lower()
	    print(newios, newiosmd5)
	``

(*continued*)

CHAPTER 9   CISCO IOS-XE UPGRADE TOOLS DEVELOPMENT PART 1

#	Task
⑬	Run the script now. You will get each device's new IOS file name and MD5 value. Because you have the same device types, the new IOS name and MD5 values will be the same. jdoe@u22s1:~/ch9_tools_dev1/tool3_read_csv$ **python read_info5.py** c8000v-universalk9.17.06.05a.SPA.bin 13f0161a50210f2f21618fc59c5f5343 c8000v-universalk9.17.06.05a.SPA.bin 13f0161a50210f2f21618fc59c5f5343
| ⑭ | Initially, when you created the CSV file, you used more information than required, and there is a reason behind this. Using the `pandas` file read method, you also want to create a list containing the `netmiko` dictionary format. You can pass the values of a dictionary to `netmiko ConnectHandler` for SSH connections while logging into the routers. Let's study the simple `netmiko` dictionary format and use IP address and `devicetype` information from `device_list` to make another dictionary for SSH logins. A device dictionary looks like the following example. You can also add the logging and delay factors options, but this lab keeps it simple.<br><br>```
device1 = {
'device_type': 'cisco_ios',
'host': '10.10.10.1',
'username': 'username',
'password': 'password',
'secret': 'secret',
}
```<br><br>So, from `tool2_login`'s previous development, you can collect the username, password, and secret from the user via an interactive input session. Now you have read the `device_type` and `host` (IP address) values from a CSV file. You can use the collected information to turn it into a `netmiko` compatible dictionary format for the SSH connection. Note that the username and password collection tool are added to teach you how they can be collected interactively. If you are running Python scripts in an extremely secure environment, you can feed all the information through the file `read` method, including the username, password, and secret. |

(continued)
```

#	Task

```
jdoe@u22s1:~/ch9_tools_dev1/tool3_read_csv$ cp read_info5.py read_
info6.py
jdoe@u22s1:~/ch9_tools_dev1/tool3_read_csv$ nano read_info6.py
jdoe@u22s1:~/ch9_tools_dev1/tool3_read_csv$ cat read_info6.py
import pandas as pd

df = pd.read_csv(r'./devices_info.csv')
number_of_rows = len(df.index)

# Read the values and save as a list, read the column as df and save
it as a list
devicename = list(df['devicename'])
device = list(df['device'])
devicetype = list(df['devicetype'])
ip = list(df['host'])
newios = list(df['newios'])
newiosmd5 = list(df['newiosmd5'])

# Convert the list into a device_list
device_list = []
for index, rows in df.iterrows():
    device_append = [rows.devicename, rows.device, rows.devicetype,
    rows.host, rows.newios, rows.newiosmd5]
    device_list.append(device_append)

i = 0
for x in device_list:
    if len(x) != 0:
        i += 1
        name = f'device{str(i)}'
        devicetype, host = x[2], x[3]
        device = {
            'device_type': devicetype,
            'host': host,
```

(continued)

#	Task
	``` 'username': 'username', 'password': 'password', 'secret': 'secret', } print(name, "=", device) ```

⑮	Now when you run the script, you will see that the information is formatted into a dictionary and you can assign the dictionary to variables, in this case, device1 and device2. **Imagine you have 100 or 200 devices to turn the read information into netmiko-friendly dictionaries. Some people say that automation is all-in-the-loop commands. If a task is repetitive, that is a potential target for automation.**
	
	```jdoe@u22s1:~/ch9_tools_dev1/tool3_read_csv$ python read_info6.py```
	```device1 = {'device_type': 'cisco_xe', 'host': '192.168.127.111', 'username': 'username', 'password': 'password', 'secret': 'secret'}```
	```device2 = {'device_type': 'cisco_xe', 'host': '192.168.127.222', 'username': 'username', 'password': 'password', 'secret': 'secret'}```

⑯	Copy the previous script and create a new script called read_info7.py. You are going to modify it, so the dictionaries are now stored in a list. In other words, you will create a list with multiple dictionaries as its items. This way, you can call them out and the dictionary during the SSH connection to your devices.
	
	Notice that I have manually added the username, password, and secret to this script for testing purposes. However, when you integrate the user ID and password collection tool developed earlier, the variables will be replaced. When you work in a production environment, it is highly recommended that you remove usernames and password information or completely delete the files if they contain sensitive information.
	
	```jdoe@u22s1:~/ch9_tools_dev1/tool3_read_csv$ cp read_info6.py read_info7.py```
	```jdoe@u22s1:~/ch9_tools_dev1/tool3_read_csv$ nano read_info7.py```
	```jdoe@u22s1:~/ch9_tools_dev1/tool3_read_csv$ cat read_info7.py```
	import pandas as pd
	
	df = pd.read_csv(r'./devices_info.csv')
	number_of_rows = len(df.index)

(continued)

#	Task
| | ```python
# Read the values and save as a list, read the column as df and save
it as a list
devicename = list(df['devicename'])
device = list(df['device'])
devicetype = list(df['devicetype'])
ip = list(df['host'])
newios = list(df['newios'])
newiosmd5 = list(df['newiosmd5'])

# Convert the list into a device_list
device_list = []
for index, rows in df.iterrows():
    device_append = [rows.devicename, rows.device, rows.devicetype,
    rows.host, rows.newios, rows.newiosmd5]
    device_list.append(device_append)

device_list_netmiko = []
i = 0
for x in device_list:
    if len(x) != 0:
        i += 1
        name = f'device{str(i)}'
        devicetype, host = x[2], x[3]
        device = {
            'device_type': devicetype,
            'host': host,
            'username': 'jdoe',
            'password': 'cisco123',
            'secret': 'cisco123',
        }
        device_list_netmiko.append(device)

print(device_list_netmiko)
``` |

(continued)
```

CHAPTER 9   CISCO IOS-XE UPGRADE TOOLS DEVELOPMENT PART 1

#	Task
⑰	When you run the script, you should get a list with two dictionaries as items. You must get familiar with massaging the data if you want to parse it to Python scripts to access any information from any network device. jdoe@u22s1:~/ch9_tools_dev1/tool3_read_csv$ **python read_info7.py** [{'device_type': 'cisco_xe', 'host': '192.168.127.111', 'username': 'jdoe', 'password': 'cisco123', 'secret': 'cisco123'}, {'device_type': 'cisco_xe', 'host': '192.168.127.222', 'username': 'jdoe', 'password': 'cisco123', 'secret': 'cisco123'}]
| ⑱ | To test that this script will be working as expected, add a netmiko ConnectHandler and run the Cisco router command. Note that only one import statement (from netmiko import ConnectHandler) and the last four lines of code are added to the previous script. This Python script is available for download from my GitHub site.<br>Download URL: https://github.com/pynetauto/apress_pynetauto_ed2.0/tree/main/source_codes<br>Create the final script to use the read data and then perform a simple task; in this case, run the show clock command to display the time of the Catalyst routers.<br><br>jdoe@u22s1:~/ch9_tools_dev1/tool3_read_csv$ **cp read_info7.py read_info8.py**<br>jdoe@u22s1:~/ch9_tools_dev1/tool3_read_csv$ **nano read_info8.py**<br>jdoe@u22s1:~/ch9_tools_dev1/tool3_read_csv$ **cat read_info8.py**<br><br>```
import pandas as pd
from netmiko import ConnectHandler

df = pd.read_csv(r'./devices_info.csv')
number_of_rows = len(df.index)

# Read the values and save as a list, read the column as df and save it as a list
devicename = list(df['devicename'])
device = list(df['device'])
devicetype = list(df['devicetype'])
``` |

(continued)
```

#	Task

```
        ip = list(df['host'])
        newios = list(df['newios'])
        newiosmd5 = list(df['newiosmd5'])

        # Convert the list into a device_list
        device_list = []
        for index, rows in df.iterrows():
            device_append = [rows.devicename, rows.device, \
            rows.devicetype, rows.host, rows.newios, rows.newiosmd5]
            device_list.append(device_append)

        device_list_netmiko = []
        i = 0
        for x in device_list:
            if len(x) !=0:
                i += 1
                name = f'device{str(i)}'
                devicetype, host = x[2], x[3]
                device = {
                'device_type': devicetype,
                'host': host,
                'username': 'jdoe',
                'password': 'cisco123',
                'secret': 'cisco123',
                }
                device_list_netmiko.append(device)
        for device in device_list_netmiko:
            net_connect = ConnectHandler(**device)
            show_clock = net_connect.send_command("show clock")
            # show_clock = net_connect.send_command("show version")
            print(show_clock)
```

(continued)

CHAPTER 9   CISCO IOS-XE UPGRADE TOOLS DEVELOPMENT PART 1

#	Task
⑲	When you run the script, you should see each router's time. The `show clock` command is one of the simplest commands that you can run from your script to check if your SSH connection is working. Checking port 22 does not test your credentials, so it is worth writing code that runs a simple `show clock` command. jdoe@u22s1:~/ch9_tools_dev1/tool3_read_csv$ **python read_info8.py** *17:44:19.071 UTC Sun Nov 26 2023 *17:44:22.721 UTC Sun Nov 26 2023

You have now completed the read .csv file tool to read information from .csv files and massage the data to be used in your IOS upgrade application script. Next, you'll create a tool that checks the MD5 value of the new IOS router images. Checking the MD5 value is critical and can prevent upgrade failures due to software corruption or file manipulations.

Tools Development 4: Check the MD5 Value of the New IOS on the Server

As discussed earlier, authenticating the new IOS file is the key to a successful IOS upgrade. The MD5 value is available from Cisco's website, or you can use a tool such as WinMD5.exe or even a command line to check the MD5 value of the file after the new IOS download. Before uploading the file, you need to check if this MD5 value is correct to double-check the file's integrity you are about to upload. When you upload a new IOS file manually to a TFTP or FTP server, you will verify this manually, but in this IOS upgrade tool, you are replacing the TFTP/FTP file transfer method with Secure Copy Protocol (SCP) file transfer. Before the file transfer occurs, you'll need to check the MD5 value of the IOS file in the SCP folder against the good MD5 value, so that the file integrity is guaranteed to avoid unexpected results.

CHAPTER 9　CISCO IOS-XE UPGRADE TOOLS DEVELOPMENT PART 1

#	Task
①	First, create a directory to upload your new IOS file using WinSCP or FileZilla. Then upload the new IOS to your server, as shown in Figures 9-3 and 9-4.

jdoe@u22s1:~/ch9_tools_dev1$ **pwd**
/home/jdoe/ch9_tools_dev1
jdoe@u22s1:~/ch9_tools_dev1$ **mkdir new_ios && cd new_ios**

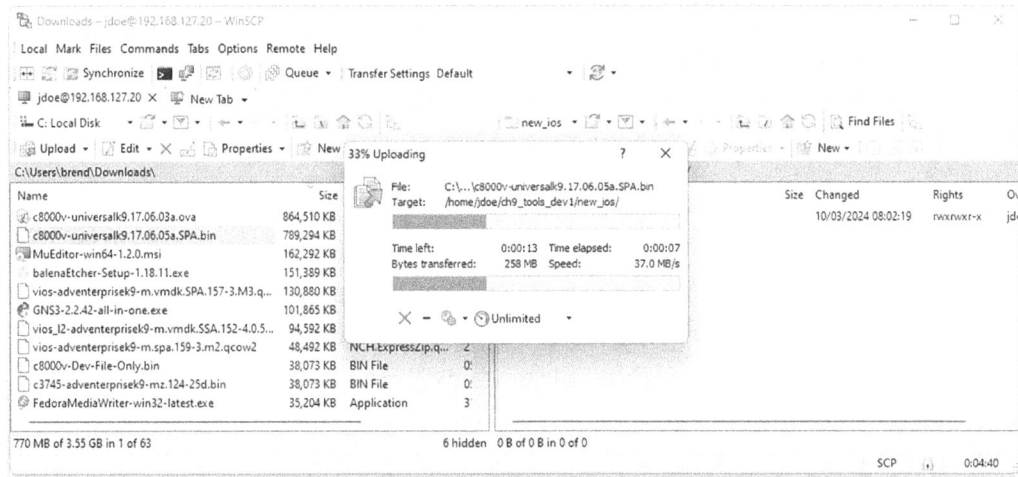

Figure 9-3. u22s1, copying the new IOS to the new_ios directory

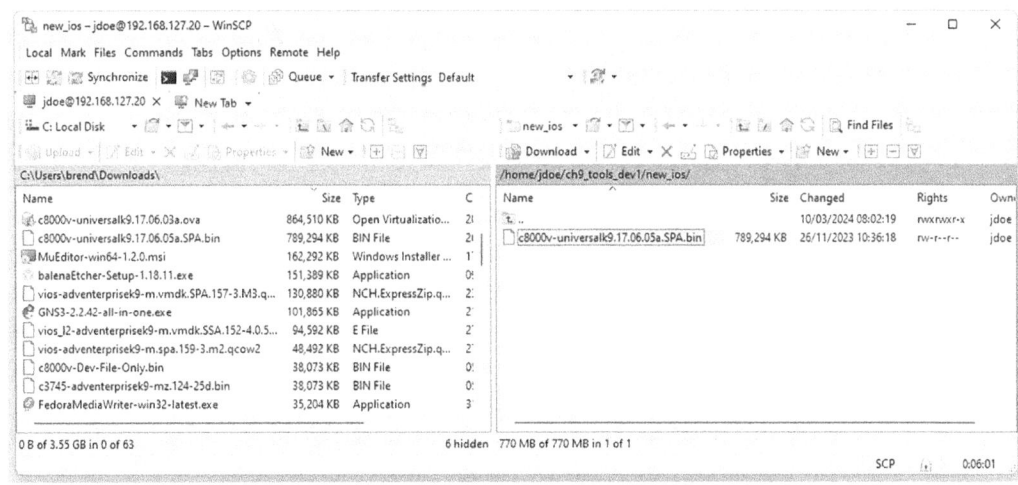

Figure 9-4. u22s1, copied new IOS file to the new_ios directory

(*continued*)

CHAPTER 9   CISCO IOS-XE UPGRADE TOOLS DEVELOPMENT PART 1

#	Task
②	After the uploading completes, check the new_ios directory. ``` jdoe@u22s1:~/ch9_tools_dev1/new_ios$ ls -lh total 771M -rw-r--r-- 1 jdoe jdoe 771M Nov 26 10:36 c8000v-universalk9.17.06.05a.SPA.bin ```
③	You already know the MD5 value from previous exercises, but to check the MD5 value of a file from the Linux server, you can use the md5sum filename command, as shown here. You should expect to see the same MD5 value as the original file uploaded. ``` jdoe@u22s1:~/ch9_tools_dev1/new_ios$ md5sum c8000v-universalk9.17.06.05a.SPA.bin 13f0161a50210f2f21618fc59c5f5343 c8000v-universalk9.17.06.05a.SPA.bin ```
④	Create the tool4_md5_linux directory; then copy the read_info5.py file as the basis of the md5_validate1.py script. Also, to help with development, copy the devices_info.csv file. Both the read_info5.py and devices_info.csv files are from the tool3 development. ``` jdoe@u22s1:~/ch9_tools_dev1/new_ios$ cd .. jdoe@u22s1:~/ch9_tools_dev1$ mkdir tool4_md5_linux && cd tool4_md5_linux jdoe@u22s1:~/ch9_tools_dev1/tool4_md5_linux$ cp /home/jdoe/ch9_tools_dev1/tool3_read_csv/read_info5.py ./md5_validate1.py jdoe@u22s1:~/ch9_tools_dev1/tool4_md5_linux$ cp /home/jdoe/ch9_tools_dev1/tool3_read_csv/devices_info.csv ./devices_info.csv jdoe@u22s1:~/ch9_tools_dev1/tool4_md5_linux$ pwd /home/jdoe/ch9_tools_dev1/tool4_md5_linux jdoe@u22s1:~/ch9_tools_dev1/tool4_md5_linux$ ls -lh total 8.0K -rw-rw-r-- 1 jdoe jdoe 254 Nov 26 18:05 devices_info.csv -rw-rw-r-- 1 jdoe jdoe 686 Nov 26 18:05 md5_validate1.py ```

(continued)

#	Task
| ⑤ | Now you need to modify the last `for` loop of the code from `read_info5.py`, so the task is to modify this `for` loop and add a function to check the MD5 value of the just-copied IOS file and check the file size to be used during the flash size check later.<br><br>`jdoe@u22s1:~/ch9_tools_dev1/tool4_md5_linux$` **`nano md5_validate1.py`**<br>`jdoe@u22s1:~/ch9_tools_dev1/tool4_md5_linux$` **`cat md5_validate1.py`**<br>```
import pandas as pd

df = pd.read_csv(r'./devices_info.csv')
number_of_rows = len(df.index)

# Read the values and save as a list, read the column as df and save
it as a list
devicename = list(df['devicename'])
device = list(df['device'])
devicetype = list(df['devicetype'])
ip = list(df['host'])
newios = list(df['newios'])
newiosmd5 = list(df['newiosmd5'])

# Convert the list into a device_list
device_list = []
for index, rows in df.iterrows():
    device_append = [rows.devicename, rows.device, rows.devicetype,
    rows.host, rows.newios, rows.newiosmd5]
    device_list.append(device_append)

for x in device_list:
    newios, newiosmd5 = x[4], x[5].lower()
    print(newios, newiosmd5)
``` |

(continued)
```

CHAPTER 9   CISCO IOS-XE UPGRADE TOOLS DEVELOPMENT PART 1

#	Task
⑥	Now add the import code for the built-in `os.path` and `hashlib` modules. `hashlib` is used to check the MD5 of the file on the Linux server and `os.path` is used to calculate the actual file size. The middle part of the script is the same as the copied script. Still, you will change the `for x in device_list:` so, the script returns the MD5 value calculated on the server side and the newios size, which will be used later to check if the free size on the router's flash can accommodate the new IOS size. After the change, `md5_validate1.py` should look similar to this:

```
jdoe@u22s1:~/ch9_tools_dev1/tool4_md5_linux$ cp md5_validate1.py md5_validate2.py
jdoe@u22s1:~/ch9_tools_dev1/tool4_md5_linux$ nano md5_validate2.py
jdoe@u22s1:~/ch9_tools_dev1/tool4_md5_linux$ cat md5_validate2.py
import pandas as pd
import os.path
import hashlib

[... omitted for brevity, same as read_info5.py]

for x in device_list:
    print(x[0])
    newios = x[4]
    newiosmd5 = x[5].lower()
    newiosmd5hash = hashlib.md5()
    file = open(f'/home/jdoe/ch9_tools_dev1/new_ios/{newios}', 'rb')
    content = file.read()
    newiosmd5hash.update(content)
    newiosmd5server = newiosmd5hash.hexdigest()
    print(newiosmd5server)
    newiossize = round(os.path.getsize(f'/home/jdoe/ch9_tools_dev1/new_ios/{newios}')/1000000,2)
    print(newiossize, "MB")
```

(continued)

CHAPTER 9   CISCO IOS-XE UPGRADE TOOLS DEVELOPMENT PART 1

#	Task
⑦	Run the script. It should return the server-side IOS MD5 value and the actual IOS size in megabytes. You will also learn how to check the router's used, free, and total size to make sure that the flash has enough free space to accommodate the new IOS. ``` jdoe@u22s1:~/ch9_tools_dev1/tool4_md5_linux$ python md5_validate2.py c8kv01 13f0161a50210f2f21618fc59c5f5343 808.24 MB # Warning! Calculation discrepancy, the actual file size is 771 MB c8kv02 13f0161a50210f2f21618fc59c5f5343 808.24 MB # Warning! Calculation discrepancy, the actual file size is 771 MB ```

⑧	Once you are happy with the result in Step 3, add an `if-else` statement to the end of the last print line to see how you can control the script flow by testing the two MD5 values.
	
	``
	jdoe@u22s1:~/ch9_tools_dev1/tool4_md5_linux$ cp md5_validate2.py md5_validate3.py
	jdoe@u22s1:~/ch9_tools_dev1/tool4_md5_linux$ nano md5_validate3.py
	jdoe@u22s1:~/ch9_tools_dev1/tool4_md5_linux$ cat md5_validate3.py
	import pandas as pd
	import os.path
	import hashlib
	
	df = pd.read_csv(r'./devices_info.csv')
	number_of_rows = len(df.index)
	
	# Read the values and save as a list, read the column as df and save it as a list
	devicename = list(df['devicename'])
	device = list(df['device'])
	devicetype = list(df['devicetype'])
	ip = list(df['host'])
	newios = list(df['newios'])
	newiosmd5 = list(df['newiosmd5'])
	``

(continued)

#	Task
| | ```python
# Convert the list into a device_list
device_list = []
for index, rows in df.iterrows():
    device_append = [rows.devicename, rows.device, \
    rows.devicetype, rows.host, rows.newios, rows.newiosmd5]
    device_list.append(device_append)

for x in device_list:
    print(x[0])
    newios = x[4]
    newiosmd5 = str(x[5].lower()).strip()
    print(newiosmd5)
    newiosmd5hash = hashlib.md5()
    file = open(f'/home/jdoe/ch9_tools_dev1/new_ios/{newios}', 'rb')
    content = file.read()
    newiosmd5hash.update(content)
    newiosmd5server = newiosmd5hash.hexdigest()
    print(newiosmd5server.strip())
    newiossize = round(os.path.getsize(f'/home/jdoe/ch9_tools_dev1/
    new_ios/{newios}')/1000000, 2)
    print(newiossize, "MB")

    if newiosmd5server == newiosmd5:
        print("MD5 values matched!")
    else:
        print("Mismatched MD5 values. Exit")
        exit()
``` |

(*continued*)
```

CHAPTER 9   CISCO IOS-XE UPGRADE TOOLS DEVELOPMENT PART 1

#	Task
⑨	When you run the script, you will see a result like the following output. You have now confirmed that the information read has been turned into a Python variable for use in your script. You should get the two MD5 values, one provided by the user in the device_info.csv file and the other calculated by the Linux server. ``` jdoe@u22s1:~/ch9_tools_dev1/tool4_md5_linux$ python md5_validate3.py c8kv01 13f0161a50210f2f21618fc59c5f5343 # read from the .csv file, user provided 13f0161a50210f2f21618fc59c5f5343 # calculated by Linux server 808.24 MB # Warning! Calculation discrepancy, the actual file size is 771 MB MD5 values matched! c8kv02 13f0161a50210f2f21618fc59c5f5343 # read from the .csv file, user provided 13f0161a50210f2f21618fc59c5f5343 # calculated by Linux server 808.24 MB # Warning! Calculation discrepancy, the actual file size is 771 MB MD5 values matched! ```

You have completed the MD5 checking tool for your new IOS file, and this tool is now ready to be integrated into the main IOS upgrade application. Despite the increased flash memory sizes in later models of Cisco routers, unlike modern-day computers, networking devices still possess limited flash memory to store primary files. Therefore, understanding whether the available free flash size can accommodate the new IOS file remains vital during IOS upgrades. Let's proceed to develop a tool that checks the remaining flash size on your Cisco routers.

Tools Development 5: Check the Flash Size on Cisco Routers

You learned how to get the IOS size on the server in the previous script, and the new IOS file size was 808.24MB (Warning! Calculation discrepancy! The actual file size is 771MB). Before you can upload this new IOS file to the router's flash memory, you must check if there is enough free space in the flash memory. If the flash memory has enough free space, uploading can occur immediately. Still, if the flash size is not big enough, you will have to delete old or redundant files, or for older IOS devices, you will need to delete the currently running IOS file from the flash. During the boot process, the IOS is copied and decompressed in random access memory (RAM) from flash. On some of the latest Cisco platforms, the files are already decompressed to save time during the Power-On-Self-Test (POST) process. Either way, you need to make enough room for a new IOS file. To get the free flash size on a Cisco device, you can use a show flash: or dir command and then dissect the information using a Regular Expression. If you know what command to run and which information you need, it's easy to get it. Still, many vendor network devices do not provide API support. Also, SNMP has its limitations while trying to work out this type of information. Although this is not the smartest or the most sophisticated way to collect the data you want, this could be the only option for some devices.

After working out the free flash size, if the IOS size exceeds the flash-free size, you want to give the user (or the script) an option to locate the old IOS file and remove it. Sometimes, there might be a large file on the router's flash memory, so in this case, you also must give the user (or the script) an option to search the file and delete a large file. These steps show how this can be achieved.

#	Task
①	Create another working directory called tool5_fsize_cisco, and then create a new script to run and capture either the dir or show flash: command.

Write the following script to capture the output of the dir command on the Cisco router. On IOS XE routers, it is easier to capture the information using the dir command, but the show flash: command works on most devices. For this practice, use the dir command because the output is shorter.

```
jdoe@u22s1:~/ch9_tools_dev1/tool4_md5_linux$ cd ..
jdoe@u22s1:~/ch9_tools_dev1$ mkdir tool5_fsize_cisco && cd tool5_fsize_cisco
jdoe@u22s1:~/ch9_tools_dev1/tool5_fsize_cisco$ nano check_flash1.py
jdoe@u22s1:~/ch9_tools_dev1/tool5_fsize_cisco$ cat check_flash1.py
import time
from netmiko import ConnectHandler

# Borrowed from read_info7.py result in tool3_read_csv directory.
devices_list = [
    {
        'device_type': 'cisco_xe',
        'host': '192.168.127.111',
        'username': 'jdoe',
        'password': 'cisco123',
        'secret': 'cisco123'
    },
    {
        'device_type': 'cisco_xe',
        'host': '192.168.127.222',
        'username': 'jdoe',
        'password': 'cisco123',
        'secret': 'cisco123'
    }
]
```

(continued)

#	Task
|   | ```python
for device in devices_list:
    net_connect = ConnectHandler(**device)
    net_connect.send_command("terminal length 0")
    showdir = net_connect.send_command("dir")
    # showflash = net_connect.send_command("show flash:") #
    Alternatively use 'show flash:'
    print(showdir)
    print("-" * 80)
    time.sleep(2)
``` |
| ② | When you run the check_flash1.py script, the output displays the file details and flash memory usage. You are interested in the free size of the router's flash memory. Here, you are interested in "11575476224 bytes total (10065227776 bytes free)" for c8kv01 and "11575476224 bytes total (10064220160 bytes free)" c8kv02 in the last line of the output.

```
jdoe@u22s1:~/ch9_tools_dev1/tool5_fsize_cisco$ python check_flash1.py
Directory of bootflash:/

262146   drwx           4096  Nov 26 2023 19:31:51 +00:00  tracelogs
[...output omitted for brevity]
11       drwx          16384  Nov 26 2023 12:24:33 +00:00  lost+found

11575476224 bytes total (10065227776 bytes free)
--------------------------------------------------------------------

Directory of bootflash:/
131076   drwx          12288  Nov 26 2023 19:45:16 +00:00  tracelogs
[...output omitted for brevity]
11       drwx          16384  Nov 25 2023 16:09:35 +00:00  lost+found

11575476224 bytes total (10064220160 bytes free)
--------------------------------------------------------------------
``` |

(continued)
```

#	Task
③	So, this is the `dir` output from the first router. You are putting your Regular Expression knowledge to the test here. See how you can use a Regular Expression to get the "bytes free."

```
c8kv01#dir
Directory of bootflash:/

262146  drwx             4096  Nov 26 2023 19:31:51 +00:00  tracelogs
[...output omitted for brevity]
11      drwx            16384  Nov 26 2023 12:24:33 +00:00  lost+found

11575476224 bytes total (10065227776 bytes free)
```

Because there are many digits, you must locate the number with pinpoint accuracy, which can be achieved by using a positive lookahead method. Because the digits are in front of `bytes free`, you can use this as the search handle and simply find the string of digits ahead of this specific string. So, the required Regular Expression is as shown here:

\d+(?=\sbytes\sfree\))

\d+: Matches one or more digits.
(?=): Denotes a positive lookahead assertion in Regular Expressions. It's used to match a group of characters only if they are followed by another specified pattern.
\sbytes\sfree\): This looks for the literal string "`bytes free)`" preceded and followed by whitespace (\s), indicating a specific sequence of characters to match. The \s represents any whitespace character, and "`bytes free)`" is the string to match.
So, the previous Regular Expression should match any digits occurring in front of `bytes free`, which is 10065227776 in this example. |

(continued)

CHAPTER 9   CISCO IOS-XE UPGRADE TOOLS DEVELOPMENT PART 1

#	Task
| ④ | Make a copy of the first script and create the following script to apply the Regular Expression:<br><br>jdoe@u22s1:~/ch9_tools_dev1/tool5_fsize_cisco$ **cp check_flash1.py check_flash2.py**<br>jdoe@u22s1:~/ch9_tools_dev1/tool5_fsize_cisco$ **nano check_flash2.py**<br>jdoe@u22s1:~/ch9_tools_dev1/tool5_fsize_cisco$ **cat check_flash2.py**<br>```
import time
from netmiko import ConnectHandler
import re # Import re module

# This is borrowed from read_info7.py result
devices_list = [
    {
        'device_type': 'cisco_xe',
        'host': '192.168.127.111',
        'username': 'jdoe',
        'password': 'cisco123',
        'secret': 'cisco123'
    },
    {
        'device_type': 'cisco_xe',
        'host': '192.168.127.222',
        'username': 'jdoe',
        'password': 'cisco123',
        'secret': 'cisco123'
    }
]

for device in devices_list:
    net_connect = ConnectHandler(**device)
    net_connect.send_command("terminal length 0")
    showdir = net_connect.send_command("dir")
``` |

(continued)
```

CHAPTER 9   CISCO IOS-XE UPGRADE TOOLS DEVELOPMENT PART 1

#	Task
|   | ```
# showflash = net_connect.send_command("show flash:")
# print(showdir)   # Hash out the line
print("-" * 80)
time.sleep(2)
p1 = re.compile("\d+(?=\sbytes\sfree\))")  # Compiled Regular
                                              Expression
m1 = p1.findall(showdir)  # Match Regular Expression
flashfree = (int(m1[0]) / 1000000)  # convert bytes into MB
print(flashfree)
``` |
| ⑤ | Run the previous script, and now you have the free bytes in megabytes. The first value is from c8kv01, and the second value is from c8kv02.<br><br>```
jdoe@u22s1:~/ch9_tools_dev1/tool5_fsize_cisco$ python check_flash2.py
--------------------------------------------------------------------
10065.227776 # free flash space in MB, c8kv01
--------------------------------------------------------------------
10064.22016 # free flash space in MB, c8kv02
``` |

When combining the previous script with md5_validate2.py from earlier exercises, you can verify if the free space on the flash memory can accommodate the size of the new IOS file. The application flow can adapt based on this comparison. Both routers have over 10GB of free flash memory, while the new IOS file size is 808.24MB (#Warning! Calculation discrepancy; the actual file size is 771MB). You don't have to remove any files before uploading the new file to the Flash. However, as mentioned, older Cisco devices often have constrained flash memory, requiring the script to offer users the option to delete files from the flash memory to accommodate the new IOS upload.

Usually, when upgrading IOS on mission-critical devices such as edge routers or core/backbone switches, engineers spend over 80 percent of their time preparing for such changes. Addressing flash space limitations is a crucial part of this preparation and is typically detected early in the change planning process. While most flash space issues must be resolved before the actual upgrade, at times, engineers might attempt to expedite this preparation.

Contemporary IT-driven companies often adhere to the ITIL process, guiding change management. Owners of such changes conduct end-to-end checks before approval from stakeholders, ensuring meticulous preparation before implementation. Consequently, to streamline this chapter, I'm excluding the script that provides users with options to remove or search for files in a directory.

Tools Development 6: Make Backups of running-config, Interface Status, and Routing Table

Assuming your routers possess sufficient flash memory for the new IOS, the next step involves backing up the running configuration. In most production environments, routers' configurations are routinely backed up, either daily or weekly, on a dedicated configuration backup server. Here, it's crucial to ensure that the last backed-up configuration of the device being upgraded isn't older than a couple of hours. In essence, the most recent `running-config` backup, just before the router undergoes a reload, holds utmost importance.

Using the Python script, there are two methods to back up the routers' running-config. The first involves saving the configuration on the local disk of your Python server, while the second method utilizes the `copy running-config tftp/ftp:` command. The latter method is preferable for extended storage, while backing up to your server suffices when the system loses its configuration unexpectedly. Hence, you'll learn to save running configurations using the SSH connection's `show` command.

Before implementing any changes, it's imperative to create backups of the running configuration before initiating the router reload.

#	Task
①	Create a new directory to develop a fresh tool. As I've covered similar content in previous chapters, you're likely familiar with file handling. This tool will be written in one go, without reiteration. Follow along by typing each word and referring to the explanations in the code lines.

The script generates dictionaries for each device with respective variables to simplify the process. Using the IP address of each device as part of the file names ensures unique backups for individual devices. This prevents multiple devices' configurations from being stored within a single file, aiding in future troubleshooting. You have the flexibility to execute various show commands to back up your router's configurations. Essential commands for Cisco routers include show running-config, show ip route, and show ip interface brief. Similarly, when working with Cisco switches, commands like show running-config, show ip interface brief, and show switch can be beneficial. Let's proceed to write and execute this code.

```
jdoe@u22s1:~/ch9_tools_dev1/tool5_fsize_cisco$ cd ..
jdoe@u22s1:~/ch9_tools_dev1$ mkdir tool6_make_backup && cd tool6_make_backup
jdoe@u22s1:~/ch9_tools_dev1/tool6_make_backup$ nano make_backup1.py
jdoe@u22s1:~/ch9_tools_dev1/tool6_make_backup$ cat make_backup1.py
import time
from netmiko import ConnectHandler

# Converted the list of dictionaries back to each variable
device1 = {
    'device_type': 'cisco_xe',
    'host': '192.168.127.111',
    'username': 'jdoe',
    'password': 'cisco123',
    'secret': 'cisco123'
}
```
|

(continued)

#	Task
| | ```
device2 = {
    'device_type': 'cisco_xe',
    'host': '192.168.127.222',
    'username': 'jdoe',
    'password': 'cisco123',
    'secret': 'cisco123'
}
devices_list = [device1, device2] # Make a list of devices

for device in devices_list: # Read each device from device_list
    print(device) # Print out each device information during
                  development
    ip = str(device['host']) # To make unique file names, assign IP
                             address of device as variable, ip
    f1 = open(ip + '_show_run_1.txt', 'w+') # Create and open f1
    net_connect = ConnectHandler(**device) # Connect to device, **
                                            denotes parsing of a
                                            dictionary
    net_connect.send_command("terminal length 0") # Change terminal
                                                   length to 0
    showrun = net_connect.send_command("show running-config")
    # Run show run and save to showrun variable
    f1.write(showrun) # Write showrun to f1. showrun is in the memory.
    time.sleep(1) # Pause for 1 second
    f1.close() # Close f1

    # The remaining code is the same as above except for the file name
    and actual command
    f2 = open(ip + '_show_ip_route_1.txt', 'w') # Create and open f2
    howiproute = net_connect.send_command("show ip route")
    f2.write(howiproute)
    time.sleep(1)
    f2.close()
``` |

(continued)
```

#	Task
| | ```
f3 = open(ip + '_show_ip_int_bri_1.txt', 'w')
showiproute = net_connect.send_command("show ip interface brief")
f3.write(showiproute)
time.sleep(1)
f3.close()

print("All tasks completed successfully") # Information to let you
                                          know that all tasks completed
``` |
| ② | Now execute the script. It will make backups by running the show commands of your choice. Here you are capturing three show command outputs to separate files. Later, the post-upgrade check will rerun the same commands and create post-reload files. Then, the script will perform the post-upgrade check to ensure that the before and after configurations remain the same and the IOS upgrade did not cause any failures.

Once you have written the make_backup1.py Python code, run it to make backups of the show commands.

```
jdoe@u22s1:~/ch9_tools_dev1/tool6_make_backup$ ls -lh
total 4.0K
-rw-rw-r-- 1 jdoe jdoe 1.1K Nov 26 20:26 make_backup1.py
jdoe@u22s1:~/ch9_tools_dev1/tool6_make_backup$ python make_backup1.py
All tasks completed successfully
jdoe@u22s1:~/ch9_tools_dev1/tool6_make_backup$ ls -lh
total 36K
-rw-rw-r-- 1 jdoe jdoe  323 Nov 26 20:26 192.168.127.111_show_ip_int_bri_1.txt
-rw-rw-r-- 1 jdoe jdoe 1.1K Nov 26 20:26 192.168.127.111_show_ip_route_1.txt
-rw-rw-r-- 1 jdoe jdoe 5.9K Nov 26 20:26 192.168.127.111_show_run_1.txt
-rw-rw-r-- 1 jdoe jdoe  323 Nov 26 20:26 192.168.127.222_show_ip_int_bri_1.txt
-rw-rw-r-- 1 jdoe jdoe 1.1K Nov 26 20:26 192.168.127.222_show_ip_route_1.txt
-rw-rw-r-- 1 jdoe jdoe 5.9K Nov 26 20:26 192.168.127.222_show_run_1.txt
-rw-rw-r-- 1 jdoe jdoe 1.1K Nov 26 20:26 make_backup1.py
``` |
```

CHAPTER 9 CISCO IOS-XE UPGRADE TOOLS DEVELOPMENT PART 1

Optionally, if you want to be more thorough with your pre-checks and post-checks, you can capture even more router status information using the following show commands. These commands work across IOS-XE routers and most also can be used in IOS-XE-based switches (see Table 9-1). For any other show commands, refer to Cisco's official documentation.

Table 9-1. Useful Cisco show Commands for Pre-Check and Post-Check

Command	Description
show clock	Displays the current system time and date.
show ver | in uptime	Shows the system's uptime since the last reload.
show ip route	Provides routing table entries.
show version	Offers detailed system information.
show running-config	Displays the current running configuration.
show cdp	Exhibits Cisco Discovery Protocol (CDP) info.
show cdp neighbors	Shows directly connected devices via CDP.
show ip interface brief	Summarizes IP interfaces' status.
show interfaces	Provides detailed information about interfaces.
show run | be line vty	Displays configuration lines related to VTY.
show run | in username	Filters configuration lines containing "username."
show run | in aaa	Filters configuration lines related to AAA.
show run license summary	Summarizes licensing information.
show run license status	Shows the current status of licenses.
show run license all	Displays all licensing information.
show logging	Shows system log messages.

Source: https://wordpress.com/post/italchemy.wordpress.com/5777

Congratulations on completing the sixth and final tool in this chapter! By utilizing a variety of show commands, you've empowered the tool to capture running configurations and operational status. This step is critical as it establishes a benchmark for the most recent stable configuration and device status. With comprehensive backups of crucial devices' configurations, you can swiftly recover from any unexpected failures during a change window, ensuring minimal downtime and efficient troubleshooting. As mentioned, these tools can be used as individual miniature tools, but when you combine them in Chapter 11, you will see the true power of these tools.

Chapter 9 concludes with the introduction of six fundamental tools. The journey doesn't end here; in the upcoming chapter, you'll delve deeper into IOS upgrading tools, exploring functionalities such as IOS uploading and router reloading tools. Stay focused as you continue to enhance your arsenal for managing and optimizing network operations.

Summary

This chapter draws to a close, leaving behind a trail of invaluable tools that converted engineers' manual tasks into logical Python-based tools. Each tool crafted within this chapter had specific uses, enhancing the capabilities of network engineers in the IOS upgrading tasks. From validating network connectivity and collecting user inputs to ensuring the integrity of the new IOS file and checking flash sizes on Cisco Routers, these tools collectively form a robust arsenal. The last Tools Development provided a crucial safeguard through comprehensive backups of running configurations, interface status, and routing tables using Cisco IOS's robust show commands. These tools, specifically designed and discussed, will serve as beacons of efficiency, empowering engineers to navigate the complexities of IOS upgrading with finesse. They lay the groundwork for future endeavors, setting the stage for more advanced tools development and strategies in the remaining chapters. As you bid farewell to Chapter 9, the journey continues. These tools marked only the first half of the IOS upgrading tools development, equipping engineers with the necessary skills and resources to tackle evolving challenges in the ever-expanding landscape of network device OS upgrading and patch management. In the next chapter, you will aim to complete your tools development.

CHAPTER 10

Cisco IOS-XE Upgrade Tools Development: Part 2

This chapter marks the penultimate stage of the IOS upgrade automation journey, focusing on the development of the remaining four crucial applications. These applications serve as the foundational elements that will eventually converge into an end-to-end IOS upgrade application in the subsequent chapter. Throughout this phase, you will carefully write a set of tools with distinct functionalities. First, you will establish a connectivity validation tool, enabling you to verify network connections. Additionally, you will build a tool for an interactive collection of usernames and passwords, along with a file information reading tool. Further, you will write an MD5 check tool specifically tailored for file integrity verification on a Linux server. To complement these, you will develop a configuration backup tool for network devices, an IOS file uploading tool, and a router flash IOS MD5 check tool. In addition to these, you will create a user input tool, which will allow you to alter the application flow as needed. To round it off, you will develop a router reload tool with an accompanying post-check tool. Along the way, you will also learn to write the basic sequential Python code and enhance the code with parallel code using the threading module, which enables your Python code to control multiple devices using multiple processes. Each tool in this chapter has undergone rigorous testing to ensure robustness and to validate their functionalities. These tools lay the foundation for the final application, shaping it into a comprehensive and efficient IOS upgrade application.

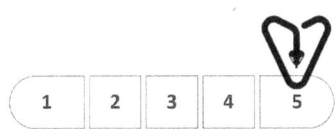

CHAPTER 10   CISCO IOS-XE UPGRADE TOOLS DEVELOPMENT: PART 2

Cisco IOS Upgrade Application Development (Continued)

In this chapter, you continue to delve deeper into writing mini-applications crucial to your IOS upgrading toolset. You will develop applications for uploading IOS files, for verifying new IOS MD5 values on Cisco flash, for providing an option to stop or reload the router upgrade, and for checking and configuring post-upgrade settings. These applications align with the objective of streamlining IOS upgrading tasks. Once you complete these mini-applications, you will see the culmination of your efforts in Chapter 11. You will merge the initial six applications from Chapter 9 with the four from this chapter to create an end-to-end Cisco IOS upgrading application. This robust toolset aims to automate and replace manual IOS upgrading processes typically managed by network engineers. Let's dive into the development, applying your newfound knowledge to create these essential applications.

> **Tip**
>
> **Looking for the source code used in this chapter?**
>
> All the source code featured in this book is available for access and download via the author's GitHub or Apress's GitHub repositories.
>
> Author's GitHub:
>
> https://github.com/pynetauto/apress_pynetauto_ed2.0/tree/main/source_codes/Part2/ch10_tools_dev2
>
> Apress GitHub:
>
> https://github.com/Apress/

Tools Development 7: IOS Uploading and More Pre-check Tools Development

After completing the comprehensive pre-checks and securing the current running configurations, the next step involves uploading the new IOS from a server to the router's flash memory. As previously discussed, the file-transfer process utilizes SCP due to the built-in module in the `netmiko` library, ensuring efficient handling of various file sizes.

CHAPTER 10   CISCO IOS-XE UPGRADE TOOLS DEVELOPMENT: PART 2

Gratitude is owed to Kirk Byers for crafting and sharing such a useful library. While FTP and TFTP methods were functional during testing, they didn't match the reliability of netmiko's SCPConn file-transfer method.

The new IOS file size stands at 771MB, and uploading such a large file to a networking device's flash memory typically takes between 10 to 20 minutes using a reliable production network. Even in a lab environment on a computer, it can take over 15 minutes. During tool development and testing phases, opting for a smaller file size saves considerable time. In this instance, I'm using a 37.1MB file—an old c3745-adventerprisek9-mz.124-25d.bin—renamed c8000v-Dev-File-Only.bin (37.1MB). **This substitution allows swift development without enduring lengthy upload (waiting) times. Using the full 771MB file during development isn't necessary and can lead to time wastage.** When it's time to upload the authentic file and reload the router, this placeholder or the dummy file will be replaced. Note that network speeds vary, influencing the duration of file transfers across different networks. Let's kick off the practical part of this.

#	Task
| ① | In Chapter 9, you copied a new IOS file to the /home/jdoe/ch9_tools_dev1/new_ios/ directory. Now copy the new_ios directory to the /home/jdoe/ch10_tools_dev2 directory. Include the IOS file, because you are going to need this file later. Use the recursive copy command to copy the directory as well as the actual IOS file. <br><br> ```
jdoe@u22s1:~$ mkdir ch10_tools_dev2 && cd ch10_tools_dev2
jdoe@u22s1:~/ch10_tools_dev2$ ls -lh ../ch9_tools_dev1/new_ios/
total 771M
-rw-r--r-- 1 jdoe jdoe 771M Nov 26 10:36 c8000v-universalk9.17.06.05a.SPA.bin
jdoe@u22s1:~/ch10_tools_dev2$ cp -r ../ch9_tools_dev1/new_ios . # use recursive option to copy the whole directory
jdoe@u22s1:~/ch_tools_dev2$ ls
new_ios  tool7_upload_ios
jdoe@u22s1:~/ch_tools_dev2$ ls -lh new_ios
total 771M
-rw-r--r-- 1 jdoe jdoe 771M Nov 27 21:23 c8000v-universalk9.17.06.05a.SPA.bin
``` |

(*continued*)
```

#	Task
②	At this stage, you can simply use this file, but waiting for uploading to complete is a boring task. So, let's copy any small file and use it during the application development phase so you can save time. Here, I made a copy of the old Cisco `c3745-adventerprisek9-mz.124-25d.bin` file and renamed it `c8000v-Dev-File-Only.bin` for development purposes. Use WinSCP to upload the file to the /home/jdoe/ch10_tools_dev2/new_ios directory. You should have two files available from this location to move to the next step. Also, run the `md5sum` Linux command to get ready to write the IOS uploading script. You will need the file name and the MD5 value. ``` jdoe@u22s1:~/ch10_tools_dev2$ ls -lh new_ios total 808M -rw-r--r-- 1 jdoe jdoe 38M Aug 5 08:41 c8000v-Dev-File-Only.bin -rw-r--r-- 1 jdoe jdoe 771M Nov 27 21:23 c8000v-universalk9.17.06.05a.SPA.bin jdoe@u22s1:~/ch10_tools_dev2$ md5sum new_ios/c8000v-Dev-File-Only.bin 563797308a3036337c3dee9b4ab54649 new_ios/c8000v-Dev-File-Only.bin ```
③	It is time to create another working directory for the IOS uploading tool development. This time, copy the script created in the last section of Chapter 9 and modify the code to test the IOS file uploading. Type the Linux commands and follow along. ``` jdoe@u22s1:~/ch10_tools_dev2$ mkdir tool7_upload_ios && cd tool7_upload_ios jdoe@u22s1:~/ch10_tools_dev2/tool7_upload_ios$ cp ../../ch10_tools_dev2/tool6_make_backup/make_backup1.py ./upload_ios1.py # notice the use of ../../ to move up two parent directories jdoe@u22s1:~/ch10_tools_dev2/tool7_upload_ios$ ls -lh total 4.0K -rw-rw-r-- 1 jdoe jdoe 1.1K Nov 27 21:09 upload_ios1.py ```

(continued)

#	Task
④	Before diving into modifying the upload_ios1.py file, let's cover the prerequisites for this code. First, three key configurations are necessary on Cisco routers (or switches) for successful SCP file transfers: the user must possess level 15 privilege, aaa authentication login must be set up, and authorization exec should be preconfigured. In a production environment where device access is controlled by a TACACS server, a slight modification to the script will be necessary. The login process remains the same, but authentication occurs through the TACACS server. However, in environments without TACACS server usage, local aaa authorization and authentication can be leveraged, as demonstrated in this example. The script first checks for level 15 privileges under your username and then validates the setup of aaa authentication and authorization. Additionally, for SCP file transfer to function, the router must act as an SCP server. The script ensures this by verifying if the ip scp server command is configured. If it is not, it enables the SCP service before commencing the IOS file-transfer process, disabling it once the transfer completes. Another critical consideration in a production environment is the SSH idle timeout on line vty 0 15. SCP file transfer uses the same port as SSH, hence if the default ten-minute timer is in use, you might have to adjust this setting if you are transferring a large file that will take more than ten minutes. For the sake of convenience in this book, exec-timeout 0 0 is configured on both routers to enable an infinite SSH idle timeout without session revocation. As this is your development environment, you can relax the security measures a bit to ensure your solutions function smoothly. Let's enhance the upload_ios1.py Python file. After completing your SCP IOS upload application, your code should resemble the following structure. Should you desire additional features, feel free to expand and explore various options. **The beauty of Python scripting lies in its freedom—unlike YAML code in Ansible, Python grants you the role of a designer, mechanic, and driver of your car. In contrast, Ansible limits you to the role of a mere driver.** With Python scripting, the canvas is yours to craft your tools. As always, detailed explanations accompany critical lines of code to aid in comprehension. ``` jdoe@u22s1:~/ch10_tools_dev2/tool7_upload_ios$ nano upload_ios1.py jdoe@u22s1:~/ch10_tools_dev2/tool7_upload_ios$ cat upload_ios1.py # Import necessary libraries ```

(*continued*)

#	Task
| | ```python
import time
from netmiko import ConnectHandler, SCPConn
# File paths and names
source_newios = "/home/jdoe/ch10_tools_dev2/new_ios/c8000v-Dev-File-Only.bin"
destination_newios = "c8000v-Dev-File-Only.bin"

# Device configurations
device1 = {
    'device_type': 'cisco_xe',
    'host': '192.168.127.111',
    'username': 'jdoe',
    'password': 'cisco123',
    'secret': 'cisco123'
}

device2 = {
    'device_type': 'cisco_xe',
    'host': '192.168.127.222',
    'username': 'jdoe',
    'password': 'cisco123',
    'secret': 'cisco123'
}

devices_list = [device1, device2]

# Iterating through devices for configuration and file transfer
for device in devices_list:
    # Assigning necessary variables
    ip = str(device['host'])
    username = str(device['username'])

    # Establishing connection
    net_connect = ConnectHandler(**device)
    net_connect.send_command("terminal length 0")
``` |

(continued)
```

#	Task
| | ```
# Retrieving running configuration
showrun = net_connect.send_command("show running-config")

# Checking privilege and authentication/authorization settings
check_priv15 = (f'username {username} privilege 15')
aaa_authenication = "aaa authentication login default local enable"
aaa_authorization = "aaa authorization exec default local"

# Verifying privilege level and authentication/authorization settings
if check_priv15 in showrun:
    print(f"{username} has level 15 privilege - OK")
    if aaa_authenication in showrun:
        print("check_aaa_authentication - OK")
        if aaa_authorization in showrun:
            print("check_aaa_authorization - OK")
        else:
            print("aaa_authorization - FAILED ")
            exit()
    else:
        print("aaa_authentication - FAILED ")
        exit()
else:
    print(f"{username} has not enough privilege - FAILED")
    exit()

# Enabling SCP server and transferring file
net_connect.enable(cmd='enable 15')
net_connect.config_mode()
net_connect.send_command('ip scp server enable')
net_connect.exit_config_mode()
time.sleep(1)
print("New IOS uploading in progress! Please wait...")
``` |

(continued)
```

#	Task
| | ```
scp_conn = SCPConn(net_connect)
scp_conn.scp_transfer_file(source_newios, destination_newios)
scp_conn.close()
time.sleep(1)

# Disabling SCP server
net_connect.config_mode()
net_connect.send_command('no ip scp server enable')
net_connect.exit_config_mode()
print("-"*79)
``` |
| ⑤ | Note that the jdoe user is correctly configured with level 15 admin privileges, but the aaa new model isn't enabled on the c8kv01 and c8kv02 routers. First, ensure that both of your routers are up and reachable from the Python server. Then, execute your script to identify the error message returned. As you've learned from other experiences in Python, encountering and troubleshooting various errors are fundamental aspects of a coder's drill.
Use fping to check the connectivity to routers.
jdoe@u22s1:~/ch10_tools_dev2/tool7_upload_ios$ **fping 192.168.127.111 192.168.127.222**
192.168.127.111 is alive
192.168.127.222 is alive

Alternatively, use the ping_1.py tool for connectivity testing.
jdoe@u22s1:~/ch10_tools_dev2/tool7_upload_ios$ **python ../../ch10_tools_dev2/tool1_ping/ping_1.py**
Sending ICMP packets to 192.168.127.111
[...output omitted for brevity]
Sending ICMP packets to 192.168.127.222
[...output omitted for brevity]

Now run the upload_ios1.py command. You should expect an error relating to the aaa configuration.
jdoe@u22s1:~/ch10_tools_dev2/tool7_upload_ios$ **python upload_ios1.py**
jdoe has level 15 privilege - OK
aaa_authentication - FAILED |

(continued)
```

#	Task

⑥	Now copy the upload_ios1.py file and create another file to configure aaa configurations on your routers. Because you have only two devices, you could open the console to configure these, but it's better to work on the assumption that there are 200 routers to configure in inventory. After you complete your configuration script, it should look like this:
	
	jdoe@u22s1:~/ch10_tools_dev2/tool7_upload_ios$ **cp upload_ios1.py config_aaa.py**
	jdoe@u22s1:~/ch10_tools_dev2/tool7_upload_ios$ **nano config_aaa.py**
	jdoe@u22s1:~/ch10_tools_dev2/tool7_upload_ios$ **cat config_aaa.py**
	``python
	from netmiko import ConnectHandler
	
	# Device information
	device1 = {
	    'device_type': 'cisco_ios',
	    'host': '192.168.127.111',
	    'username': 'jdoe',
	    'password': 'cisco123',
	    'secret': 'cisco123'
	}
	
	device2 = {
	    'device_type': 'cisco_ios',
	    'host': '192.168.127.222',
	    'username': 'jdoe',
	    'password': 'cisco123',
	    'secret': 'cisco123'
	}
	
	devices_list = [device1, device2]
	
	for device in devices_list:
	    net_connect = ConnectHandler(**device)
	    net_connect.enable()
	``

(continued)

Task

```
# AAA configurations
commands = [
'aaa new-model',
'aaa authentication login default local enable'
]

output = net_connect.send_config_set(commands)
print(output)

# Save configuration
save_output = net_connect.save_config()
print(save_output)

net_connect.disconnect()
```

(7) Note that the netmiko scripts may encounter issues if you have not configured the enable secret cisco123 or enable password cisco123 command in Chapter 8. You must configure the enable password/secret on both routers. This security configuration is often overlooked by many network engineers. To strengthen security measures, it's crucial to configure this setting. In your demonstration lab, you hadn't initially configured the enable secret/password, so I have set them up manually through the c8kv01 and c8kv02 consoles.

On the c8kv01 console:
c8kv01#**conf t**
c8kv01(config)#**enable secret cisco123**

On the c8kv02 console:
c8kv02#**conf t**
c8kv02(config)#**enable secret cisco123**

(continued)

#	Task
| ⑧ | Now back to the u22s1 server. Execute the Python script to add the required aaa configurations and save the file. If your script worked, the output should look like the following.<br><br>```
jdoe@u22s1:~/ch10_tools_dev2/tool7_upload_ios$ python config_aaa.py
configure terminal
Enter configuration commands, one per line.  End with CNTL/Z.
c8kv01(config)#aaa new-model
c8kv01(config)#aaa authentication login default local enable
c8kv01(config)#end
c8kv01#
write mem
Building configuration...
[OK]
c8kv01#
configure terminal
Enter configuration commands, one per line.  End with CNTL/Z.
c8kv02(config)#aaa new-model
c8kv02(config)#aaa authentication login default local enable
c8kv02(config)#end
c8kv02#
write mem
Building configuration...
[OK]
c8kv02#
```<br><br>Alternatively, you can add the following aaa configurations to both routers via SSH or console connection:<br><br>On the c8kv01 console:<br>c8kv01(config)#**aaa new-model**<br>c8kv01(config)#**aaa authentication login default local enable**<br><br>On the c8kv02 console:<br>c8kv02(config)#**aaa new-model**<br>c8kv02(config)#**aaa authentication login default local enable** |

(continued)
```

CHAPTER 10   CISCO IOS-XE UPGRADE TOOLS DEVELOPMENT: PART 2

#	Task
⑨	Now run the IOS upload application one more time and observe where the script stops. Although the admin user, `jdoe`, has a level 15 privilege, the script complains that the user does not have the right privilege because it has now moved to aaa local authentication. **The last aaa configuration,** `aaa authorization exec default local`**, is needed to fix this problem.** `jdoe@u22s1:~/ch10_tools_dev2/tool7_upload_ios$` **python upload_ios1.py** `jdoe has not enough privilege - FAILED`

⑩	To counter this problem, add the last aaa configuration to both routers.
	
	`jdoe@u22s1:~/ch10_tools_dev2/tool7_upload_ios$` **cp config_aaa.py config_aaa_all.py**
	`jdoe@u22s1:~/ch10_tools_dev2/tool7_upload_ios$` **nano config_aaa_all.py**
	`jdoe@u22s1:~/ch10_tools_dev2/tool7_upload_ios$` **cat config_aaa_all.py**
	`[...output omitted for brevity]`
	`for device in devices_list:`
	`    net_connect = ConnectHandler(**device)`
	`    net_connect.enable()`
	
	`    # AAA configurations`
	`    commands = [`
	`        'aaa new-model',`
	`        'aaa authentication login default local enable',`
	`        `**`'aaa authorization exec default local'`**
	`    ]`
	`[...output omitted for brevity]`
	
	You can use the router console, but perhaps it would be easier to add this command to the commands list in your `config_aaa.py` file. I am going to make a copy of this file, name it `config_aaa_all.py`, and then execute the script.

(continued)

#	Task
	jdoe@u22s1:~/ch10_tools_dev2/tool7_upload_ios$ **python config_aaa_all.py** *[...output omitted for brevity]* c8kv01(config)#aaa authorization exec default local *[...output omitted for brevity]* c8kv02(config)#aaa authorization exec default local *[...output omitted for brevity]* Alternatively, you can configure this on each router through a console connection. On the c8kv01 console: c8kv01(config)# **aaa authorization exec default local** On the c8kv02 console: c8kv02(config)# **aaa authorization exec default local**
(11)	Wait! Hold on before running the script. Access the console of each router and enable SCP debugging. This step allows you to monitor and analyze the file transfers through the console. Debugging such traffic is crucial during development to ensure the proper functioning of your application. On the c8kv01 console: c8kv01#**debug ip scp** Incoming SCP debugging is on c8kv01#**terminal monitor** # if connected via SSH connection On the c8kv02 console: c8kv02#**debug ip scp** Incoming SCP debugging is on c8kv02#**terminal monitor** # if connected via SSH connection

(continued)

CHAPTER 10   CISCO IOS-XE UPGRADE TOOLS DEVELOPMENT: PART 2

#	Task
⑫	Now, re-run the upload_ios1.py script. If all the configuration verifications have passed, the transfer of the new IOS file will commence for the first router. Once the file transfer completes on c8kv01, the same script will execute for the second router, c8kv02. This demonstration involves IOS uploading on only two routers, but consider the scenario with hundreds of routers requiring IOS images. Moreover, now you have the flexibility to schedule this script to run outside of business hours, eliminating the need to monitor the screen for extended periods. **This is perhaps the most rewarding aspect of automating mundane and repetitive tasks like these.** The username, password, and secret are printed on the screen during development to check that the developing applications are working as they were designed, but in the production implementations, you must disable any redundant print() statements. As mentioned, the print() function is for you; the computer does not have to output the information to the screen (see Figure 10-1). ``` jdoe@u22s1:~/ch10_tools_dev2/tool7_upload_ios$ python upload_ios1.py ``` ``` jdoe@u22s1:~/ch21/tool7_upload_ios$ python upload_ios1.py jdoe has level 15 privilege - OK check_aaa_authentication - OK check_aaa_authorization - OK New IOS uploading in progress! Please wait... --- jdoe has level 15 privilege - OK check_aaa_authentication - OK check_aaa_authorization - OK New IOS uploading in progress! Please wait... --- jdoe@u22s1:~/ch21/tool7_upload_ios$ python upload_ios1.py jdoe has level 15 privilege - OK check_aaa_authentication - OK check_aaa_authorization - OK New IOS uploading in progress! Please wait... --- jdoe has level 15 privilege - OK check_aaa_authentication - OK check_aaa_authorization - OK New IOS uploading in progress! Please wait... --- ``` *Figure 10-1. upload_ios1.py execution output*

(continued)

CHAPTER 10　CISCO IOS-XE UPGRADE TOOLS DEVELOPMENT: PART 2

#	Task
⑬	While the script is running, check the terminal screens of both routers. If you observe the following on-screen logs, you know that the new IOS file transfer has been completed successfully. Check the uploaded file on both routers and save the configuration to complete the task (see Figure 10-2). *Figure 10-2. c8kv01 and c8kv02 sequential test IOS uploading completed*
⑭	The first IOS upgrade script worked fine, but it uploaded the file one device at a time, in other words, in sequence. The second router had to wait for the first upload to complete before it started to receive the test IOS file. **To enable parallel file uploads, you can use Python's threading module. Python's threading module enables concurrent execution by creating threads, independent sequences of code execution. These threads run concurrently, sharing the same memory space but having their own execution path**. It's used to perform multiple tasks simultaneously, improving efficiency and responsiveness in Python programs. Interestingly, although Ansible is mainly based on Python, it doesn't use Python's threading module directly. Instead, it uses a task-based system through modules like `async` or `async_task` that execute tasks concurrently, but it's not based on Python's native threading. Ansible employs its execution engine for managing parallelism and async operations (see Figure 10-3).

(*continued*)

Task

```python
import time
from netmiko import ConnectHandler, SCPConn

source_newios = "/home/jdoe/ch21/new_ios/c8000v-Dev-File-Or
destination_newios = "c8000v-Dev-File-Only.bin"

device1 = {
    'device_type': 'cisco_xe',
    'host': '192.168.127.111',
    'username': 'jdoe',
    'password': 'cisco123',
    'secret': 'cisco123'
}

device2 = {
    'device_type': 'cisco_xe',
    'host': '192.168.127.222',
    'username': 'jdoe',
    'password': 'cisco123',
    'secret': 'cisco123'
}

devices_list = [device1, device2]

for device in devices_list:
    ip = str(device['host'])
    username = str(device['username'])
    net_connect = ConnectHandler(**device)
    net_connect.send_command("terminal length 0")
    showrun = net_connect.send_command("show running-config
    check_priv15 = (f'username {username} privilege 15')
    aaa_authenication = "aaa authentication login default l
    aaa_authorization = "aaa authorization exec default loc

    if check_priv15 in showrun:
        print(f"{username} has level 15 privilege - OK")
        if aaa_authenication in showrun:
            print("check_aaa_authentication - OK")
            if aaa_authorization in showrun:
                print("check_aaa_authorization - OK")
            else:
                print("aaa_authorization - FAILED ")
                exit()
        else:
            print("aaa_authentication - FAILED ")
            exit()
    else:
        print(f"{username} has not enough privilege - FAILE
        exit()

    net_connect.enable(cmd='enable 15')
    net_connect.config_mode()
    net_connect.send_command('ip scp server enable')
    net_connect.exit_config_mode()
    time.sleep(1)
    print("New IOS uploading in progress! Please wait...")
    scp_conn = SCPConn(net_connect)
    scp_conn.scp_transfer_file(source_newios, destination_r
    scp_conn.close()
    time.sleep(1)
    net_connect.config_mode()
    net_connect.send_command('no ip scp server enable')
    net_connect.exit_config_mode()

    print("-"*80)
```

```python
import time
from netmiko import ConnectHandler, SCPConn
import threading

source_newios = "/home/jdoe/ch21/new_ios/c8000v-Dev-File-Or
destination_newios = "c8000v-Dev-File-Only.bin"

device1 = {
    'device_type': 'cisco_xe',
    'host': '192.168.127.111',
    'username': 'jdoe',
    'password': 'cisco123',
    'secret': 'cisco123'
}

device2 = {
    'device_type': 'cisco_xe',
    'host': '192.168.127.222',
    'username': 'jdoe',
    'password': 'cisco123',
    'secret': 'cisco123'
}

devices_list = [device1, device2]

def upload_file(device):
    ip = str(device['host'])
    username = str(device['username'])
    net_connect = ConnectHandler(**device)
    net_connect.send_command("terminal length 0")
    showrun = net_connect.send_command("show running-config
    check_priv15 = (f'username {username} privilege 15')
    aaa_authenication = "aaa authentication login default l
    aaa_authorization = "aaa authorization exec default loc

    if check_priv15 in showrun:
        print(f"{username} has level 15 privilege - OK")
        if aaa_authenication in showrun:
            print("check_aaa_authentication - OK")
            if aaa_authorization in showrun:
                print("check_aaa_authorization - OK")
            else:
                print("aaa_authorization - FAILED ")
                exit()
        else:
            print("aaa_authentication - FAILED ")
            exit()
    else:
        print(f"{username} has not enough privilege - FAILE
        exit()

    net_connect.enable(cmd='enable 15')
    net_connect.config_mode()
    net_connect.send_command('ip scp server enable')
    net_connect.exit_config_mode()
    time.sleep(1)
    print(f"New IOS uploading in progress to {ip}! Please w
    scp_conn = SCPConn(net_connect)
    scp_conn.scp_transfer_file(source_newios, destination_r
    scp_conn.close()
    time.sleep(1)
    net_connect.config_mode()
    net_connect.send_command('no ip scp server enable')
    net_connect.exit_config_mode()
    print(f"Upload to {ip} completed.")
    print("-"*79)

# Create threads for each device and start uploading files
threads = []
for device in devices_list:
    thread = threading.Thread(target=upload_file, args=(dev
    thread.start()
    threads.append(thread)

# Wait for all threads to complete before proceeding
for thread in threads:
    thread.join()
```

Figure 10-3. *upload_ios1.py vs upload_ios2_parallel.py compared on Notepad++*

(*continued*)

#	Task

Putting the first and second IOS upload Python scripts side by side, as in Figure 10-3, most of the code is nearly identical, apart from the additional code added for threading at the end. You will copy the last script and create the parallel IOS uploading script, which uploads the files in parallel using the Python's threading module. You only need to modify a few lines of code, making the main uploading script into a function, and then add the threading using script. Time to execute and start applying.

```
jdoe@u22s1:~/ch10_tools_dev2/tool7_upload_ios$ cp upload_ios1.py upload_ios2_parallel.py
jdoe@u22s1:~/ch10_tools_dev2/tool7_upload_ios$ nano upload_ios2_parallel.py
jdoe@u22s1:~/ch10_tools_dev2/tool7_upload_ios$ cat upload_ios2_parallel.py
import time
from netmiko import ConnectHandler, SCPConn
import threading

source_newios = "/home/jdoe/ch10_tools_dev2/new_ios/c8000v-Dev-File-Only.bin"
destination_newios = "c8000v-Dev-File-Only.bin"

device1 = {
    'device_type': 'cisco_xe',
    'host': '192.168.127.111',
    'username': 'jdoe',
    'password': 'cisco123',
    'secret': 'cisco123'
}

device2 = {
    'device_type': 'cisco_xe',
    'host': '192.168.127.222',
    'username': 'jdoe',
    'password': 'cisco123',
    'secret': 'cisco123'
}
```

(*continued*)

#	Task
| | ```
devices_list = [device1, device2]
def upload_file(device):
    ip = str(device['host'])
    username = str(device['username'])
    net_connect = ConnectHandler(**device)
    net_connect.send_command("terminal length 0")
    showrun = net_connect.send_command("show running-config")
    check_priv15 = (f'username {username} privilege 15')
    aaa_authenication = "aaa authentication login default local
    enable"
    aaa_authorization = "aaa authorization exec default local"

    if check_priv15 in showrun:
        print(f"{username} has level 15 privilege - OK")
        if aaa_authenication in showrun:
            print("check_aaa_authentication - OK")
            if aaa_authorization in showrun:
                print("check_aaa_authorization - OK")
            else:
                print("aaa_authorization - FAILED ")
                exit()
        else:
            print("aaa_authentication - FAILED ")
            exit()
    else:
        print(f"{username} has not enough privilege - FAILED")
        exit()

    net_connect.enable(cmd='enable 15')
    net_connect.config_mode()
    net_connect.send_command('ip scp server enable')
    net_connect.exit_config_mode()
    time.sleep(1)
``` |

(continued)
```

#	Task

```
        print(f"New IOS uploading in progress to {ip}! Please wait...")
        scp_conn = SCPConn(net_connect)
        scp_conn.scp_transfer_file(source_newios, destination_newios)
        scp_conn.close()
        time.sleep(1)
        net_connect.config_mode()
        net_connect.send_command('no ip scp server enable')
        net_connect.exit_config_mode()
        print(f"Upload to {ip} completed.")
        print("-"*79)

    # Create threads for each device and start uploading files in parallel
    threads = []
    for device in devices_list:
        thread = threading.Thread(target=upload_file, args=(device,))
        thread.start()
        threads.append(thread)

    # Wait for all threads to complete before proceeding
    for thread in threads:
        thread.join()

    print("All file uploads completed.")
```

(continued)

#	Task
⑮	Now execute the new Python script. It will use two separate processes to upload the testing IOS file in parallel. Now you can upload the same file to multiple devices with fewer delays and save time, thus increasing efficiency. jdoe@u22s1:~/ch10_tools_dev2/tool7_upload_ios$ **python upload_ios2_parallel.py** jdoe has level 15 privilege - OK check_aaa_authentication - OK check_aaa_authorization - OK jdoe has level 15 privilege - OK check_aaa_authentication - OK check_aaa_authorization - OK New IOS uploading in progress to 192.168.127.222! Please wait... New IOS uploading in progress to 192.168.127.111! Please wait... Upload to 192.168.127.222 completed. -- Upload to 192.168.127.111 completed. -- All file uploads completed.

(*continued*)

#	Task
⑯	When you observe the debugging consoles, you can see that the uploading takes place almost at the same time. There is a slight delay in the file trigger, but it is almost in parallel (see Figure 10-4).

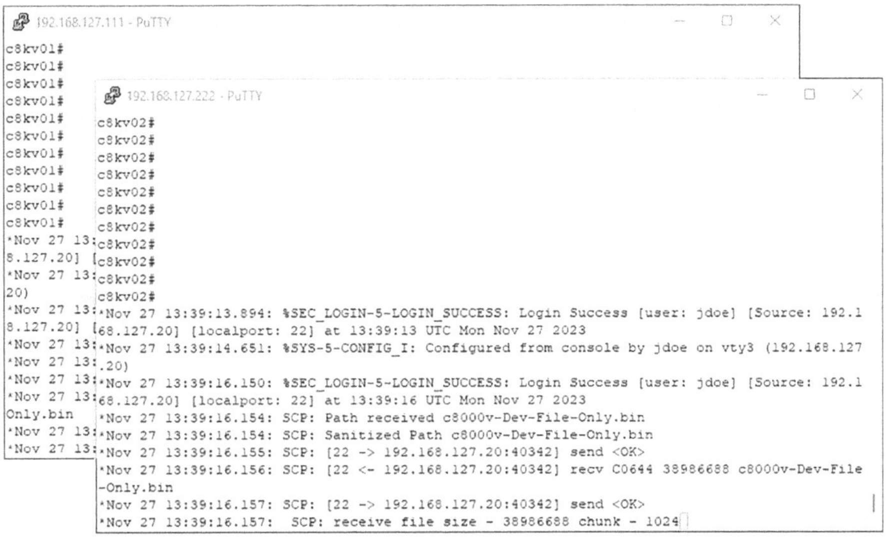

Figure 10-4. c8kv01 and c8kv02 parallel test IOS uploading in progress

You have completed writing Tool 7, learning to build a Cisco IOS file-uploading tool with SCP on Cisco routers. The initial script operated sequentially, using a single thread for all tasks, which limits concurrent actions. In the revamped version, you transformed it into a parallel file upload application, enhancing efficiency and reducing upload time significantly. Python development often involves code refactoring to optimize performance and streamline execution for smoother operations. Next, you'll learn to develop a tool to check the MD5 value of the uploaded IOS file on the router flash.

Tools Development 8: Check the New IOS MD5 Value on the Cisco Device's Flash

In the previous development phase of the Cisco IOS uploading tool, you successfully uploaded the temporary IOS version for DevOps to the c8kv01 and c8kv02 routers. The next step is to create a script that validates the `verify IOS_file` command and confirms the MD5 value of the IOS copy stored in the router's flash memory. For this iteration, the script will execute only if the actual IOS file is located in the `flash:/` root directory. It will proceed to run the verification command and check for an expected `'Verified'` output using a Regular Expression to confirm the successful IOS verification. Upon success, the script will display the MD5 values of both the original and the new IOS files in the flash memory. Additionally, you'll aim to ensure the application continues running even if the first router is unreachable on the network. To address this, you will incorporate a `netmiko` timeout exception.

Although the script should persist in its execution upon encountering an exception until it finishes processing the last device on the list, **it's crucial to acknowledge that while Python threading significantly boosts efficiency, there are specific scenarios where employing a single thread for sequential tasks is more beneficial. This is particularly true when dealing with mission-critical network and security devices. Taking a cautious, step-by-step approach helps mitigate the risk of simultaneous disruptions to multiple services across various devices. In certain cases, a slower but more deliberate approach proves more beneficial than giving precedence to speed.**

In the domain of network automation, a predominant focus lies in reducing the time required for specific tasks as resource time is often considered an overhead. However, there are rare instances where precision holds far more significance than the speed of the automation process. These two concepts might seem contradictory, but determining the speed and logic of your application must align with the objectives you aim to achieve in your automation task.

Let's create and test the next application in question.

#	Task
①	As always, begin by creating a new directory and then create your new Python script. `jdoe@u22s1:~/ch10_tools_dev2/tool7_upload_ios$` **`cd ..`** `jdoe@u22s1:~/ch10_tools_dev2$` **`mkdir tool8_md5_cisco && cd tool8_md5_cisco`**

(continued)

#	Task
| ② | Much of this script mirrors the previous one. Let's swiftly refactor the code to validate the command and compare the MD5 provided by the user with the MD5 calculated on the router. The application's flow will pivot on the outcome of these two MD5 value comparisons. Your script, upon completing the IOS MD5 checker application for Cisco IOS routers, should resemble the following. Remember, your script need not match this one precisely, but it should be optimized and functional. Writing the code line by line is beneficial practice, offering insight into the professional coding experience. If you relish handling every aspect independently, then you'll find pleasure in crafting lines of code.<br><br>```
jdoe@u22s1:~/ch10_tools_dev2/tool8_md5_cisco$ nano md5_verify1.py
jdoe@u22s1:~/ch10_tools_dev2/tool8_md5_cisco$ cat md5_verify1.py
from netmiko import ConnectHandler, SCPConn
import threading
import time
import re
from netmiko import NetMikoTimeoutException

# Verification parameters
destination_newios = "c8000v-Dev-File-Only.bin" # From development 7
newiosmd5 = "563797308a3036337c3dee9b4ab54649" # From development 7

# Device configurations
device1 = {
    'device_type': 'cisco_xe',
    'host': '192.168.127.111',
    'username': 'jdoe',
    'password': 'cisco123',
    'secret': 'cisco123',
    'global_delay_factor': 2, # Run netmiko commands twice slower
    'read_timeout_override': 90, # This is important & required for a big output commands
}
``` |

(continued)
```

CHAPTER 10   CISCO IOS-XE UPGRADE TOOLS DEVELOPMENT: PART 2

#	Task

```python
device2 = {
    'device_type': 'cisco_xe',
    'host': '192.168.127.222',
    'username': 'jdoe',
    'password': 'cisco123',
    'secret': 'cisco123',
    'global_delay_factor': 2,
    'read_timeout_override': 90, # This is important & required for a
    big output commands
}

devices_list = [device1, device2]

def upload_file(device):
    ip = str(device['host'])
    try:
        net_connect = ConnectHandler(**device)
        net_connect.send_command("terminal length 0")
        locate_newios = net_connect.send_command(f"show flash: | in
        {destination_newios}")

        # If new IOS is found on the router's flash, run this script
        if destination_newios in locate_newios:
            # Cisco IOS/IOS XE verify command, run and assign variable
            result to the output
            result = net_connect.send_command(f"verify /md5
            flash:{destination_newios} {newiosmd5}", read_timeout=120)
            print(f"Connecting to {ip}...") # Print the connected
            device host IP
            print(result)
            net_connect.disconnect()
            p1 = re.compile(r'Verified') # Regular Expression (re)
            compiler for word 'Verified'
            p2 = re.compile(r'[a-fA-F0-9]{31}[a-fA-F0-9]') # re
            compiler for MD5 value
```

(continued)

#	Task
| | ```
            verified = p1.findall(result) # If 'Verified' was found in
            result, run this part of the script
            newiosmd5flash = p2.findall(result)

            if verified:
                result = True
                print(f"Connecting to {ip}...") # Print the connected
                device host IP
                print("MD5 values MATCH! Continue")
                print("MD5 of new IOS on Server : ", newiosmd5)
                print("MD5 of new IOS on flash  : ",
                newiosmd5flash[0])
                print("-" * 79)
            else: # If 'Verified' was not found in result, print and
            exit the application
                result = False
                print(f"Connecting to {ip}...") # Print the connected
                device host IP
                print("MD5 values DO NOT MATCH! Exiting.")
                print("-" * 79)
                exit()
        else: # If no new IOS file was found on the router's flash:/.
        Print the statement
            print(f"Connecting to {ip}...") # Print the connected device
            host IP
            print("No new IOS found on router's flash. Continue to next
            device...")
            print("-" * 79)
    except NetMikoTimeoutException as e: # Handle Timeout error due to
    network issue
        print(f'Timeout error to : {ip}')
        print("-" * 79)
``` |

(continued)
```

#	Task
| | ```
        except Exception as unknown_error: # Handle other errors as
                                              exception
            print(f'Unknown error occurred: {unknown_error}')

            print("-" * 79)

    def perform_verification_and_upload(devices):
        threads = []
        for device in devices:
            thread = threading.Thread(target=upload_file, args=(device,))
            thread.start()
            threads.append(thread)

    for thread in threads:
    thread.join()
        print("All file uploads and verifications completed.")

    perform_verification_and_upload(devices_list)
``` |
| ③ | Make sure that both routers are powered on and in normal operational mode. Run the application to check the MD5 value of the newly uploaded IOS image files stored in flash: of c8kv01 and c8kv02. If everything worked correctly, you should get the same output as shown here. Some of the output has been truncated to save space.<br><br>```
jdoe@u22s1:~/ch10_tools_dev2/tool8_md5_cisco$ python md5_verify1.py
Connecting to 192.168.127.222...
... truncated ...
.......................................................Done!
Verified (bootflash:c8000v-Dev-File-Only.bin) =
563797308a3036337c3dee9b4ab54649

Connecting to 192.168.127.111...
... truncated ...
.......................................................Done!
Verified (bootflash:c8000v-Dev-File-Only.bin) =
563797308a3036337c3dee9b4ab54649
``` |

(continued)

#	Task
	Connecting to 192.168.127.222... MD5 values MATCH! Continue MD5 of new IOS on Server : 563797308a3036337c3dee9b4ab54649 MD5 of new IOS on flash : 563797308a3036337c3dee9b4ab54649 -- Connecting to 192.168.127.111... MD5 values MATCH! Continue MD5 of new IOS on Server : 563797308a3036337c3dee9b4ab54649 MD5 of new IOS on flash : 563797308a3036337c3dee9b4ab54649 -- All file uploads and verifications completed. **WARNING**! The file used for this testing is only 37.1MB. However, if you replace it with the real IOS-XE file, which is more than 700+MB, you are likely to run into "Unknown error occurred: Pattern not detected:" error. This is a known issue with netmiko and users will encounter this problem when running a command that takes a long time, such as the show tech and verify /md5 commands. The workaround is to add read_timeout=90 to the command or overwrite the Netmiko ConnectHandler value at the device level, using read_timeout_override': 90. I am adding these to the scripts, but this is not shown here. Refer to this URL for further reading about this netmiko issue: https://pynet.twb-tech.com/blog/netmiko-read-timeout.html

(continued)

CHAPTER 10   CISCO IOS-XE UPGRADE TOOLS DEVELOPMENT: PART 2

#	Task
④	To simulate a scenario where c8kv01 is not on the network (it's unreachable), disable GigabitEthernet1. You have to disable the port via VMware Workstation's main console (see Figure 10-5). c8kv01#**configure terminal** c8kv01(config)#**interface GigabitEthernet1** c8kv01(config-if)#**shutdown** 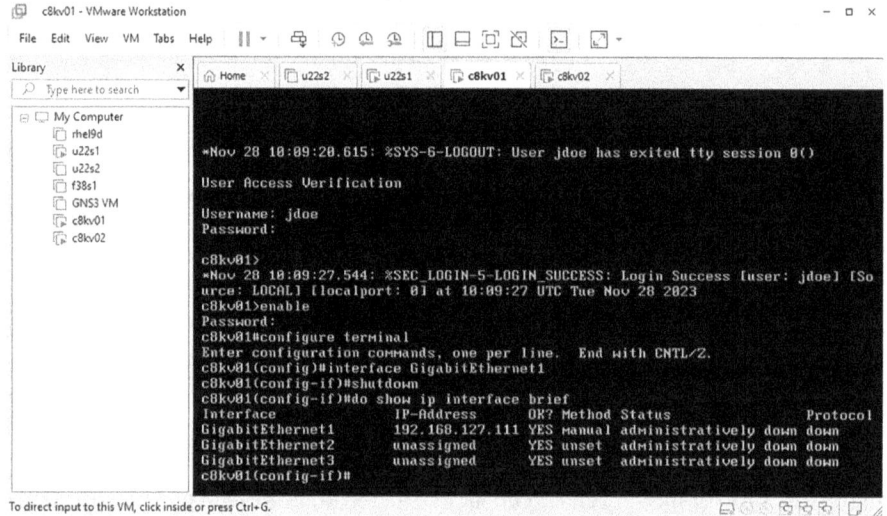 *Figure 10-5. The shutdown interface GigabitEthernet1 on c8kv01 on VMware console*
⑤	Now re-run the Python script to check the result. Because you have already added an exception to counter this problem, the script will continue to run until the end. If the exception handler was not added, you will see the following errors: jdoe@u22s1:~/ch10_tools_dev2/tool8_md5_cisco$ **python md5_verify1.py** Connecting to 192.168.127.222... ... truncated ... .Done! Verified (bootflash:c8000v-Dev-File-Only.bin) = 563797308a3036337 c3dee9b4ab54649

(continued)

#	Task
	``
Connecting to 192.168.127.222...
MD5 values MATCH! Continue
MD5 of new IOS on Server :   563797308a3036337c3dee9b4ab54649
MD5 of new IOS on flash :   563797308a3036337c3dee9b4ab54649

Timeout error to : 192.168.127.111

All file uploads and verifications completed.
```
As you can see, the Python script handled the connection error nicely. Note that, when troubleshooting connection-related issues, it's crucial to consider various common root causes that might lead to connectivity problems. These issues often stem from factors such as incorrect configurations of hostnames or IP addresses, incorrect usage of TCP ports during connection attempts, or potential hindrances caused by intermediate firewalls blocking access. Addressing these factors is vital to ensure seamless connectivity. Specifically, in the context of device settings, there is a `cisco_xe` device located at the IP address 192.168.127.111 utilizing port 22 for communication, where meticulous attention to configuration accuracy and firewall permissions is essential to maintain a reliable connection. |

(*continued*)
```

CHAPTER 10   CISCO IOS-XE UPGRADE TOOLS DEVELOPMENT: PART 2

#	Task
⑥	To simulate a scenario with a missing IOS file, delete the Dev-File-Only IOS file from c8kv01. Afterward, you will unshut the GigabitEthernet1 port on the router and verify that your script operates smoothly. Ensure that c8kv01 is connected to the network before executing your script in the next step. You can streamline this process by continuing to work from the VMware Workstation console (see Figure 10-6). c8kv01#**delete flash:c8000v-Dev-File-Only.bin** Delete filename [c8000v-Dev-File-Only.bin]? Delete bootflash:/c8000v-Dev-File-Only.bin? [confirm] c8kv01#**configure terminal** c8kv01(config)#**interface GigabitEthernet1** c8kv01(config-if)#**no shutdown**

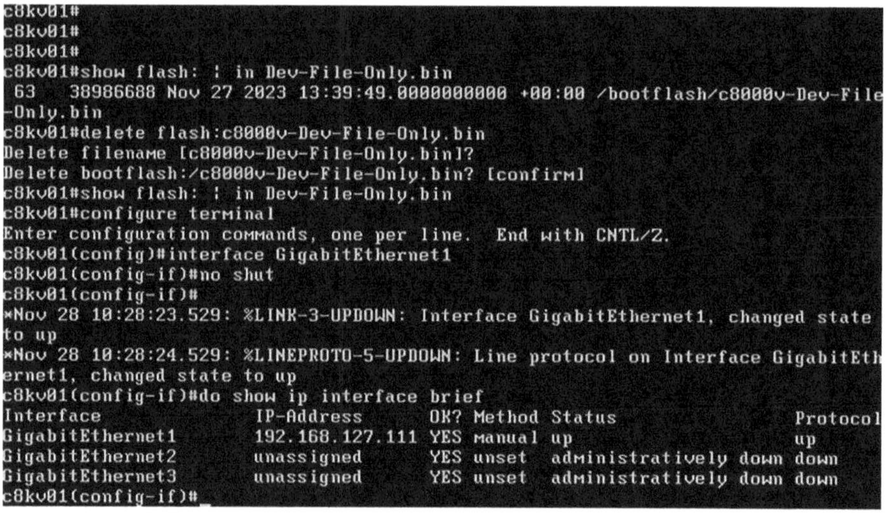

Figure 10-6. *Delete the IOS file and bring up the GigabitEthernet1 interface on c8kv01 on VMware console*

(*continued*)

#	Task
| ⑦ | Re-run the script. It should run smoothly and persist until completion. Your Python script should gracefully handle any missing file errors, ensuring they don't halt the script's progress or become a showstopper.<br><br>```
jdoe@u22s1:~/ch10_tools_dev2/tool8_md5_cisco$ python md5_verify1.py
Connecting to 192.168.127.222...
... truncated ...
........................................................Done!
Verified (bootflash:c8000v-Dev-File-Only.bin) = 563797308a3036337c3dee9b4ab54649
Connecting to 192.168.127.222...
MD5 values MATCH! Continue
MD5 of new IOS on Server :   563797308a3036337c3dee9b4ab54649
MD5 of new IOS on flash  :   563797308a3036337c3dee9b4ab54649
-----------------------------------------------------------------------
Connecting to 192.168.127.111...
No new IOS found on router's flash. Continue to next device...
-----------------------------------------------------------------------
All file uploads and verifications completed.
``` |

(*continued*)
```

Task

(8) To prepare for the next lab, upload the real IOS file by modifying the IOS upload script.

```
jdoe@u22s1:~/ch10_tools_dev2$ ls new_ios
c8000v-Dev-File-Only.bin   c8000v-universalk9.17.06.05a.SPA.bin
jdoe@u22s1:~/ch10_tools_dev2$ md5sum new_ios/c8000v-universalk9.17.06.05a.SPA.bin
13f0161a50210f2f21618fc59c5f5343  new_ios/c8000v-universalk9.17.06.05a.SPA.bin
jdoe@u22s1:~/ch10_tools_dev2$ cd tool7_upload_ios/
jdoe@u22s1:~/ch10_tools_dev2/tool7_upload_ios$ ls
config_aaa.py  config_aaa_all.py  upload_ios1.py  upload_ios1_tmp.py
upload_ios2_parallel.py
jdoe@u22s1:~/ch10_tools_dev2/tool7_upload_ios$ cp upload_ios2_parallel.py upload_ios3_parallel.py
jdoe@u22s1:~/ch10_tools_dev2/tool7_upload_ios$ nano upload_ios3_parallel.py
jdoe@u22s1:~/ch10_tools_dev2/tool7_upload_ios$ cat upload_ios3_parallel.py
```

[...omitted for brevity]
```
source_newios = "/home/jdoe/ch10_tools_dev2/new_ios/c8000v-universalk9.17.06.05a.SPA.bin"
destination_newios = "c8000v-universalk9.17.06.05a.SPA.bin"
```
[...omitted for brevity]

You only need to change the file names to the genuine IOS-XE file name for Cisco c8000v.

(*continued*)

CHAPTER 10 CISCO IOS-XE UPGRADE TOOLS DEVELOPMENT: PART 2

#	Task
⑨	Optionally, enable debug `ip scp` on both routers (see Figure 10-7).

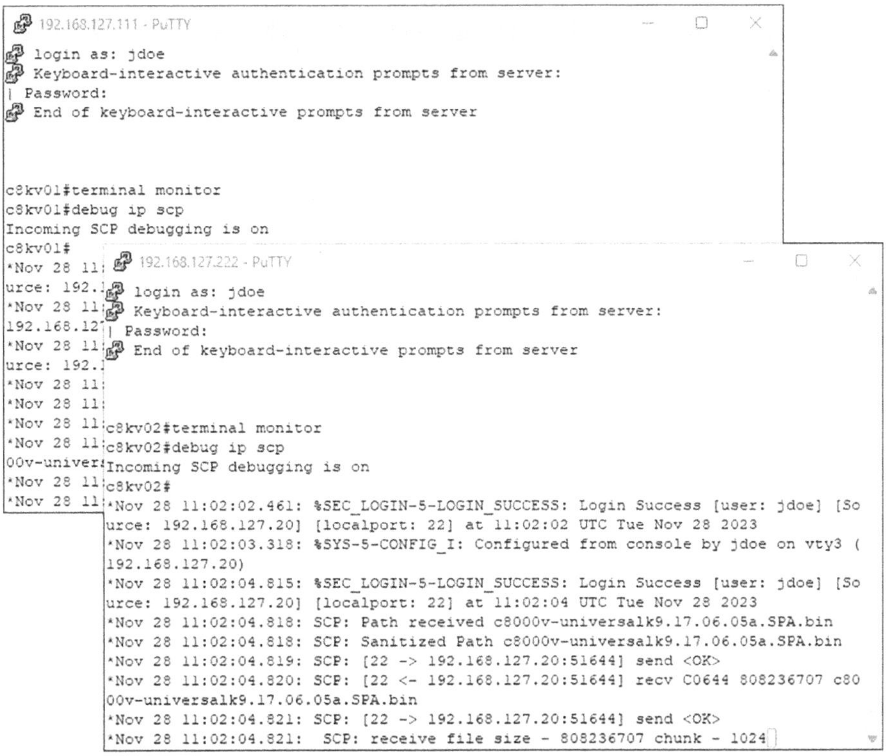

```
192.168.127.111 - PuTTY                                    −  □  ×
login as: jdoe
Keyboard-interactive authentication prompts from server:
Password:
End of keyboard-interactive prompts from server


c8kv01#terminal monitor
c8kv01#debug ip scp
Incoming SCP debugging is on
c8kv01#
*Nov 28 11:
urce: 192.1    192.168.127.222 - PuTTY                    −  □  ×
*Nov 28 11:   login as: jdoe
192.168.12    Keyboard-interactive authentication prompts from server:
*Nov 28 11:   Password:
urce: 192.1   End of keyboard-interactive prompts from server
*Nov 28 11:
*Nov 28 11:
*Nov 28 11: c8kv02#terminal monitor
*Nov 28 11: c8kv02#debug ip scp
00v-univers Incoming SCP debugging is on
*Nov 28 11: c8kv02#
*Nov 28 11: *Nov 28 11:02:02.461: %SEC_LOGIN-5-LOGIN_SUCCESS: Login Success [user: jdoe] [So
            urce: 192.168.127.20] [localport: 22] at 11:02:02 UTC Tue Nov 28 2023
            *Nov 28 11:02:03.318: %SYS-5-CONFIG_I: Configured from console by jdoe on vty3 (
            192.168.127.20)
            *Nov 28 11:02:04.815: %SEC_LOGIN-5-LOGIN_SUCCESS: Login Success [user: jdoe] [So
            urce: 192.168.127.20] [localport: 22] at 11:02:04 UTC Tue Nov 28 2023
            *Nov 28 11:02:04.818: SCP: Path received c8000v-universalk9.17.06.05a.SPA.bin
            *Nov 28 11:02:04.818: SCP: Sanitized Path c8000v-universalk9.17.06.05a.SPA.bin
            *Nov 28 11:02:04.819: SCP: [22 -> 192.168.127.20:51644] send <OK>
            *Nov 28 11:02:04.820: SCP: [22 <- 192.168.127.20:51644] recv C0644 808236707 c80
            00v-universalk9.17.06.05a.SPA.bin
            *Nov 28 11:02:04.821: SCP: [22 -> 192.168.127.20:51644] send <OK>
            *Nov 28 11:02:04.821:  SCP: receive file size - 808236707 chunk - 1024
```

Figure 10-7. c8kv01 and ckv02 SSH PuTTY session, debug ip scp enabled example

(*continued*)

CHAPTER 10 CISCO IOS-XE UPGRADE TOOLS DEVELOPMENT: PART 2

#	Task
⑩	Run the python upload_ios3_parallel.py command from the u22s1 server. jdoe@u22s1:~/ch10_tools_dev2/tool7_upload_ios$ **python upload_ios3_parallel.py** jdoe has level 15 privilege - OK check_aaa_authentication - OK check_aaa_authorization - OK jdoe has level 15 privilege - OK check_aaa_authentication - OK check_aaa_authorization - OK New IOS uploading in progress to 192.168.127.222! Please wait... New IOS uploading in progress to 192.168.127.111! Please wait...
⑪	From your router console, run the dir or show flash: command to check that the correct IOS file is being uploaded to both routers in parallel (see Figure 10-8).

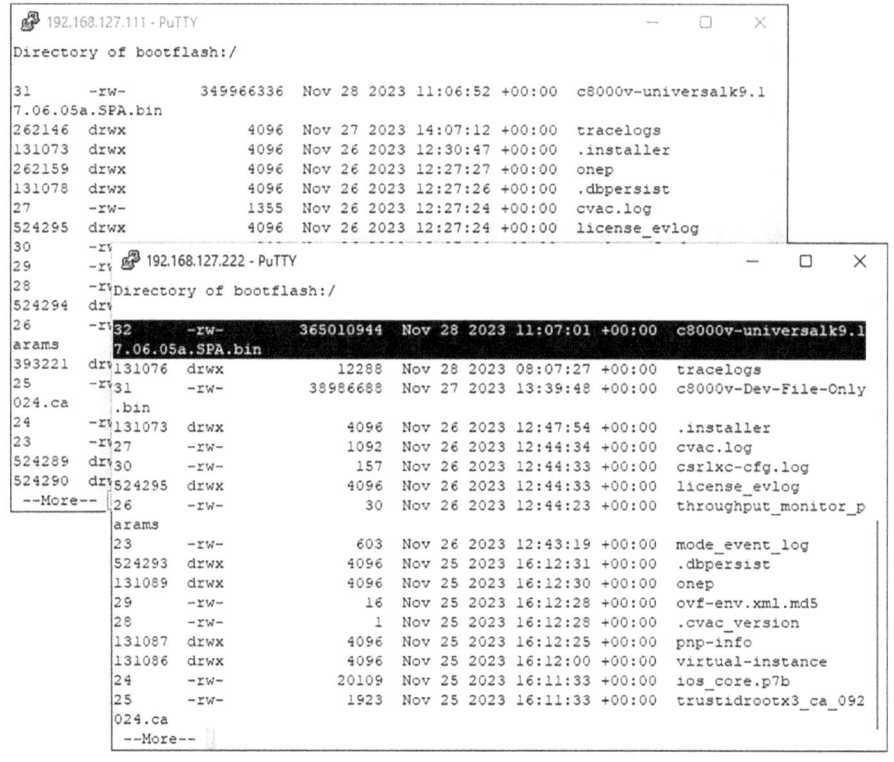

Figure 10-8. *Upload genuine c8000v IOS .bin file to c8kv01 and c8kv02*

(continued)

#	Task
⑫	When you receive the following messages from the Python file on u22s1, your file upload has been completed successfully and you are ready for the next development lab. ``` jdoe@u22s1:~/ch10_tools_dev2/tool7_upload_ios$ python upload_ios3_parallel.py [...omitted for brevity] Upload to 192.168.127.222 completed. --- Upload to 192.168.127.111 completed. --- All file uploads completed. ```

During the development of your Python tools, it's crucial to repeatedly test your scripts to verify the repeatability of automated tasks and ensure consistent expected outcomes. In the event of a failure, it's essential to thoroughly investigate and pinpoint the root cause, and then correct the problem accordingly. This will prevent any exposure of software bugs or shortcomings of your work in the production network.

Tools Development 9: Options to Stop or Reload the Routers

In this lab, your task is to create a script offering users two options: one to modify the boot system, back up the running configuration, and perform a reload, and another to exit the application for a reload later. To execute the first option, you'll need to code the process for altering the boot system, saving the changes, and optionally creating a backup of the running configuration for future validation of the modifications.

CHAPTER 10   CISCO IOS-XE UPGRADE TOOLS DEVELOPMENT: PART 2

#	Task
①	Navigate to the ch10_tools_dev2 directory, create a new folder specifically for this section, and generate a new base script to begin your development. jdoe@u22s1:~/ch10_tools_dev2/tool8_md5_cisco$ **cd ..** jdoe@u22s1:~/ch10_tools_dev2$ **mkdir tool9_yes_no && cd tool9_yes_no**
| ② | In this iteration, you'll write a clean yes/no function that anticipates input in the format of yes/y or no/n. If the user inputs anything else, the yes_or_no function will re-prompt them until it receives the correct input. This approach maintains command over the input, effectively guiding the script's progression. A yes/y response initiates the configuration adjustment of routers followed by a reload. Conversely, choosing no/n exits the application, postponing the reload action for a later time.<br><br>jdoe@u22s1:~/ch10_tools_dev2/tool9_yes_no$ **nano yes_no1.py**<br>jdoe@u22s1:~/ch10_tools_dev2/tool9_yes_no$ **cat yes_no1.py**<br>```python
yes = ['yes', 'y']
no = ['no', 'n']
def yes_or_no():
    resp = input("Would you like to reload your devices? (y/n)? ")
    .lower()
    if resp in yes:
        print("YES")
    elif resp in no:
        print("NO")
    else:
        yes_or_no()

yes_or_no()

print("All tasks completed.")
``` |

(continued)
```

#	Task
③	After completing the initial Yes or No script, execute the script and perform a test, by following these steps: ``` jdoe@u22s1:~/ch10_tools_dev2/tool9_yes_no$ python yes_no1.py Would you like to reload your devices? (y/n)? y YES All tasks completed. jdoe@u22s1:~/ch10_tools_dev2/tool9_yes_no$ python yes_no1.py Would you like to reload your devices? (y/n)? yes YES All tasks completed. jdoe@u22s1:~/ch10_tools_dev2/tool9_yes_no$ python yes_no1.py Would you like to reload your devices? (y/n)? N NO All tasks completed. jdoe@u22s1:~/ch10_tools_dev2/tool9_yes_no$ python yes_no1.py Would you like to reload your devices? (y/n)? NO NO All tasks completed. jdoe@u22s1:~/ch10_tools_dev2/tool9_yes_no$ python yes_no1.py Would you like to reload your devices? (y/n)? sure Would you like to reload your devices? (y/n)? why not? Would you like to reload your devices? (y/n)? 12345 Would you like to reload your devices? (y/n)? OK Would you like to reload your devices? (y/n)? YeS YES All tasks completed. ```

(*continued*)

CHAPTER 10   CISCO IOS-XE UPGRADE TOOLS DEVELOPMENT: PART 2

#	Task

④	Optionally, you could write the initial tool, as demonstrated next, and it appears well-structured. However, crafting complex code over simplicity might lead to challenges. While this code stands alone as a decent tool, integrating it revealed an excessive indentation depth, breaching the Pythonic guidelines laid out in the PEP 8 style guide. Hence, the goal should focus on creating tools that suit the scenario while adhering to the PEP 8 style guide.
	
	``
	jdoe@u22s1:~/ch10_tools_dev2/tool9_yes_no$ cp yes_no1.py yes_no2.py
	jdoe@u22s1:~/ch10_tools_dev2/tool9_yes_no$ nano yes_no2.py
	jdoe@u22s1:~/ch10_tools_dev2/tool9_yes_no$ cat yes_no2.py
	def yes_or_no():
	    # Dictionary to map user responses to boolean values
	    valid_responses = {'yes': True, 'y': True, 'no': False, 'n': False}
	
	    while True: # Loop until a valid response is received
	        resp = input("Would you like to reload your devices? (y/n)? ")
	            .lower()
	        # Check if the response is in the valid_responses dictionary
	        if resp in valid_responses:
	            return valid_responses[resp] # Return the corresponding
	            boolean value
	        print("Please enter 'yes' or 'no'.")
	
	# Call the function to get the user's choice
	user_choice = yes_or_no()
	
	# Output based on the user's choice
	if user_choice:
	    print("You have chosen to reload your devices.")
	else:
	    print("You chose not to reload the devices.")
	
	# Completion message
	print("All tasks completed.")
	``

(continued)

CHAPTER 10   CISCO IOS-XE UPGRADE TOOLS DEVELOPMENT: PART 2

#	Task
⑤	Run the Yes or No application several times, providing your responses. Verify that the tool functions precisely as you've designed it to. This application version will not be used for your integration—the first version will be used. However, if you want to check out this version in the integration, download the file called `reload_yes_no2.py` for your reference. ``` jdoe@u22s1:~/ch10_tools_dev2/tool9_yes_no$ python yes_no2.py Would you like to reload your devices? (y/n)? N You chose not to reload the devices. All tasks completed. jdoe@u22s1:~/ch10_tools_dev2/tool9_yes_no$ python yes_no2.py Would you like to reload your devices? (y/n)? NO You chose not to reload the devices. All tasks completed. jdoe@u22s1:~/ch10_tools_dev2/tool9_yes_no$ python yes_no2.py Would you like to reload your devices? (y/n)? Kangaroo Please enter 'yes' or 'no'. Would you like to reload your devices? (y/n)? Sure Please enter 'yes' or 'no'. Would you like to reload your devices? (y/n)? 54321 Please enter 'yes' or 'no'. Would you like to reload your devices? (y/n)? OK Please enter 'yes' or 'no'. Would you like to reload your devices? (y/n)? Yes You have chosen to reload your devices. All tasks completed. jdoe@u22s1:~/ch10_tools_dev2/tool9_yes_no$ python yes_no2.py Would you like to reload your devices? (y/n)? y You have chosen to reload your devices. All tasks completed. ```

(continued)

CHAPTER 10   CISCO IOS-XE UPGRADE TOOLS DEVELOPMENT: PART 2

#	Task
⑥	It's time to leverage one of your base scripts for this scenario. You will utilize the yes_no1.py script as a foundation to develop your application for reloading. Begin by making a copy and naming it reload_yes_no1.py. This script will verify the presence of the new IOS file in the router's flash memory by comparing it with the show dir output. If the new IOS is detected, it will proceed to modify the boot system statement on the device, save the configuration, execute specific show commands to capture the pre-reload operational status, and ultimately reload the router. Conversely, if no matching IOS file name is found in the flash, the application will encounter an error and exit.

```
jdoe@u22s1:~/ch10_tools_dev2/tool9_yes_no$ cp yes_no1.py reload_yes_no1.py
jdoe@u22s1:~/ch10_tools_dev2/tool9_yes_no$ nano reload_yes_no1.py
jdoe@u22s1:~/ch10_tools_dev2/tool9_yes_no$ cat reload_yes_no1.py
from netmiko import ConnectHandler
import time

# Device information
destination_newios = "c8000v-universalk9.17.06.05a.SPA.bin"
# newiosmd5 = "563797308a3036337c3dee9b4ab54649" # Placeholder only,
md5 check tool

device1 = {
    'device_type': 'cisco_xe',
    'host': '192.168.127.111',
    'username': 'jdoe',
    'password': 'cisco123',
    'secret': 'cisco123',
    'global_delay_factor': 2  # Used to slow down the script
}

device2 = {
    'device_type': 'cisco_xe',
    'host': '192.168.127.222',
    'username': 'jdoe',
    'password': 'cisco123',
```

(continued)

```python
    'secret': 'cisco123',
    'global_delay_factor': 2
}

devices_list = [device1, device2]
yes_list = ['yes', 'y']
no_list = ['no', 'n']

# Function to reload a specific device, check if the destination
exists and follow the logic.
def reload_device(device):
    ip = str(device['host'])
    net_connect = ConnectHandler(**device)
    net_connect.enable(cmd='enable 15')
    config_commands1 = ['no boot system', 'boot system flash:/' +
    destination_newios, 'do write memory']
    output = net_connect.send_config_set(config_commands1)
    print(output)
    net_connect.send_command('terminal length 0\n')
    show_boot = net_connect.send_command('show boot\n')
    show_dir = net_connect.send_command('dir\n')
    if destination_newios not in show_dir:
        print(f'Unable to locate new IOS on the flash: {ip}. Exiting.')
        print("-" * 79)
        raise Exception("IOS not found")
    elif destination_newios not in show_boot:
        print(f'Boot system was not correctly configured on {ip}. Exiting.')
        print("-" * 79)
        raise Exception("Boot system misconfiguration")
    elif destination_newios in show_boot and destination_newios in show_dir:
    print(f'Found {destination_newios} in show boot on {ip}')
    print("-" * 79)
    net_connect.send_command("terminal length 0")
```

(continued)

#	Task

```
            time.sleep(1)
            with open(f'{ip}_showver_pre.txt', 'w+') as f1:
                print(f"Capturing pre-reload 'show version' on {ip}")
                showver_pre = net_connect.send_command("show version")
                f1.write(showver_pre)
                time.sleep(1)
            with open(f'{ip}_showrun_pre.txt', 'w+') as f2:
                print(f"Capturing pre-reload 'show running-config' on {ip}")
                showrun_pre = net_connect.send_command("show running-config")
                f2.write(showrun_pre)
                time.sleep(1)
            with open(f'{ip}_showint_pre.txt', 'w+') as f3:
                print(f"Capturing pre-reload 'show ip interface brief' on
                {ip}")
                showint_pre = net_connect.send_command("show ip interface
                brief")
                f3.write(showint_pre)
    time.sleep(1)
            with open(f'{ip}_showroute_pre.txt', 'w+') as f4:
                print(f"Capturing pre-reload 'show ip route' on {ip}")
                showroute_pre = net_connect.send_command("show ip route")
                f4.write(showroute_pre)
                time.sleep(1)
            # ... Add more devices as needed

            print("-" * 79)
            # Trigger the device reload
            print(f"Your device {ip} is now reloading.")
            net_connect.send_command('reload', expect_string='[confirm]')
            net_connect.send_command('yes\n')
            net_connect.send_command('\n')
            net_connect.disconnect()
            print("-" * 79)
```

(continued)

#	Task
| | ```
# Function to prompt user for reload decision
def yes_or_no():
    resp = input("Would you like to reload your devices? (y/n)? ")
    .lower()
    if resp in yes_list:
        print("Reloading devices")
        for device in devices_list:
            try:
                reload_device(device)
            except Exception as e:
                print(f"Exception occurred for {device['host']}: {e}")
    elif resp in no_list:
        print("You have chosen to reload the devices later. Exiting
        the application.")
    else:
        yes_or_no()

# Initiating the script
yes_or_no()
print("All tasks completed.")
``` |

(continued)
```

#	Task
⑦	After completing the script, run it for verification purposes. You'll observe the capture of the show commands, and the modification of the boot system variables, followed by the execution of a reload command. However, this script might cause a socket error. To handle this, you can utilize except Exception as e to capture the socket error, allowing the script to proceed and loop through to the second device, c8kv02. This sequential or linear execution is deliberate, used to slow down the process, thereby causing an outage on a single device at a time, particularly when initially implementing this script. ``` jdoe@u22s1:~/ch10_tools_dev2/tool9_yes_no$ python reload_yes_no1.py Would you like to reload your devices? (y/n)? y Reloading devices configure terminal Enter configuration commands, one per line. End with CNTL/Z. c8kv01(config)#no boot system c8kv01(config)#boot system flash:/c8000v-universalk9.17.06.05a.SPA.bin c8kv01(config)#do write memory Building configuration... [OK] c8kv01(config)#end c8kv01# Found c8000v-universalk9.17.06.05a.SPA.bin in show boot on 192.168.127.111 --- Capturing pre-reload 'show version' on 192.168.127.111 Capturing pre-reload 'show running-config' on 192.168.127.111 Capturing pre-reload 'show ip interface brief' on 192.168.127.111 Capturing pre-reload 'show ip route' on 192.168.127.111 --- Your device 192.168.127.111 is now reloading. Exception occurred for 192.168.127.111: Socket is closed [...output omitted for brevity, 192.168.127.222's output is same with the IP ending with .222] Exception occurred for 192.168.127.222: Socket is closed ```

(continued)

#	Task
	In the first edition of this book, accepting the end-user license agreement was necessary to perform reloading on the router platform. However, with the c8000v platform, no such requirement was found. It's advisable to verify any additional requirements by consulting the official vendor documentation.
⑧	While your routers are reloading, run the `ls -lh 192.*` command to confirm the show commands in your working folder. ``` jdoe@u22s1:~/ch10_tools_dev2/tool9_yes_no$ ls -lh 192.* -rw-rw-r-- 1 jdoe jdoe 323 Dec 2 10:39 192.168.127.111_showint_pre.txt -rw-rw-r-- 1 jdoe jdoe 1.1K Dec 2 10:39 192.168.127.111_showroute_pre.txt -rw-rw-r-- 1 jdoe jdoe 6.1K Dec 2 10:39 192.168.127.111_showrun_pre.txt -rw-rw-r-- 1 jdoe jdoe 2.5K Dec 2 10:39 192.168.127.111_showver_pre.txt -rw-rw-r-- 1 jdoe jdoe 323 Dec 2 10:39 192.168.127.222_showint_pre.txt -rw-rw-r-- 1 jdoe jdoe 1.1K Dec 2 10:39 192.168.127.222_showroute_pre.txt -rw-rw-r-- 1 jdoe jdoe 6.1K Dec 2 10:39 192.168.127.222_showrun_pre.txt -rw-rw-r-- 1 jdoe jdoe 2.5K Dec 2 10:39 192.168.127.222_showver_pre.txt ```

(*continued*)

#	Task
| ⑨ | This was a sequential or linear application utilizing a single process. Now, let's use the threading module to execute the reloading jobs in parallel. This approach allows your script to utilize multiple threads, saving time when performing this task in a large batch.<br><br>jdoe@u22s1:~/ch10_tools_dev2/tool9_yes_no$ **cp reload_yes_no1.py reload_yes_no2.py**<br>jdoe@u22s1:~/ch10_tools_dev2/tool9_yes_no$ **nano reload_yes_no2.py**<br>jdoe@u22s1:~/ch10_tools_dev2/tool9_yes_no$ **cat reload_yes_no2.py**<br>```python
from netmiko import ConnectHandler
import threading
import time

# List of devices to perform operations on
devices = [
    {
    'device_type': 'cisco_ios',
    'host': '192.168.127.111',
    'username': 'jdoe',
    'password': 'cisco123',
    'secret': 'cisco123',
    'global_delay_factor': 2,
    'timeout' : 0,
    },
    {
    'device_type': 'cisco_ios',
    'host': '192.168.127.222',
    'username': 'jdoe',
    'password': 'cisco123',
    'secret': 'cisco123',
    'global_delay_factor': 2,
    'timeout' : 0,
    },
    # ... Add more devices as needed
    ]
``` |

(*continued*)
```

#	Task
| | ```python
# Function to capture show command outputs before device reload
def show_and_capture(device):
    ip = str(device['host'])
    try:
        net_connect = ConnectHandler(**device)
        net_connect.enable()
        # Capturing 'show' commands before reload
        with open(f'{ip}_showver_pre.txt', 'w+') as f1:
            showver_pre = net_connect.send_command("show version")
            f1.write(showver_pre)
            time.sleep(1)
        with open(f'{ip}_showrun_pre.txt', 'w+') as f2:
            showrun_pre = net_connect.send_command("show running-
            config")
            f2.write(showrun_pre)
            time.sleep(1)
        with open(f'{ip}_showint_pre.txt', 'w+') as f3:
            showint_pre = net_connect.send_command("show ip interface
            brief")
            f3.write(showint_pre)
            time.sleep(1)
        with open(f'{ip}_showroute_pre.txt', 'w+') as f4:
            showroute_pre = net_connect.send_command("show ip route")
            f4.write(showroute_pre)
            time.sleep(1)
        net_connect.disconnect()
        # ... Add more show commands to capture as needed
    except Exception as e:
        print(f"Connection error with {device['host']}: {e}")
``` |

(continued)
```

#	Task
| | ```python
# Function to reload an individual device
def reload_device(device):
    try:
        net_connect = ConnectHandler(**device)
        net_connect.enable()
        # Reload the device
        output = net_connect.send_command_timing("reload", strip_
        prompt=False, strip_command=False)
        if "confirm" in output:
            output += net_connect.send_command_timing("\n", strip_
            prompt=False, strip_command=False)
        if "Reload scheduled" in output:
            print(f"Reload command successful for {device['host']}")
        else:
            print(f"Reload command failed for {device['host']}")
        net_connect.disconnect()
    except Exception as e:
        print(f"Connection error with {device['host']}: {e}")

# Function to display the current time on a device
def show_clock(device):
    try:
        net_connect = ConnectHandler(**device)
        net_connect.enable()
        # Display the current time
        output = net_connect.send_command("show clock")
        print(f"Time on {device['host']}:")
        print(output)
        net_connect.disconnect()
    except Exception as e:
        print(f"Connection error with {device['host']}: {e}")
``` |

(*continued*)
```

#	Task
| | ```python
# Function to reload the devices
def reload_devices(devices):

    threads = []
    # Capture 'show' commands before reloading devices
    for device in devices:
        print(f"{device['host']} running show commands and capture")
        thread = threading.Thread(target=show_and_capture,
        args=(device,))
    thread.start()
    threads.append(thread)
    for thread in threads:
        thread.join()

    threads = []
    # Show current time on device before the reload
    for device in devices:
        print(f"{device['host']} Showing time after routers has been
        reloaded")
        thread = threading.Thread(target=show_clock, args=(device,))
        thread.start()
        threads.append(thread)
    for thread in threads:
        thread.join()

threads = []
# Reload devices
for device in devices:
print(f"{device['host']} Reloading the device")
thread = threading.Thread(target=reload_device, args=(device,))
thread.start()
threads.append(thread)
``` |

(continued)
```

#	Task
| | ```
for thread in threads:
thread.join()
time.sleep(300)   # Sleep for 5 minutes, until the routers reload and
normalize

threads = []
# Show current time on device after the reload
for device in devices:
print(f"{device['host']} Showing time after router has been reloaded")
thread = threading.Thread(target=show_clock, args=(device,))
thread.start()
threads.append(thread)
for thread in threads:
thread.join()
# Main function to initiate operations based on user input
def main():
yes = ['yes', 'y']
no = ['no', 'n']
resp = input("Would you like to reload your devices? (y/n): ").lower()
if resp in yes:
reload_devices(devices)
elif resp in no:
print("You chose not to reload the devices.")
else:
print("Invalid input. Please enter 'y' or 'n'.")
main()
if __name__ == "__main__":
main()
print("All tasks completed.")
``` |

(continued)
```

#	Task
⑩	Once you have written the reload_yes_no2.py script, execute the command. Immediately after executing the application, make sure you open the VMware Workstation consoles for c8kv01 and c8kv02 and observe the full reloading process. If the application runs smoothly, it will reload your devices in parallel using the threading method (see Figure 10-9). ``` jdoe@u22s1:~/ch10_tools_dev2/tool9_yes_no$ python reload_yes_no2.py Would you like to reload your devices? (y/n): y [...omitted for brevity] Time on 192.168.127.111: *13:50:25.811 UTC Sat Dec 2 2023 Time on 192.168.127.222: *13:50:25.733 UTC Sat Dec 2 2023 192.168.127.111 Reloading the device 192.168.127.222 Reloading the device [...omitted for brevity] ``` You can look at the netmiko session_log or debug log for more information. ``` # Here your script will go to sleep for 5 minutes or 300 seconds, you have to adjust this timer depending on your environment. [...Wait for 300 seconds, until you see the next lines] 192.168.127.111 Showing time after router has been reloaded 192.168.127.222 Showing time after router has been reloaded Time on 192.168.127.111: *13:56:36.275 UTC Sat Dec 2 2023 Time on 192.168.127.222: *13:56:36.234 UTC Sat Dec 2 2023 All tasks completed. ```

(continued)

CHAPTER 10 CISCO IOS-XE UPGRADE TOOLS DEVELOPMENT: PART 2

#	Task

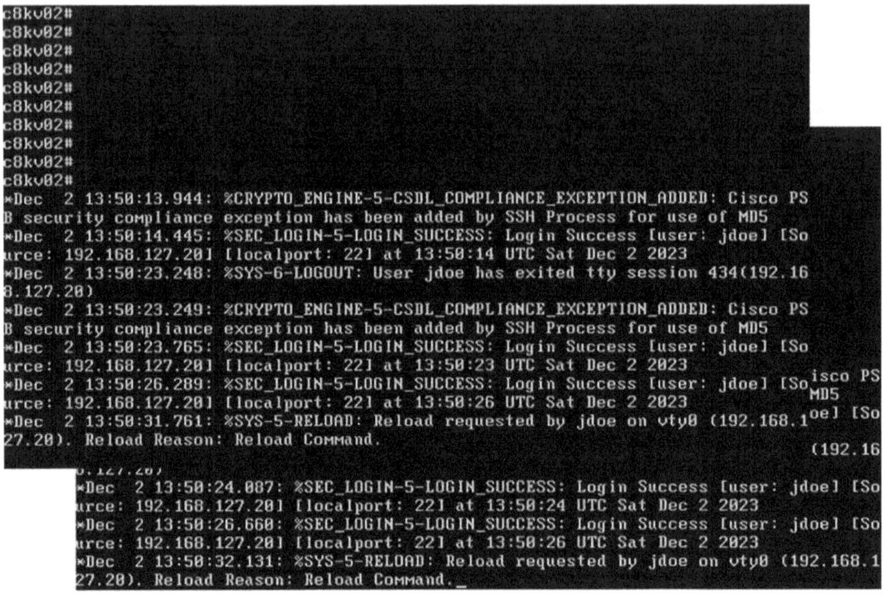

Figure 10-9. c8kv01 and c8kv02 reloading devices in parallel console logs

⑪ You might wonder how I came up with the 300-second value. The c8000v took approximately 3 minutes and 22 seconds in my lab environment (see Figure 10-10).

time.sleep(300) # Sleep for 5 minutes, until the routers reload and normalize

(*continued*)

CHAPTER 10   CISCO IOS-XE UPGRADE TOOLS DEVELOPMENT: PART 2

#	Task
	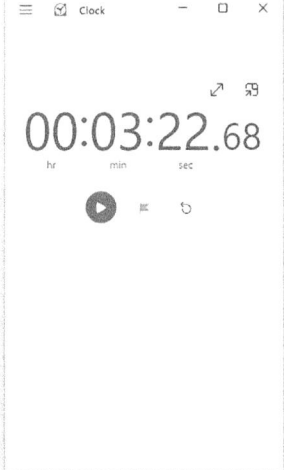

Figure 10-10. Router reload time manual measure

Also, I added an extra timer in the script to measure the end-to-end script run. You need to consider giving the routers enough time to stabilize and operate normally. Looking at the two time intervals of c8kv02 at the time the reload command was issued and the time after it fully came up and another show clock was run, it took about 6 minutes 11 seconds. Every environment and every platform will differ, and your test runs will require different times. Just ensure that you give enough time for the network devices to come up and stabilize.

```
Reload command execution time on c8kv02
Time on 192.168.127.222:
*13:50:25.733 UTC Sat Dec 2 2023
Time on 192.168.127.222:
*13:56:36.234 UTC Sat Dec 2 2023
```

You have completed the development of the Yes or No application, enabling the reloading of your devices. This single application required a significant amount of Python code, incorporating various networking and Python concepts, automating and replacing your manual tasks into lines of code. Ensure you obtain the desired outputs from your test runs. You've explored both sequential and parallel approaches to reloading the routers. Choosing between a more efficient yet powerful tool and a slower but more stable one often presents a programmer's dilemma. However, **with the task of enterprise networking device OS upgrade and patch management, speed** isn't usually the primary concern. **Many enterprise network devices are aggregated devices, serving tens or hundreds of IP services and connecting numerous IP endpoints, IP applications, and users. Therefore, a slower yet more sure-footed and reliable approach is typically preferred by most network administrators and businesses in the domain of enterprise network automation.** The speed and efficiency of an application vary across platforms and technologies, depending on the IT infrastructure domains. As highlighted in earlier chapters, each customer environment is unique and demands a different set of priorities.

Now that both routers have been reloaded with the newest IOS XE images, the final touch involves logging back into the devices and performing a post-upgrade check using the captured information. Proceed to the next section.

Tools Development 10: Check the Reloading Device and Perform a Post-Reload Configuration Verification

You have arrived at the final tool you will develop in Chapter 10. This tool will teach you how to scan and verify that port 22 remains open after your script initiates a router reload. Upon detecting an open port 22, the application will SSH into the device and execute a post-reload capture of running configurations. The pre-reload capture can be compared to leveraging Python's `difflib` library. You can now proceed with the step-by-step process to develop this final tool.

#	Task
①	To keep things simple, copy the previous script and modify it to create the last mini-script for this chapter. `jdoe@u22s1:~/ch10_tools_dev2/tool9_yes_no$` **cd ..** `jdoe@u22s1:~/ch10_tools_dev2$` **mkdir tool10_post_check && cd tool10_post_check** `jdoe@u22s1:~/ch10_tools_dev2/tool10_post_check$` **nano post_check1.py**

(continued)

#	Task
| ② | You will be creating a socket application to monitor port 22. Upon detecting an open port 22, it will initiate a login to that device and execute the show clock command. If the port remains closed, the application will wait ten seconds before rechecking. This check will repeat 60 times, running for ten minutes, aligning with the router reboot and the boot process into the new IOS XE image. In real-world scenarios, this value must be adjusted based on network speed and the actual hardware CPU processor speed, requiring estimation for the wait time. This exercise assumes that failure to restore service on port 22 after the reload command might necessitate troubleshooting.<br><br>jdoe@u22s1:~/ch10_tools_dev2/tool10_post_check$ **cat post_check1.py**<br>```
import socket
import time
from netmiko import ConnectHandler

t1 = time.mktime(time.localtime()) # Timer start to measure script running time

device1 = {
    'device_type': 'cisco_xe',
    'host': '192.168.127.111',
    'username': 'jdoe',
    'password': 'cisco123',
    'secret': 'cisco123',
    'global_delay_factor': 2
}
device2 = {
    'device_type': 'cisco_xe',
    'host': '192.168.127.222',
    'username': 'jdoe',
    'password': 'cisco123',
    'secret': 'cisco123',
'global_delay_factor': 2
}
devices_list = [device1, device2]
``` |

(*continued*)
```

#	Task
| | ```
# Function to check if the port is open
def isOpen(ip, port):
    s = socket.socket(socket.AF_INET, socket.SOCK_STREAM)
    s.settimeout(3)
    try:
        s.connect((ip, int(port)))
        s.shutdown(socket.SHUT_RDWR)
        return True
    except:
        return False
    finally:
        s.close()

for device in devices_list:
    ip = str(device['host'])
    port = 22
    retry = 60
    delay = 10
    t1 = time.mktime(time.localtime())
    ipup = False
    for i in range(retry):
        if isOpen(ip, port):
            ipup = True
            print(f"{ip} is online. Logging into device to perform post
            reload check")
            net_connect = ConnectHandler(**device)
            print(net_connect.send_command("show clock"))
            break
        else:
            print("The device is still reloading. Please wait...")
            time.sleep(delay)
    t2 = time.mktime(time.localtime()) - t1
    print("Total wait time : {0} seconds".format(t2))
``` |

(continued)
```

#	Task
③	When you run the previous script, it will check for port 22, and if port 22 is opened on the target device, it should SSH into the router and run the show clock command. Then it will break out of the for loop and move to the next device, so your result should look like this output. Now you know that the port 22 post-checkers are working fine, so you are ready to write your full post-check application. jdoe@u22s1:~/ch10_tools_dev2/tool10_post_check$ **python post_check1.py** 192.168.127.111 is online. Logging into device to perform post reload check *14:33:23.424 UTC Sat Dec 2 2023 Total wait time : 3.0 seconds 192.168.127.222 is online. Logging into device to perform post reload check *14:33:25.597 UTC Sat Dec 2 2023 Total wait time : 2.0 seconds
④	From the VMware Workstation user interface, go to c8kv01's console and enable an access list to block port 22 to test the application. Enable access-list 100 to block traffic on port 22 on c8kv01. Username: **jdoe** Password: ******** c8kv01> **enable** Password: ******** c8kv01#**configure terminal** c8kv01(config)#**access-list 100 deny tcp any any eq 22** c8kv01(config)#**access-list 100 permit ip any any** c8kv01(config)#**interface GigabitEthernet 1** c8kv01(config-if)#**ip access-group 100 in**

(continued)

CHAPTER 10   CISCO IOS-XE UPGRADE TOOLS DEVELOPMENT: PART 2

#	Task
⑤	Use the following Python command to run the base post-check script. Because you blocked port 22 to simulate an unreachable device, it will return the "Device is still reloading. Please wait..." message every ten seconds. jdoe@u22s1:~/ch10_tools_dev2/tool10_post_check$ **python post_check1.py** The device is still reloading. Please wait... The device is still reloading. Please wait... The device is still reloading. Please wait... [...omitted for brevity]
⑥	On VMware Workstation's main console, log in to c8kv01 and remove access-group 100 from the GigabitEthernet1 interface; this will allow the script to detect when port 22 is open for SSH. c8kv01#**configure terminal** c8kv01(config)#**interface GigabitEthernet 1** c8kv01(config-if)# **no ip access-group 100 in** As soon as the access list is removed from interface GigabitEthernet1, your script will detect an open port 22, log in to the router, and run the show clock command. It then continues to the second router and completes the loop. Device is still reloading. Please wait... Device is still reloading. Please wait... Device is still reloading. Please wait... [...omitted for brevity] Device is still reloading. Please wait... 192.168.127.111 is online. Logging into device to perform post reload check *14:42:55.365 UTC Sat Dec 2 2023 Total wait time : 153.0 seconds 192.168.127.222 is online. Logging into device to perform post reload check *14:42:57.557 UTC Sat Dec 2 2023 Total wait time : 2.0 seconds

(continued)

#	Task
⑦	To test the previous application, you are going to need the files created from the previous development. Copy all of the pre-reload show files from the tool9 development. ``` jdoe@u22s1:~/ch10_tools_dev2/tool10_post_check$ pwd /home/jdoe/ch10_tools_dev2/tool10_post_check jdoe@u22s1:~/ch10_tools_dev2/tool10_post_check$ ls -lh ../tool9_yes_no/192* -rw-rw-r-- 1 jdoe jdoe 323 Dec 2 13:50 ../tool9_yes_no/192.168.127.111_showint_pre.txt -rw-rw-r-- 1 jdoe jdoe 1.1K Dec 2 13:50 ../tool9_yes_no/192.168.127.111_showroute_pre.txt -rw-rw-r-- 1 jdoe jdoe 6.1K Dec 2 13:50 ../tool9_yes_no/192.168.127.111_showrun_pre.txt -rw-rw-r-- 1 jdoe jdoe 2.5K Dec 2 13:50 ../tool9_yes_no/192.168.127.111_showver_pre.txt -rw-rw-r-- 1 jdoe jdoe 323 Dec 2 13:50 ../tool9_yes_no/192.168.127.222_showint_pre.txt -rw-rw-r-- 1 jdoe jdoe 1.1K Dec 2 13:50 ../tool9_yes_no/192.168.127.222_showroute_pre.txt -rw-rw-r-- 1 jdoe jdoe 6.1K Dec 2 13:50 ../tool9_yes_no/192.168.127.222_showrun_pre.txt -rw-rw-r-- 1 jdoe jdoe 2.5K Dec 2 13:50 ../tool9_yes_no/192.168.127.222_showver_pre.txt jdoe@u22s1:~/ch10_tools_dev2/tool10_post_check$ cp ../tool9_yes_no/192* ./ jdoe@u22s1:~/ch10_tools_dev2/tool10_post_check$ ls -lh total 44K -rw-rw-r-- 1 jdoe jdoe 323 Dec 2 14:49 192.168.127.111_showint_pre.txt -rw-rw-r-- 1 jdoe jdoe 1.1K Dec 2 14:49 192.168.127.111_showroute_pre.txt -rw-rw-r-- 1 jdoe jdoe 6.1K Dec 2 14:49 192.168.127.111_showrun_pre.txt -rw-rw-r-- 1 jdoe jdoe 2.5K Dec 2 14:49 192.168.127.111_showver_pre.txt -rw-rw-r-- 1 jdoe jdoe 323 Dec 2 14:49 192.168.127.222_showint_pre.txt ```

(continued)

#	Task
| | ```
-rw-rw-r-- 1 jdoe jdoe 1.1K Dec  2 14:49 192.168.127.222_showroute_pre.txt
-rw-rw-r-- 1 jdoe jdoe 6.1K Dec  2 14:49 192.168.127.222_showrun_pre.txt
-rw-rw-r-- 1 jdoe jdoe 2.5K Dec  2 14:49 192.168.127.222_shower_pre.txt
-rw-rw-r-- 1 jdoe jdoe 1.5K Dec  2 14:33 post_check1.py
``` |
| ⑧ | Make a copy of the first Python file and name it post_check2.py. Use this as the base for your next iteration.

jdoe@u22s1:~/ch10_tools_dev2/tool10_post_check$ **cp post_check1.py post_check2.py** |
| ⑨ | Write the following code and complete the application. This application will check port 22 and then log in to each router and capture the four show commands from each router. Then, using Python's difflib library, you can compare the pre- and post-capture configurations line by line. Next, you save each comparison set in HTML format for easy viewing. Read and study each line of code carefully; the explanations are embedded in the following code for your reference:<br><br>jdoe@u22s1:~/ch10_tools_dev2/tool10_post_check$ **nano post_check2.py**<br>```
from netmiko import ConnectHandler
import socket
import time
import difflib

destination_newios = "c8000v-universalk9.17.06.05a.SPA.bin"

device1 = {
    'device_type': 'cisco_xe',
    'host': '192.168.127.111',
    'username': 'jdoe',
    'password': 'cisco123',
    'secret': 'cisco123',
    'global_delay_factor': 2
}
``` |

(continued)

#	Task
| | ```
device2 = {
    'device_type': 'cisco_xe',
    'host': '192.168.127.222',
    'username': 'jdoe',
    'password': 'cisco123',
    'secret': 'cisco123',
    'global_delay_factor': 2
}

devices_list = [device1, device2]

# Checks SSH port and then logs back in to complete the post-upgrade
check.
def post_check():
    for device in devices_list:
        ip = str(device['host'])
        port = 22
        retry = 120
        delay = 10

        def isOpen(ip, port):
            s = socket.socket(socket.AF_INET, socket.SOCK_STREAM)
            s.settimeout(3)
            try:
                s.connect((ip, int(port)))
                s.shutdown(socket.SHUT_RDWR)
                return True
            except:
                return False
            finally:
                s.close()

        t1 = time.mktime(time.localtime())
        ipup = False
``` |

(continued)
```

#	Task
| | ```
for i in range(retry):
    if isOpen(ip, port):
        ipup = True
        print(f"{ip} is online. Logging into device to perform
        post reload check")
        # Capture four show commands result from each router
        print("Performing post upgrade check.")
        net_connect = ConnectHandler(**device)
        net_connect.enable(cmd='enable 15')
        config_commands1 = ['no boot system', 'boot system
        flash:/' + destination_newios, 'do write memory']
        output = net_connect.send_config_set(config_commands1)
        print(output)
        net_connect.send_command('terminal length 0\n')

        with open(f'{ip}_showver_post.txt', 'w+') as f1:
            print("Capturing post-reload 'show version'")
            showver_post = net_connect.send_command("show
            version")
            f1.write(showver_post)
            time.sleep(1)

        with open(f'{ip}_showrun_post.txt', 'w+') as f2:
            print("Capturing post-reload 'show running-
            config'")
            showrun_post = net_connect.send_command("show
            running-config")
            f2.write(showrun_post)
            time.sleep(1)
``` |

(*continued*)
```

#	Task

```python
        with open(f'{ip}_showint_post.txt', 'w+') as f3:
            print("Capturing post-reload 'show ip interface
            brief'")
            showint_post = net_connect.send_command("show ip
            interface brief")
            f3.write(showint_post)
            time.sleep(1)

        with open(f'{ip}_showroute_post.txt', 'w+') as f4:
            print("Capturing post-reload 'show ip route'")
            showroute_post = net_connect.send_command("show ip
            route")
            f4.write(showroute_post)
            time.sleep(1)

        # Compare pre vs post configurations
        showver_pre = "showver_pre"
        showver_post = "showver_post"
        showver_pre_lines = open(f"{ip}_showver_pre.txt").
        readlines()
        showver_post_lines = open(f"{ip}_showver_post.txt").
        readlines()
        difference = difflib.HtmlDiff(wrapcolumn=70).make_
        file(showver_pre_lines, showver_post_lines, showver_
        pre, showver_post)
        difference_report = open(f"{ip}_show_ver_compared.
        html", "w+")
        difference_report.write(difference) # Writes the
        differences to html file
        difference_report.close()
        time.sleep(1)
        showrun_pre = "showrun_pre"
        showrun_post = "showrun_post"
```

(continued)

#	Task
| | ```
showrun_pre_lines = open(f"{ip}_showrun_pre.txt").
readlines()
showrun_post_lines = open(f"{ip}_showrun_post.txt").
readlines()
difference = difflib.HtmlDiff(wrapcolumn=70).make_
file(showrun_pre_lines, showrun_post_lines, showrun_
pre, showrun_post)
difference_report = open(f"{ip}_show_run_compared.
html", "w+")
difference_report.write(difference)
difference_report.close()
time.sleep(1)

showint_pre = "showint_pre"
showint_post = "showint_post"
showint_pre_lines = open(f"{ip}_showint_pre.txt").
readlines()
showint_post_lines = open(f"{ip}_showint_post.txt").
readlines()
difference = difflib.HtmlDiff(wrapcolumn=70).make_
file(showint_pre_lines, showint_post_lines, showint_
pre, showint_post)
difference_report = open(f"{ip}_show_int_compared.
html", "w+")
difference_report.write(difference)
difference_report.close()
time.sleep(1)
showroute_pre = "showroute_pre"
showroute_post = "showroute_post"
showroute_pre_lines = open(f"{ip}_showroute_pre.txt").
readlines()
showroute_post_lines = open(f"{ip}_showroute_post.
txt").readlines()
``` |

(continued)
```

#	Task

```
                difference = difflib.HtmlDiff(wrapcolumn=70).
                make_file(showroute_pre_lines, showroute_post_lines,
                showroute_pre, showroute_post)
    difference_report = open(f"{ip}_show_route_compared.html", "w+")
                difference_report.write(difference)
                difference_report.close()
                time.sleep(1)

                break
            else:
                print("Device is still reloading. Please wait...")
                time.sleep(delay)

    t2 = time.mktime(time.localtime()) - t1
    print("Total wait time : {0} seconds".format(t2))
    print("=" * 80)
    time.sleep(1)

post_check()
print("All tasks completed. Check pre and post configuration comparison
html files.")
```

(continued)

#	Task
⑩	Run the script; it should create backup files post-reload and then create the HTML files after comparing each line. This script will not trigger the routers to reload as part of the previous script. Here you are interested in creating the files, and then pre- and post-file contents are compared. In the final script, you will need to add this script to reload_yes_no2.py and turn it into a fully working application. So, don't be alarmed if the HTML is different from what you expected. ``` jdoe@u22s1:~/ch10_tools_dev2/tool10_post_check$ python post_check2.py 192.168.127.111 is online. Logging into device to perform post reload check Performing post upgrade check. configure terminal Enter configuration commands, one per line. End with CNTL/Z. c8kv01(config)#no boot system c8kv01(config)#boot system flash:/csr1000v-universalk9.16.09.06.SPA.bin c8kv01(config)#do write memory Building configuration... [OK] c8kv01(config)#end c8kv01# Capturing post-reload 'show version' Capturing post-reload 'show running-config' Capturing post-reload 'show ip interface brief' Capturing post-reload 'show ip route' Total wait time : 15.0 seconds == 192.168.127.222 is online. Logging into device to perform post reload check [...omitted for brevity] == All tasks completed. Check pre and post configuration comparison html files. ```

(*continued*)

#	Task
⑪	Run the `ls -lh 192*` command on your Python server. You should observe 16 new files—eight post-show capture files, and eight HTML files containing the differences between the _pre.txt files and _post.txt files. ``` jdoe@u22s1:~/ch10_tools_dev2/tool10_post_check$ ls -lh 192* -rw-rw-r-- 1 jdoe jdoe 5.3K Dec 2 14:59 192.168.127.111_show_int_compared.html -rw-rw-r-- 1 jdoe jdoe 11K Dec 2 14:59 192.168.127.111_show_route_compared.html -rw-rw-r-- 1 jdoe jdoe 75K Dec 2 14:59 192.168.127.111_show_run_compared.html -rw-rw-r-- 1 jdoe jdoe 26K Dec 2 14:59 192.168.127.111_show_ver_compared.html -rw-rw-r-- 1 jdoe jdoe 323 Dec 2 14:59 192.168.127.111_showint_post.txt -rw-rw-r-- 1 jdoe jdoe 323 Dec 2 14:49 192.168.127.111_showint_pre.txt -rw-rw-r-- 1 jdoe jdoe 1.1K Dec 2 14:59 192.168.127.111_showroute_post.txt -rw-rw-r-- 1 jdoe jdoe 1.1K Dec 2 14:49 192.168.127.111_showroute_pre.txt -rw-rw-r-- 1 jdoe jdoe 6.2K Dec 2 14:59 192.168.127.111_showrun_post.txt -rw-rw-r-- 1 jdoe jdoe 6.1K Dec 2 14:49 192.168.127.111_showrun_pre.txt -rw-rw-r-- 1 jdoe jdoe 2.5K Dec 2 14:59 192.168.127.111_showver_post.txt -rw-rw-r-- 1 jdoe jdoe 2.5K Dec 2 14:49 192.168.127.111_showver_pre.txt -rw-rw-r-- 1 jdoe jdoe 5.3K Dec 2 14:59 192.168.127.222_show_int_compared.html -rw-rw-r-- 1 jdoe jdoe 11K Dec 2 14:59 192.168.127.222_show_route_compared.html -rw-rw-r-- 1 jdoe jdoe 74K Dec 2 14:59 192.168.127.222_show_run_compared.html -rw-rw-r-- 1 jdoe jdoe 26K Dec 2 14:59 192.168.127.222_show_ver_compared.html ```

(continued)

#	Task

```
-rw-rw-r-- 1 jdoe jdoe  323 Dec  2 14:59 192.168.127.222_showint_post.txt
-rw-rw-r-- 1 jdoe jdoe  323 Dec  2 14:49 192.168.127.222_showint_pre.txt
-rw-rw-r-- 1 jdoe jdoe 1.1K Dec  2 14:59 192.168.127.222_showroute_post.txt
-rw-rw-r-- 1 jdoe jdoe 1.1K Dec  2 14:49 192.168.127.222_showroute_pre.txt
-rw-rw-r-- 1 jdoe jdoe 6.1K Dec  2 14:59 192.168.127.222_showrun_post.txt
-rw-rw-r-- 1 jdoe jdoe 6.1K Dec  2 14:49 192.168.127.222_showrun_pre.txt
-rw-rw-r-- 1 jdoe jdoe 2.5K Dec  2 14:59 192.168.127.222_showver_post.txt
-rw-rw-r-- 1 jdoe jdoe 2.5K Dec  2 14:49 192.168.127.222_showver_pre.txt
```

(12) Connect to the u22s1 server using WinSCP or FileZilla and locate the backup and comparison files under the /home/jdoe/ch10_tools_dev2/tool10_post_check directory. Copy the eight HTML files to your Windows Host (see Figure 10-11).

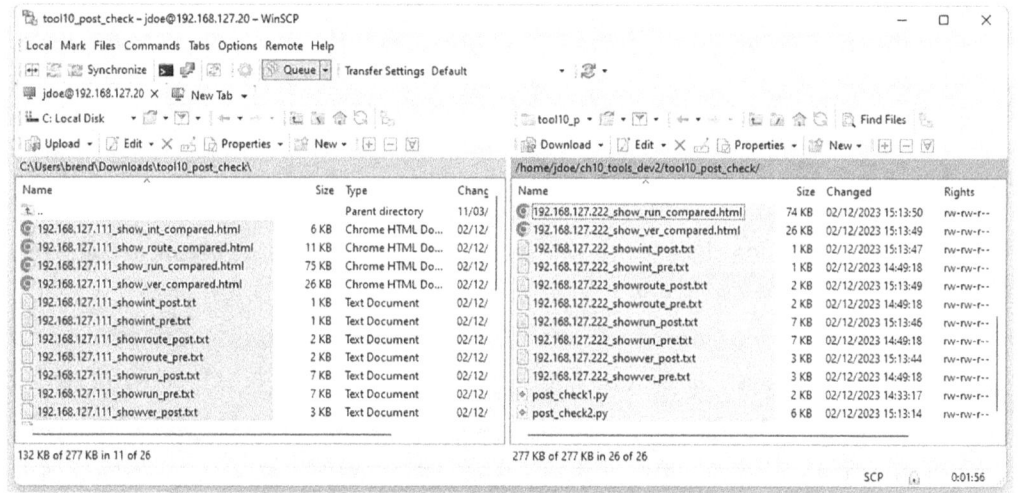

Figure 10-11. WinSCP, checking and copying HTML compared files

(*continued*)

CHAPTER 10 CISCO IOS-XE UPGRADE TOOLS DEVELOPMENT: PART 2

#	Task
⑬	Open the HTML files to check the results. If you get a similar result as shown in Figure 10-12, you have completed the last task of this lab.

Figure 10-12. *Open HTML-compared files on Google Chrome*

Congratulations on completing the development of the last four mini tools. I hope you have successfully finished all the necessary tasks and are eager to integrate these applications, much like assembling LEGO blocks, into a comprehensive tool in Chapter 11. Soon, you'll combine these elements to create a complete tool and test its functionality. I hope you're excited to progress to the next chapter!

Summary

This chapter marked a pivotal phase in your journey toward automating IOS upgrades, focusing on the development of the final set of four essential applications. These applications form the building blocks that culminate in an end-to-end IOS upgrade application in the subsequent chapter. Throughout this phase, a series of unique tools were written, each serving a distinct function. Beginning with a connectivity validation tool for seamless network connection verification, you progressed to tools for interactive username and password collection, alongside a file information read tool. Expanding your toolkit, a specific MD5 check tool was tailored for file integrity verification on a Linux server. This complemented the creation of a configuration backup tool for network devices, an IOS file uploading tool, and a router flash IOS MD5 check tool. Additionally, a user input tool offered adaptability, allowing the application flow to be adjusted as needed. This phase culminated in the development of a router reload tool, coupled with a post-check tool. You also looked at the Python parallelism using the threading module. Through this journey, knowledge grew from creating basic sequential Python code to enhancing it with parallel code using the threading module. This empowered Python scripts to efficiently manage multiple devices through multiple processes. Every tool underwent careful testing to ensure resilience and validate functionality. These tools will form the foundation for your final application, shaping it into a comprehensive and efficient IOS upgrade solution. Leveraging a range of Python modules tailored to your needs showcased their adaptability. Not only can these tools merge into a unified application, but they also enable the creation of diverse workflows and varied Python network applications.

The process of writing Python programs remains consistent across networking devices from different vendors. Leveraging Python modules compatible with specific vendor products allows for similar automation strategies across various platforms. With the six applications from the previous chapter and four applications from this chapter in hand, the final phase begins—amalgamating them into a comprehensive Cisco IOS XE upgrade tool, signifying the next stride in your Python Network Automation journey.

Storytime 3: IT Evolution: Unraveling the Cloud's Impact with Chicken Littles

Once upon a time in the tech world, there were fervent advocates of cloud computing, known as "chicken littles." They passionately declared that traditional network engineers were on the brink of doom due to the ascent of the cloud. In this world of service as everything, giants like AWS, Azure, and GCP reigned as influential leaders in cloud computing. The chicken littles warned that the cloud would render traditional network engineering obsolete, claiming it signaled the end of the hardware era.

However, an insightful owl raised questions about these assertions, pondering how one could connect to the cloud without relying on networking. While the Internet seemed like magic to most users, it relied on networks managed by ISPs and routing protocols, anchored by the age-old BGP, quietly hidden from plain sight. Despite the fervor of the chicken littles, reality stood firm: the Internet and networks continued to lean on traditional networking protocols. Despite claims that the cloud would make physical connections and routing protocols irrelevant, networking vendors like Cisco, Juniper, HPE, Palo Altos, Fortinet, and Huawei thrived. They knew the cloud wasn't some magical abstraction but someone else's computer humming away in a server farm, hardware-based with network and SAN adapters, and physical storage—nothing more than a clever abstraction.

The owl reminded everyone that the Internet relied on the robust backbone of ISP's BGP and enterprise networks. Beneath the cloud's glossy facade, traditional networking infrastructure provided its unshakable support. The chicken littles overlooked the fact that the connections underpinning the cloud were firmly tethered by traditional networks. While roles might evolve, the significance of networking persists. The owl's sagacity triumphed. The Ethernet switch markets surged, ensuring the sustained demand for network engineers, albeit under different titles. The cloud, far from being the demise of networks, marked a new chapter in their enduring story. Learn to differentiate between private, hybrid, and multi-cloud setups, and understand the differences between private and public clouds. Private clouds offer dedicated resources for a single organization, hybrid clouds combine private and public cloud resources, and multi-cloud involves using multiple cloud services from different providers. As more Enterprise IT workloads migrate to the public cloud, which provides shared resources over the internet, network engineers with a solid grasp of infrastructure will be needed to support all cloud models.

CHAPTER 11

Upgrading the Application and Routers

In this penultimate chapter, you will assemble the ten Python stand-alone applications developed in the last two chapters. This application, an exploratory Cisco IOS/IOS-XE upgrade application, encompasses pre-checks, IOS uploads, pre-reload checks, pre-reload configuration backups, reloads, and post-upgrade verification checks. Traditional network engineers have manually upgraded IOS on Cisco network devices for years. However, now you can develop and amalgamate various Python tools, enabling your Python code to automate these tasks. You will integrate one application at a time and test the code during the integration process. By the chapter's end, you will have learned how to combine smaller applications to create a sophisticated and highly exploratory application. Python provides programmers with the freedom to write almost any code, limited only by the programmer's imagination. Conversely, configuration and orchestration tools like Terraform, Ansible, Jenkins, and Puppet offer non-programming engineers control over devices using declarative languages, with several limitations that restrict the freedom and flexibility depending on the tool sets available to users. While writing Python code grants you freedom, it's essential to operate within the imaginary framework established by company policies and agreements among business stakeholders. Upon completing this chapter, you'll have achieved a significant milestone in your Python network programming journey. Are you ready to give it one big push?

CHAPTER 11   UPGRADING THE APPLICATION AND ROUTERS

Python Mastery for Network Automation: Building the Final IOS Upgrade Tool

In the previous two chapters, you made significant progress by crafting ten crucial applications for Cisco IOS upgrades. It's now time to bring these tools together and create an all-inclusive Cisco IOS XE upgrade tool. As you dive deeper into Python projects, you'll continuously encounter uncharted territories—an essential aspect of effective learning. For beginners in Python network automation, starting with simpler tasks is advisable. Document your progress, automate easier tasks, and dedicate time to research via online resources, courses, books, or by seeking guidance from seasoned professionals. Additionally, learning to code has become more accessible with conversational AIs like ChatGPT, Claude AI, BARD, Bing Chat, Perplexity AI, and others. These AI-powered assistants have demystified coding, enabling even those without computer science degrees to develop software in various languages. The paradigm has shifted; now, even primary school children can learn Python and Scratch, creating functional applications in mere days or weeks.

This transformation suggests that the advantage of certain software leans less on coding and more on user interfaces and innovative features. As engineers from diverse fields engage in coding, the competition among software products will shift to different arenas. Despite this transformation, my roots lie in infrastructure engineering. I've been exploring software engineering for some time, understanding the synergy and gaps between the two worlds. While technology silos are evolving, the demand for network engineers with robust infrastructure knowledge and hands-on experience remains high. Software engineers cannot fully replace traditional network engineers due to their lack of understanding of true infrastructure technologies. All they see, think, and read is software, software, software—they don't see that this software still requires platforms, connections, and storage and that these still run on physical hardware, called the "IT infrastructure". Thus, technology silos are akin to peeling an onion, with each layer revealing another silo that may take significant time to dismantle.

I introduced you briefly to object-oriented programming (OOP) in Chapter 8. While it's pivotal, you won't utilize it in your final IOS XE upgrade application due to increased complexity. Notably, Python allows coding without the need for custom OOP classes. Almost everything in Python functions as an object, even without explicit knowledge of OOP; you inherently use its principles. I intentionally excluded the OOP segment from the final script, keeping it distinct from the Python application.

CHAPTER 11   UPGRADING THE APPLICATION AND ROUTERS

You've arrived at the concluding exercise of the IOS XE upgrading lab, marking the book's penultimate chapter—a surprise Chapter 12 awaits. Congratulations on your unwavering dedication throughout this journey. Reflecting on my initial steps into network automation with Python, tasks like automating a Cisco IOS upgrade felt like distant aspirations. Yet, the momentum surrounding network automation and Python coding has surged, gaining significant traction across IT and networking teams. For the entire chapter, you will only be required to use the devices listed in Figure 11-1. Without further delay, let's proceed to complete your ultimate application: a functional IOS upgrade tool.

Tip

Looking for the source code used in this chapter?

All the source code featured in this book is available for access and download via the author's GitHub or Apress's GitHub repositories.

Author's GitHub:https://github.com/pynetauto/apress_pynetauto_ed2.0/tree/main/source_codes/Part2/ch11_upgrade_ios

Apress GitHub:

https://github.com/Apress/

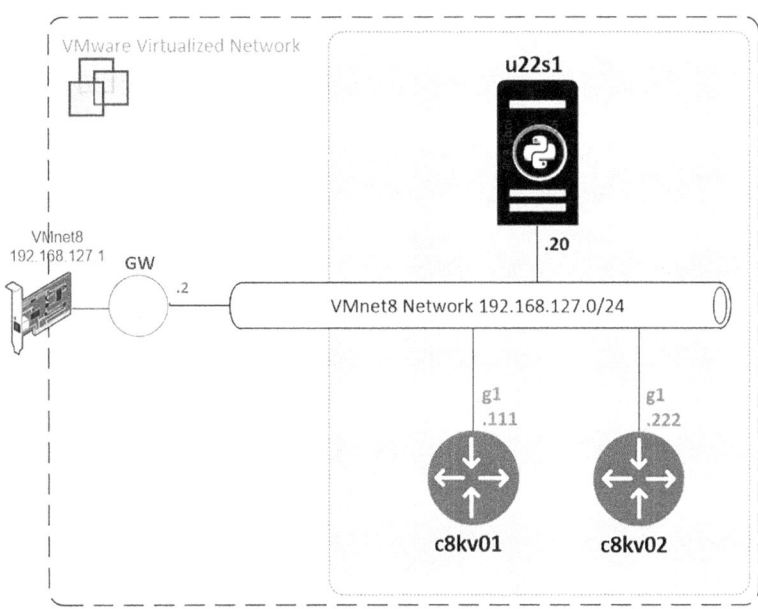

Figure 11-1. Devices for the IOS upgrade lab

CHAPTER 11   UPGRADING THE APPLICATION AND ROUTERS

For simplicity, you will create a single Python file (upgrade_c8000v.py). In a real-world production environment, the best practice is to create and save tools into smaller files (modules) to reduce the number of lines written to the main script. However, you are still exploring and learning, so it is more important to see the whole picture of how each application works. After you have finished this book, you can move some parts of the code into separate modules, which was discussed, and you have already practiced in the previous chapters. As discussed earlier, you will make this application an interactive tool, getting the user credentials and user inputs while the script runs. The device information will be written to a .csv file and read from the file using the pandas module as df (dataframe); this is a more convenient way to read large text-based data than to make the user enter the data.

#	Task
| ① | You will continue the tasks in the newly created /home/jdoe/ch11_upgrade_ios directory. Copy and move all the relevant files developed in Chapters 9 and 10 to this directory to work from a single location. Create the ch11_upgrade_ios directory and initiate the upgrade_c8000v.py IOS upgrade Python application to start integrating the ten applications.<br><br>You'll soon begin filling in the final upgrade_c8000v.py script. For now, leave it there as a placeholder.<br><br>```
jdoe@u22s1:~$ pwd
/home/jdoe
jdoe@u22s1:~$ mkdir ch11_upgrade_ios && cd ch11_upgrade_ios
jdoe@u22s1:~/ch11_upgrade_ios$ touch upgrade_c8000v.py
jdoe@u22s1:~/ch11_upgrade_ios$ ls
upgrade_c8000v.py
``` |

(*continued*)
```

CHAPTER 11   UPGRADING THE APPLICATION AND ROUTERS

#	Task

②	Now gather all the required files in one place to handle the tasks more easily. Move the new_ios directory, which contains the new IOS, to the ch11_upgrade_ios directory.
	
	`# new IOS .bin file location`
	`jdoe@u22s1:~/ch11_upgrade_ios$ pwd`
	`/home/jdoe/ch11_upgrade_ios`
	`jdoe@u22s1:~/ch11_upgrade_ios$ ls ../ch10_tools_dev2/new_ios`
	`c8000v-universalk9.17.06.05a.SPA.bin`
	`jdoe@u22s1:~/ch11_upgrade_ios$ mv ../ ch10_tools_dev2/new_ios .`
	`# Move whole dir current working dir`
	`jdoe@u22s1:~/ch11_upgrade_ios$ ls`
	`new_ios   upgrade_c8000v.py`
	`jdoe@u22s1:~/ch11_upgrade_ios$ ls new_ios`
	`c8000v-universalk9.17.06.05a.SPA.bin`

③	Move all of the tools developed in the last two chapters to the ch11_upgrade_ios directory, as shown here.
	
	Chapter 9 tools:
	
	Tool 1:
	`jdoe@u22s1:~/ch11_upgrade_ios$ cp ../ ch9_tools_dev1/tool1_ping/ping_6_module.py .`
	`jdoe@u22s1:~/ch11_upgrade_ios$ cp ../ ch9_tools_dev1/tool1_ping/ping_6_modular.py .`
	Tool 2:
	`jdoe@u22s1:~/ch11_upgrade_ios$ cp ../ ch9_tools_dev1/tool2_login/get_cred3.py .`
	Tool 3:
	`jdoe@u22s1:~/ch11_upgrade_ios$ cp ../ ch9_tools_dev1/tool3_read_csv/read_info8.py .`

(continued)

#	Task
	jdoe@u22s1:~/ch11_upgrade_ios$ **cp ../ ch9_tools_dev1/tool3_read_csv/ devices_info.csv .** # device info csv file **Tool 4:** jdoe@u22s1:~/ch11_upgrade_ios$ **cp ../ ch9_tools_dev1/tool4_md5_linux/ md5_validate3.py .** **Tool 5:** jdoe@u22s1:~/ch11_upgrade_ios$ **cp ../ ch9_tools_dev1/tool5_fsize_ cisco/check_flash2.py .** **Tool 6:** jdoe@u22s1:~/ch11_upgrade_ios$ **cp ../ ch9_tools_dev1/tool6_make_ backup/make_backup1.py .** Chapter 10 tools: **Tool 7:** jdoe@u22s1:~/ch11_upgrade_ios$ **cp ../ ch10_tools_dev2/tool7_upload_ ios/upload_ios3_parallel.py .** **Tool 8:** jdoe@u22s1:~/ch11_upgrade_ios$ **cp ../ ch10_tools_dev2/tool8_md5_ cisco/md5_verify1.py .** **Tool 9:** jdoe@u22s1:~/ch11_upgrade_ios$ **cp ../ ch10_tools_dev2/tool9_yes_no/ reload_yes_no_in_parallel.py .** **Tool 10:** jdoe@u22s1:~/ch11_upgrade_ios$ **cp ../ ch10_tools_dev2/tool10_post_ check/post_check2.py .**

(continued)

#	Task
④	Once you have copied all the relevant files, run the ls -lh command to check all the files you will be referencing and using to complete the main IOS upgrade application. ``` jdoe@u22s1:~/ch11_upgrade_ios$ ls -lh total 56K -rw-rw-r-- 1 jdoe jdoe 761 Dec 3 01:32 check_flash2.py # tool 5 -rw-rw-r-- 1 jdoe jdoe 262 Dec 3 01:50 devices_info.csv # device info csv file -rw-rw-r-- 1 jdoe jdoe 1.4K Dec 3 01:26 get_cred3.py # tool 2 -rw-rw-r-- 1 jdoe jdoe 1.1K Dec 3 01:33 make_backup1.py # tool 6 -rw-rw-r-- 1 jdoe jdoe 1.3K Dec 3 01:29 md5_validate3.py # tool 4 -rw-rw-r-- 1 jdoe jdoe 3.4K Dec 3 01:38 md5_verify1.py # tool 8 drwxrwxr-x 2 jdoe jdoe 4.0K Nov 26 17:54 new_ios -rw-rw-r-- 1 jdoe jdoe 765 Dec 3 01:23 ping_6_modular.py # tool 1 -rw-rw-r-- 1 jdoe jdoe 880 Dec 3 01:23 ping_6_module.py # tool 1 module -rw-rw-r-- 1 jdoe jdoe 6.0K Dec 3 01:43 post_check2.py # tool 10 -rw-rw-r-- 1 jdoe jdoe 1.2K Dec 3 01:28 read_info8.py # tool 3 -rw-rw-r-- 1 jdoe jdoe 2.5K Dec 3 01:41 reload_yes_no_in_parallel.py # tool 7 -rw-rw-r-- 1 jdoe jdoe 0 Dec 3 00:28 upgrade_c8000v.py # main IOS upgrade tool -rw-rw-r-- 1 jdoe jdoe 2.5K Dec 3 01:37 upload_ios3_parallel.py # tool 7 ```

(*continued*)

CHAPTER 11   UPGRADING THE APPLICATION AND ROUTERS

#	Task

⑤	As a precaution, check the md5 value of your IOS XE and cross-check the `devices_info.csv` file to ensure that the corresponding information matches your target devices, c8kv01 and c8kv02.
	
	`jdoe@u22s1:~/ch11_upgrade_ios$ `**`ls -lh new_ios`**
	`total 771M`
	`-rw-r--r-- 1 jdoe jdoe 771M Nov 26 10:36 c8000v-universalk9.17.06.05a.SPA.bin`
	`jdoe@u22s1:~/ch11_upgrade_ios$ `**`md5sum new_ios/c8000v-universalk9.17.06.05a.SPA.bin`**
	`13f0161a50210f2f21618fc59c5f5343  new_ios/c8000v-universalk9.17.06.05a.SPA.bin`
	
	`jdoe@u22s1:~/ch11_upgrade_ios$ `**`more devices_info.csv`**
	`devicename,device,devicetype,host,newios,newiosmd5`
	`c8kv01,RT,cisco_xe,192.168.127.111,c8000v-universalk9.17.06.05a.SPA.bin,13f0161a50210f2f21618fc59c5f5343`
	`c8kv02,RT,cisco_xe,192.168.127.222,c8000v-universalk9.17.06.05a.SPA.bin,13f0161a50210f2f21618fc59c5f5343`

⑥	Now that the standard procedures are out of the way, you can begin the final application build. You will now add the series of scripts to the `upgrade_c8000v.py` Python application in a linear fashion, and once everything has been strung together, it will work as a single application that can be directly used in most production environments for c8000v and other Cisco ISR platforms. Also, the application can be easily refactored to include other Cisco IOS routers and switches. First, add all the Python modules that were used in your ten applications and then bolt them on and modify the scripts one at a time. Follow these instructions to add the required modules to the `upgrade_c8000v.py` file. In the next step, you will append and modify sections of code from each tool and slowly build up the main application, `upgrade_c8000v.py`.

(continued)

#	Task
	jdoe@u22s1:~/ch11_upgrade_ios$ **cp upgrade_c8000v.py upgrade_c8000v_bt1.py** jdoe@u22s1:~/ch11_upgrade_ios$ **nano upgrade_c8000v_bt1.py** **import socket** # For socket networking **import os** # For Python Server OS **import time** # Python time module **import pandas as pd** # Pandas for reading data into a data frame **import re** # Python Regular Expression module **from getpass import getpass** # For uid & password collection **import os.path** # For Python Server OS directory **import hashlib** # For MD5 checks **from netmiko import ConnectHandler, SCPConn** # netmiko SSH connection, and SCP file transfer **from netmiko import NetMikoTimeoutException** # To catch netmiko timeout exceptions **import threading** # For multitasking using threads **import difflib** # For analyzing two files and differences **t1 = time.mktime(time.localtime())** # Timer(t1) start to measure script running time # Add your ten applications here! **tt = time.mktime(time.localtime()) - t1** **print("Total time : {0} seconds".format(tt))** # Timer finish to show total time (tt)

(*continued*)

CHAPTER 11 UPGRADING THE APPLICATION AND ROUTERS

#	Task
⑦	Utilize get_cred3.py from tool2 development as the foundation and make slight modifications to the tool. This script will interactively acquire the network administrator's ID and password, prompt for the enable secret, and provide the option for the user to confirm if the password and enable secret match. You've integrated Regular Expressions to control the input strings for the username, password, and secret. These regex patterns were initially defined as global variables during development but have now been transitioned to local variables within the get_credentials() function. Additionally, note that the p2 or pattern 2 regex for password validation is passed into the get_secret() function, eliminating the need to declare it again in another function. For p1 or pattern 1, the expected user ID characters range from 4 to 30 characters in length, commencing and concluding with a letter or a number. Regarding p2 or pattern 2, it anticipates characters from 8 to 50 in length, commencing with a letter and ending with any character, including special characters. Understanding Regular Expressions might pose a challenge, especially for individuals without a programming background. I recommend reviewing and referencing Chapter 9 of the Part I book to initiate your understanding of Regular Expressions in Python. It's crucial to invest sufficient time to grasp their usage and relevance in your work environment. While I'm adept at memorization, I too can't recall all the Regular Expressions and their applications. Therefore, I often refer to Chapter 9 of the Part I book as my go-to resource whenever I encounter difficulties with Regular Expressions. That chapter alone consumed around three months of reading across multiple books and sources, and it's unlikely you'll find books or training videos that delve as deeply into Regular Expressions in Python. It's vital to type this Python code yourself for ample coding practice. Let's now dive into the practical side of things. ``` jdoe@u22s1:~/ch11_upgrade_ios$ cp upgrade_c8000v_bt1.py upgrade_c8000v_bt2.py jdoe@u22s1:~/ch11_upgrade_ios$ nano upgrade_c8000v_bt2.py [...omitted for brevity] t1 = time.mktime(time.localtime()) # Timer(t1) start to measure script running time ``` *(continued)*

#	Task

```python
# <<< Continue Your Build
def get_secret(p2):
  global secret # declare secret as a global variable
  resp = input("Is secret same as password? (y/n) : ")
  resp = resp.lower()
  if resp == "yes" or resp == "y":
    secret = pwd
  elif resp == "no" or resp == "n":
    secret = None
    while not secret:
      secret = getpass("Enter the secret : ")
      while not p2.match(secret):
        print("User ID: min. 4 letters, starts with a letter")
        secret = getpass(r"*Enter the secret : ")
      secret_verify = getpass("Confirm the secret : ")
      if secret != secret_verify:
        print("Secrets Mismatch! Retry.")
        secret = None
  else:
    get_secret(p2)

def get_credentials():
  p1 = re.compile(r'^[a-zA-Z][a-zA-Z0-9_-]{2,28}[a-zA-Z0-9]$')
  # local variable to the function
  p2 = re.compile(r'^[a-zA-Z][a-zA-Z0-9!@#$%^&*()_+=\-[\]
  {};:\'",.<>?]{7,49}') # local variable to the function
  global uid # declare uid as a global variable
  uid = input("Enter Network Admin ID : ")
  while not p1.match(uid):
    print("User ID: min. 4 letters, starts with a letter")
    uid = input(r"*Enter Network Admin ID : ")
  global pwd # declare pwd as a global variable
```

(continued)

CHAPTER 11   UPGRADING THE APPLICATION AND ROUTERS

#	Task
| | ```
    pwd = None
    while not pwd:
      pwd = getpass("Enter Network Admin PWD : ")
      while not p2.match(pwd):
        print("Password: min. 7 chars, starts with a letter.")
        pwd = getpass(r"*Enter Network Admin PWD : ")
      pwd_verify = getpass("Confirm Network Admin PWD : ")
      if pwd != pwd_verify:
        print("Passwords Mismatch! Retry.")
        pwd = None
      get_secret(p2) # Trigger get_secret function to run
      return uid, pwd, secret

  get_credentials() # Trigger get_Credential function to run

  tt = time.mktime(time.localtime()) - t1

  print("Total time : {0} seconds".format(tt)) # Timer finish to show
  total time (tt)
``` |
| ⑧ | The next step involves testing and confirming the recent changes in the Python code. Execute the Python command to test the application. Test scenarios involving incorrect password entry, including a scenario where the password and secret don't match, necessitating the use of different credentials. This practice reflects standard security measures in a production network.

```
jdoe@u22s1:~/ch11_upgrade_ios$ python upgrade_c8000v_bt2.py
Enter Network Admin ID : jd
User ID: min. 4 letters, starts with a letter
*Enter Network Admin ID : jdoe
Enter Network Admin PWD : cisco1
Password: min. 7 chars, starts with a letter.
*Enter Network Admin PWD : cisco123
```
 |

(continued)
```

#	Task
	Confirm Network Admin PWD : **cisco555** Passwords Mismatch! Retry. Enter Network Admin PWD : **cisco123** Confirm Network Admin PWD : **cisco123** Is secret same as password? (y/n) : **n** Enter the secret : **cisco2** User ID: min. 4 letters, starts with a letter *Enter the secret : **cisco123** Confirm the secret : **cisco777** Secrets Mismatch! Retry. Enter the secret : **cisco123** Confirm the secret : **cisco123** Total time : 68.0 seconds
	Re-run the Python code to test a scenario where the password and secret match. To simplify the lab setup, you're using identical password and secret values. However, if they are different, as in most production environments, you've already addressed this, as shown in previous output. jdoe@u22s1:~/ch11_upgrade_ios$ **python upgrade_c8000v_bt2.py** Enter Network Admin ID : **jdoe** Enter Network Admin PWD : **cisco123** Confirm Network Admin PWD : **cisco123** Is secret same as password? (y/n) : **y** Total time : 11.0 seconds

(continued)

#	Task
⑨	Duplicate and insert the excerpts from read_info8.py (tool3_read_csv) toward the end of the script, immediately after the get_credentials() line. This script reads the devices_info.csv file to provide input for your script, converting the retrieved items into variables used throughout the script. Transform this process into a function for enhanced portability and readability. Make sure to pass the three arguments received from the user input in the preceding step, preventing script errors. Remember, this application requires three arguments from the prior workflow: uid, pwd, and secret. Additionally, include print(x) to verify the .csv file read, highlighted in the following code excerpt. You can comment out this print statement in the subsequent script build by using the # symbol." ```\njdoe@u22s1:~/ch11_upgrade_ios$ cp upgrade_c8000v_bt2.py upgrade_c8000v_bt3.py\njdoe@u22s1:~/ch11_upgrade_ios$ nano upgrade_c8000v_bt3.py\n[...omitted for brevity]\nget_credentials()\n# <<< Continue Your Build\ndef read_info(uid, pwd, secret):\n df = pd.read_csv(r'./devices_info.csv') # ensure the correct file location\n number_of_rows = len(df.index)\n # Read the values and save as a list, read the column as df and save it as a list\n devicename = list(df['devicename'])\n device = list(df['device'])\n devicetype = list(df['devicetype'])\n ip = list(df['host'])\n newios = list(df['newios'])\n newiosmd5 = list(df['newiosmd5'])\n # Append the items and convert to a list, device_list\n global device_list # For md5_validate3.py\n device_list = []\n```

(*continued*)

#	Task

```
    for index, rows in df.iterrows():
      device_append = [rows.devicename, rows.device, \
      rows.devicetype, rows.host, rows.newios, rows.newiosmd5]
      device_list.append(device_append)
    # Using device_list, create a netmiko friendly list device_list_
    netmiko
    global device_list_netmiko
    device_list_netmiko = []
    i = 0
    for x in device_list:
    print(x) # Check device info read
      if len(x) != 0: # As long as number of items in device_list is
                     not 0 (empty)
        i += 1
      name = f'device{str(i)}' # Each for loop, the name is updated to
      device1, device2, device3, ...
        devicetype, host = x[2], x[3]
      device = {
      'device_type': devicetype,
      'host': host,
      'username': uid,
      'password': pwd,
      'secret': secret,
      }
      device_list_netmiko.append(device)
  # Trigger read_info function to run
  read_info(uid, pwd, secret)
  tt = time.mktime(time.localtime()) - t1
  print("Total time : {0} seconds".format(tt))
```

(continued)

CHAPTER 11   UPGRADING THE APPLICATION AND ROUTERS

#	Task
⑩	Testing and validating the freshly updated Python code is your next step. Run `python upgrade_c8000v_bt3.py` and test that your Python application is working as it should. jdoe@u22s1:~/ch11_upgrade_ios$ **python upgrade_c8000v_bt3.py** Enter Network Admin ID : **jdoe** Enter Network Admin PWD : **cisco123** Confirm Network Admin PWD : **cisco123** Is secret same as password? (y/n) : **y** ['c8kv01', 'RT', 'cisco_xe', '192.168.127.111', 'c8000v-universalk9.17.06.05a.SPA.bin', '13f0161a50210f2f21618fc59c5f5343'] ['c8kv02', 'RT', 'cisco_xe', '192.168.127.222', 'c8000v-universalk9.17.06.05a.SPA.bin', '13f0161a50210f2f21618fc59c5f5343'] Total time : 8.0 seconds
⑪	The next step involves examining the code excerpts from `ping_6_modular.py` and `ping_6_module.py` to combine them into a single function. This integration eliminates the need to call the `check_port` module from `ping_6_module.py`. Copy these scripts and adapt them as outlined in the provided code excerpts. Before executing this build and test script, it's imperative to reassess the application logic. During the connectivity test, this Python code should utilize both the Linux OS `ping` command and socket networking to verify an open socket status on port 22. In cases where an IP address is unreachable, the script will exit the application, necessitating troubleshooting to resolve why the specific device is unreachable on the network. This script is designed to generate three distinct files reflecting the reachability status. If port 22 is open, the device's IP address will be written to `f1`. In situations where port 22 is closed, the script will additionally examine the status of port 23 (telnet port) and log the IP address to `f2`. Should the device prove unreachable, the IP address will be recorded in `f3` for further troubleshooting. jdoe@u22s1:~/ch11_upgrade_ios$ **cp upgrade_c8000v_bt3.py upgrade_c8000v_bt4.py** jdoe@u22s1:~/ch11_upgrade_ios$ **nano upgrade_c8000v_bt4.py** [...omitted for brevity] read_info(uid, pwd, secret)

(*continued*)

#	Task
| | ```
# <<< Continue Your Build
def test_connectivity(device_list_netmiko):
    f1 = open('reachable_ips_ssh.txt', 'w+')
    f2 = open('reachable_ips_telnet.txt', 'w+')
    f3 = open('unreachable_ips.txt', 'w+')
    for device in device_list_netmiko:
        ip = device['host'].strip()
        print(ip)
        resp = os.system('ping -c 4 ' + ip)
        if resp == 0:
            for port in range(22, 23):
                destination = (ip, port)
                try:
                    with socket.socket(socket.AF_INET, socket.SOCK_STREAM) as s:
                        s.settimeout(3)
                        connection = s.connect(destination)
                        print(f"{ip} {port} open")
                        f1.write(f"{ip}\n")
                except:
                    print(f"{ip} {port} closed")
                    f3.write(f"{ip} {port} closed\n")
            for port in range(23, 24):
                destination = (ip, port)
                try:
                    with socket.socket(socket.AF_INET, socket.SOCK_STREAM) as s:
                        s.settimeout(3)
                        connection = s.connect(destination)
                        print(f"{ip} {port} open")
                        f2.write(f"{ip}\n")
                except:
                    print(f"{ip} {port} closed")
                    f3.write(f"{ip} {port} closed\n")
``` |

(continued)
```

#	Task
| | ```python
    else:
      print(f"{ip} unreachable")
      f3.write(f"{ip} unreachable\n")
f1.close()
f2.close()
f3.close()

# Trigger test_connectivity function to run
test_connectivity(device_list_netmiko)

tt = time.mktime(time.localtime()) - t1
print("Total time : {0} seconds".format(tt))
``` |
| ⑫ | Now verify and validate the updated Python code by running tests. Once you are happy with your build, execute the latest application by issuing python upgrade_c8000v_bt4.py. Enter the username and password. You should then observe output like this:<br><br>```
jdoe@u22s1:~/ch11_upgrade_ios$ python upgrade_c8000v_bt4.py
[...omitted for brevity]
192.168.127.111
PING 192.168.127.111 (192.168.127.111) 56(84) bytes of data.
64 bytes from 192.168.127.111: icmp_seq=2 ttl=255 time=0.740 ms
64 bytes from 192.168.127.111: icmp_seq=3 ttl=255 time=0.608 ms
64 bytes from 192.168.127.111: icmp_seq=4 ttl=255 time=1.16 ms

--- 192.168.127.111 ping statistics ---
4 packets transmitted, 3 received, 25% packet loss, time 3023ms
rtt min/avg/max/mdev = 0.608/0.836/1.162/0.236 ms
192.168.127.111 22 open
192.168.127.111 23 closed
192.168.127.222
PING 192.168.127.222 (192.168.127.222) 56(84) bytes of data.
64 bytes from 192.168.127.222: icmp_seq=2 ttl=255 time=0.700 ms
64 bytes from 192.168.127.222: icmp_seq=3 ttl=255 time=0.474 ms
64 bytes from 192.168.127.222: icmp_seq=4 ttl=255 time=0.699 ms
``` |

(*continued*)

#	Task
	``
--- 192.168.127.222 ping statistics ---
4 packets transmitted, 3 received, 25% packet loss, time 3089ms
rtt min/avg/max/mdev = 0.474/0.624/0.700/0.106 ms
192.168.127.222 22 open
192.168.127.222 23 closed
Total time : 25.0 seconds
```

Confirm that three files have been created as expected.

```
jdoe@u22s1:~/ch11_upgrade_ios$ ls -lh *.txt
-rw-rw-r-- 1 jdoe jdoe 32 Dec 3 14:22 reachable_ips_ssh.txt
-rw-rw-r-- 1 jdoe jdoe 0 Dec 3 14:22 reachable_ips_telnet.txt
-rw-rw-r-- 1 jdoe jdoe 52 Dec 3 14:22 unreachable_ips.txt
``` |
| ⑬ | This excerpt originates from md5_validate3.py (tool4_md5_linux), as previously mentioned. Its purpose is to validate the MD5 value of the new IOS file on the server. Given that both the file and script operate from a unified Python automation server, you can leverage Python's hashlib to verify the IOS's MD5 value within a directory. Rigorous testing of the MD5 values across your workflow is crucial to ensuring a successful upgrade to a newer IOS XE version. Additionally, it's vital to confirm that your new IOS resides in the correct directory.

```
jdoe@u22s1:~/ch11_upgrade_ios$ cp upgrade_c8000v_bt4.py upgrade_c8000v_bt5.py
jdoe@u22s1:~/ch11_upgrade_ios$ nano upgrade_c8000v_bt5.py
[...omitted for brevity]
# Trigger test_connectivity function to run
test_connectivity(device_list_netmiko)
# <<< Continue Your Build
def validate_md5(device_list):
for x in device_list:
print(x[3], x[0], x[1], x[2])
newios = x[4]
newiosmd5 = str(x[5].lower()).strip()
``` |

(continued)
```

#	Task
| | ```python
print(newiosmd5)
newiosmd5hash = hashlib.md5()
file = open(f'./new_ios/{newios}', 'rb') # Update your IOS directory
here
content = file.read()
newiosmd5hash.update(content)
newiosmd5server = newiosmd5hash.hexdigest()
print(newiosmd5server.strip())
global newiossize
newiossize = round(os.path.getsize(f'./new_ios/{newios}') / 1000000, 2)
# Update your IOS directory here
print(newiossize, "MB")
if newiosmd5server == newiosmd5:
print("MD5 values matched!")
else:
print("Mismatched MD5 values. Exit")
exit()
return newiossize

# Trigger validate_md5 function to run
validate_md5(device_list)

tt = time.mktime(time.localtime()) - t1
print("Total time : {0} seconds".format(tt))
``` |
| ⑭ | It's time to execute the latest build and validate that your tool is working as designed.<br><br>```
jdoe@u22s1:~/ch11_upgrade_ios$ python upgrade_c8000v_bt5.py
[...omitted for brevity]
192.168.127.111 c8kv01 RT cisco_xe
13f0161a50210f2f21618fc59c5f5343
13f0161a50210f2f21618fc59c5f5343
808.24 MB
``` |

(*continued*)

CHAPTER 11 UPGRADING THE APPLICATION AND ROUTERS

#	Task
|  | ```
MD5 values matched!
192.168.127.222 c8kv02 RT cisco_xe
13f0161a50210f2f21618fc59c5f5343
13f0161a50210f2f21618fc59c5f5343
808.24 MB
MD5 values matched!
Total time : 35.0 seconds
``` |
| (15) | In this task, the script closely resembles check_flash2.py (tool5_fsize_cisco). While you're creating the function, ensure accurate parsing of variables from the preceding workflow. Here, you're parsing the device_list_netmiko list for the SSH connection and using the newiossize variable to verify sufficient free space in the router's flash memory for the new IOS file size. If the available space is less than 1.5 times the new IOS file size, the script will terminate.<br><br>On newer Cisco devices, accommodating multiple larger IOS and IOS XE files isn't typically problematic due to ample flash free size. However, on older devices, managing space might necessitate removing old IOS files from the flash. If your IOS upgrade fails, this could result in a disastrous situation where the device boots into ROMMON mode without a valid image. The only recovery method in such a scenario involves working on the device via a physical console connection, occasionally resorting to xmodem for file transfer.<br><br>You're fortunate to witness dropping prices of flash drives, yet Cisco's approach to physical storage of flash and licensing remains puzzling. It might be beneficial to devise a method for automatically identifying the largest or oldest file if needed. Nonetheless, consistently assessing free space during the change planning phase remains critical.<br><br>```
jdoe@u22s1:~/ch11_upgrade_ios$ cp upgrade_c8000v_bt5.py upgrade_c8000v_bt6.py
jdoe@u22s1:~/ch11_upgrade_ios$ nano upgrade_c8000v_bt6.py
[...omitted for brevity]
# Trigger validate_md5 function to run
validate_md5(device_list)
# <<< Continue Your Build
``` |

(continued)

#	Task
|  | ```python
def check_flash(device_list_netmiko, newiossize):
    for device in device_list_netmiko:
        ip = str(device['host'])
        net_connect = ConnectHandler(**device)
        net_connect.send_command("terminal length 0")
        showdir = net_connect.send_command("dir")
        # showflash = net_connect.send_command("show flash:") # For Cisco switches
        time.sleep(2)
        p1 = re.compile("\d+(?=\sbytes\sfree\))")
        m1 = p1.findall(showdir)
        flashfree = ((int(m1[0])/1000000))
        print(f"{ip} Free flash size : ", flashfree, "MB")
        if flashfree < (newiossize * 1.5):
            print(f"Not enough space on {ip}'s flash! Exiting")
            exit()
        else:
            print(f"{ip} has enough space for new IOS.")

# Trigger check_flash function to run
check_flash(device_list_netmiko, newiossize)

tt = time.mktime(time.localtime()) - t1
print("Total time : {0} seconds".format(tt))
``` |
| 16 | It's time to put the newly updated Python code through testing and validation. Execute the following Python command and follow the prompt as before.<br><br>```
jdoe@u22s1:~/ch11_upgrade_ios$ python upgrade_c8000v_bt6.py
[...omitted for brevity]
192.168.127.111 Free flash size :   8418.267136 MB
192.168.127.111 has enough space for new IOS.
192.168.127.222 Free flash size :   8380.002304 MB
192.168.127.222 has enough space for new IOS.
Total time : 39.0 seconds
``` |

(continued)

#	Task
⑰	This script, extracted from upload_ios3_parallel.py (tool7_upload_ios), represents a substantial portion of the main script. As its name suggests, following the completion of the preceding task, the application prepares to upload the new IOS image to the initial router's flash memory. However, before commencing the file upload, certain preparatory steps are essential. First, the logged-in user must possess a level 15 privilege to execute IOS uploading tasks using SCP. Secondly, the router must be configured appropriately to enable aaa authentication and authorization for this user. Disregarding these initial requirements indicates a failure to adequately prepare for the change, necessitating the resolution of access issues before proceeding with the application. If the access configurations are correct for this device, the application proceeds to verify the status of `ip scp server enable`, enabling the SCP server service on the Cisco device if necessary. If the service is disabled, the script automatically activates this feature before initiating the upload of the new IOS file, subsequently deactivating it upon completion of the upload. Before integrating the next application, it is important to note that you are utilizing a set of device information. The first piece of information, `device_list`, is a Python list extracted and converted from the `pandas device_info.csv` file. This data is structured as a list of lists, where each inner list represents a device. The items within each inner list represent the device's attributes, and the order of these items is consistent across all devices (see Table 11-1). ``` device_list = [['c8kv01', 'RT', 'cisco_xe', '192.168.127.111', 'c8000v-universalk9.17.06.05a.SPA.bin', '13f0161a50210f2f21618fc59c5f5343'], ['c8kv02', 'RT', 'cisco_xe', '192.168.127.222', 'c8000v-universalk9.17.06.05a.SPA.bin', '13f0161a50210f2f21618fc59c5f5343']] ```

(continued)

#	Task

Table 11-1. The device_list Is a List of Lists

Index	Item	Data Type	Description
0	hostname	String	The hostname of the device, as c8kv01 and c8kv02.
1	role	String	The role of the device, such as "RT" for router.
2	device_type	String	The type of device, such as "cisco_xe" for Cisco XE router.
3	ip_address	String	The IP address of the device.
4	image_filename	String	The file name of the IOS image.
5	MD5 value	String	The MD5 value of the device.

The second set of data is device_list_netmiko, which is presented in JSON format, where each dictionary represents a device. This JSON-formatted information has been partially derived from the aforementioned device_list list. The keys within the dictionaries correspond to the device's attributes, and the values represent the specific values of those attributes. For instance, the first dictionary contains a key named device_type with a value of cisco_xe. This indicates that the device is a Cisco XE router. The logic follows suit, as detailed in Table 11-2.

Here's a breakdown of the changes:

device_list_netmiko = [
{
'device_type': 'cisco_xe',
'host': '192.168.127.111',
'username': 'jdoe',
'password': 'cisco123',
'secret': 'cisco123'
},

(continued)

#	Task
| | ```
{
'device_type': 'cisco_xe',
'host': '192.168.127.222',
'username': 'jdoe',
'password': 'cisco123',
'secret': 'cisco123'
}
]
``` |

Table 11-2. JSON Data in device_list_netmiko

Key	Value	Data Type	Description
device_type	cisco_xe	String	Device type
host	192.168.127.111	String	Device IP address
username	jdoe	String	Network admin username
password	cisco123	String	Network admin password
secret	cisco123	String	Cisco device enable password

Let's dive back into coding—time to begin writing more code. To get practice in coding, type this Python code on your own.

jdoe@u22s1:~/ch11_upgrade_ios$ **cp upgrade_c8000v_bt6.py upgrade_c8000v_bt7.py**
jdoe@u22s1:~/ch11_upgrade_ios$ **nano upgrade_c8000v_bt7.py**
[...omitted for brevity]
# Trigger check_flash function to run
check_flash(device_list_netmiko, newiossize)
# <<< Continue Your Build

def ios_upload(device_list_netmiko, device_list):
def upload_file(device):
ip = str(device['host'])

(continued)
```

#	Task
| | ```
username = str(device['username'])
net_connect = ConnectHandler(**device)
net_connect.send_command("terminal length 0")
showrun = net_connect.send_command("show running-config")
check_priv15 = (f'username {username} privilege 15')
aaa_authentication = "aaa authentication login default local enable"
aaa_authorization = "aaa authorization exec default local"

if check_priv15 in showrun:
print(f"{username} has level 15 privilege - OK")
if aaa_authentication in showrun:
print("check_aaa_authentication - OK")
if aaa_authorization in showrun:
print("check_aaa_authorization - OK")
else:
print("aaa_authorization - FAILED ")
exit()
else:
print("aaa_authentication - FAILED ")
exit()
else:
print(f"{username} has not enough privilege - FAILED")
exit()

net_connect.enable(cmd='enable 15')
net_connect.config_mode()
net_connect.send_command('ip scp server enable')
net_connect.exit_config_mode()
time.sleep(1)
print(f"New IOS uploading in progress to {ip}! Please wait...")
scp_conn = SCPConn(net_connect)
scp_conn.scp_transfer_file(source_newios, destination_newios)
scp_conn.close()
``` |

(continued)
```

#	Task

```python
    time.sleep(1)
    net_connect.config_mode()
    net_connect.send_command('no ip scp server enable')
    net_connect.exit_config_mode()
    print(f"Upload to {ip} completed.")
    print("-"*79)

    # Create threads for each device and start uploading files in
    parallel
    threads = []
    for device in device_list:
    thread = threading.Thread(target=upload_file, args=(device,))
    thread.start()
    threads.append(thread)

    # Wait for all threads to complete before proceeding
    for thread in threads:
    thread.join()

    print("All file uploads completed.")

    # Trigger ios_upload() application.
    ios_upload(device_list_netmiko, device_list)

    tt_ios_upload = time.mktime(time.localtime()) - t2
    print("Total Time : {0} seconds".format(tt_ios_upload)) # Time taken
    to upload IOS file

    tt = time.mktime(time.localtime()) - t1
    print("Total time : {0} seconds".format(tt)) # Timer finish to show
    total time (tt)
```

(continued)

#	Task
⑱	Now, it's time for comprehensive testing and validation of the updated Python code. The new IOS image is already present on your router's flash memory, and executing the Python code will overwrite it. While deleting the existing IOS image is not necessary, you may choose to remove it before commencing the validation process. Before initiating the Python command, enable `debug ip scp` on both routers to thoroughly observe the end-to-end file transfer verification. During the upload of the new IOS image to the routers' flash memory, the process may appear to stall (see Figure 11-2). This is due to the large file size exceeding 700MB and the parallel upload to both routers running on the same PC. It's best to refrain from interrupting the file transfer process and allow the upload to complete seamlessly. ``` jdoe@u22s1:~/ch11_upgrade_ios$ python upgrade_c8000v_bt7.py [...omitted for brevity] New IOS uploading in progress to 192.168.127.222! Please wait... New IOS uploading in progress to 192.168.127.111! Please wait... ``` ``` 64 bytes from 192.168.127.222: icmp_seq=4 ttl=255 time=0.434 ms --- 192.168.127.222 ping statistics --- 4 packets transmitted, 4 received, 0% packet loss, time 3044ms rtt min/avg/max/mdev = 0.423/1.904/6.230/2.497 ms 192.168.127.222 22 open 192.168.127.222 23 closed 192.168.127.111 c8kv01 RT cisco_xe 13f0161a50210f2f21618fc59c5f5343 13f0161a50210f2f21618fc59c5f5343 808.24 MB MD5 values matched! 192.168.127.222 c8kv02 RT cisco_xe 13f0161a50210f2f21618fc59c5f5343 13f0161a50210f2f21618fc59c5f5343 808.24 MB MD5 values matched! 192.168.127.111 Free flash size : 8956.862464 MB 192.168.127.111 has enough space for new IOS. 192.168.127.222 Free flash size : 8918.994944 MB 192.168.127.222 has enough space for new IOS. New IOS uploading in progress to 192.168.127.111! Please wait... New IOS uploading in progress to 192.168.127.222! Please wait... ``` *Figure 11-2. python upgrade_c8000v_bt7.py - PuTTY output*

(continued)

CHAPTER 11   UPGRADING THE APPLICATION AND ROUTERS

#	Task
⑲	I have enabled debug ip scp on c8kv01 and c8kv02, allowing me to observe and learn about the file transfer process using the SCP protocol. Your output should resemble Figures 11-3 and 11-4. Additionally, you can verify the file being uploaded using the standard dir | in command, as shown here. On c8kv01: c8kv01#**terminal monitor** # For SSH connection only c8kv01#**debug ip scp** c8kv01#**dir | in c8000v-universalk9.17.06.05a.SPA.bin** 31 -rw- 234426368 Dec 3 2023 07:20:23 +00:00 c8000v-universalk9.17.06.05a.SPA.bin *Figure 11-3. c8kv01 debug ip scp output* On c8kv02: c8kv02#**terminal monitor** # For SSH connection only c8kv02#**debug ip scp** c8kv02#**dir | in c8000v-universalk9.17.06.05a.SPA.bin** 32 -rw- 239680512 Dec 3 2023 07:20:27 +00:00 c8000v-universalk9.17.06.05a.SPA.bin

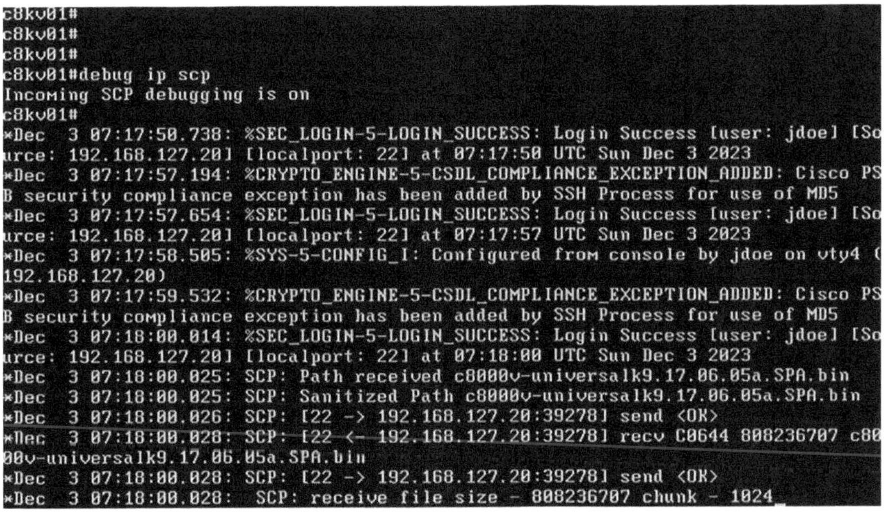

(*continued*)

#	Task

```
c8kv02#
c8kv02#
c8kv02#
c8kv02#debug ip scp
Incoming SCP debugging is on
c8kv02#
*Dec  3 07:17:53.863: %SEC_LOGIN-5-LOGIN_SUCCESS: Login Success [user: jdoe] [So
urce: 192.168.127.20] [localport: 22] at 07:17:53 UTC Sun Dec 3 2023
*Dec  3 07:17:57.177: %CRYPTO_ENGINE-5-CSDL_COMPLIANCE_EXCEPTION_ADDED: Cisco PS
B security compliance exception has been added by SSH Process for use of MD5
*Dec  3 07:17:57.792: %SEC_LOGIN-5-LOGIN_SUCCESS: Login Success [user: jdoe] [So
urce: 192.168.127.20] [localport: 22] at 07:17:57 UTC Sun Dec 3 2023
*Dec  3 07:17:59.488: %SYS-5-CONFIG_I: Configured from console by jdoe on vty4 (
192.168.127.20)
*Dec  3 07:18:00.553: %CRYPTO_ENGINE-5-CSDL_COMPLIANCE_EXCEPTION_ADDED: Cisco PS
B security compliance exception has been added by SSH Process for use of MD5
*Dec  3 07:18:01.075: %SEC_LOGIN-5-LOGIN_SUCCESS: Login Success [user: jdoe] [So
urce: 192.168.127.20] [localport: 22] at 07:18:01 UTC Sun Dec 3 2023
*Dec  3 07:18:01.106: SCP: Path received c8000v-universalk9.17.06.05a.SPA.bin
*Dec  3 07:18:01.106: SCP: Sanitized Path c8000v-universalk9.17.06.05a.SPA.bin
*Dec  3 07:18:01.108: SCP: [22 -> 192.168.127.20:48018] send <OK>
*Dec  3 07:18:01.121: SCP: [22 <- 192.168.127.20:48018] recv C0644 808236707 c80
00v-universalk9.17.06.05a.SPA.bin
*Dec  3 07:18:01.122: SCP: [22 -> 192.168.127.20:48018] send <OK>
*Dec  3 07:18:01.122: SCP: receive file size = 808236707 chunk = 1024
```

Figure 11-4. c8kv02 debug ip scp output

(20)	Figures 11-5, 11-6, and 11-7 illustrate the immediate output upon the completion of the IOS file upload. As is evident from the figures, uploading the new IOS file to both routers took 879 minutes, while the entire script execution took 933 seconds. In other words, the file upload took over 14 minutes, and the script execution completed all required tasks on both routers in over 15 minutes. Because you only have two routers in the lab setting, you are testing this under optimal conditions. In the production network, it might take slightly longer; however, the increase in the number of devices does not necessarily increase the script execution time with network automation. Python loops handle repetition, eliminating human error and leaving only machine errors, which is the key to efficiency.

(continued)

CHAPTER 11 UPGRADING THE APPLICATION AND ROUTERS

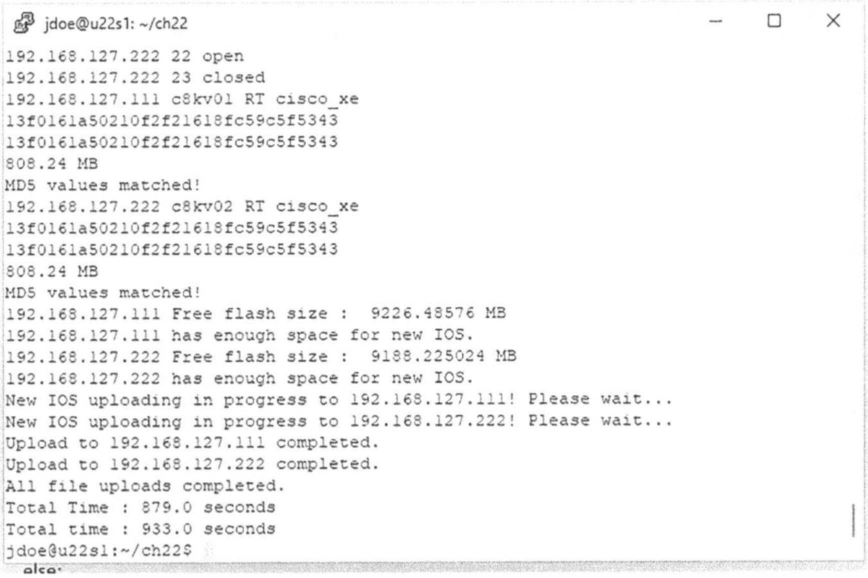

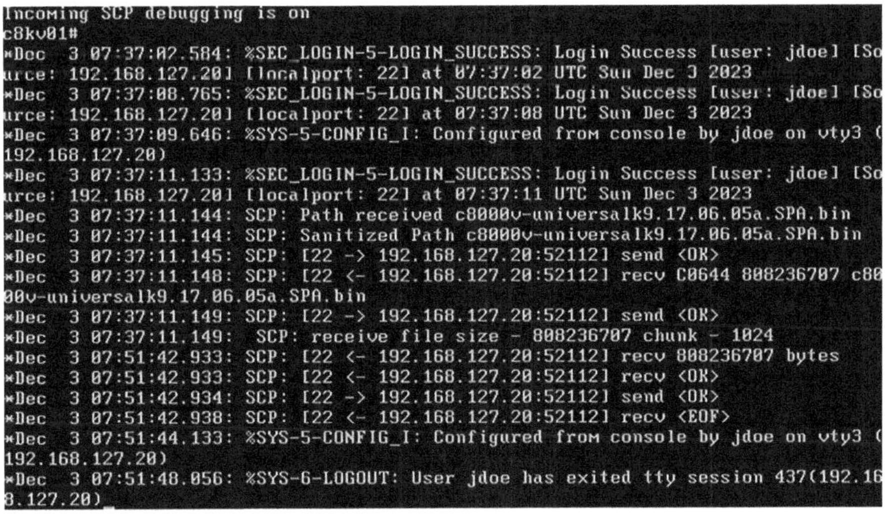

#	Task
	``
jdoe@u22s1:~/ch22
192.168.127.222 22 open
192.168.127.222 23 closed
192.168.127.111 c8kv01 RT cisco_xe
13f0161a50210f2f21618fc59c5f5343
13f0161a50210f2f21618fc59c5f5343
808.24 MB
MD5 values matched!
192.168.127.222 c8kv02 RT cisco_xe
13f0161a50210f2f21618fc59c5f5343
13f0161a50210f2f21618fc59c5f5343
808.24 MB
MD5 values matched!
192.168.127.111 Free flash size : 9226.48576 MB
192.168.127.111 has enough space for new IOS.
192.168.127.222 Free flash size : 9188.225024 MB
192.168.127.222 has enough space for new IOS.
New IOS uploading in progress to 192.168.127.111! Please wait...
New IOS uploading in progress to 192.168.127.222! Please wait...
Upload to 192.168.127.111 completed.
Upload to 192.168.127.222 completed.
All file uploads completed.
Total Time : 879.0 seconds
Total time : 933.0 seconds
jdoe@u22s1:~/ch22$
```

*Figure 11-5.* *u22s1 IOS upload completed*

On c8kv01:

c8kv01#**dir | in c8000v-universalk9.17.06.05a.SPA.bin**

31      -rw-       808236707    Dec 3 2023 06:17:36 +00:00   c8000v-universalk9.17.06.05a.SPA.bin

```
Incoming SCP debugging is on
c8kv01#
*Dec 3 07:37:02.584: %SEC_LOGIN-5-LOGIN_SUCCESS: Login Success [user: jdoe] [Source: 192.168.127.20] [localport: 22] at 07:37:02 UTC Sun Dec 3 2023
*Dec 3 07:37:08.765: %SEC_LOGIN-5-LOGIN_SUCCESS: Login Success [user: jdoe] [Source: 192.168.127.20] [localport: 22] at 07:37:08 UTC Sun Dec 3 2023
*Dec 3 07:37:09.646: %SYS-5-CONFIG_I: Configured from console by jdoe on vty3 (192.168.127.20)
*Dec 3 07:37:11.133: %SEC_LOGIN-5-LOGIN_SUCCESS: Login Success [user: jdoe] [Source: 192.168.127.20] [localport: 22] at 07:37:11 UTC Sun Dec 3 2023
*Dec 3 07:37:11.144: SCP: Path received c8000v-universalk9.17.06.05a.SPA.bin
*Dec 3 07:37:11.144: SCP: Sanitized Path c8000v-universalk9.17.06.05a.SPA.bin
*Dec 3 07:37:11.145: SCP: [22 -> 192.168.127.20:52112] send <OK>
*Dec 3 07:37:11.148: SCP: [22 <- 192.168.127.20:52112] recv C0644 808236707 c8000v-universalk9.17.06.05a.SPA.bin
*Dec 3 07:37:11.149: SCP: [22 -> 192.168.127.20:52112] send <OK>
*Dec 3 07:37:11.149: SCP: receive file size - 808236707 chunk - 1024
*Dec 3 07:51:42.933: SCP: [22 <- 192.168.127.20:52112] recv 808236707 bytes
*Dec 3 07:51:42.933: SCP: [22 <- 192.168.127.20:52112] recv <OK>
*Dec 3 07:51:42.934: SCP: [22 -> 192.168.127.20:52112] send <OK>
*Dec 3 07:51:42.938: SCP: [22 <- 192.168.127.20:52112] recv <EOF>
*Dec 3 07:51:44.133: %SYS-5-CONFIG_I: Configured from console by jdoe on vty3 (192.168.127.20)
*Dec 3 07:51:48.056: %SYS-6-LOGOUT: User jdoe has exited tty session 437(192.168.127.20)
``` |

Figure 11-6. *c8kv01 debug ip scp output, IOS upload completed*

(*continued*)
```

CHAPTER 11 UPGRADING THE APPLICATION AND ROUTERS

#	Task
	On c8kv02: c8kv02#**dir	in c8000v-universalk9.17.06.05a.SPA.bin** 32 -rw- 808236707 Dec 3 2023 06:17:43 +00:00 c8000v-universalk9.17.06.05a.SPA.bin *Figure 11-7. c8kv02 debug ip scp output, IOS upload completed*
㉑	This task involves utilizing the excerpt from md5_verify1.py (located in tool8_md5_cisco) to perform MD5 verification of the newly uploaded IOS image on the router's flash memory and compare it to the MD5 value calculated on the server. Once again, you leverage the power of Regular Expressions to test if Cisco IOS XE's verify command has been successfully executed by examining the "Verified" characters in the command's output. Subsequently, you extract the 32-character MD5 value using the re (Regular Expression) compiler and use it as a variable to validate the MD5 value. The try/except block effectively captures most potential errors that may arise during the process. jdoe@u22s1:~/ch11_upgrade_ios$ **cp upgrade_c8000v_bt7.py upgrade_c8000v_bt8.py** jdoe@u22s1:~/ch11_upgrade_ios$ **nano upgrade_c8000v_bt8.py** [...omitted for brevity]

(continued)

#	Task
| | # Trigger ios_upload() application.<br># ios_upload(device_list_netmiko, device_list) # Disable the last IOS uploading app to save time, and enable it later for end-to-end testing.<br><br>```python
ios_upload(device_list_netmiko, device_list)
def verify_ios_md5(device, device_info):
ip = device['host']
newios = device_info[4]
newiosmd5 = device_info[5]
net_connect = ConnectHandler(**device)

try:
locate_newios = net_connect.send_command(f"show flash: | in {newios}")
if newios in locate_newios:
result = net_connect.send_command(f"verify /md5 flash:{newios} {newiosmd5}")
print(result)
p1 = re.compile(r'Verified')
p2 = re.compile(r'[a-fA-F0-9]{31}[a-fA-F0-9]')
verified = p1.findall(result)
newiosmd5flash = p2.findall(result)
if verified:
result = True
print(f"{ip} - MD5 values MATCH! Continue")
print("MD5 of new IOS on Server : ", newiosmd5)
print("MD5 of new IOS on flash  : ", newiosmd5flash[0])
return True
else:
result = False
print(f"{ip} - MD5 values DO NOT MATCH! Exiting.")
exit()
``` |

(continued)
```

#	Task
| | ```
        else:
            print("No new IOS found on router's flash. Continue to next
            device...")
    except (NetMikoTimeoutException):
        print(f'Timeout error to : {ip}')
    except unknown_error:
        print('Unknown error occurred : ' + str(unknown_error))
        return False

def verify_md5(device_list_netmiko, device_list):
    for device in device_list_netmiko:
        device["read_timeout_override"] = 90 # Must add this netmiko
                                        attribute to counter delay issue
    # print(device_list_netmiko)
    threads = []
    for device, device_info in zip(device_list_netmiko, device_list):
        thread = threading.Thread(target=verify_ios_md5, args=(device,
        device_info))
        thread.start()
        threads.append(thread)

verify_md5(device_list_netmiko, device_list)

# tt_ios_upload = time.mktime(time.localtime()) - t2
# print("Total Time : {0} seconds".format(tt_ios_upload)) # Time
taken to upload IOS file

tt = time.mktime(time.localtime()) - t1
print("Total time : {0} seconds".format(tt)) # Timer finish to show
total time (tt)
``` |

(continued)
```

#	Task
㉒	Let's ensure the integrity of the updated Python code by testing and validating it. Execute python upgrade_c8000v_bt8.py to verify the changes. This update will invoke the verify command on the routers to validate the MD5 values and compare them to the original values you gathered earlier. ``` jdoe@u22s1:~/ch11_upgrade_ios$ python upgrade_c8000v_bt8.py Enter Network Admin ID : jdoe Enter Network Admin PWD : cisco123 Confirm Network Admin PWD : cisco123 Is secret same as password? (y/n) : y [...omitted for brevity] ....Done! Verified (bootflash:c8000v-universalk9.17.06.05a.SPA.bin) = 13f0161a50210f2f21618fc59c5f5343 192.168.127.111 - MD5 values MATCH! Continue MD5 of new IOS on Server : 13f0161a50210f2f21618fc59c5f5343 MD5 of new IOS on flash : 13f0161a50210f2f21618fc59c5f5343 [...omitted for brevity] ....Done! Verified (bootflash:c8000v-universalk9.17.06.05a.SPA.bin) = 13f0161a50210f2f21618fc59c5f5343 192.168.127.222 - MD5 values MATCH! Continue MD5 of new IOS on Server : 13f0161a50210f2f21618fc59c5f5343 MD5 of new IOS on flash : 13f0161a50210f2f21618fc59c5f5343 ```

(continued)

#	Task
| ㉓ | Under normal circumstances, this step may not be necessary. However, to enhance the application's realism, I have decided to introduce an interactive element where, upon completing a new IOS upload, the user is prompted to choose whether to reload the router by responding with y or n. This constitutes a significant portion of the script. As soon as the user responds with "y" and presses the Enter key, the script proceeds to reconfigure the router's boot system configuration and save the running configuration to the startup configuration.<br><br>The excerpts of this section have been borrowed from reload_yes_no_in_parallel.py (tool9_yes_no). Also, to make this script work properly, you must update the read_timout_override value to 360 or more. You originally set this value to 90 in upgrade_c8000v_bt8.py, but you can increase this time to give the routers enough time to go through the normal POST, load the new IOS XE image into the RAM, and settle down. In my environment, it took around 300 seconds, so I have increased this value by another minute, which is 360 seconds. Alternatively, you can leave this value to 90 seconds like before and use time.sleep(270), which will have the identical effect.<br><br>```
jdoe@u22s1:~/ch11_upgrade_ios$ cp upgrade_c8000v_bt8.py upgrade_c8000v_bt9.py
jdoe@u22s1:~/ch11_upgrade_ios$ nano upgrade_c8000v_bt9.py
[...omitted for brevity]
verify_md5(device_list_netmiko, device_list)
[...omitted for brevity]
  def verify_md5(device_list_netmiko, device_list):
    for device in device_list_netmiko:
      device["read_timeout_override"] = 360 # changed this value to
        increase the timeout to 6 minutes, routers take about 5 minutes
        to reload in my lab settings, so adjust this time accordingly.
    print(device_list_netmiko)
    threads = []
    for device, device_info in zip(device_list_netmiko, device_list):
      thread = threading.Thread(target=verify_ios_md5, args=(device,
        device_info))
      thread.start()
      threads.append(thread)
``` |

(continued)
```

#	Task
| | `verify_md5(device_list_netmiko, device_list)`<br>`##################################################################`<br>```
def reload_device(device):
  try:
    net_connect = ConnectHandler(**device)
    net_connect.enable()
    print(f"Reloading {device['host']}...")
    output = net_connect.send_command_timing("reload", strip_
    prompt=False, strip_command=False)
    if "confirm" in output:
      output += net_connect.send_command_timing("\n", strip_
      prompt=False, strip_command=False)
    if "Reload scheduled" in output:
      print(f"Reload command successful for {device['host']}")
    else:
       print(f"Reload command failed for {device['host']}")
    net_connect.disconnect()
  except Exception as e:
    print(f"Connection error with {device['host']}: {e}")

def show_clock(device):
  try:
    net_connect = ConnectHandler(**device)
    net_connect.enable()
    print(f"Connecting to {device['host']}...")
    output = net_connect.send_command("show clock")
    print(f"Time on {device['host']}:")
    print(output)
    net_connect.disconnect()
  except Exception as e:
    print(f"Connection error with {device['host']}: {e}")
``` |

(continued)
```

#	Task

```
def reload_devices(device_list_netmiko):
  threads = []
  for device in device_list_netmiko:
    thread = threading.Thread(target=reload_device, args=(device,))
    thread.start()
    threads.append(thread)

  for thread in threads:
    thread.join()

#     time.sleep(270)  # Sleep for 2 minutes, use this sleep timer to
                       add more delays for router reboot as needed

  threads = []
  for device in device_list_netmiko:
    thread = threading.Thread(target=show_clock, args=(device,))
    thread.start()
    threads.append(thread)

  for thread in threads:
    thread.join()

def reload_yes_no_in_parallel(device_list_netmiko):
  time.sleep(10) # Slow down the script for 10 seconds
  print("*"*79)
  resp = input("Would you like to reload your devices? (y/n): ")
  .lower()
    if resp == 'y':
      reload_devices(device_list_netmiko)
    elif resp == 'n':
      print("You chose not to reload the devices.")
    else:
      print("Invalid input. Please enter 'y' or 'n'.")
```

(continued)

#	Task
	time.sleep(10) **reload_yes_no_in_parallel(device_list_netmiko)** **###** # tt_ios_upload = time.mktime(time.localtime()) - t2 # print("Total Time : {0} seconds".format(tt_ios_upload)) # Time taken to upload IOS file tt = time.mktime(time.localtime()) - t1 print("Total time : {0} seconds".format(tt)) # Timer finish to show total time (tt)
24	jdoe@u22s1:~/ch11_upgrade_ios$ **python upgrade_c8000v_bt9.py** [...omitted for brevity] Would you like to reload your devices? (y/n): **y** Reloading 192.168.127.222... Reloading 192.168.127.111... Connection error with 192.168.127.111: read_channel_timing's absolute timer expired. The network device was continually outputting data for longer than 360 seconds. If this is expected (the command you are executing is continually generating data for a long time), you can set read_timeout=x' seconds. If you want netmiko to keep reading indefinitely (i.e., to stop only when there is no new data), you can set read_timeout=0. You can look at the netmiko session_log or debug log for more information. Connection error with 192.168.127.222: read_channel_timing's absolute timer expired.

(continued)

#	Task
	The network device was continually outputting data for longer than 360 seconds.

If this is expected (i.e., the command you are executing is continually emitting data for a long time), you can set read_timeout=x' seconds. If you want netmiko to keep reading indefinitely (i.e., to stop only when there is no new data), you can set read_timeout=0.

You can look at the netmiko session_log or debug log for more information.

```
Connecting to 192.168.127.111...
Connecting to 192.168.127.222...
Time on 192.168.127.111:
*21:22:28.268 UTC Mon Dec 4 2023
Time on 192.168.127.222:
*21:22:27.435 UTC Mon Dec 4 2023
Total time : 430.0 seconds
``` |
| ㉕ | Now you'll add the function to capture the show commands. You will be using the excerpt from reload_yes_no1.py:

```
jdoe@u22s1:~/ch11_upgrade_ios$ cp upgrade_c8000v_bt9.py upgrade_c8000v_bt10.py
jdoe@u22s1:~/ch11_upgrade_ios$ nano upgrade_c8000v_bt10.py
[...omitted for brevity]
for thread in threads:
    thread.join()
######################################################################
# Function to capture and show command outputs before the device reload
def show_and_capture(device):
    ip = device['host']
``` |

(continued)

#	Task

```python
    try:
        net_connect = ConnectHandler(**device)
        net_connect.enable()
        with open(f'{ip}_showver_pre.txt', 'w+') as f1:
            showver_pre = net_connect.send_command("show version")
            f1.write(showver_pre)
            time.sleep(1)
        with open(f'{ip}_showrun_pre.txt', 'w+') as f2:
            showrun_pre = net_connect.send_command("show running-config")
            f2.write(showrun_pre)
            time.sleep(1)
        with open(f'{ip}_showint_pre.txt', 'w+') as f3:
            showint_pre = net_connect.send_command("show ip interface
            brief")
            f3.write(showint_pre)
            time.sleep(1)
        with open(f'{ip}_showroute_pre.txt', 'w+') as f4:
            showroute_pre = net_connect.send_command("show ip route")
            f4.write(showroute_pre)
            time.sleep(1)
        net_connect.disconnect()
    except Exception as e:
        print(f"Connection error with {device['host']}: {e}")

# Function to run threads for each device
def run_show_and_capture(device_list_netmiko):
    threads = []
    for device in device_list_netmiko:
        thread = threading.Thread(target=show_and_capture,
        args=(device,))
        threads.append(thread)
        thread.start()
```

(continued)

#	Task
| | ```
# Wait for all threads to complete
for thread in threads:
  thread.join()
#######################################################
def reload_yes_no_in_parallel(device_list_netmiko):
  time.sleep(10)
  print("*"*79)
  # Run the threads using the function
  run_show_and_capture(device_list_netmiko) # Running this function
                                            before asking for a
                                            reload
  print("*"*79)
  resp = input("Would you like to reload your devices? (y/n): ")
  .lower()
  if resp == 'y':
    # Run the router reload command
    reload_devices(device_list_netmiko)
  elif resp == 'n':
    print("You chose not to reload the devices.")
  else:
    print("Invalid input. Please enter 'y' or 'n'.")

time.sleep(10)
reload_yes_no_in_parallel(device_list_netmiko)

tt = time.mktime(time.localtime()) - t1
print("Total time : {0} seconds".format(tt)) # Timer finish to show
total time (tt)
``` |

(continued)
```

#	Task
㉖	Running tests to validate the updated Python code is your immediate task. At the end of the test run, you should have eight files, four from each router, as shown here: ``` jdoe@u22s1:~/ch11_upgrade_ios$ python upgrade_c8000v_bt10.py [...omitted for brevity] jdoe@u22s1:~/ch11_upgrade_ios$ ls -lh 192* -rw-rw-r-- 1 jdoe jdoe 323 Dec 4 22:27 192.168.127.111_showint_pre.txt -rw-rw-r-- 1 jdoe jdoe 1.1K Dec 4 22:27 192.168.127.111_showroute_pre.txt -rw-rw-r-- 1 jdoe jdoe 6.2K Dec 4 22:27 192.168.127.111_showrun_pre.txt -rw-rw-r-- 1 jdoe jdoe 2.5K Dec 4 22:27 192.168.127.111_showver_pre.txt -rw-rw-r-- 1 jdoe jdoe 323 Dec 4 22:27 192.168.127.222_showint_pre.txt -rw-rw-r-- 1 jdoe jdoe 1.1K Dec 4 22:27 192.168.127.222_showroute_pre.txt -rw-rw-r-- 1 jdoe jdoe 6.1K Dec 4 22:27 192.168.127.222_showrun_pre.txt -rw-rw-r-- 1 jdoe jdoe 2.5K Dec 4 22:27 192.168.127.222_showver_pre.txt ```

(continued)

CHAPTER 11 UPGRADING THE APPLICATION AND ROUTERS

#	Task

㉗	As a best practice, change the boot system command right at the end to avoid any problems.
	
	`jdoe@u22s1:~/ch11_upgrade_ios$ `**`cp upgrade_c8000v_bt10.py upgrade_c8000v_bt11.py`**
	`jdoe@u22s1:~/ch11_upgrade_ios$ `**`nano upgrade_c8000v_bt11.py`**
	[...omitted for brevity]
	`#######################################################`
	`def change_boot_var(device, device_info):`
	`  ip = device['host']`
	`  newios = device_info[4]`
	`  net_connect = ConnectHandler(**device)`
	
	`  try:`
	`    net_connect = ConnectHandler(**device)`
	`    net_connect.enable(cmd='enable 15')`
	`    config_commands1 = ['no boot system', 'boot system flash:/' + newios, 'do write memory']`
	`    output = net_connect.send_config_set(config_commands1)`
	`    print(output)`
	`  except (NetMikoTimeoutException):`
	`    print(f'Timeout error to : {ip}')`
	`  except unknown_error:`
	`    print('Unknown error occurred : ' + str(unknown_error))`
	`  return False`
	
	`def run_change_boot_var(device_list_netmiko, device_list):`
	`  print(device_list_netmiko)`
	`  threads = []`
	`  for device, device_info in zip(device_list_netmiko, device_list):`
	`    thread = threading.Thread(target=change_boot_var, args=(device, device_info))`
	`    thread.start()`
	`    threads.append(thread)`

(continued)

#	Task

```
###########################################################
def reload_yes_no_in_parallel(device_list_netmiko, device_list):
# Add the device_list list
  time.sleep(10)
  print("*"*79)
  # Run the threads using the function
  run_show_and_capture(device_list_netmiko)
  print("*"*79)
  resp = input("Would you like to reload your devices? (y/n): ").lower()
  if resp == 'y':
    # Change the boot variable to the new IOS version
    run_change_boot_var(device_list_netmiko, device_list)
    # Add the new function
    # Run the router reload command
    reload_devices(device_list_netmiko)
  elif resp == 'n':
    print("You chose not to reload the devices.")
  else:
    print("Invalid input. Please enter 'y' or 'n'.")

time.sleep(10)
reload_yes_no_in_parallel(device_list_netmiko, device_list)
# Add the device_list list

tt = time.mktime(time.localtime()) - t1
print("Total time : {0} seconds".format(tt)) # Timer finish to show
                                               total time (tt)
```

(continued)

#	Task
| ㉘ | After you run this command, when you get to the y or n for the router reload, the run_change_boot_var() will execute first when you run y, before running the reload_devices() function. The output should look like the following.<br><br>```
jdoe@u22s1:~/ch11_upgrade_ios$ python upgrade_c8000v_bt11.py
[...omitted for brevity]
configure terminal
Enter configuration commands, one per line.  End with CNTL/Z.
c8kv01(config)#no boot system
c8kv01(config)#boot system flash:/c8000v-universalk9.17.06.05a.SPA.bin
c8kv01(config)#do write memory
Building configuration...
[OK]
c8kv01(config)#end
c8kv01#
configure terminal
Enter configuration commands, one per line.  End with CNTL/Z.
c8kv02(config)#no boot system
c8kv02(config)#boot system flash:/c8000v-universalk9.17.06.05a.SPA.bin
c8kv02(config)#do write memory
Building configuration...
[OK]
c8kv02(config)#end
c8kv02#
[...omitted for brevity]
``` |

(*continued*)
```

#	Task
29	Now copy the last build and test file and name it upgrade_c8000v_bt12.py.

```
jdoe@u22s1:~/ch11_upgrade_ios$ cp upgrade_c8000v_bt11.py upgrade_
c8000v_bt12.py
jdoe@u22s1:~/ch11_upgrade_ios$ nano upgrade_c8000v_bt12.py
[...omitted for brevity]
##############################################################
def post_check(device):
  try:
    ip = device['host']
    net_connect = ConnectHandler(**device)
    net_connect.enable(cmd='enable 15')
    net_connect.send_command('terminal length 0\n')

    with open(f'{ip}_showver_post.txt', 'w+') as f1:
      print("Capturing post-reload 'show version'")
      showver_post = net_connect.send_command("show version")
      f1.write(showver_post)
      time.sleep(1)

    with open(f'{ip}_showrun_post.txt', 'w+') as f2:
      print("Capturing post-reload 'show running-config'")
      showrun_post = net_connect.send_command("show running-config")
      f2.write(showrun_post)
      time.sleep(1)

    with open(f'{ip}_showint_post.txt', 'w+') as f3:
      print("Capturing post-reload 'show ip interface brief'")
      showint_post = net_connect.send_command("show ip interface
      brief")
      f3.write(showint_post)
      time.sleep(1)
``` |

(continued)

#	Task

```python
    with open(f'{ip}_showroute_post.txt', 'w+') as f4:
      print("Capturing post-reload 'show ip route'")
      showroute_post = net_connect.send_command("show ip route")
      f4.write(showroute_post)
      time.sleep(1)

    # Compare pre vs post configurations
    showver_pre = "showver_pre"
    showver_post = "showver_post"
    showver_pre_lines = open(f"{ip}_showver_pre.txt").readlines()
    showver_post_lines = open(f"{ip}_showver_post.txt").readlines()
    difference = difflib.HtmlDiff(wrapcolumn=70).make_file(showver_
    pre_lines, showver_post_lines, showver_pre, showver_post)
    difference_report = open(f"{ip}_show_ver_compared.html", "w+")
    difference_report.write(difference) # Writes the differences to
    html file
    difference_report.close()
    time.sleep(1)

    showrun_pre = "showrun_pre"
    showrun_post = "showrun_post"
    showrun_pre_lines = open(f"{ip}_showrun_pre.txt").readlines()
    showrun_post_lines = open(f"{ip}_showrun_post.txt").readlines()
    difference = difflib.HtmlDiff(wrapcolumn=70).make_file(showrun_
    pre_lines, showrun_post_lines, showrun_pre, showrun_post)
    difference_report = open(f"{ip}_show_run_compared.html", "w+")
    difference_report.write(difference)
    difference_report.close()
    time.sleep(1)

    showint_pre = "showint_pre"
    showint_post = "showint_post"
    showint_pre_lines = open(f"{ip}_showint_pre.txt").readlines()
```

(continued)

#	Task

```python
    showint_post_lines = open(f"{ip}_showint_post.txt").readlines()
    difference = difflib.HtmlDiff(wrapcolumn=70).make_file(showint_
    pre_lines, showint_post_lines, showint_pre, showint_post)
    difference_report = open(f"{ip}_show_int_compared.html", "w+")
    difference_report.write(difference)
    difference_report.close()
    time.sleep(1)

    showroute_pre = "showroute_pre"
    showroute_post = "showroute_post"
    showroute_pre_lines = open(f"{ip}_showroute_pre.txt").readlines()
    showroute_post_lines = open(f"{ip}_showroute_post.txt").
    readlines()
    difference = difflib.HtmlDiff(wrapcolumn=70).make_file(showroute_
    pre_lines, showroute_post_lines, showroute_pre, showroute_post)
    difference_report = open(f"{ip}_show_route_compared.html", "w+")
    difference_report.write(difference)
    difference_report.close()
    time.sleep(1)
############################################################
def change_boot_var(device, device_info):
    ip = device['host']
    newios = device_info[4]
    net_connect = ConnectHandler(**device)

    try:
        net_connect = ConnectHandler(**device)
        net_connect.enable(cmd='enable 15')
        config_commands1 = ['no boot system', 'boot system flash:/' +
        newios, 'do write memory']
        output = net_connect.send_config_set(config_commands1)
        print(output)
    except (NetMikoTimeoutException):
        print(f'Timeout error to : {ip}')
```

(continued)

#	Task

```
      except unknown_error:
        print('Unknown error occurred : ' + str(unknown_error))
      return False

  def run_change_boot_var(device_list_netmiko, device_list):
    print(device_list_netmiko)
    threads = []
    for device, device_info in zip(device_list_netmiko, device_list):
      thread = threading.Thread(target=change_boot_var, args=(device,
      device_info))
      thread.start()
      threads.append(thread)
############################################################
```
[...omitted for brevity]

run_change_boot_var(device_list_netmiko, device_list)
[...omitted for brevity]

```
time.sleep(10)
reload_yes_no_in_parallel(device_list_netmiko, device_list) # Add the
device_list list

tt = time.mktime(time.localtime()) - t1
print("Total time : {0} seconds".format(tt)) # Timer finish to show
total time (tt)
```

(continued)

#	Task
| ㉚ | You are almost there. Before moving forward, let's test and validate the recent modifications in the Python code. Execute the second last script. You should get 16 more files, eight files for the post-show command captures and eight comparison HTML files, which is the same result you saw in the last development lab of Chapter 10.<br><br>```
jdoe@u22s1:~/ch11_upgrade_ios$ python upgrade_c8000v_bt12.py
[...omitted for brevity]
Capturing post-reload 'show version'
Capturing post-reload 'show version'
Capturing post-reload 'show running-config'
Capturing post-reload 'show running-config'
Capturing post-reload 'show ip interface brief'
Capturing post-reload 'show ip interface brief'
Capturing post-reload 'show ip route'
Capturing post-reload 'show ip route'
Total Time for IOS upload: 706.0 seconds # 11.75 minutes to upload
IOS in my env.
Total time to reload and compare: 393.0 seconds # 6.6 minutes to
reload to completion
Total time to run all applications: 1128.0 seconds # 18.82 minutes
end-to-end
```<br><br>To expedite the testing process, you can comment out all the code sections associated with IOS uploading. You can choose to execute the full script, but bear in mind that each test script will take longer to run. In my testing, the end-to-end runs consistently clocked in under 19 minutes. However, given the variability of different environments, your runtime might differ. Use your discretion to adjust the timers accordingly. Now, execute the ls -lh 192* command to verify the newly created files.<br><br>```
jdoe@u22s1:~/ch11_upgrade_ios$ ls -lh 192*
-rw-rw-r-- 1 jdoe jdoe 5.3K Dec  5 00:13 192.168.127.111_show_int_
compared.html
-rw-rw-r-- 1 jdoe jdoe  11K Dec  5 00:13 192.168.127.111_show_route_
compared.html
``` |

(continued)

#	Task
| | ```
-rw-rw-r-- 1 jdoe jdoe  75K Dec  5 00:13 192.168.127.111_show_run_
compared.html
-rw-rw-r-- 1 jdoe jdoe  26K Dec  5 00:13 192.168.127.111_show_ver_
compared.html
-rw-rw-r-- 1 jdoe jdoe  323 Dec  5 00:13 192.168.127.111_showint_
post.txt
-rw-rw-r-- 1 jdoe jdoe  323 Dec  5 00:06 192.168.127.111_showint_pre.txt
-rw-rw-r-- 1 jdoe jdoe 1.1K Dec  5 00:13 192.168.127.111_showroute_
post.txt
-rw-rw-r-- 1 jdoe jdoe 1.1K Dec  5 00:06 192.168.127.111_showroute_
pre.txt
-rw-rw-r-- 1 jdoe jdoe 6.2K Dec  5 00:13 192.168.127.111_showrun_
post.txt
-rw-rw-r-- 1 jdoe jdoe 6.2K Dec  5 00:06 192.168.127.111_showrun_pre.txt
-rw-rw-r-- 1 jdoe jdoe 2.5K Dec  5 00:13 192.168.127.111_showver_
post.txt
-rw-rw-r-- 1 jdoe jdoe 2.5K Dec  5 00:06 192.168.127.111_showver_pre.txt
-rw-rw-r-- 1 jdoe jdoe 5.3K Dec  5 00:13 192.168.127.222_show_int_
compared.html
-rw-rw-r-- 1 jdoe jdoe  11K Dec  5 00:13 192.168.127.222_show_route_
compared.html
-rw-rw-r-- 1 jdoe jdoe  74K Dec  5 00:13 192.168.127.222_show_run_
compared.html
-rw-rw-r-- 1 jdoe jdoe  26K Dec  5 00:13 192.168.127.222_show_ver_
compared.html
-rw-rw-r-- 1 jdoe jdoe  323 Dec  5 00:13 192.168.127.222_showint_
post.txt
-rw-rw-r-- 1 jdoe jdoe  323 Dec  5 00:06 192.168.127.222_showint_pre.txt
``` |

(continued)
```

CHAPTER 11   UPGRADING THE APPLICATION AND ROUTERS

#	Task
	-rw-rw-r-- 1 jdoe jdoe 1.1K Dec 5 00:13 192.168.127.222_showroute_post.txt -rw-rw-r-- 1 jdoe jdoe 1.1K Dec 5 00:06 192.168.127.222_showroute_pre.txt -rw-rw-r-- 1 jdoe jdoe 6.1K Dec 5 00:13 192.168.127.222_showrun_post.txt -rw-rw-r-- 1 jdoe jdoe 6.1K Dec 5 00:06 192.168.127.222_showrun_pre.txt -rw-rw-r-- 1 jdoe jdoe 2.5K Dec 5 00:13 192.168.127.222_showver_post.txt -rw-rw-r-- 1 jdoe jdoe 2.5K Dec 5 00:06 192.168.127.222_showver_pre.txt
31	Use the VMware Workstation snapshot to revert to the original Cisco IOS XE version (Cisco IOS XE Software, Version 17.06.03a). Your version may be different if you used other IOS XE software. Both c8kv01 and c8kv02 must be reverted to the original version and ready for the ultimate IOS upgrade test run (see Figures 11-8 and 11-9).

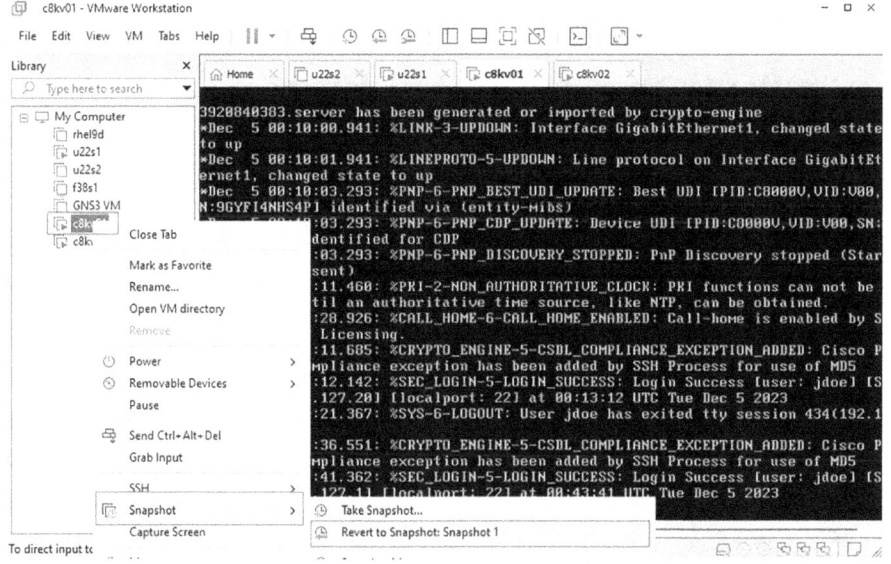

Figure 11-8. Revert to the original IOS XE using the snapshot

(*continued*)

#	Task

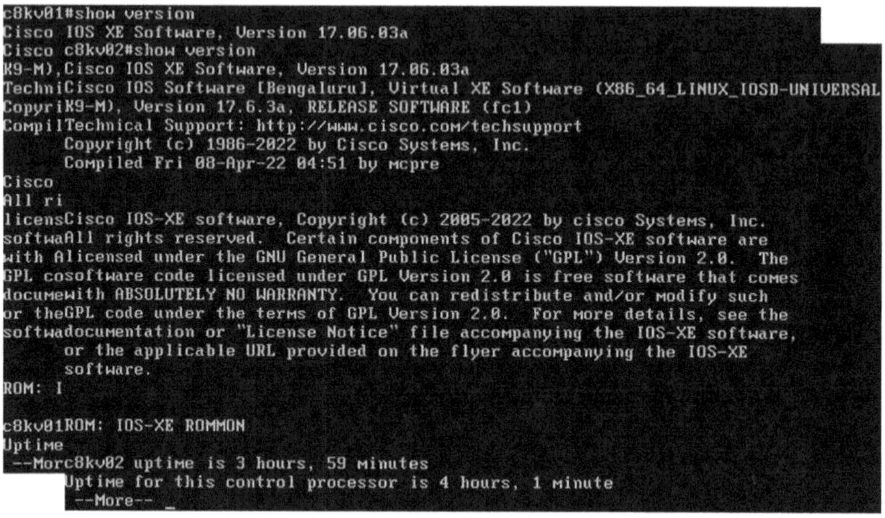

```
c8kv01#show version
Cisco IOS XE Software, Version 17.06.03a
Cisco c8kv02#show version
K9-M),Cisco IOS XE Software, Version 17.06.03a
TechniCisco IOS Software [Bengaluru], Virtual XE Software (X86_64_LINUX_IOSD-UNIVERSAL
CopyriK9-M), Version 17.6.3a, RELEASE SOFTWARE (fc1)
CompilTechnical Support: http://www.cisco.com/techsupport
      Copyright (c) 1986-2022 by Cisco Systems, Inc.
      Compiled Fri 08-Apr-22 04:51 by mcpre
Cisco
All ri
licensCisco IOS-XE software, Copyright (c) 2005-2022 by cisco Systems, Inc.
softwAll rights reserved. Certain components of Cisco IOS-XE software are
with Alicensed under the GNU General Public License ("GPL") Version 2.0. The
GPL cosoftware code licensed under GPL Version 2.0 is free software that comes
documewith ABSOLUTELY NO WARRANTY. You can redistribute and/or modify such
or theGPL code under the terms of GPL Version 2.0. For more details, see the
softwadocumentation or "License Notice" file accompanying the IOS-XE software,
      or the applicable URL provided on the flyer accompanying the IOS-XE
      software.
ROM: I

c8kv01ROM: IOS-XE ROMMON
Uptime
  --Morc8kv02 uptime is 3 hours, 59 minutes
      Uptime for this control processor is 4 hours, 1 minute
      --More--
```

Figure 11-9. c8kv01 and c8kv02 reverted IOS XE

(32) Because you reverted, you go back to the point where you still have not configured the vital aaa configurations. For this script to work, you need to configure the three missing aaa configurations on your devices.

aaa new-model
aaa authentication login default local enable
aaa authorization exec default local

Write a quick interactive configuration script to configure the aaa settings. Refer to the following URL for more information about this problem. https://italchemy.wordpress.com/2020/08/17/netmiko-scp-ios-upload-error-troubleshooting/

jdoe@u22s1:~/ch11_upgrade_ios$ **nano configure_aaa.py**
from netmiko import ConnectHandler
import getpass

def get_credentials():
username = input("Enter username: ")
password = getpass.getpass("Enter password: ")
secret = getpass.getpass("Enter secret: ")
return username, password, secret

(continued)

#	Task
| | ```python
device_list_netmiko = [
{'device_type': 'cisco_xe', 'host': '192.168.127.111'},
{'device_type': 'cisco_xe', 'host': '192.168.127.222'}
]

def configure_aaa(device, username, password, secret):
try:
device['username'] = username
device['password'] = password
device['secret'] = secret

net_connect = ConnectHandler(**device)
net_connect.enable()
config_commands = [
'aaa new-model',
'aaa authentication login default local enable',
'aaa authorization exec default local'
]
output = net_connect.send_config_set(config_commands)

# Save configuration after applying AAA configuration
save_config = net_connect.save_config()

print(f"Configuration successful on {device['host']}:")
print(output)
print(save_config)
net_connect.disconnect()
except Exception as e:
print(f"Failed to configure AAA on {device['host']}: {str(e)}")
``` |

(continued)
```

CHAPTER 11   UPGRADING THE APPLICATION AND ROUTERS

#	Task
| | ```
def main():
    username, password, secret = get_credentials()
    for device in device_list_netmiko:
    configure_aaa(device, username, password, secret)

if __name__ == "__main__":
    main()
``` |
| (33) | When you run this script, your aaa configurations will be configured on your routers. You are now ready to run the end-to-end IOS upgrade script.<br><br>```
jdoe@u22s1:~/ch11_upgrade_ios$ python configure_aaa.py
Enter username: jdoe
Enter password: cisco123
Enter secret: cisco123
Configuration successful on 192.168.127.111:
configure terminal
Enter configuration commands, one per line.  End with CNTL/Z.
c8kv01(config)#aaa new-model
c8kv01(config)#aaa authentication login default local enable
c8kv01(config)#aaa authorization exec default local
c8kv01(config)#end
c8kv01#
write mem
Building configuration...
[OK]
c8kv01#
Configuration successful on 192.168.127.222:
configure terminal
Enter configuration commands, one per line.  End with CNTL/Z.
c8kv02(config)#aaa new-model
c8kv02(config)#aaa authentication login default local enable
c8kv02(config)#aaa authorization exec default local
c8kv02(config)#end
``` |

(continued)

#	Task
	c8kv02# write mem Building configuration... [OK] c8kv02#
34	Now copy (overwrite) the build and test file 12 as upgrade_c8000v.py. Reorder the applications so the functions are declared first. Then place all function triggers at the end of the script. Open the upgrade_c8000v.py file and you will see the re-ordered final script. Save the updated Python file. The two files are identical and upgrade_c8000v_bt12.py is in the final build and test file. jdoe@u22s1:~/ch11_upgrade_ios$ **cp upgrade_c8000v_bt12.py upgrade_c8000v.py** jdoe@u22s1:~/ch11_upgrade_ios$ **nano upgrade_c8000v.py** [...omitted for brevity]
35	Finally, execute the final Cisco IOS Upgrade application and follow the prompts. jdoe@u22s1:~/ch11_upgrade_ios$ **python upgrade_c8000v_clean.py** Enter Network Admin ID : **jdoe** Enter Network Admin PWD : **cisco123** Confirm Network Admin PWD : **cisco123** Is secret same as password? (y/n) : **y** ['c8kv01', 'RT', 'cisco_xe', '192.168.127.111', 'c8000v-universalk9.17.06.05a.SPA.bin', '13f0161a50210f2f21618fc59c5f5343'] ['c8kv02', 'RT', 'cisco_xe', '192.168.127.222', 'c8000v-universalk9.17.06.05a.SPA.bin', '13f0161a50210f2f21618fc59c5f5343'] 192.168.127.111 PING 192.168.127.111 (192.168.127.111) 56(84) bytes of data. 64 bytes from 192.168.127.111: icmp_seq=1 ttl=255 time=14.5 ms 64 bytes from 192.168.127.111: icmp_seq=2 ttl=255 time=0.627 ms 64 bytes from 192.168.127.111: icmp_seq=3 ttl=255 time=0.731 ms 64 bytes from 192.168.127.111: icmp_seq=4 ttl=255 time=0.532 ms

(continued)

CHAPTER 11 UPGRADING THE APPLICATION AND ROUTERS

#	Task
| | ```
--- 192.168.127.111 ping statistics ---
4 packets transmitted, 4 received, 0% packet loss, time 3022ms
rtt min/avg/max/mdev = 0.532/4.107/14.541/6.024 ms
192.168.127.111 22 open
192.168.127.111 23 closed
192.168.127.222
PING 192.168.127.222 (192.168.127.222) 56(84) bytes of data.
64 bytes from 192.168.127.222: icmp_seq=1 ttl=255 time=1.52 ms
64 bytes from 192.168.127.222: icmp_seq=2 ttl=255 time=0.896 ms
64 bytes from 192.168.127.222: icmp_seq=3 ttl=255 time=0.472 ms
64 bytes from 192.168.127.222: icmp_seq=4 ttl=255 time=0.705 ms

--- 192.168.127.222 ping statistics ---
4 packets transmitted, 4 received, 0% packet loss, time 3029ms
rtt min/avg/max/mdev = 0.472/0.898/1.522/0.389 ms
192.168.127.222 22 open
192.168.127.222 23 closed
192.168.127.111 c8kv01 RT cisco_xe
13f0161a50210f2f21618fc59c5f5343
13f0161a50210f2f21618fc59c5f5343
808.24 MB
MD5 values matched!
192.168.127.222 c8kv02 RT cisco_xe
13f0161a50210f2f21618fc59c5f5343
13f0161a50210f2f21618fc59c5f5343
808.24 MB
MD5 values matched!
192.168.127.111 Free flash size :   10065.34656 MB
192.168.127.111 has enough space for new IOS.
192.168.127.222 Free flash size :   10064.326656 MB
192.168.127.222 has enough space for new IOS.
``` |

(continued)
```

#	Task
	New IOS uploading in progress to 192.168.127.111! Please wait... New IOS uploading in progress to 192.168.127.222! Please wait... Upload to 192.168.127.111 completed. Upload to 192.168.127.222 completed. All file uploads completed. [{'device_type': 'cisco_xe', 'host': '192.168.127.111', 'username': 'jdoe', 'password': 'cisco123', 'secret': 'cisco123', 'read_timeout_override': 360}, {'device_type': 'cisco_xe', 'host': '192.168.127.222', 'username': 'jdoe', 'password': 'cisco123', 'secret': 'cisco123', 'read_timeout_override': 360}] [{'device_type': 'cisco_xe', 'host': '192.168.127.111', 'username': 'jdoe', 'password': 'cisco123', 'secret': 'cisco123', 'read_timeout_override': 360}, {'device_type': 'cisco_xe', 'host': '192.168.127.222', 'username': 'jdoe', 'password': 'cisco123', 'secret': 'cisco123', 'read_timeout_override': 360}] configure terminal Enter configuration commands, one per line. End with CNTL/Z. c8kv02(config)#no boot system c8kv02(config)#boot system flash:/c8000v-universalk9.17.06.05a.SPA.bin c8kv02(config)#do write memory Building configuration... [OK] c8kv02(config)#end c8kv02# configure terminal Enter configuration commands, one per line. End with CNTL/Z. c8kv01(config)#no boot system c8kv01(config)#boot system flash:/c8000v-universalk9.17.06.05a.SPA.bin c8kv01(config)#do write memory

(*continued*)

#	Task
	Building configuration... [OK] c8kv01(config)#end c8kv01# *** 192.168.127.222 - MD5 values MATCH! Continue MD5 of new IOS on Server : 13f0161a50210f2f21618fc59c5f5343 MD5 of new IOS on flash : 13f0161a50210f2f21618fc59c5f5343 192.168.127.111 - MD5 values MATCH! Continue MD5 of new IOS on Server : 13f0161a50210f2f21618fc59c5f5343 MD5 of new IOS on flash : 13f0161a50210f2f21618fc59c5f5343 *** Would you like to reload your devices? (y/n): **y** # You must answer y to reload! Reloading 192.168.127.111... Reloading 192.168.127.222... Connection error with 192.168.127.111: read_channel_timing's absolute timer expired. The network device was continually outputting data for longer than 360 seconds. If this is expected (i.e., the command you are executing is continually emitting data for a long time), you can set read_timeout=x seconds. If you want netmiko to keep reading indefinitely (i.e., to stop only when there is no new data), you can set read_timeout=0. You can look at the netmiko session_log or debug log for more information.

(continued)

#	Task
	Connection error with 192.168.127.222:

read_channel_timing's absolute timer expired.

The network device was continually outputting data for longer than 360 seconds.

If this is expected (i.e., the command you are executing is continually emitting data for a long time), you can set read_timeout=x seconds. If you want netmiko to keep reading indefinitely (i.e., to stop only when there is no new data), you can set read_timeout=0.

You can look at the netmiko session_log or debug log for more information.

```
Capturing post-reload 'show version'
Capturing post-reload 'show version'
Capturing post-reload 'show running-config'
Capturing post-reload 'show running-config'
Capturing post-reload 'show ip interface brief'
Capturing post-reload 'show ip interface brief'
Capturing post-reload 'show ip route'
Capturing post-reload 'show ip route'
Total Time for IOS upload: 705.0 scconds # 11.75 minutes to upload IOS in my env.
Total time to reload and compare: 394.0 seconds # 6.6 minutes to reload to completion
Total time to run all applications: 1129.0 seconds # 18.82 minutes end-to-end
```

These times can vary from one environment to another. In the production environment, it may be faster if you remove the interactive code and simply run the end-to-end application. Note that your times will be different from my tests runs and that is normal. |

(continued)

CHAPTER 11   UPGRADING THE APPLICATION AND ROUTERS

#	Task
㊱	Now you can run the `ls -lh 192*` command. You will see 16 newly created files, eight of which are the post-capture files and eight of which are comparison files between the pre-captured and post-captured files. ``` jdoe@u22s1:~/ch11_upgrade_ios$ ls -lh 192* -rw-rw-r-- 1 jdoe jdoe 5.6K Dec 6 18:19 192.168.127.111_show_int_compared.html -rw-rw-r-- 1 jdoe jdoe 11K Dec 6 18:19 192.168.127.111_show_route_compared.html -rw-rw-r-- 1 jdoe jdoe 74K Dec 6 18:19 192.168.127.111_show_run_compared.html -rw-rw-r-- 1 jdoe jdoe 27K Dec 6 18:19 192.168.127.111_show_ver_compared.html -rw-rw-r-- 1 jdoe jdoe 323 Dec 6 18:19 192.168.127.111_showint_post.txt -rw-rw-r-- 1 jdoe jdoe 323 Dec 6 18:13 192.168.127.111_showint_pre.txt -rw-rw-r-- 1 jdoe jdoe 1.1K Dec 6 18:19 192.168.127.111_showroute_post.txt -rw-rw-r-- 1 jdoe jdoe 1.1K Dec 6 18:13 192.168.127.111_showroute_pre.txt -rw-rw-r-- 1 jdoe jdoe 6.1K Dec 6 18:19 192.168.127.111_showrun_post.txt -rw-rw-r-- 1 jdoe jdoe 6.1K Dec 6 18:13 192.168.127.111_showrun_pre.txt -rw-rw-r-- 1 jdoe jdoe 2.5K Dec 6 18:19 192.168.127.111_showver_post.txt -rw-rw-r-- 1 jdoe jdoe 2.5K Dec 6 18:13 192.168.127.111_showver_pre.txt -rw-rw-r-- 1 jdoe jdoe 5.3K Dec 6 18:19 192.168.127.222_show_int_compared.html ```

(*continued*)

#	Task
| | ```
-rw-rw-r-- 1 jdoe jdoe  11K Dec  6 18:19 192.168.127.222_show_route_compared.html
-rw-rw-r-- 1 jdoe jdoe  74K Dec  6 18:19 192.168.127.222_show_run_compared.html
-rw-rw-r-- 1 jdoe jdoe  27K Dec  6 18:19 192.168.127.222_show_ver_compared.html
-rw-rw-r-- 1 jdoe jdoe  323 Dec  6 18:19 192.168.127.222_showint_post.txt
-rw-rw-r-- 1 jdoe jdoe  323 Dec  6 18:13 192.168.127.222_showint_pre.txt
-rw-rw-r-- 1 jdoe jdoe 1.1K Dec  6 18:19 192.168.127.222_showroute_post.txt
-rw-rw-r-- 1 jdoe jdoe 1.1K Dec  6 18:13 192.168.127.222_showroute_pre.txt
-rw-rw-r-- 1 jdoe jdoe 6.1K Dec  6 18:19 192.168.127.222_showrun_post.txt
-rw-rw-r-- 1 jdoe jdoe 6.1K Dec  6 18:13 192.168.127.222_showrun_pre.txt
-rw-rw-r-- 1 jdoe jdoe 2.5K Dec  6 18:19 192.168.127.222_showver_post.txt
-rw-rw-r-- 1 jdoe jdoe 2.5K Dec  6 18:13 192.168.127.222_showver_pre.txt
``` |

(continued)
```

CHAPTER 11   UPGRADING THE APPLICATION AND ROUTERS

#	Task
㊲	Open WinSCP on your host PC, connect to your Python server (u22s1), and copy the contents of the ch11_upgrade_ios directory to your Windows Downloads folder (see Figure 11-10).

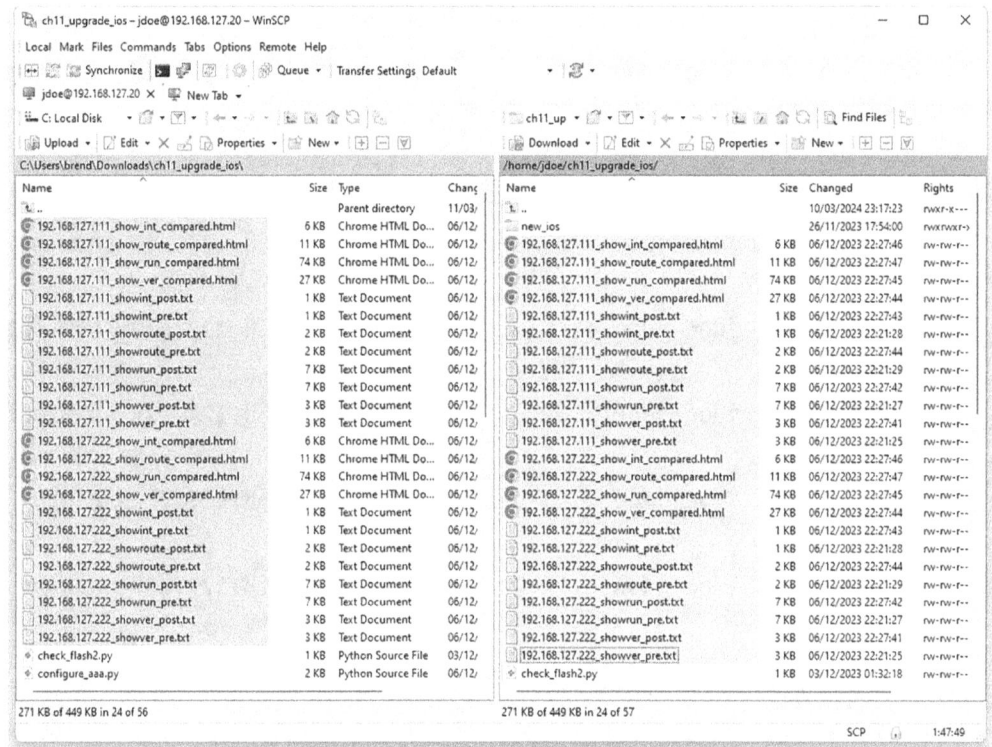

Figure 11-10. WinSCP copy ch11_upgrade_ios files to Windows PC

(continued)

CHAPTER 11   UPGRADING THE APPLICATION AND ROUTERS

#	Task
38	Open the HTML file on your favorite web browser and confirm that the upgrade was successful. If your IOS upgrade was successful, you should see the version changes between the pre-captured and post-captured `show version` output files (see Figure 11-11).

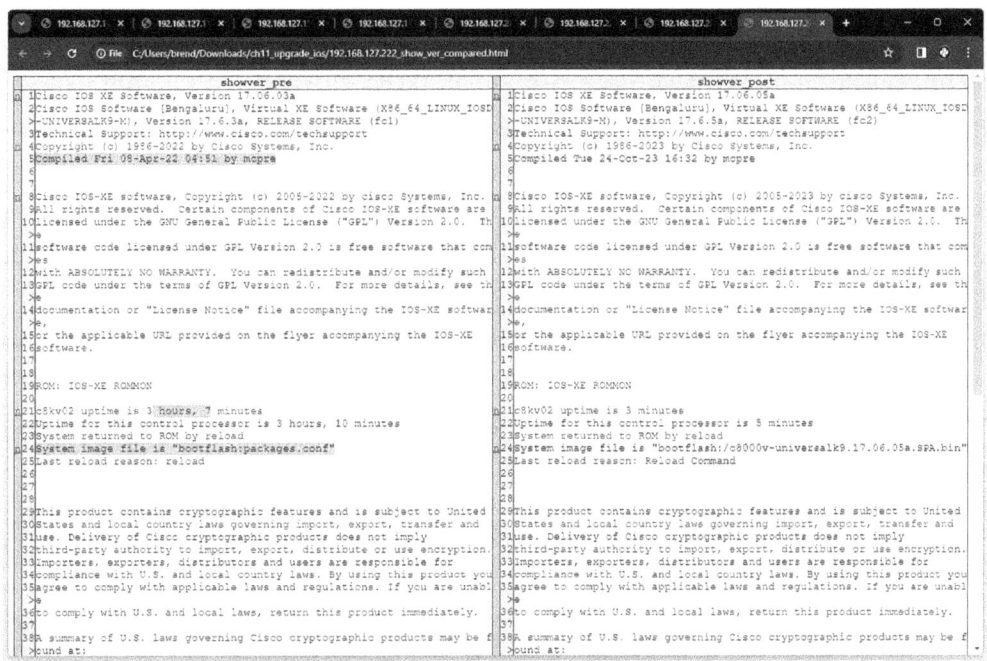

Figure 11-11. *Open HTML files for final validation*

You have tackled one of the most challenging chapters of this book, following all the tasks detailed in this chapter, and successfully upgraded your IOS images on your routers. You can revisit the snapshots and repeatedly run your tests for validation and revalidation of your IOS upgrade Python script. You have worked up to this point and conquered a small mountain. Now, as you approach the bonus chapter, take a moment to unwind and contemplate your journey through the last 11 chapters. Dive into Chapter 12 at your own pace, relax, and enjoy the process.

Summary

In real-world enterprise production environments, scripts are divided into distinct functional units known as *modules*. This modular approach streamlines the main script by separating individual applications into distinct modules stored in separate files. When these module files are combined, they form libraries. Essentially, you created a library of modules in Chapters 9 and 10, which you've unpacked to form a single Cisco IOS XE upgrading application with approximately 500 lines of code. In real-life scenarios, developing 500-line applications is impractical due to their complexity and reader-unfriendliness. Code is written once but read more often. To facilitate your understanding, I have consolidated the entire application into a single Python script, which makes the application a little bit like Frankenstein. However, you've likely noticed the inefficiency of working with lengthy code. While I could split the code into separate modules and re-import functions as modules, I will leave this challenge for you as your homework. Move to the bonus chapter, where you'll construct an IPAM/DCIM (IP Address Management/Data Center Infrastructure Management) server on an Ubuntu Server for your company using a Python script.

CHAPTER 12

Installing NetBox (IPAM/DCIM) with Python

This chapter delves into the fundamental process of installing NetBox, a comprehensive IP Address Management (IPAM) and Data Center Infrastructure Management (DCIM) tool built on the Python-based, Django Web framework. The chapter begins by outlining the significance of NetBox and its essential features while clarifying what it isn't. The chapter elaborates on server specifications, dependencies, and core functions vital for NetBox operations. It focuses on the central concept of the NetBox "site" and provides a high-level overview of its IP address management capabilities. Additionally, it explores the transformation of NetBox into multi-tenancy IPAM, offering insights into manual and automated installations of NetBox 3.6 on Ubuntu 22.04 LTS through Python scripting. This chapter serves as a comprehensive guide to understanding, configuring, and deploying NetBox effectively on Ubuntu systems, thereby enhancing network management capabilities. This will set you up for success in grasping REST API concepts by utilizing the tools you've built from the ground up.

What Is NetBox, and Why Do You Need It?

IP Address Management (IPAM) and Data Center Infrastructure Management (DCIM) software tools are crucial for simplifying IP address and data center infrastructure management for any enterprise network. They enable network administrators to automatically discover unallocated and assignable IP addresses, making it easy to provision IP addresses for devices on a network. If your company has been managing IP addresses using a single Excel spreadsheet and is seeking a viable IP address management solution that caters to its engineers and managers, NetBox serves as the single source of truth for all IP address needs. Some IPAMs only support single tenancy, while others can accommodate multi-tenancy. A single tenancy is typically designed for a single organization, whereas the multi-tenancy feature is intended to support multiple organizations under a single brand or cater to multiple customers of a Managed Service Provider (MSP).

In 2015, a network engineer named Jeremy Stretch developed NetBox to automate network provisioning at Digital Ocean. Recognizing its potential, Digital Ocean released NetBox as an open-source project in June, 2016 (`https://netbox.dev/`). Yes, this amazing tool is at no cost. In other words, NetBox is a tool created by an engineer for engineers. Since then, thousands of organizations worldwide have adopted NetBox as their central source of truth for networking, empowering both network operators and automation processes. NetBox is built on the Django Python framework and utilizes a PostgreSQL database. It operates as a WSGI service behind the HTTP server of your choice. At the time of writing this book, the latest release of NetBox from the developer was version 3.7.0, with a release date of 29/12/2023. It offers several valuable features suitable for MSP use cases. The latest iteration of NetBox, version 3.6.3 or newer, requires Python version 3.8.6 or newer, PostgreSQL version 11 or newer, and Redis version 4.0 or newer.

According to Digital Ocean's home page, NetBox is the leading solution for modeling and documenting modern networks. It combines the traditional disciplines IPAM and DCIM with powerful APIs and extensions. NetBox provides the ideal "source of truth" to power network automation. The first page (the dashboard) is shown in Figure 12-1 and a populated rack diagram example is shown in Figure 12-2, for your reference.

CHAPTER 12 INSTALLING NETBOX (IPAM/DCIM) WITH PYTHON

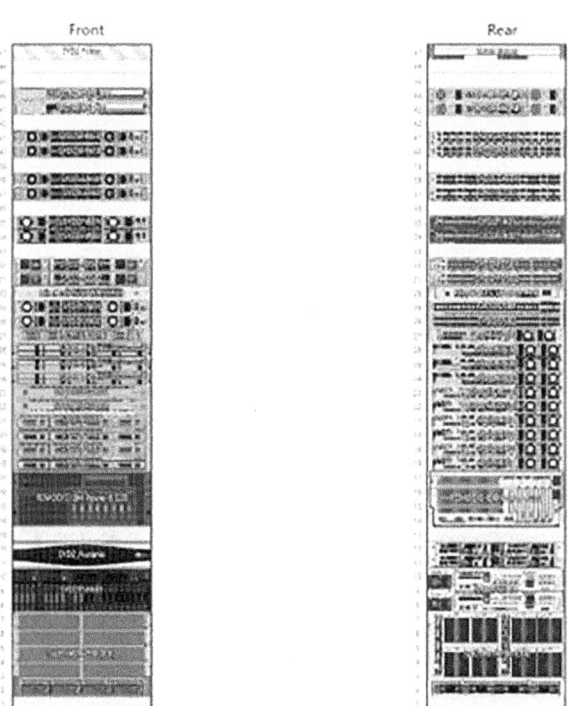

Figure 12-1. *NetBox dashboard example*

Figure 12-2. *A fully populated DC rack diagram*

717

CHAPTER 12   INSTALLING NETBOX (IPAM/DCIM) WITH PYTHON

Some NetBox Features

NetBox was purpose-built to cater to the specific needs of engineers and administrators. For a comprehensive list of the features offered by NetBox, refer to the official NetBox documentation at this link: https://docs.netbox.dev/en/stable/introduction/

NetBox's core design principles include:

- Real-world replication
- Serving as a "source of truth"
- Simplicity in use and design
- Application stack optimization

What NetBox Is Not

While NetBox supports numerous features, it is important to clarify what it does not intend to be:

- Network monitoring tool
- DNS server
- RADIUS server
- Configuration management system
- Facilities management platform

Server Specifications, Dependencies, and Functions

To install and deploy the latest NetBox version for your organization, your Linux server needs to meet all of the minimum dependencies. Although you can use various Linux distributions to set up your NetBox Server, for the convenience of this chapter, you will use an Ubuntu Server, specifically u22s1, to host the NetBox services. Both Ubuntu and NetBox Servers are open-source and do not require any commercial service fees. Administrators have the freedom to upgrade both the operating system and software whenever necessary.

First, create a table to document some important server specifications for your reference and to determine whether your servers meet the minimum requirements. Refer to Tables 12-1, 12-2, and 12-3 for the details of your IPAM server build.

Table 12-1. Host Details

Hostname	IP	OS Distribution	OS Version	Kernel Version
u22s1	192.168.127.20	Ubuntu Server (Jammy Jellyfish)	22.04.3 LTS	5.15.0-84-generic

Table 12-2. NetBox 3.6.3 Dependencies

Dependency	Minimum Version	Role
Python	3.8	The latest Django is a web framework built on Python 3.8.
PostgreSQL	11	PostgreSQL 11 is used to save data for NetBox 3.6.3
Redis	4.0	Redis 4.0 is used for an in-memory key-value store, background task, and queuing.

Table 12-3. NetBox 3.6.3 Functions and Components

Function	Component
HTTP service	Nginx or Apache
WSGI service	Gunicon or uWSGI
Application	Django/Python
Database	PostgreSQL 11+
Task queuing	Redis/Django-rq
Live device access	NAPALM (Optional)

NetBox Concept: The "Site" Is the Core of NetBox

From the perspective of IP addressing design and network architecting, there are different ways to model your network depending on the nature of your organization. Generally, a site will equate to a building or campus. Each site requires a unique name and may be assigned to a region and/or tenant. The "site" is the core of NetBox, as you can see in Figure 12-3. The site will usually be associated with geological attributes such as the region and location, tenancy or customers, and physical racks. Also, on the other side, the site is associated with Configuration Information (CI) related issues such as devices, network interfaces, VLANs, and other general IP address management information, as shown in Figure 12-3.

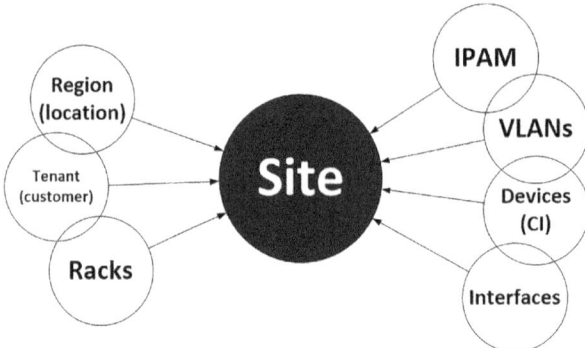

Figure 12-3. *The site is the core of NetBox*

NetBox IP Address Management (IPAM): High-Level Overview

NetBox's primary feature is IPAM, and it should serve as the source of the truth for all IP addresses used on the network. As represented in Figure 12-4, an IP address is typically associated directly with an interface (physical/virtual), and parent and child prefixes, where the interface of an IP address belongs to an IP device and the parent prefixes are associated with the aggregated network.

CHAPTER 12   INSTALLING NETBOX (IPAM/DCIM) WITH PYTHON

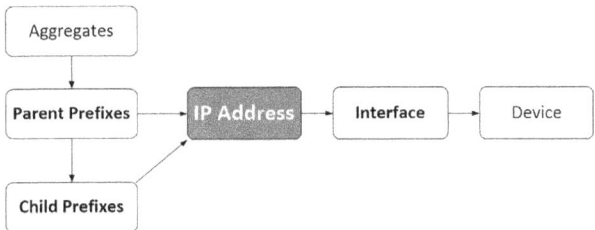

Figure 12-4. *IP address attributes in NetBox*

Changing NetBox into Multi-tenancy IPAM

If your organization is a single entity with an autonomous network and IP addresses are consistent throughout the network, but even within the same organizational network, keeping things under an acceptable level of standard becomes a challenge. Now, if your organization is an entity consisting of multiple companies or different organizations, then this will complicate things even more. The most challenging of all would be a situation where you are working for an MSP who manages multiple clients' networks, and you must keep up with tens and hundreds of IP address schemes with clashing IP addresses and subnets. With a little bit of imagination, you can tweak the regions to host several tenants/organizations. Using an MSP for an example, as depicted by Figure 12-5, you can use a customer as the region and then apply each country (or each state) as the sub-regions, then these regions will be associated with the sites. In a MSP environment, this model is suitable, as multiple organizational information will be held in the same database. If NetBox is used in a single enterprise environment, this model can be simplified based on the geolocational information.

CHAPTER 12   INSTALLING NETBOX (IPAM/DCIM) WITH PYTHON

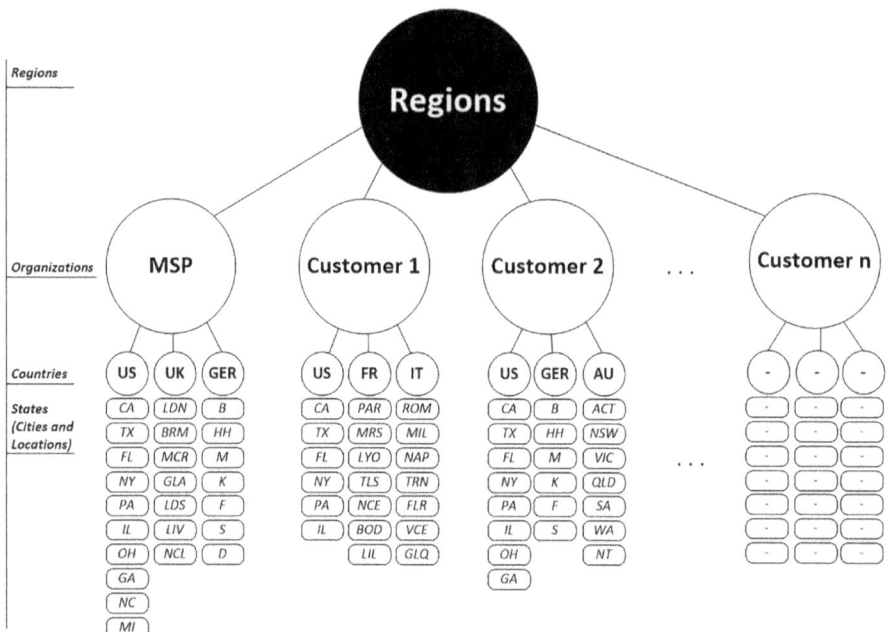

Figure 12-5. MSP modeling using regions in NetBox

It's now time to install NetBox on your Ubuntu Server. Power on your u22s1 (192.168.127.20) server and follow each task step-by-step to set up your first NetBox Server for your team and organization.

NetBox 3 Manual Installation on Ubuntu 22.04 LTS

To simplify the explanation and enhance your understanding of the NetBox installation process, you will be working on an Ubuntu 22.04.1 LTS server with Internet access. However, the installation process for NetBox on other Linux systems is nearly identical. The subsequent steps outline the complete NetBox installation process, which you can use to rebuild or troubleshoot any NetBox issues. Realistically, you should be able to build this server in under 30 minutes if you are familiar with Linux or under two hours if you are new to the Linux environment. For any upgrade procedures, follow the official NetBox documents (https://docs.netbox.dev/en/stable/introduction/).

Figure 12-6 shows that a single Ubuntu Server connected to the Internet is all you need for this lab. In this chapter, you will install NetBox on the u22s1 server using the manual way and a Python automated way. Optionally, should you prefer to set NetBox up on a dedicated server, you can power down the current server, create a complete

CHAPTER 12 INSTALLING NETBOX (IPAM/DCIM) WITH PYTHON

clone, rename it to u22s2 or your preferred server name, and then build a dedicated IPAM/DCIM server. Installing NetBox in the production environment is almost identical to the processes shown here, with more security hardening around your Linux server.

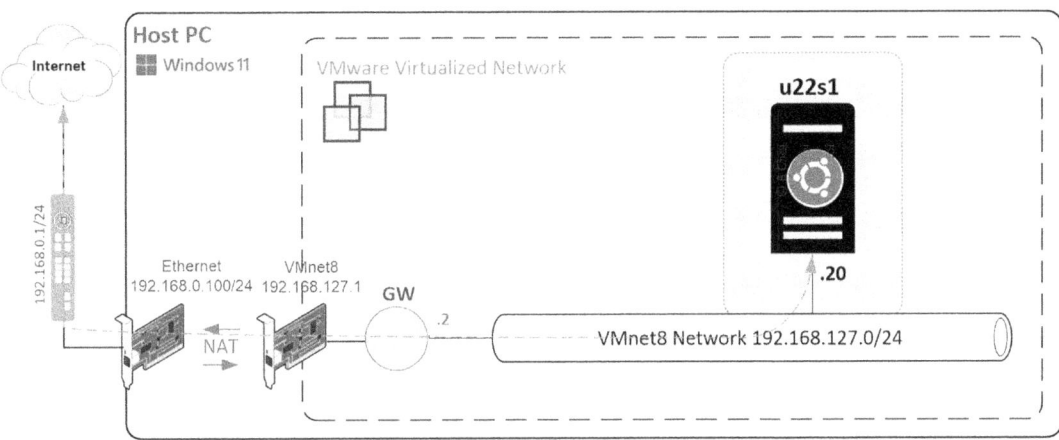

Figure 12-6. *NetBox (IPAM/DCIM) build lab device(s) in use*

Before you begin, take a snapshot of the current state of your Ubuntu Server. If you want to practice NetBox installation multiple times, it is always a good idea to take a snapshot before you begin the NetBox installation to reserve the initial state (see Figure 12-7).

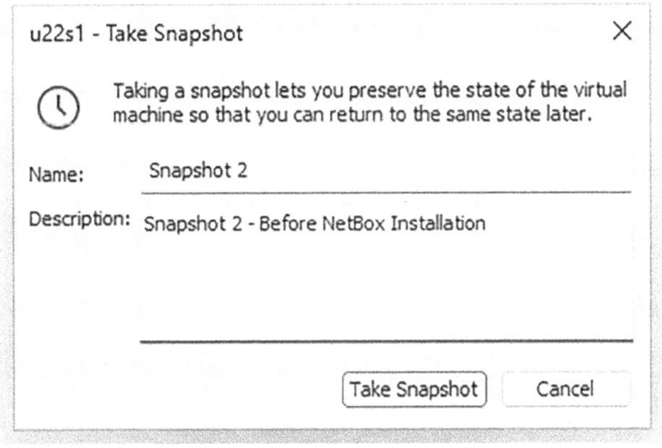

Figure 12-7. *Take a snapshot of your Ubuntu Server*

CHAPTER 12 INSTALLING NETBOX (IPAM/DCIM) WITH PYTHON

#	Task

①	Verify your Python version before you begin and then run the `apt update` command to update the package repository. Your Python version must be a minimum of 3.8 before you can proceed.
	
	`jdoe@u22s1:~$ `**`python -V`**
	`Python 3.10.12`
	`jdoe@u22s1:~$ `**`sudo apt update`**
	`[sudo] password for jdoe: ***************`

②	Install PostgreSQL on your Linux server. `jdoe@u22s1:~$ `**`sudo apt install -y postgresql`** `[sudo] password for jdoe: ***************`
③	Check the PostgreSQL version. `jdoe@u22s1:~$ `**`psql -V`** `psql (PostgreSQL) 14.9 (Ubuntu 14.9-0ubuntu0.22.04.1)`

④	Start, enable, and check the PostgreSQL service.
	
	`jdoe@u22s1:~$ `**`sudo systemctl start postgresql`**
	`jdoe@u22s1:~$ `**`sudo systemctl enable postgresql`**
	`jdoe@u22s1:~$ `**`sudo systemctl status postgresql`**
	
	Synchronizing state of postgresql.service with SysV service script with /lib/systemd/systemd-sysv-install.
	
	`Executing: /lib/systemd/systemd-sysv-install enable postgresql`
	`jdoe@u22s1:~$ sudo systemctl status postgresql`
	`• postgresql.service - PostgreSQL RDBMS`
	`Loaded: loaded (/lib/systemd/system/postgresql.service; enabled;`
	`vendor preset: enabled)`
	`Active: active (exited) since Mon 2023-10-02 15:29:33 AEDT; 3min 10s`
	`ago`
	`Main PID: 5389 (code=exited, status=0/SUCCESS)`
	`CPU: 11ms`

(continued)

CHAPTER 12   INSTALLING NETBOX (IPAM/DCIM) WITH PYTHON

#	Task
| | ```
Oct 02 15:29:33 u22s1.pynetauto.com systemd[1]: Starting PostgreSQL RDBMS...
Oct 02 15:29:33 u22s1.pynetauto.com systemd[1]: Finished PostgreSQL RDBMS.
``` |
| ⑤ | Create a database and a SQL user called netbox with a password, and then grant the database privileges to this user.<br><br>```
jdoe@u22s1:~$ sudo -u postgres psql
postgres=# CREATE DATABASE netbox;
CREATE DATABASE
postgres=# CREATE USER netbox WITH PASSWORD 'P0stgr3SQLP@ss!';
CREATE ROLE
postgres=# GRANT ALL PRIVILEGES ON DATABASE netbox TO netbox;
GRANT
postgres=# exit
``` |
| ⑥ | Test the login and confirm its functionality.<br><br>```
jdoe@u22s1:~$ psql --username netbox --password --host localhost netbox
Password: P0stgr3SQLP@ss!
psql (14.9 (Ubuntu 14.9-0ubuntu0.22.04.1))
SSL connection (protocol: TLSv1.3, cipher: TLS_AES_256_GCM_SHA384, bits: 256, compression: off)
Type "help" for help.

netbox=>
``` |

(continued)

725
```

CHAPTER 12　INSTALLING NETBOX (IPAM/DCIM) WITH PYTHON

#	Task

⑦	Next, install Redis (used for an in-memory key-value store, background task, and queuing).
	
	`jdoe@u22s1:~$ `**`sudo apt install -y redis-server`**
	`[...omitted for brevity]`
	`jdoe@u22s1:~$ `**`redis-server -v`**
	`Redis server v=6.0.16 sha=00000000:0 malloc=jemalloc-5.2.1 bits=64 build=a3fdef44459b3ad6`
	`jdoe@u22s1:~$ `**`sudo service redis-server start`**
	`jdoe@u22s1:~$ `**`redis-cli ping`**
	`PONG`
	`jdoe@u22s1:~$ `**`sudo service redis-server status`**
	`● redis-server.service - Advanced key-value store`
	`Loaded: loaded (/lib/systemd/system/redis-server.service; enabled; vendor preset: enabled)`
	`Active: active (running) since Mon 2023-10-02 15:42:11 AEDT; 1min 45s ago`
	`Docs: http://redis.io/documentation,`
	`man:redis-server(1)`
	`Main PID: 6024 (redis-server)`
	`Status: "Ready to accept connections"`
	`Tasks: 5 (limit: 4509)`
	`Memory: 2.6M`
	`CPU: 374ms`
	`CGroup: /system.slice/redis-server.service`
	`└─6024 "/usr/bin/redis-server 127.0.0.1:6379" "" "" "" "" "" "" "" "" "" "" "" "" "" "" "" "" "" "" "" "" "" "" "" "" ""`
	
	`Oct 02 15:42:11 u22s1.pynetauto.com systemd[1]: Starting Advanced key-value store...`
	`Oct 02 15:42:11 u22s1.pynetauto.com systemd[1]: Started Advanced key-value store.`

(continued)

#	Task
⑧	Install Netbox dependencies for NetBox. jdoe@u22s1:~$ **sudo apt install -y python3 python3-pip python3-venv python3-dev build-essential libxml2-dev libxslt1-dev libffi-dev libpq-dev libssl-dev zlib1g-dev**
⑨	Create a new directory under the /opt directory, named netbox. Change the working directory to /opt/netbox. Download NetBox (the git method is used here, but you can do the .tar method, which is easier). jdoe@u22s1:~$ **sudo mkdir -p /opt/netbox/** jdoe@u22s1:~$ **cd /opt/netbox/** Install git: jdoe@u22s1:/opt/netbox$ **sudo apt install -y git** Install the current version of NetBox in the current path: jdoe@u22s1:/opt/netbox$ **sudo git clone -b master --depth 1 https://github.com/netbox-community/netbox.git .** jdoe@u22s1:/opt/netbox$ **ls** CHANGELOG.md CONTRIBUTING.md LICENSE.txt NOTICE README.md SECURITY.md base_requirements.txt contrib docs mkdocs.yml netbox pyproject.toml requirements.txt scripts upgrade.sh
⑩	Create a NetBox system user and change the ownership of the media directory for any device images to be used. jdoe@u22s1:/opt/netbox$ **sudo adduser --system --group netbox** *[...omitted for brevity]* jdoe@u22s1:/opt/netbox$ **sudo chown --recursive netbox /opt/netbox/netbox/media/** #<<<holds all the uploaded files like images etc.

(continued)

#	Task

⑪	Change the directory, copy the example configuration file, and then modify it.
	
	jdoe@u22s1:/opt/netbox$ **cd /opt/netbox/netbox/netbox/**
	jdoe@u22s1:/opt/netbox/netbox/netbox$ **sudo cp configuration_example.py configuration.py**
	jdoe@u22s1:/opt/netbox/netbox/netbox$ **sudo nano configuration.py**
	...
	ALLOWED_HOSTS
	jdoe@u22s1:/opt/netbox/netbox/netbox$ sudo nano configuration.py
	ALLOWED_HOSTS = ['*']
	...
	DATABASE = {
	'NAME': 'netbox', # Database name
	'USER': '**netbox**', # PostgreSQL username
	'PASSWORD': '**P0stgr3SQLP@ss!**', # PostgreSQL password
	'HOST': 'localhost', # Database server
	'PORT': '', # Database port (leave blank for default)
	'CONN_MAX_AGE': 300, # Max database connection age (seconds)
	}
	...

⑫	Use Python to generate a secret key for your use. jdoe@u22s1:/opt/netbox/netbox/netbox$ **python3 ../generate_secret_key.py** RDqld)zethPs2$hx@iVM$cT7sH=d@)OrL=n87W@x!V8x$ux%nw This needs to be entered into configuration.py at 'SECRET_KEY', it is blank and you need to update it. jdoe@u22s1:/opt/netbox/netbox/netbox$ **sudo nano configuration.py** SECRET_KEY = 'RDqld)zethPs2$hx@iVM$cT7sH=d@)OrL=n87W@x!V8x$ux%nw'

(continued)

#	Task
⑬	Optionally, add a package you will need to install in the local_requirements.txt file. This will install any required packages in the next stage. Here, I am adding a napalm module, as I might need this at a later stage of network automation. jdoe@u22s1:~$ **sudo sh -c "echo 'napalm' >> /opt/netbox/local_requirements.txt"**
⑭	Run the upgrade script. This may take a while to complete. jdoe@u22s1:~$ **sudo /opt/netbox/upgrade.sh** OR jdoe@u22s1:~$ **sudo PYTHON=/usr/bin/python3.10 /opt/netbox/upgrade.sh**
⑮	Activate the virtual environment, change the working directory, and then create a superuser with a password (admin/admin). You will be running NetBox in a virtual environment for convenience. jdoe@u22s1:~$ **source /opt/netbox/venv/bin/activate** (venv) jdoe@u22s1:~$ **cd /opt/netbox/netbox** (venv) jdoe@u22s1:/opt/netbox/netbox$ **python3 manage.py createsuperuser** Username (leave blank to use 'jdoe'): **admin** Email address: **admin@pynetauto.com** Password: **admin** Password (again): **admin** Superuser created successfully.
⑯	Schedule the housekeeping task (still in the virtual environment) and provide the sudo password for jdoe if prompted. (venv) jdoe@u22s1:/opt/netbox/netbox$ **sudo ln -s /opt/netbox/contrib/netbox-housekeeping.sh /etc/cron.daily/netbox-housekeeping** [sudo] password for jdoe: **MySuperSecretPassword1**

(*continued*)

CHAPTER 12   INSTALLING NETBOX (IPAM/DCIM) WITH PYTHON

#	Task

⑰	To allow clients to connect to the Django Server, you must modify the settings.py file's ALLOWED_HOSTS setting to allow all connections. Disable the original line and add the new line as ALLOWED_HOSTS = ['*'], indicating that you want to allow all traffic to this NetBox Server.
	
	(venv) jdoe@u22s1:/opt/netbox/netbox$ **sudo nano /opt/netbox/netbox/netbox/settings.py**
	...
	`# Set required parameters`
	`#ALLOWED_HOSTS = getattr(configuration, 'ALLOWED_HOSTS')`
	`ALLOWED_HOSTS = ['*']`
	...

⑱	At this point, NetBox is ready to run. Test the basic functionality of the web using the runserver (Django) command for testing. (For development only, at this stage.) (venv) jdoe@u22s1:/opt/netbox/netbox$ **python3 manage.py runserver 0.0.0.0:8000 --insecure** Performing system checks... System check identified no issues (0 silenced). October 02, 2023 - 05:20:07 Django version 4.2.5, using settings 'netbox.settings' Starting development server at http://0.0.0.0:8000/ Quit the server with CONTROL-C. [02/Oct/2023 05:20:37] "GET / HTTP/1.1" 200 29120 [02/Oct/2023 05:20:37] "GET /static/setmode.js HTTP/1.1" 200 3506 [02/Oct/2023 05:20:37] "GET /static/netbox-light.css?v=3.6.3 HTTP/1.1" 200 232824 [02/Oct/2023 05:20:37] "GET /static/netbox-external.css?v=3.6.3 HTTP/1.1" 200 349159 [02/Oct/2023 05:20:37] "GET /static/netbox-dark.css?v=3.6.3 HTTP/1.1" 200 375651

(continued)

#	Task
| | ```
[02/Oct/2023 05:20:37] "GET /static/netbox.js?v=3.6.3 HTTP/1.1" 200 529868
[02/Oct/2023 05:20:37] "GET /static/netbox_logo.svg HTTP/1.1" 200 4719
[02/Oct/2023 05:20:37] "GET /static/netbox_icon.svg HTTP/1.1" 200 835
[02/Oct/2023 05:20:37] "GET /static/netbox-print.css?v=3.6.3 HTTP/1.1" 200 728021
[02/Oct/2023 05:20:38] "GET /static/materialdesignicons-webfont-ER2MFQKM.woff2?v=7.0.96 HTTP/1.1" 200 385360
[02/Oct/2023 05:20:38] "GET /static/netbox.ico HTTP/1.1" 200 1174
^C(venv) jdoe@u22s1:/opt/netbox/netbox^C
(venv) jdoe@u22s1:/opt/netbox/netbox$ sudo nano /opt/netbox/netbox/netbox/settings.py
(venv) jdoe@u22s1:/opt/netbox/netbox$ python3 manage.py runserver 0.0.0.0:8000 --insecure
Performing system checks...

System check identified no issues (0 silenced).
October 02, 2023 - 05:28:07
Django version 4.2.5, using settings 'netbox.settings'
Starting development server at http://0.0.0.0:8000/
Quit the server with CONTROL-C.
``` |

(continued)
```

CHAPTER 12   INSTALLING NETBOX (IPAM/DCIM) WITH PYTHON

#	Task
⑲	Open the web browser and the NetBox development server page. If everything goes as planned, you should be able to log in with `admin/admin` as your username and password. See Figure 12-8. `http://192.168.127.20:8000/` 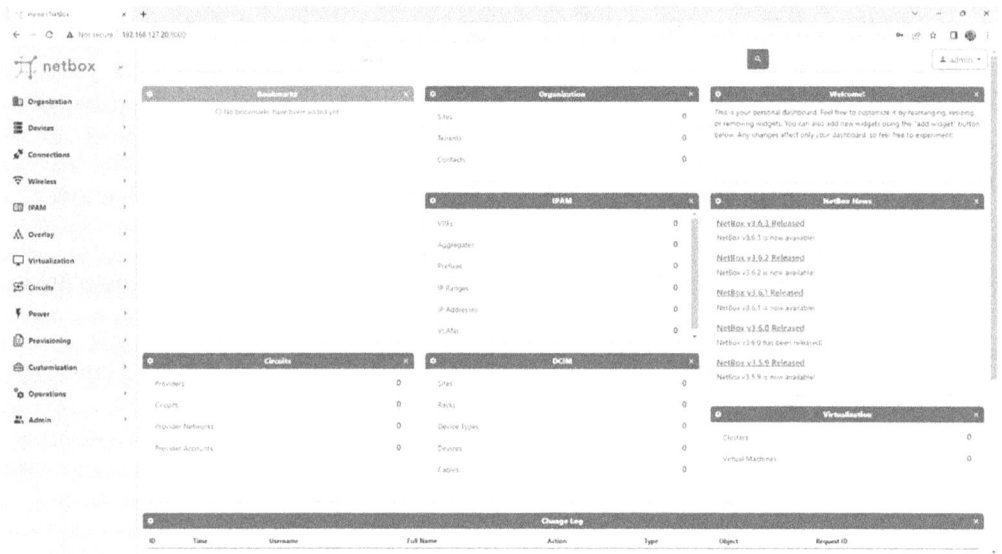 *Figure 12-8. NetBox test login page*
⑳	You still have some more work to do to make this server fully functional. Use the Ctrl+C (^C) to stop the NetBox DevOps server.

(*continued*)

#	Task

21	In the Gunicorn server (uWSGI server sitting behind HTTP and allowing long-running servers), copy the gunicorn.py file from /opt/netbox/contrib/gunicorn.py to /opt/netbox/gunicorn.py.
	
	(venv) jdoe@u22s1:/opt/netbox/netbox$ **sudo cp /opt/netbox/contrib/gunicorn.py /opt/netbox/gunicorn.py**
	(venv) jdoe@u22s1:/opt/netbox/netbox$ **cd ..**
	(venv) jdoe@u22s1:/opt/netbox$ **cat gunicorn.py**
	
	``
	# The IP address (typically localhost) and port that the NetBox WSGI
	process should listen on
	bind = '127.0.0.1:8001'
	# Number of gunicorn workers to spawn. This should typically be 2n+1, where
	# n is the number of CPU cores present.
	workers = 5
	
	# Number of threads per worker process
	threads = 3
	
	# Timeout (in seconds) for a request to complete
	timeout = 120
	
	# The maximum number of requests a worker can handle before being respawned
	max_requests = 5000
	max_requests_jitter = 500
	``

(continued)

CHAPTER 12   INSTALLING NETBOX (IPAM/DCIM) WITH PYTHON

#	Task
㉒	Use the sudo systemd (systemctl) commands setup, start, and enable with NetBox. (venv) jdoe@u22s1:/opt/netbox$ **sudo cp -v /opt/netbox/contrib/*.service /etc/systemd/system/** '/opt/netbox/contrib/netbox-housekeeping.service' -> '/etc/systemd/system/netbox-housekeeping.service' '/opt/netbox/contrib/netbox-rq.service' -> '/etc/systemd/system/netbox-rq.service' '/opt/netbox/contrib/netbox.service' -> '/etc/systemd/system/netbox.service' (venv) jdoe@u22s1:/opt/netbox$ **sudo systemctl daemon-reload** (venv) jdoe@u22s1:/opt/netbox$ **sudo systemctl start netbox netbox-rq** (venv) jdoe@u22s1:/opt/netbox$ **sudo systemctl enable netbox netbox-rq**
㉓	Now quickly check the status of the NetBox service. (venv) jdoe@u22s1:/opt/netbox$ **sudo systemctl status netbox.service** ●netbox.service - NetBox WSGI Service Loaded: loaded (/etc/systemd/system/netbox.service; enabled; vendor preset: enabled) Active: active (running) since Mon 2023-10-02 16:35:15 AEDT; 17s ago Docs: https://docs.netbox.dev/ Main PID: 7636 (gunicorn) Tasks: 6 (limit: 4509) Memory: 513.6M CPU: 9.900s CGroup: /system.slice/netbox.service ├─7636 /opt/netbox/venv/bin/python3 /opt/netbox/venv/bin/gunicorn --pid /var/tmp/netbox.pid --pythonpath /opt/netbox/netbox --config /opt/netbox/gunicorn.py netbox.wsgi ├─7638 /opt/netbox/venv/bin/python3 /opt/netbox/venv/bin/gunicorn --pid /var/tmp/netbox.pid --pythonpath /opt/netbox/netbox --config /opt/netbox/gunicorn.py netbox.wsgi

(continued)

#	Task
	├─7639 /opt/netbox/venv/bin/python3 /opt/netbox/venv/bin/gunicorn --pid /var/tmp/netbox.pid --pythonpath /opt/netbox/netbox --config /opt/netbox/gunicorn.py netbox.wsgi ├─7640 /opt/netbox/venv/bin/python3 /opt/netbox/venv/bin/gunicorn --pid /var/tmp/netbox.pid --pythonpath /opt/netbox/netbox --config /opt/netbox/gunicorn.py netbox.wsgi ├─7641 /opt/netbox/venv/bin/python3 /opt/netbox/venv/bin/gunicorn --pid /var/tmp/netbox.pid --pythonpath /opt/netbox/netbox --config /opt/netbox/gunicorn.py netbox.wsgi └─7642 /opt/netbox/venv/bin/python3 /opt/netbox/venv/bin/gunicorn --pid /var/tmp/netbox.pid --pythonpath /opt/netbox/netbox --config /opt/netbox/gunicorn.py netbox.wsgi Oct 02 16:35:15 u22s1.pynetauto.com systemd[1]: Started NetBox WSGI Service. [... omitted for brevity]
24	Install HTTP for the frontend. (Self-signed, you can get the paid SSL certificate from Let's Encrypt or Verisign.) (venv) jdoe@u22s1:/opt/netbox$ **sudo openssl req -x509 -nodes -days 365 -newkey rsa:2048 -keyout /etc/ssl/private/netbox.key -out /etc/ssl/certs/netbox.crt**

(continued)

CHAPTER 12 INSTALLING NETBOX (IPAM/DCIM) WITH PYTHON

#	Task

	`.+....++++++++++++++++++++++++++++++++++++++++++++++++++++++++++++++++`
	`++*...+....+.....+.........+..............+...........+....+.......`
	`....+...+.....+++++++++++++++++++++++++++++++++++++++++++++++++++++++`
	`++++++++++*.+.................+....+................+..+...+.+...+...+..`
	`.+...........+.........+.........+........+...........+..........+.+...`
	`+............+...+..+............+.....+....+...+.........+.+........`
	`.......+.....+...+......+.+.....+...+.......+.+...+..+....+.+........+`
	`...+..+...............+.......+....+........+.............+..........`
	`..+.+...........+.......+.+...+..+..............+..+.+.........+.`
	`.+...+.+..+..........+..+...+.+......+.........+.....+..........+.`
	`..+.....+...+.......+..+.+..+......+.+...................+......+`
	`....+...+.......+.........+......+..+.+............+..+........`
	`+...+.......+..+........+.+.....+...................+.........+..+..`
	`.+.......+...+..............................+..+.+.........+.`
	`.....+.+.................+...+....+...+...+...+..+...+.+++++++++++++`
	`+++++++++++++++++++++++++++++++++++++++++++++++++++++`
	`.+...+.......+........++++++++++++++++++++++++++++++++++++++++++++++`
	`++++++++++++++++*.........+.+.........+.+.........+.+..+..........+...+..`
	`.....+......+.+++++++++++++++++++++++++++++++++++++++++++++++++++++`
	`+++++++++*.+.+..+.....+............+...+.+......+++++++++++++++++++++`
	`++++++++++++++++++++++++++++++++++++++++++`
	`-----`
	You are about to be asked to enter information that will be incorporated
	into your certificate request.
	What you are about to enter is what is called a Distinguished Name or a DN.
	There are quite a few fields but you can leave some blank
	For some fields there will be a default value,
	If you enter '.', the field will be left blank.
	`-----`

(continued)

#	Task
	Country Name (2 letter code) [AU]: State or Province Name (full name) [Some-State]:**NSW** Locality Name (eg, city) []:Sydney Organization Name (eg, company) [Internet Widgets Pty Ltd]:**pynetauto.com** Organizational Unit Name (eg, section) []:**DevOps** Common Name (e.g. server FQDN or YOUR name) []: Email Address []:**admin@pynetauto.com**
25	Install the nginx server for the GUI. (venv) jdoe@u22s1:/opt/netbox$ **sudo apt install -y nginx**

26	Create a new site file called netbox under the /etc/ngix/sites-available directory and add the following content.
	
	(venv) jdoe@u22s1:/opt/netbox$ **sudo nano /etc/nginx/sites-available/netbox**
	
	server {
	listen [::]:443 ssl ipv6only=off;
	
	# CHANGE THIS TO YOUR SERVER'S NAME
	server_name **192.168.127.20**;
	
	ssl_certificate /etc/ssl/certs/netbox.crt;
	ssl_certificate_key /etc/ssl/private/netbox.key;
	
	client_max_body_size 25m;
	
	location /static/ {
	alias /opt/netbox/netbox/static/;
	}

(*continued*)

CHAPTER 12 INSTALLING NETBOX (IPAM/DCIM) WITH PYTHON

#	Task
| | ```
location / {
proxy_pass http://127.0.0.1:8001;
proxy_set_header X-Forwarded-Host $http_host;
proxy_set_header X-Real-IP $remote_addr;
proxy_set_header X-Forwarded-Proto $scheme;
}
}

server {
# Redirect HTTP traffic to HTTPS
listen [::]:80 ipv6only=off;
server_name _;
return 301 https://$host$request_uri;
}
``` |
| ㉗ | Delete the Nginx default and create a symlink to the new active file.

(venv) jdoe@u22s1:/opt/netbox$ **sudo rm /etc/nginx/sites-enabled/default**
(venv) jdoe@u22s1:/opt/netbox$ **sudo ln -s /etc/nginx/sites-available/netbox /etc/nginx/sites-enabled/netbox** |
| ㉘ | Restart the Nginx service to complete the manual installation.

(venv) **bchoi@u2204s:**/opt/netbox$ **sudo systemctl restart nginx** |

(*continued*)
```

CHAPTER 12   INSTALLING NETBOX (IPAM/DCIM) WITH PYTHON

#	Task
29	The latest NetBox version 3.6.3 has now been installed successfully. As shown in Figure 12-9, log in using `https`. Your IPAM server is using the self-signed certificate, so you will be prompted to accept the risk. Because this is your environment and you're using HTTPS, you can trust this site and proceed to 192.168.127.20. URL: `https://192.168.127.20/` 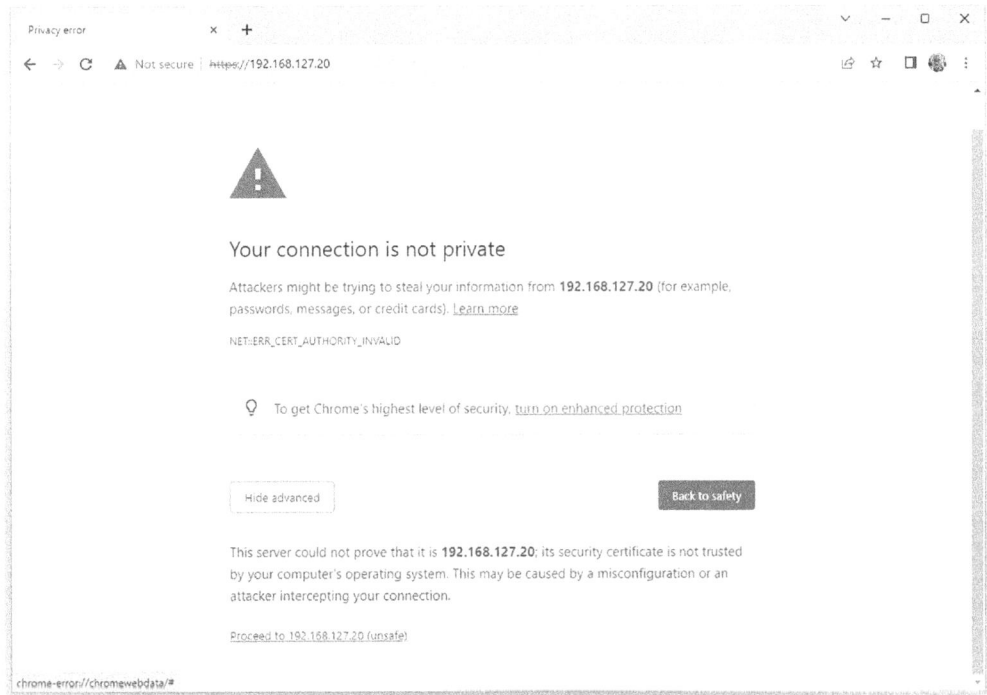 ***Figure 12-9.*** *Accept and proceed to 192.168.127.20 (unsafe)*

(*continued*)

CHAPTER 12   INSTALLING NETBOX (IPAM/DCIM) WITH PYTHON

#	Task
	Use the `admin` username and password to log in to the dashboard of your NetBox. Figure 12-10 shows the first login page. 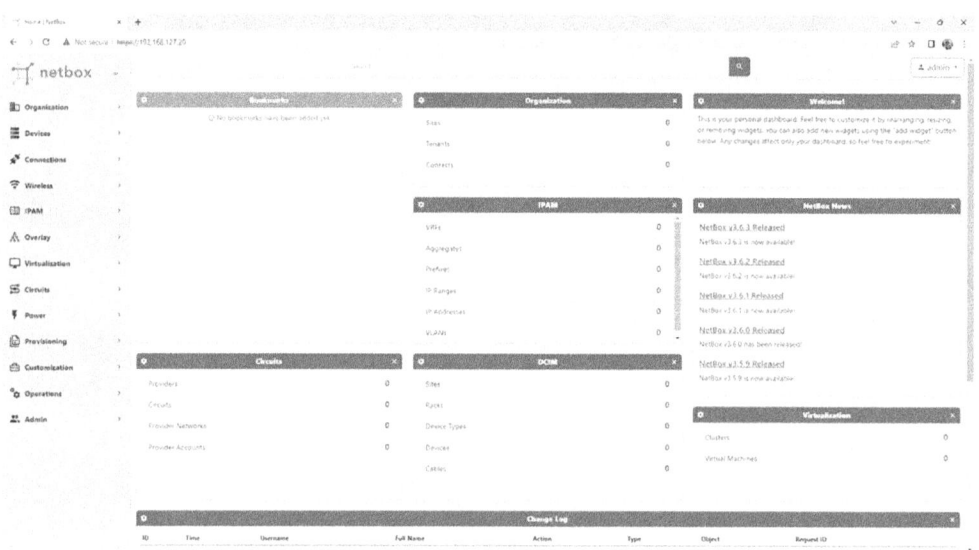 *Figure 12-10. Netbox dashboard, first page*
30	At this stage, go to your DNS server and add the hostname and IP address for DNS resolution. https://github.com/netbox-community/devicetype-library

You have successfully learned how to install your organization's first IPAM/DCIM server. NetBox, an open-source software favored by seasoned engineers for efficiently managing network and data center infrastructure, has been your tool of choice. NetBox offers API integrations with other configuration items (CI) and customer relationship management (CRM) software for data sharing. With your grasp of the manual installation method firmly in place, the next step is to transition these intricate tasks into Python scripts. The goal? To streamline the process and accomplish the provisioning of your IPAM server within a fraction of the time—minutes instead of the painstaking hours typically required.

Tip

Behind the curtain: The dual facets of working in IT

The Network Academy's LinkedIn post, "3 Things I Wish I Knew Before Working in IT," echoed a familiar sentiment in December 2023. It emphasized the importance of lifelong learning, professional skills development, and gaining experience through volunteering, projects, and certifications. This was nothing new and you all know this by now.

However, the idealistic portrayal conveniently sidestepped the harsher truths embedded within IT careers. Certification exams merely mark the outset—a phase demanding extensive personal time for studying, leaving scant room for personal lives and hobbies. The insatiable demand for dedication surpasses reasonable limits, shackling IT professionals to a perpetual 24/7 on-call existence, eroding family ties and personal well-being. The toll taken by this relentless pressure extends beyond work, often fracturing relationships and resulting in divorces. Moreover, the facade of corporate camaraderie masks a stark reality: companies, despite lauding human connections, often view employees as replaceable components. In times of illness or burnout, scant support is proffered, leaving employees discarded like a malfunctioning piece of machinery. Adding to the complexity, the human aspect of the industry carries a dual edge. While many IT professionals epitomize collaboration and support, a significant faction embodies the antithesis, fostering a toxic environment. Beyond this, the landscape of IT harbors additional perils—burnout culture pervades, diversity and inclusion remain elusive, and constant technological upheavals demand perpetual learning. Navigating this multifaceted domain demands resilience and adaptability, urging for a reevaluation of work cultures and support mechanisms. The glamour of IT often overshadows these intricate challenges, underscoring the necessity for a balanced perspective and robust support structures within the industry.

This is the reality that Network Academy failed to mention. Working in IT can be challenging, demanding, and sometimes even heartbreaking. But for those who are passionate, resilient, and possess a thick skin, it can also be incredibly rewarding. So, if you are considering an IT career, proceed with open eyes and realistic expectations. Remember, the path ahead may be difficult, but the rewards can be worth the struggle. Try to learn, grow, and share with others.

CHAPTER 12   INSTALLING NETBOX (IPAM/DCIM) WITH PYTHON

NetBox 3 Automated Python Script Installation on Ubuntu 22.04 LTS

This section begins the final coding session in this book and explains how to write the last code. Here, you will translate the manual NetBox installation tasks into simple Python code using the shell module. If you have been following along, this should be slightly less challenging than when you started your Python automation journey. Let's proceed with the step-by-step procedures to create functional Python code for installing NetBox on your Ubuntu Server.

Tip

Looking for the source code used in this chapter?

All the source code featured in this book is available for access and download via the author's GitHub or Apress's GitHub repositories. Look under the ch12_ipam_dcim folder.

Author's GitHub: https://github.com/pynetauto/apress_pynetauto_ed2.0/tree/main/source_codes/Part2/ch12_ipam_dcim

Apress GitHub:

https://github.com/Apress/

#	Task
①	First, ensure that you have saved all your work and you have reverted to Snapshot 2, taken at the beginning of this chapter. You must go back to Snapshot 2 to begin your final lab.
②	Log in to your Ubuntu Server as the root user. You must run the script as the root user to install NetBox successfully.

```
jdoe@u22s1:~$ python -V
Python 3.10.12
jdoe@u22s1:~$ su -
Password: ***************
root@u22s1:~# pwd
/root
```

(continued)

#	Task
③	Now create a working directory to save your NetBox installation Python file. `root@u22s1:~#` **`mkdir ch12_ipam_dcim && cd ch12_ipam_dcim`** `root@u22s1:~/ch12_ipam_dcim#` **`nano install_netbox.py`**
| ④ | For your installation, you need two mandatory modules, os and subprocess, and another to measure the time and time module. Add the start_time variable to mark the start of the installation.<br><br>```
import os
import subprocess
import time

# Record the start time
start_time = time.time()
``` |
| ⑤ | Next, you have to rely on the Linux Shell to run your commands, which were used in the manual installation processes earlier. To make your code more efficient, make this into a function, named run_command(command).<br><br>```
# Function to run a shell command and print its output
def run_command(command):
  try:
    result = subprocess.run(command, shell=True, stdout=subprocess.
    PIPE, stderr=subprocess.PIPE, text=True, check=True)
    output = result.stdout.strip()  # Get the command output
    print(output)
    print(f"Command executed successfully: {command}")
    return output  # Return the output
  except subprocess.CalledProcessError as e:
    print(f"Error executing command: {command}")
    print(f"Error details: {e.stderr}")
    return None
  except Exception as e:
    print(f"An error occurred: {e}")
    return None
``` |

(*continued*)

#	Task
⑥	From here, you only need to add the Linux commands in the correct sequence, so the required software gets installed in the right order. First, update the `apt` package manager and run it. ``` update_command = "sudo apt update" run_command(update_command) ```
⑦	Install PostgreSQL on your Ubuntu Server. ``` install_postgresql_command = "sudo apt install -y postgresql" run_command(install_postgresql_command) ```
⑧	Start and enable the PostgreSQL service. ``` start_postgresql_command = "sudo systemctl start postgresql" enable_postgresql_command = "sudo systemctl enable postgresql" run_command(start_postgresql_command) run_command(enable_postgresql_command) ```
⑨	Create a database and user and grant privileges. ``` create_db_user_command = ''' sudo -u postgres psql -c "CREATE DATABASE netbox;" sudo -u postgres psql -c "CREATE USER netbox WITH PASSWORD 'P0stgr3SQLP@ss!';" sudo -u postgres psql -c "GRANT ALL PRIVILEGES ON DATABASE netbox TO netbox;" ''' run_command(create_db_user_command) ```

(continued)

#	Task
⑩	Install Redis.
| | ```
install_redis_command = "sudo apt install -y redis-server"
run_command(install_redis_command)
# 5a. Start Redis service
start_redis_command = "sudo service redis-server start"
run_command(start_redis_command)
# 5b. Enable Redis service to start on boot
enable_redis_service_command = "sudo systemctl enable redis-server"
run_command(enable_redis_service_command)
# 5c. Check Redis version
check_redis_version_command = "redis-server -v"
run_command(check_redis_version_command)
``` |
| ⑪ | Install the NetBox dependencies. |
| | ```
install_dependencies_command = '''
sudo apt install -y python3 python3-pip python3-venv python3-dev
build-essential libxml2-dev libxslt1-dev libffi-dev libpq-dev libssl-
dev zlib1g-dev
'''
run_command(install_dependencies_command)
``` |
| ⑫ | Create the NetBox directory and clone the repository. |
| | ```
create_netbox_directory_command = "sudo mkdir -p /opt/netbox/"
run_command(create_netbox_directory_command)
# 7a. Change the working directory to '/opt/netbox'
os.chdir("/opt/netbox/")
# 7b. Install Git
install_git_command = "sudo apt install -y git"
run_command(install_git_command)
# 7c. Clone the NetBox repository (latest version)
clone_netbox_command = "sudo git clone -b master --depth 1 https://github.com/netbox-community/netbox.git ."
run_command(clone_netbox_command)
``` |

(continued)
```

#	Task
| ⑬ | Create a NetBox system user and change the ownership of the media directory for any device images to be used.<br><br>```
create_netbox_user_command = "sudo adduser --system --group netbox"
change_media_ownership_command = "sudo chown --recursive netbox /opt/netbox/netbox/media/"
run_command(create_netbox_user_command)
run_command(change_media_ownership_command)
``` |
| ⑭ | Change the directory, copy the example configuration file, and then modify it.<br><br>```
os.chdir("/opt/netbox/netbox/netbox/")
copy_config_file_command = "sudo cp configuration_example.py configuration.py"
run_command(copy_config_file_command)
``` |
| ⑮ | Update the configuration.py file with the SQL user and password.<br><br>```
print("# 10a. Update configuration.py file with the SQL user and password.")
# Define the path to the configuration.py file
config_file_path = '/opt/netbox/netbox/netbox/configuration.py'
# Read the existing configuration.py file
with open(config_file_path, 'r') as config_file:
lines = config_file.readlines()
# Update 'USER' and 'PASSWORD' values (only the first instance)
updated_lines = []
user_updated = False
password_updated = False

for line in lines:
if not user_updated and line.strip().startswith("'USER':"):
updated_lines.append(f"    'USER': 'netbox',\n")
user_updated = True
``` |

(continued)
```

#	Task
|  | ```python
elif not password_updated and line.strip().startswith("'PASSWORD':"):
    updated_lines.append(f"    'PASSWORD': 'P0stgr3SQLP@ss!',\n")
    password_updated = True
else:
    updated_lines.append(line)

# Write the updated lines back to the configuration.py file
with open(config_file_path, 'w') as config_file:
    config_file.writelines(updated_lines)

# Use Python to generate a secret key and append it to the
configuration.py file
print("# 10b. Generating a secret key...")
generate_secret_key_command = "python3 /opt/netbox/netbox/generate_secret_key.py"
generated_secret_key = run_command(generate_secret_key_command)
if generated_secret_key is not None:
    print("# Generated Secret Key:", generated_secret_key)
# Append the "SECRET_KEY = generated_secret_key" to the end of
configuration.py file
print("# Appending the SECRET_KEY to the configuration.py file...")
# Replace with the actual generated secret key
append_secret_key_command = f"sudo echo 'SECRET_KEY = \"{generated_secret_key}\"' >> /opt/netbox/netbox/netbox/configuration.py"
run_command(append_secret_key_command)
``` |
| ⑯ | Append napalm to local_requirements.txt. |
|  | ```python
append_napalm_command = "sudo sh -c \"echo 'napalm' >> /opt/netbox/local_requirements.txt\""
run_command(append_napalm_command)
``` |
| ⑰ | Run the upgrade script. |
|  | ```python
run_upgrade_script_command = "sudo /opt/netbox/upgrade.sh"
run_command(run_upgrade_script_command)
``` |

(continued)
```

CHAPTER 12   INSTALLING NETBOX (IPAM/DCIM) WITH PYTHON

#	Task
| ⑱ | Activate the virtual environment.<br><br>```python
activate_venv_command = "source /opt/netbox/venv/bin/activate"
subprocess.run(activate_venv_command, shell=True, text=True)
change_dir = "cd /opt/netbox/netbox"
subprocess.run(change_dir, shell=True, text=True)
``` |
| ⑲ | Define the command to create a superuser.<br><br>```python
createsuperuser_command = "python3 manage.py createsuperuser"

# Define the responses
responses = {
"Username (leave blank to use 'root'): ": "admin\n",
"Email address: ": "admin@pynetauto.com\n",
"Password: ": "admin\n",
"Password (again): ": "admin\n",
}

try:
# Run the command and provide responses interactively
process = subprocess.Popen(
createsuperuser_command,
stdin=subprocess.PIPE,
stdout=subprocess.PIPE,
stderr=subprocess.PIPE,
text=True,
shell=True,
)

for prompt, response in responses.items():
response_str = f"{response}"
process.stdin.write(response_str)
process.stdin.flush()
``` |

(continued)

#	Task
|  | ```
stdout, stderr = process.communicate()

if process.returncode == 0:
    print("Superuser created successfully.")
else:
    print(f"Error creating superuser: {stderr}")
except Exception as e:
    print(f"An error occurred: {e}")
``` |
| 20 | Define the path to the settings.py file. |
|  | ```
settings_file_path = '/opt/netbox/netbox/netbox/settings.py'
``` |
| 21 | Define the new ALLOWED_HOSTS line. |
|  | ```
new_allowed_hosts_line = "ALLOWED_HOSTS = ['*']\n"

try:
# Read the existing content of settings.py
with open(settings_file_path, 'r') as file:
    lines = file.readlines()

updated_lines = []
found_allowed_hosts = False
print("#17. Iterate through the lines and update ALLOWED_HOSTS.")
for line in lines:
    if line.strip().startswith('ALLOWED_HOSTS'):
        updated_lines.append(new_allowed_hosts_line)
        found_allowed_hosts = True
    else:
        updated_lines.append(line)
``` |

(continued)
```

#	Task
| | ```
# If ALLOWED_HOSTS was not found, append it to the end of the file
if not found_allowed_hosts:
    updated_lines.append(new_allowed_hosts_line)
print("#18. Write the modified content back to settings.py.")
with open(settings_file_path, 'w') as file:
    file.writelines(updated_lines)

print("ALLOWED_HOSTS updated successfully.")
except Exception as e:
    print(f"An error occurred: {e}")
``` |
| ㉒ | Copy the gunicorn.py file. |
| | ```
copy_gunicorn_command = "sudo cp /opt/netbox/contrib/gunicorn.py /opt/netbox/gunicorn.py"

try:
    # Copy the gunicorn.py file
    subprocess.run(copy_gunicorn_command, shell=True, text=True, check=True)
    print("gunicorn.py file copied successfully.")
except subprocess.CalledProcessError as e:
    print(f"Error copying gunicorn.py file: {e.stderr}")
except Exception as e:
    print(f"An error occurred: {e}")
``` |

(*continued*)

#	Task

23	Copy the .service files and then reload the systemd, start, and enable NetBox services.
	
	``python
	copy_service_files_command = "sudo cp -v /opt/netbox/contrib/*.service /etc/systemd/system/"
	reload_systemd_command = "sudo systemctl daemon-reload"
	start_netbox_command = "sudo systemctl start netbox netbox-rq"
	enable_netbox_command = "sudo systemctl enable netbox netbox-rq"
	try:
	# Copy .service files
	subprocess.run(copy_service_files_command, shell=True, text=True, check=True)
	print("Service files copied successfully.")
	
	# Reload systemd
	subprocess.run(reload_systemd_command, shell=True, text=True, check=True)
	print("Systemd reloaded successfully.")
	
	# Start NetBox services
	subprocess.run(start_netbox_command, shell=True, text=True, check=True)
	print("NetBox services started successfully.")
	
	# Enable NetBox services to start on boot
	subprocess.run(enable_netbox_command, shell=True, text=True, check=True)
	print("NetBox services enabled successfully.")
	except subprocess.CalledProcessError as e:
	print(f"Error: {e.stderr}")
	except Exception as e:
	print(f"An error occurred: {e}")
	``

(continued)

#	Task
| ㉔ | Create a local SSL certificate and define the `openssl` command with the default values for prompts.<br><br>```
openssl_command = (
"sudo openssl req -x509 -nodes -days 365 -newkey rsa:2048 "
"-keyout /etc/ssl/private/netbox.key "
"-out /etc/ssl/certs/netbox.crt"
)

# Create a dictionary of responses to provide default values for
prompts
responses = {
"Country Name (2 letter code) [AU]:": "\n",
"State or Province Name (full name) [Some-State]:": "\n",
"Locality Name (eg, city) []:": "\n",
"Organization Name (eg, company) [Internet Widgets Pty Ltd]:": "\n",
"Organizational Unit Name (eg, section) []:": "\n",
"Common Name (e.g. server FQDN or YOUR name) []:": "\n",
"Email Address []:": "\n",
}

try:
# Run the openssl command and provide responses interactively
process = subprocess.Popen(
openssl_command,
stdin=subprocess.PIPE,
stdout=subprocess.PIPE,
stderr=subprocess.PIPE,
text=True,
shell=True,
)
``` |

(continued)
```

CHAPTER 12 INSTALLING NETBOX (IPAM/DCIM) WITH PYTHON

#	Task
| | ```
for prompt, response in responses.items():
response_str = f"{response}"
process.stdin.write(response_str)
process.stdin.flush()

stdout, stderr = process.communicate()

if process.returncode == 0:
print("Local SSL certificate created successfully.")
else:
print(f"Error creating SSL certificate: {stderr}")
except Exception as e:
print(f"An error occurred: {e}")
``` |
| 25 | Install Nginx. While installing Nginx, you will be prompted to select a service restart. Select 37. None of the Above. |
| | ```
install_nginx_command = "sudo apt install -y nginx"

try:
# Run the command to install Nginx
subprocess.run(install_nginx_command, shell=True, text=True,
check=True)
print("Nginx installed successfully.")
except subprocess.CalledProcessError as e:
print(f"Error installing Nginx: {e.stderr}")
except Exception as e:
print(f"An error occurred: {e}")
``` |
| 26 | Create and configure the Nginx site file. |
| | ```
nginx_config_content = '''
server {
listen [::]:443 ssl ipv6only=off;

# CHANGE THIS TO YOUR SERVER'S NAME
server_name 192.168.127.20;
``` |

(*continued*)
```

#	Task

```
    ssl_certificate /etc/ssl/certs/netbox.crt;
    ssl_certificate_key /etc/ssl/private/netbox.key;

    client_max_body_size 25m;

    location /static/ {
    alias /opt/netbox/netbox/static/;
    }
    location / {
    proxy_pass http://127.0.0.1:8001;
    proxy_set_header X-Forwarded-Host $http_host;
    proxy_set_header X-Real-IP $remote_addr;
    proxy_set_header X-Forwarded-Proto $scheme;
    }
    }

    server {
    # Redirect HTTP traffic to HTTPS
    listen [::]:80 ipv6only=off;
    server_name _;
    return 301 https://$host$request_uri;
    }
    '''

    # Create and write the Nginx site configuration file
    nginx_config_path = '/etc/nginx/sites-available/netbox'
    with open(nginx_config_path, 'w') as nginx_config_file:
    nginx_config_file.write(nginx_config_content)

    # Print success message
    print("Nginx site configuration created successfully.")
```

(continued)

CHAPTER 12 INSTALLING NETBOX (IPAM/DCIM) WITH PYTHON

#	Task
27	If you get the outdated Damon message shown in Figure 12-11, choose Cancel and continue.

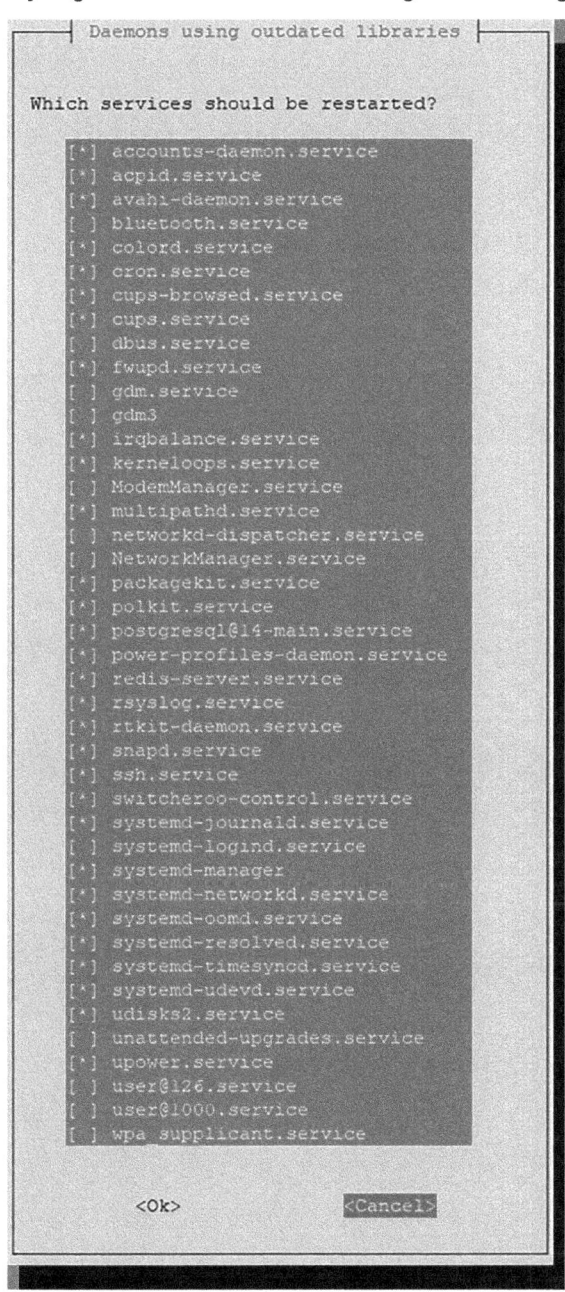

Figure 12-11. Cancel and ignore the outdated Damon message

(*continued*)

#	Task

28	Delete the Nginx default site and create a symlink to the new active file.
	
	 ``
	delete_default_nginx_site_command = "sudo rm /etc/nginx/sites-enabled/default"
	create_symlink_command = "sudo ln -s /etc/nginx/sites-available/netbox /etc/nginx/sites-enabled/netbox"
	
	# Delete the default site and create the symlink
	run_command(delete_default_nginx_site_command)
	run_command(create_symlink_command)
	
	# Print success messages
	print("Default Nginx site deleted and symlink created successfully.")
	``

29	Restart the Nginx service.
	
	 ``
	restart_nginx_command = "sudo systemctl restart nginx"
	run_command(restart_nginx_command)
	
	# Print a success message
	print("Nginx service restarted successfully.")
	``

30	Finally, write the code to print the elapsed time to run the installation script.
	
	 ``
	# Record the end time
	end_time = time.time()
	
	# Calculate the elapsed time
	elapsed_time = end_time - start_time
	
	# Print the elapsed time
	print(f"Script execution time: {elapsed_time} seconds")
	``

(continued)

#	Task
㉛	Once you have completed your script, save the file one more time. It is time to execute the script. root@u22s1:~/ch12_ipam_dcim# **python3 install_netbox.py** # You must use 'python3' #1. Update the Linux server. Hit:1 http://au.archive.ubuntu.com/ubuntu jammy InRelease Hit:2 http://au.archive.ubuntu.com/ubuntu jammy-updates InRelease Hit:3 http://au.archive.ubuntu.com/ubuntu jammy-backports InRelease Hit:4 http://au.archive.ubuntu.com/ubuntu jammy-security InRelease Reading package lists... Building dependency tree... Reading state information... 133 packages can be upgraded. Run 'apt list --upgradable' to see them. Command executed successfully: sudo apt update #2. Install PostgreSQL. Waiting for cache lock: Could not get lock /var/lib/dpkg/lock-frontend. It is held by process 3482 (unattended-upgr)... Waiting for cache lock: Could not get lock /var/lib/dpkg/lock-frontend. It is held by process 3482 (unattended-upgr)... *[...installation output omitted for brevity]* systemctl restart user@1000.service systemctl restart user@126.service systemctl restart wpa_supplicant.service No containers need to be restarted. No user sessions are running outdated binaries. No VM guests are running outdated hypervisor (qemu) binaries on this host. Nginx installed successfully. #23. Create and configure the Nginx site file. Nginx site configuration created successfully.

(continued)

CHAPTER 12   INSTALLING NETBOX (IPAM/DCIM) WITH PYTHON

#	Task
	#24. Delete the Nginx default site and create a symlink to the new active file. Command executed successfully: sudo rm /etc/nginx/sites-enabled/default Command executed successfully: sudo ln -s /etc/nginx/sites-available/netbox /etc/nginx/sites-enabled/netbox Default Nginx site deleted and symlink created successfully. #25. Restart the Nginx service. Command executed successfully: sudo systemctl restart nginx Nginx service restarted successfully. Script execution time: 443.284708738327 seconds # Approximately 7.5 minutes to install
32	This is an extremely important step. **You must activate the virtual environment, change the working directory to** /opt/netbox/netbox, **and create a superuser with a password** (admin/admin). **Then you will be exiting the NetBox virtual environment.** Ctrl+D will log you out completely from your SSH session. root@u22s1:~# **logout** jdoe@u22s1:~$ # You need to log out of the root user and log back in as jdoe user. jdoe@u22s1:~$ **source /opt/netbox/venv/bin/activate** (venv) jdoe@u22s1:~$ **cd /opt/netbox/netbox** (venv) jdoe@u22s1:/opt/netbox/netbox$ **python3 manage.py createsuperuser** Username (leave blank to use 'jdoe'): **admin** Email address: **admin@pynetauto.local** Password: **admin** Password (again): **admin** Superuser created successfully. (venv) root@u22s1:/opt/netbox/netbox# # Use Ctrl+D to logout of venv logout

(continued)

CHAPTER 12   INSTALLING NETBOX (IPAM/DCIM) WITH PYTHON

#	Task
33	You have installed NetBox successfully. The version installed for this demo installation was v3.6.6 (Release Date: November 29, 2023).Open the NetBox and select Advanced. Then choose Proceed to 192.168.127.20(unsafe) (see Figure 12-12). NetBox URL: `https://192.168.127.20/` 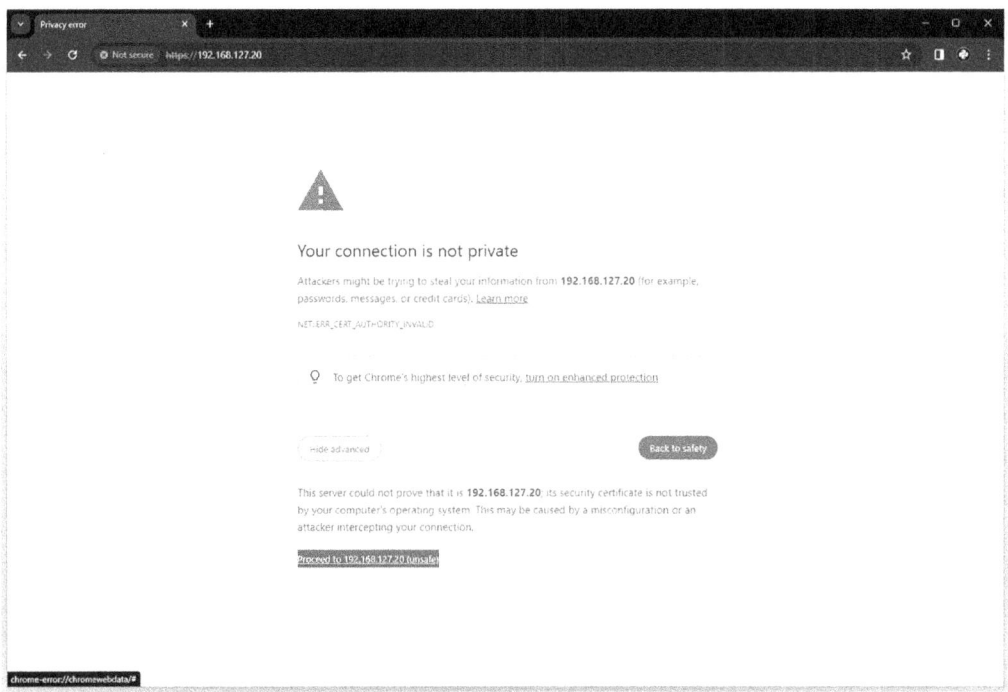 *Figure 12-12. Open NetBox on Chrome*

(continued)

CHAPTER 12 INSTALLING NETBOX (IPAM/DCIM) WITH PYTHON

#	Task
34	Click the Log In button in the top-right corner (see Figure 12-13).

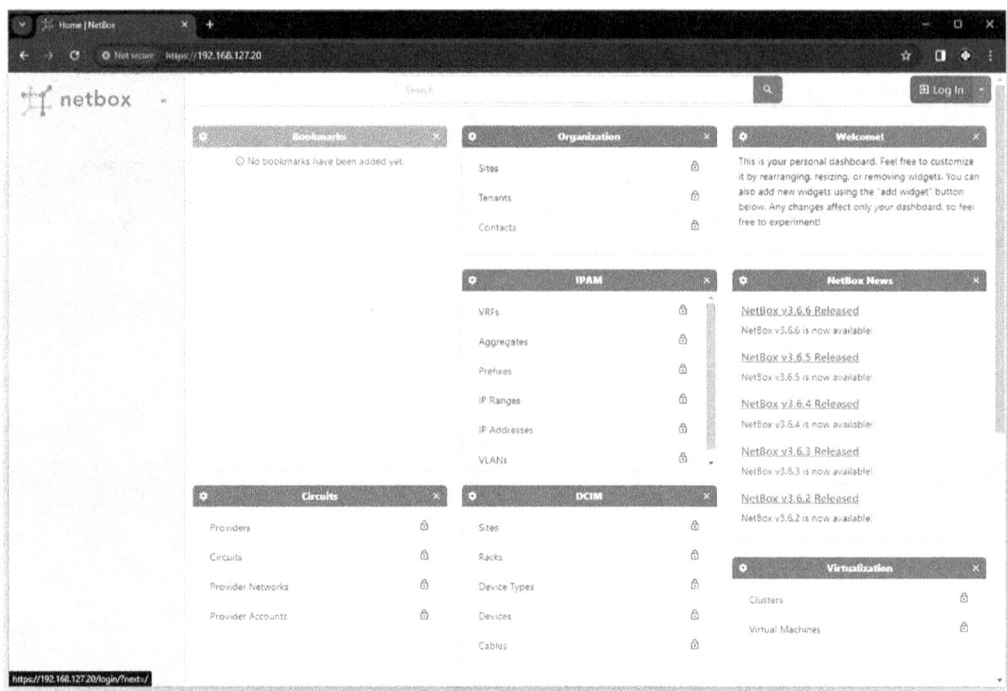

Figure 12-13. Click the Log In button

(*continued*)

#	Task
35	Enter the administrator's name and password. It was admin/admin to keep the initial login simple (see Figure 12-14).

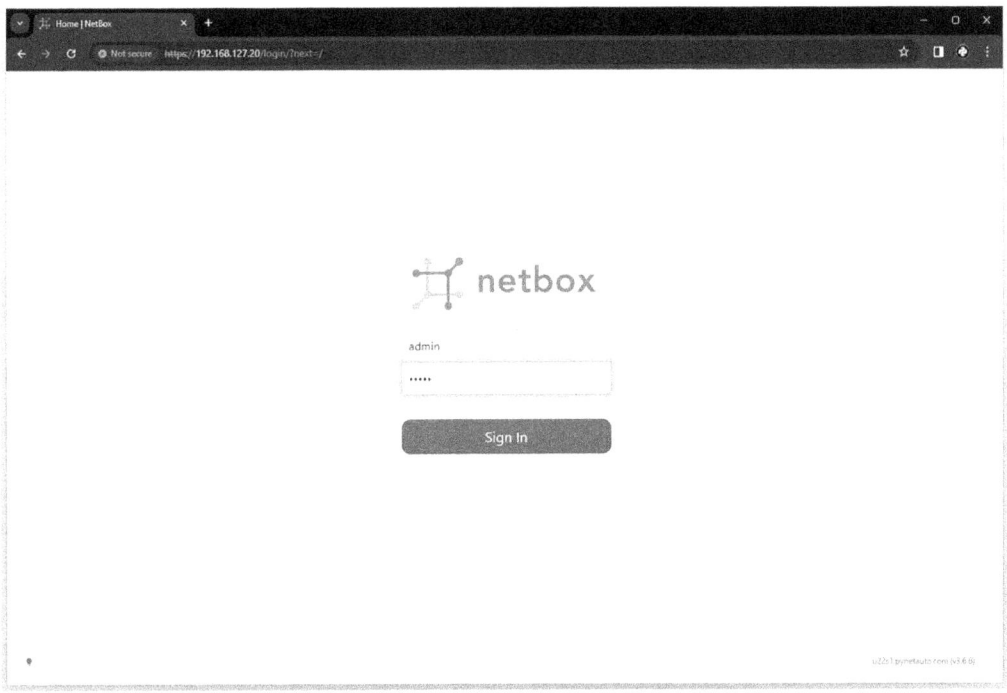

Figure 12-14. Log in with the admin username and password

(*continued*)

CHAPTER 12   INSTALLING NETBOX (IPAM/DCIM) WITH PYTHON

#	Task
㊱	Immediately after you log in, NetBox's first page should look like Figure 12-15.

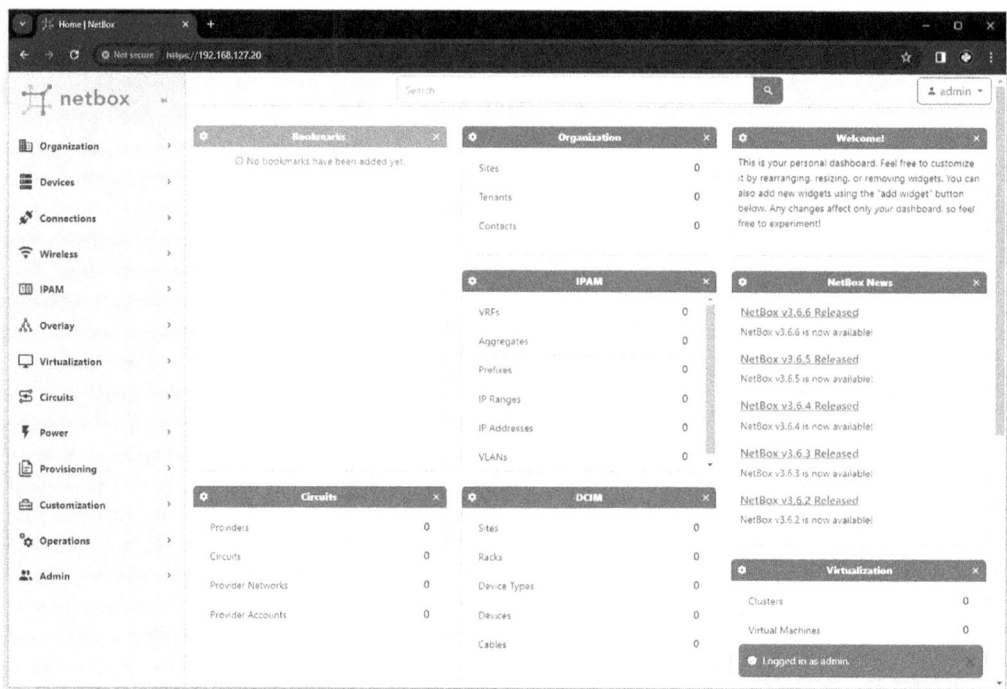

Figure 12-15. *First logged-in screen*

(*continued*)

CHAPTER 12　INSTALLING NETBOX (IPAM/DCIM) WITH PYTHON

#	Task
(37)	Choose Admin ➤ API Token from the left menu and create a new API token to get some API practice (see Figure 12-16).

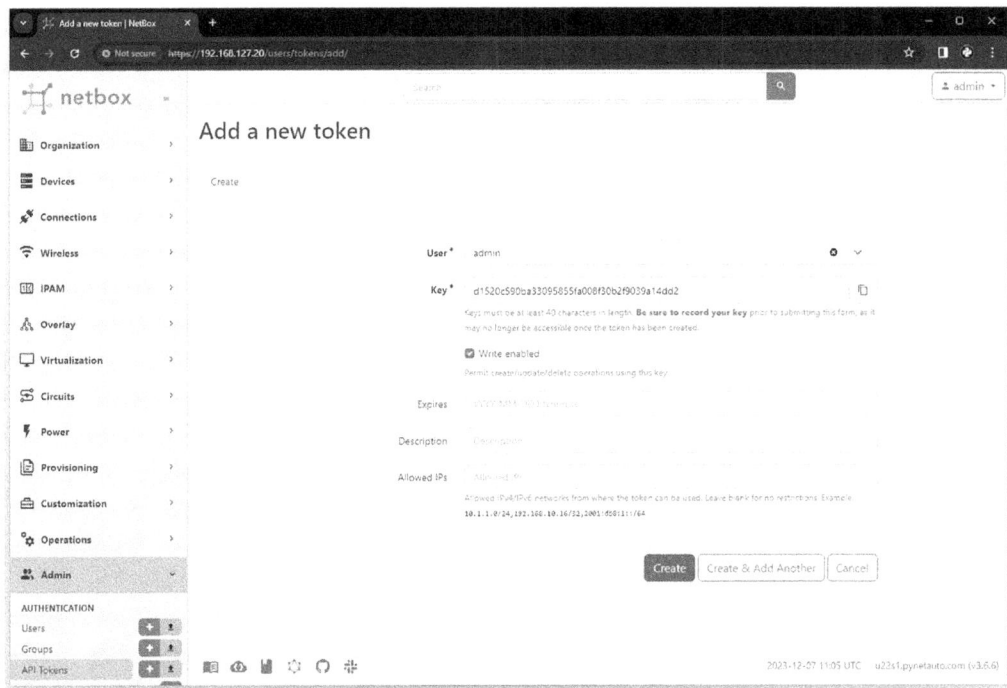

Figure 12-16. Go to Admin and select API Token. Then select +Add to add a new API token

(38)	At the bottom of the page, click the Rest API Documentation button (see Figure 12-17) to navigate to the REST API documentation page.

Figure 12-17. Click the REST API Documentation button

(*continued*)

763

CHAPTER 12　INSTALLING NETBOX (IPAM/DCIM) WITH PYTHON

#	Task
39	Now check the NetBox REST API page before studying the NetBox videos from YouTube (see Figure 12-18).

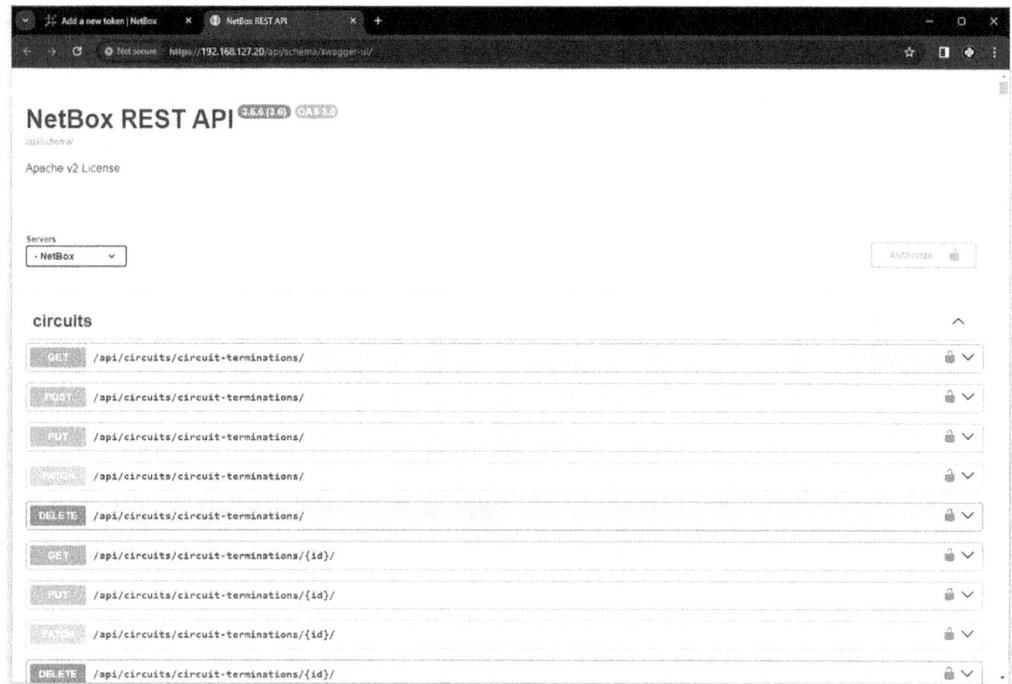

Figure 12-18. Learn how to use NetBox REST API

#	Task
40	Download Postman from the following page and install it to get ready for NetBox and API lessons. URL: https://www.postman.com/
41	Watch the NetBox Zero to Hero videos from YouTube to master how to use NetBox. Channel Link: https://www.youtube.com/@NetBoxLabs NetBox Zero to Hero video series: https://www.youtube.com/watch?v=zT82jOUCcW4&list=PL7sEPiUbBLo_iTds-NV-9Tuo5Gg2Aj8N7

This bonus chapter focused on the installation of an open-source tool for managing IP addresses and data center infrastructure (IPAM/DCIM), offered by DigitalOcean. The core of this tool is written in Python code, utilizing the Django Web framework.

My decision to guide you through NetBox installation using both manual and Python script approaches is driven by three key reasons. First, mastering both manual and automated work is crucial for maintaining relevance in your field. **Often, programmers with software backgrounds lack hands-on experience, leading to misinterpretations and hence inaccurate code. This is why I believe that network engineers who code and develop network automation scripts can achieve far superior results compared to software engineers unfamiliar with networking and security devices.** Secondly, installing an IPAM/DCIM server in the absence of one at your workplace will allow you to contribute valuable resources to your team's capabilities. There are still many teams operating from an Excel spreadsheet and Google Sheets. Finally, this practical experience equips you to learn REST API through NetBox and your network devices. You are comfortably positioned to learn REST API in a familiar environment, fostering a smoother learning process. Congratulations to everyone who has reached this page in the final chapter. **Keep at it and strive to enhance your IT skills. In the world of IT, as long as you remain technical, the learning journey must continue.**

Summary

In this final chapter, you have navigated the foundational steps of deploying NetBox—an all-encompassing IPAM and DCIM tool crafted on Python using the Django Web framework. Your journey commenced by learning about NetBox's importance and its defining features while dispelling misconceptions. You delved into critical server specifications, dependencies, and pivotal functions necessary for seamless NetBox operations. Emphasizing the pivotal 'site' concept within NetBox, the chapter explored its high-level IP Address Management capabilities. Furthermore, you ventured into the transformation of NetBox into Multi-tenancy IPAM, demystifying both manual and automated installations of NetBox 3.6 on Ubuntu 22.04 LTS through Python scripting. This chapter stands as an exhaustive roadmap, equipping you to comprehend, configure, and deploy NetBox adeptly on Ubuntu systems, thereby elevating network management capabilities. By harnessing the tools you have carefully built, this chapter sets the stage for a full grasp of REST API concepts later.

Coding, writing code, or working with code is a fundamentally different experience from configuring or troubleshooting routers, switches, or firewalls. It demands a unique mindset—a higher level of imagination and more intense concentration. To become a proficient coder, you must master coding skills and apply your general knowledge to programming tasks. While IT technicians are accustomed to reading technical books, programming success hinges on expanding your imagination by exploring a diverse range of literature and art. I hope your first Python Network Automation journey has been enjoyable thus far and that you are eager to continue your learning journey. You have covered significant ground in this book, but there's always room for improvement. Becoming proficient in anything takes time and effort. Mastering Python programming will not happen overnight; it is a journey that requires dedication and perseverance. As Joseph M. Marshall aptly stated, "Life is a circle. The end of a journey means the start of another one." It's time to bid farewell within these pages, leaving only the last storytime for your leisurely perusal. Best of luck with your Python coding and future career endeavors!

Storytime 4: Embarking on the Python Programming Odyssey: A Journey of Network Automation and Career Advancement

Embarking on a new programming language can be a daunting leap for many aspiring to elevate their careers. It is like embarking on an uncharted journey into the unknown, filled with both excitement and apprehension. **While numerous tech evangelists encourage embracing a programming language and diving into coding, often touting Python as a universal language for all, I politely but strongly disagree. Programming isn't a skill that suits everyone**. It demands a unique mindset, a blend of analytical thinking, creativity, and problem-solving prowess. However, for those who spend hours immersed in front of the computer, mastering a high-level language like Python can work wonders in their professional realm. It opens doors to a world of possibilities, enhancing productivity, streamlining tasks, and automating processes. For those involved in IT services or infrastructure support, there's hardly an excuse not to explore Python coding. It has become an indispensable tool in the IT arsenal, empowering IT engineers to manage complex infrastructure with greater efficiency and precision. While Python boasts ease of learning, not enough attention is given to what follows reaching a certain proficiency.

Just like any other programming language, Python's brilliance hinges on the person behind the code. It is not the language itself that holds the power, but the programmer's ability to wield it effectively. Learning a language should not mark the end of the journey but instead should signify the start of an ongoing odyssey without a defined finish line. It is a continuous process of learning, seeking opportunities to apply newfound skills, and constantly refining one's coding prowess. And there is no easy way out, so you must learn it the hard way. In the world of programming, the coders' locations and pay scales may vary, but the playing field remains level. The true measure of a programmer's worth lies in their ability to write elegant, efficient, and scalable code, regardless of external factors. It is a meritocracy, where talent and dedication reign supreme, enabling programmers from all corners of the globe to shape the future and enhance lives through innovative solutions.

This book was an attempt to create a hands-on Python Network Automation guide for both novice and seasoned network engineers. It aimed to cover crucial IT administration skills essential for building infrastructure from scratch and developing practical applications for efficient network management. Through hands-on exercises and real-world scenarios, I endeavored to equip readers with the knowledge and skills to navigate the ever-evolving landscape of network automation. I hope that you have gained a comprehensive understanding of diverse technologies, establishing a robust foundation for proficient enterprise network administration. Whether you're a seasoned engineer seeking to expand your skillset or a newcomer embarking on your IT career journey, I believe this book has provided a valuable stepping stone into the world of Python Network Automation. Congratulations to everyone who's reached this point after completing the final lab. **You have accomplished a great deal through this book and are now poised to accelerate your career to the next level. Embrace the challenges that lie ahead, continue to learn and grow, and let Python be your tool to revolutionize the world of network administration.**

Index

A

aaa configurations, 585, 587, 702, 704
Abstraction, 475, 647
add_lo_ospf1.py, 64–65, 67, 72
add_vlans_for_loop.py, 80, 87
add_vlans_for_range.py, 96
add_vlans_single.py, 73, 76, 80
add_vlans_while_loop.py, 87, 96
AES-type key exchange method, 361
Agent (client), 287
Agents, 281
AI ethical and accountability, 58
AI-powered assistants, 650
Ansible, 2, 303, 350
    AAP, 351
    Controller, 351
    network automation, 351, 374
    network configuration
        management, 351
    primary advantage, 351
    programming language, 351
    Red Hat, 353, 374
    virtualenv, 352–355
    virtualenv installation (*see* virtualenv
        installation, Ansible)
    YAML, 351, 374
    YAML files, endpoint automation, 351
Ansible Automation Platform (AAP),
    351, 374
ansible_venv, 357, 358
Ansible virtualenv lab, 356

API Documentation button, 763
Application Programming Interfaces
    (APIs), 57, 279
ASN.1 (Abstract Notation One), 286
Authentication methods, 124, 125, 307
AWS Lambda, 410
AWX, 351, 374
Azure Functions, 57, 410

B

~/.bashrc command, 63
/.bashrc file, 83
.bashrc file, 62
Bing AI, 516
Bytes free, 567

C

c8kv01, 519, 566, 569
c8kv01 and c8kv02, 702
c8kv01 debug, 677, 679
c8kv02, 519, 566, 569
c8kv02 debug, 678, 680
c8000v platform, 245, 621
Cacti, 288
Capital expenditure (CAPEX), 401,
    406, 408
cast() function, 334
ch11_r1_cf1.py, 77
ch11_upgrade_ios, 653, 712
ChatGPT, 58, 331, 516, 650

INDEX

check_flash1.py script, 566
Checkmk, 290
check_port() function, 530, 534
check_rsa_key_entries.py, 139
Chicken littles, 646–647
chmod command, 261
chmod +x command, 107, 261
chrony.conf file, 168, 169
chronyd service, 170
Cisco 2621XM/2651XM routers, 493
Cisco c8000v router, 245
Cisco Catalyst 8000v IOS XE Software
    versions, 494
    virtual machines, 493
    VMware Workstation 17 Pro, 494
Cisco CML Image integrations, GNS3, 4–6
Cisco CML integration, Python Network
    Automation, 3, 4
Cisco CML L2 image integration, GNS3, 6
Cisco CML L2 switch installation, GNS3
    add template pop-up, 15
    appliances, server, 10
    appliance window pops up, 14
    Cisco IOSvL2 switch, 15
    create and operate virtual
        switching labs, 8
    crucial tool, 18
    new template, 9
    pynetauto-lab topology, 16
    Qemu settings, 12
    required files, 13
    server, 11
    run show version command, SW1, 18
    Switch icon, 15
    SW1, 17
    usage, 14
Cisco CML L3 image installation, GNS3
    appliance server selection, 29

Cisco IOSv (Custom), 25, 26
cisco-iosv file, 28
cisco-iosv GNS3 appliance file, 27
CML L3 image error, 37
CML router, 33
complete message, 32
custom image file, 27
delete CML template, 38
GNU GRUB version 2.00, 36
Import Appliance, 28
Install Cisco IOSv, 31
+ New template method, 25, 27
ping test, L3 router, cml, 38–40
pynetauto-lab project
    Servers Summary window, 36
    topology summary window, 35
pynetauto-lab topology, 34
Qemu settings selection, 30
R2, pynetauto-lab topology, 34
required files, 31
router icon, 33
usage, 32
Cisco Device MIBs, 346
Cisco devices
    MIBs, 330
    SNMP agents (*see* SNMP agents, Cisco
        devices)
    SSH version 1.99, 311
Cisco device's flash
    c8kv01 and c8kv02, 602, 603
    command validation and compare
        MD5, 599–602
    creating, 598
    debug ip scp, 609
    delete the Dev-File-Only IOS file from
        c8kv01, 606
    dir or show flash, 610
    GigabitEthernet1, 604, 606

INDEX

IOS upload script, 608
MD5 value, 599
mission-critical network, 598
re-run the Python script, 604, 605
python upload_ios3_parallel.py, 610
u22s1, 611
and security devices, 598
verification command, 598
verify IOS_file command, 597
Cisco feature navigator, 330
Cisco IOS (IOS/IOS XE/IOS-XR)
    Cisco IOS upgrade, 510, 511, 513, 514
    Cisco routers, 509
    Cisco TAC recommendations, 510
    enterprise networking devices, 509
    Zero Trust security, 510
Cisco IOS Upgrade application, 705
Cisco IOS upgrades, 510, 511, 513, 514, 599, 650, 651
Cisco IOS-XE upgrading
    application, 516
    application development, 577
    backups
        interface status, 570–575
        routing tables, 570–575
        running configuration, 570–575
    Cisco routers, 515
    coding efficiency and ownership, 516
    commands for pre-check and post-check, 574
    CSV file (*see* CSV file)
    development phase, 516
    enterprise networks, 517
    flash size, 564–570
    GitHub, 578
    IaC principles, 515
    login credentials, 536–543
    MD5 value, 556–563
    mission-critical devices, 569
    network connectivity, 519–537
    network vendors, 515
    new IOS MD5 value on Cisco device's flash, 597–611
    pathways, 516
    pre-check tools, 578–597
    reloading device, 630–645
    socket validation, 519–537
    stop/reload the routers, 611–630
    u22s1 Python server, 517
    user input collector, 536–543
Cisco Modeling Labs–Personal (CML)
    Cisco Learning Network Store, 4
    Cisco's proprietary software, 4
    subscription-based CML, 3
Cisco Prime Infrastructure management platform, 281
Cisco routers, 515
    flash size, 564–570
    SCP, 597
    SCP file transfers, 581
Cisco's approach, 669
Cisco Secure ACS, 175
Cisco TAC, 93, 510
C language, 475
Clock command, 142, 388, 556, 631
Cloud computing, 277, 491, 493, 646
Cloud services, 4, 409, 509
CML lab topology
    add sw2 and r3, 41
    final topology, 49
    IP addresses, 50
    network devices, 40
    pynetauto-lab GNS3 project, 41
    R1, 44
    R2's configuration, 47
    r3, 48

INDEX

CML lab topology (*cont.*)
    servers summary, 43
    sw1 configuration, 45, 46
    topology summary, 42
CML L2 image integration, 5
CML L2 switch image, 5
CML L3 image integration, 5
CML L3 router, ping test
    output, local network, 40
    R2, 38
Coder's drill, 584
Command-line interface (CLI), 54, 56, 58, 311
Command-line methods, 294
Common Vulnerabilities and Exposures (CVEs), 492
config_aaa_all.py, 588
config_rsa_telnet.py file, 132
Configuration Information (CI), 720
Configuration items (CI), 740
Configuration validation tests, 61
Connectivity validation tool, 577
Console connection, 122, 239, 587, 669
construct_object_types function, 333
Containers, 407–409, 411
copy running-config tftp/ftp, 570
Core/backbone switches, 569
CPU utilization
    API-enabled SMS, 471
    debug all, 456
    Gauge32 digit, 468, 469
    monitoring script, 449
    OID IDs, 457
    overloading lab topology, 456
    Python application and scheduling, 470
    REST API SMS message tool, 447, 448
    R2, 448, 456, 460, 465–467

    scripts, 461–463
    an SMS message, 456–471
    unresponsive device trigged SMS message, 470
cron
    Linux, 162, 250
    Python networking, 229, 230
cron.log file, 255
crontab, 251, 267
    (*see also* Ubuntu LTS task scheduler (crontab))
crontab-e command, 252, 254, 256
crontab-l command, 255
CSV file
    MD5 values, 543–556
    new IOS names, 543–556
Custom-developed enterprise, 410
Cut-and-paste method, 59, 89, 124, 132
CVE entry, 492

D

Data Center Infrastructure Management (DCIM), 715
deactivate command, 358
debug ip scp, 609
.decode('encoding_type') method, 83
Decoding, 83
Deployment methods, 122
device_info.csv, 544
device_list, 520, 527, 549
device_list_netmiko, 672
Devices class, 478
devices_info.csv, 544
device_uptime_graph.py, 399
diffie-hellman-group1-sha1 SSH algorithm, 161
Diffie-Hellman key exchange, 161

difflib library, 636–641
dir command, 565
Django Python framework, 716
Django Web framework, 715, 765
dnf, 266
Docker, 401
    advantages, 407
    CAPEX, 406
    containers, 407–409
    CPU utilization (*see* CPU utilization)
    financial impact, 408, 409
    Hub acts, 407
    Kubernetes, 408, 409
    OPEX, 406
    resource-saving tool, 409
    sandboxed environment, 407
    Sendmail email notification, 436–447
    Sendmail lab (*see* Sendmail Docker lab)
    Sendmail Python lab, 424–436
    Twilio, 447 (*see* Twilio)
    VMs, 407
Docker boasts, 412
Docker Compose, 412
Docker Engine, 412
Docker Hub, 412
Docker images, 417
docker pull command, 417

E

Edge routers, 569
email.mime.text, 433
Encapsulation, 476
.encode ('encoding_type') method, 83
Encoding, 83
End-to-end IOS upgrade application, 577
End-user license agreement, 621

Enterprise-level network device configuration backup solutions, 175
Enterprise network equipment vendors, 311
Enterprise networking environment, 157
Enterprise networks, 517
External Python modules, 94

F

f38s1 server, 59, 259
Fedora 38, 56
Fedora task scheduler (crond)
    chmod +x command, 261
    cron.log file, 265
    /etc/crontab, 264
    f38s1, 258, 259
    IP addresses, 259
    Python script, 259, 264
    shell command/Python script, 263
    systemctl status crond command, 262
    vs. Ubuntu's crontab, 262
    user-friendly, 266
    fetch function, 333
File integrity verification, 577
FileZilla, 557, 644
Fitting analogy, 476
Flash memory, 564, 569
Flash size, on Cisco Routers, 564–570
Flow Control and User Input management
    flaws, 484
    getpass module, 488
    iterative process, 490
    optimal recommendations, 485
    Python application, 491
    Regular Expressions, 482

INDEX

Flow Control and User Input management (*cont.*)
    script collects two sets of network administrator ID and password, 486, 487
    while loop, 485
    yes_or_no1.py, 483
for loop, 559, 633
fping, 314
FTP/SFTP/TFTP server, 536

G

Gauge32 digit, 468, 469
Genie, 352
genie parse command, 382, 384
get_credentials() line, 662
Get credentials, 537
get_cred3.py, 542
get() function, 332, 337
getpass, 537, 538
GigabitEthernet1, 604, 606, 634
git clone command, 378
GitHub, 578
GitHub site, 554
GNS3
    Cisco CML image integrations, 4–6
    Cisco CML L2 image integration, 6
    Cisco CML L2 switch installation (*see* Cisco CML L2 switch installation, GNS3)
    Cisco CML L3 image installation (*see* Cisco CML L3 image installation, GNS3)
    Cisco CML L3 image integration, 6
    Marketplace
        Cisco IOSv2 appliance icon, 7
        IOSv appliance icon, 7

    Python Network Automation, 3, 4
GNS3 network devices, 174
GNS3 project, 235
    delete redundant project copy, 235
    desktop icon, 233
    increase RAM, GNS3 VM, 237
    netmiko Lab 1, 237
    netmiko library, 231
    network devices, powered off, 232
    new project, 235
    open project, disk, 234
    project export method, cloning, 230
    copy, pynetauto-lab folder, 233
    pynetauto-devops project, 236
    save_config.py script, 231
    Save Project, 234
    vRouter, 237
GNS3 topology, 59
Google Cloud Functions, 410
Google Sheets, 544
Grafana, 289
Graphical User Interfaces (GUI), 57, 311

H

hello-world command, 414
/home/jdoe/ch2_telnet directory, 108
Human network engineers, 58
Hypervisors run, 175

I

Icinga, 289
ICMP and NMAP tools, 95
ICMP application, 519
ICMP (ping), 519
ICMP tests, 94
if-else statement, 561

Infrastructure as code (IaC), 514, 515, 518
Infrastructure management, 493
Inheritance, 476
Installing Catalyst 8000v on VMware Workstation 17
  c8kv01, 504
  c8kv01 and c8kv02 virtual routers, 496
  c8kv02, 504, 507
  completion process, 501
  Deployment Options setting, 499
  development phase, 509
  file selection, 497
  interfaces, 504
  IOS upgrade application development, 508
  Network Adapter configuration, 503
  PuTTY security alert, 505
  PuTTY SSH client, 505
  RSA key, secure key exchange, 506
  Settings option, 502
  snapshot, 508
  u22s1 server, 495
  virtual router, 500
Interactive method, 293
Internet Control Management Protocol (ICMP), 276
Internetwork Operating System (IOS), 474
IOS MD5 value, Cisco device's flash, 597–611
IOS names and MD5 values, 543–556
IOS Upgrade Task Breakdown, 512
IOSv, 157
IOS verification, 598
IOSv L2 switches, 236
IOS (XE/XR), 473
ip_addresses.txt, 524
IP Address Management (IPAM), 715
ip address show command, 62
ITIL-driven organizations, 510
ITIL process, 570

J

JavaScript Object Notation (JSON), 188
jdoe, 254
jdoe user, 584
Jeremy Stretch developed NetBox, 716
JSON Data, 673
json.dumps() method, 149
JSON-formatted data, 152
Jump hosts, 59

K

Key performance indicator, 514
known_hosts file, 135, 140
Kubernetes, 407–409, 446

L

Labbing, 23, 24
Lab pre-task, delete VPCS
  add switches, 245
  configurations, 246
  extended ping/traceroute request, 248
  Google DNS, 249
  R2, ping IP, 245
Less Secure Apps, 425
LibreNMS, 289
Link flap, 437
Linux command-line interface (CLI), 56
Linux cp command, 343
Linux cron, 162
Linux NTP server, 173

INDEX

Linux Scheduler, cron
    examples, 268
        Fedora task scheduler (*see* Fedora task scheduler (crond))
        job definition, 267, 268
        job definition, example, 267–270
        off-the-shelf task schedulers, 251
        Python applications (scripts), 251
        Python scripts, 270
        Ubuntu LTS task scheduler (*see* Ubuntu LTS task scheduler (crontab))
        Wikipedia, 270
Linux server, 558, 577
Linux Shell, 743
Linux to CML Router/Switch SSH key exchange, 138
Login credentials, 536–543
Lookahead method, 567
Loops, 81
ls ansible_venv/bin/ansible*, 360

M

Managed Service Provider (MSP), 716
Management Information Base (MIB), 282, 286, 330
Matching kex (Key Exchange) algorithm, 159
matplotlib, 400
MD5, 543
md5sum Linux, 580
md5_validate2.py, 569
MD5 values
    CSV file, 543–556
    IOS on server
        directory creation, 557
        device_info.csv, 563

        double-check, 556
        for loop, 559
        if-else statement, 561
        md5sum filename, 558
        os.path and hashlib modules, 560
        run the script, 561
        SCP folder, 556
        TFTP/FTP server, 556
        tool4_md5_linux directory, 558
        uploading, 558
    user-provided, 544
MFA on Gmail account
    app password, 426–428
    home page, 429
    netstat-tuna, 431, 432
    pynetauto_ubuntu20, 430
    sendmailconfig, 432
    Sendmail installation, 431
    sent folder, 434, 435
    sign in, 425
    SMTP server, 434
Microsoft Windows-based NTP servers, 164
Mission-critical network, 598
mkdir, 357
Modular approach, 714
Multi-cloud environments, 446
Multifactor authentication (MFA), 424–436
mygenie directory, 397

N

Nagios, 288, 292
NAPALM, 376
NetBox, 723, 759, 762, 764
    ALLOWED_HOSTS line, 749
    on Chrome, 759

INDEX

concept, 715
core, 720
dashboard, 717, 740
DC rack diagram, 717
dependencies, 719
Digital Ocean's home page, 716
directory, 745, 746
features, 718
functions and components, 719
installation, 722, 723, 742
installation process, 722
IPAM, 720
login page, 732
MSP, 721
MSP modeling, 722
NetBox, IP address
    attributes, 721
PostgreSQL version 11, 716
REST API concepts, 715
service files, 751
services, 718
settings.py file, 749
version, 718
vversion 3.6.3, 739
NetBox REST API, 764
netmiko, 550
    application, 317
    dictionary, 189
    dictionary format, 550
    importing modules, 187
    installation, 187
    library, 231, 578
    Python library, 186
    scripts, 586
    timeout exception, 598
netmiko-friendly dictionaries, 552
netmiko Lab 1, 237, 238
netmiko Lab 2 scripts, 203
    netmiko's SCPConn file-transfer
        method, 579
SSH netmiko Lab 3
    data modules, 214
    difflib library, 213
    modern alternative, 215
    NAPALM library, 214
    Notepad++, comparison, 213
    practical application
        GitHub repository, 221
        .html file, 222, 223
        netmiko_compare_config.py, 216
        port-scanner tool, 215
        python command, 220
        python netmiko_compare_
          config.py, 224
        WinSCP/FileZilla, 221
    u22s1 Python server, 214
netmiko session_log, 627, 708, 709
net-snmp-utils, 293
Network automation, 53, 767
Network Automation Engineer, 58
Network Automation Engineering, 374
NetworkAutomation-1 server, 56
Network Configuration Manager
    (NCM), 175
Network connectivity, 519–537
Network DevOps Engineer, 58
Networking vendors, 58
Network outage, 58
Network Time Protocol (NTP), 164
Next-generation package management, 266
Nginx, 753, 756, 758
NMAP port, 152
Nornir, 352, 376
NTP association and synchronization, 171
NTP problems, 167
NTP servers, 164, 169

777

INDEX

O

Object, 475
Object-Identifier (OID), 282, 286, 290, 291
Object-oriented programming (OOP), 473, 650
    abstraction, 475
    class, 475
    encapsulation, 476
    enterprise network, 474
    inheritance, 476
    languages, 474
    object, 475
    polymorphism, 476
    routers and switches, 474
    theory, 476
Observium, 289
-oHostkeyAlgorithms=ssh-rsa, 133
-oKexAlgorithms=+ command, 132, 137
OOP, networking
    classes creation, 478
    hierarchical classes construction, 480
    oop_task1.py, 481
    Python Interpreter, 477
    Python script output, 481
    Router and Switch classes, 477, 478
    show() method, 479
Open HTML files, 713
Open-source task scheduler, 270
Operational expenditure (OPEX), 408
OS patching, 510
Out-of-bound (OOB), 239

P

Paessler PRTG Network Monitor, 292
Pandas, 390, 396, 398, 400, 550
pandas module, 516, 544, 545
paramiko, 358
    library, 301
    pip command, 129
    run and import library, 129
Python paramiko library, 124
Patch management, 630
PC hardware resources, 174
PEP 8 convention, 141
PEP 8 style guide, 614
Ping, 18
ping_4_socket.py, 527, 529
ping_5_pythonic.py, 534
ping_5_socket_pythonic.py, 532
ping_6_modular, 664
ping_6_modular.py, 533, 534
Ping method 1-os, 94
Ping method 2, subprocess, 94
Ping method 3, ping3, 94
Ping method 4, pythonping, 94
Ping method 5, scapy, 95
Ping test, CML L2 switch Gi0/0, SW1, 19
    PC1, ip dhcp command, 20
    labbing, 23, 24
    R1, 19
    save configuration, 23
    SW1, 19
    Tcl shell (tclsh), 21
    tclsh cut and paste command, 21, 22
    tclsh ping command output, 23
pip command, 364
pip3 commands, 358, 364, 398
Pipenv
    Django/Flask library, 354
    environment isolation, effective dependency management, 355

one-stop-shop, virtual environment creation/dependency management, 355
package and environment management solution, Python projects, 354
project-specific dependencies, 354
Python ecosystem, 354
Pipfile, 354
Platform-as-a-service (PAAS) software, 410
Polling method, 283
Polymorphism, 476
Port-scanning tool, 203
post_check2.py, 636
PostgreSQL, 744
Post Incident Reporting (PIR), 58
Post-reload configuration verification
    backup files, 642
    difflib library, 630, 636–641
    for loop, 633
    HTML files, 645
    ls-lh 192*, 643
    port 22, 633
    post-check application, 633
    pre-reload show files, 635
    socket application, 631, 632
    SSH, 630
    u22s1, 644
    VMware Workstation, 633, 634
post-upgrade check, 573
Power-On-Self-Test (POST) process, 564
Pre-check tools
    aaa configuration, 588
    async or async_task, 591
    c8kv01 and c8kv02 parallel test IOS uploading, 597
    c8kv01 and c8kv02 sequential test, 591
    config_aaa_all.py, 588
    config_aaa.py file, 588
    debugging consoles, 597
    IOS file size, 579
    jdoe user, 584
    netmiko scripts, 586
    netmiko's SCPConn file-transfer method, 579
    parallel file uploads, 591
    print() function, 590
    Python scripting, 581–584
    Python's native threading, 591
    Python's threading module, 591, 593–598
    recursive copy, 579
    router's flash memory, 578
    SCP debugging, 589
    TACACS server, 581
    and testing phases, 579
    u22s1 server, 587
    upload_ios1.py file, 581, 585
    upload_ios1.py Python file, 581
    upload_ios1.py script, 590
    upload_ios1.py *vs.* upload_ios2_parallel.py, 591, 592
    WinSCP, 580
    working directory, 580
print() function, 590
Python's print statement, 522
Proof of Concept (PoC), 24, 40
Protocol Data Unit (PDU), 285
PRTG Network Monitor, 288
pwd command, 253
pyATS (Genie)
    deactivate command, 397
    genie parse command, 382, 384
    git clone command, 378
    matplotlib libraries, 399, 400

INDEX

pyATS (Genie) (*cont.*)
    mygenie directory, 397
    mygenie and cd, 377
    pandas module, 396
    pyats run job command, 378
    show cdp neighbor command, 389
    show clock command, 388
    testbed.yml file, nano text editor, 385
pyATS (Genie), 350
    Cisco device network automation, 400
    Cisco DevNet Team, 352
    Cisco's DevNet automation environment, 375
    Cisco Systems, 375
    convert uptime information, bar graph, 394
    data collection, 390
    device-focused automation, 376
    device orchestration and streamlined network management, 376
    devices, 376
    devices' uptime, 395
    documentation, 376
    features, 375
    Host PC, 398
    install libraries, Excel, 379
    library, 377
    NAPALM, 376
    network automation, 376
    network probing and information-gathering tool, 352
    output data type, 394
    install pandas library, 390
    create testbed.yml file, 379, 380
    WinSCP, 397
    YAML testbed JSON file, device connection/authentication, 379
    zip feature, 396

pyats run job command, 378
pynetauto-lab
    GNS3 project, 232, 233
    IP addresses, 55
    main logical topology, 55
    project, 35, 54, 234
    project topology, 34
    topology, 16, 51
pynetauto_ubuntu20, 430
pysnmp, 290, 335
python3 ping_2.py, 522
Python application, 650, 664
Python-based tools, 575
Python code, 540, 649, 658, 660, 661, 673, 676, 683, 691
Python code snippets/modules, 514
Python coding, 766
Python command, 676
Python file, 705
Pythonic approach, 95
Python Integrated Development Environments (IDEs), 355
Python loops, 678
Python mastery
    application, 652
    automation server, 667
    command, 670
    get_cred3.py, 658
    IOS upgrade application, 655
    IOS upgrade lab, 651
    md5_validate3.py, 667
    network automation, 650
    OOP classes, 650
    regular expressions, 658
    ROMMON mode, 669
    and Scratch, 650
    software products, 650
    upgrade_c8000v.py, 656

upgrade_c8000v.py file, 656
Python modules, 119, 656
Python network automation, 191, 194, 203, 473, 514
  Ansible, 2
  Cisco CML integration, 3, 4
  GNS3, 3, 4
  main logical topology, 54
  network engineer's interface, 56–59
  open-source and proprietary software, 2
  problem-solving approach, 57
  streamline network engineering tasks, 57
Python network programming, 649
Python Package Index (PyPL), 291
python ping_6_modular.py, 535
Python scripting lies, 581
Python scripts, 179, 652, 713, 740
Python's native threading, 591
Python SNMP code
  run SNMP queries, interface description, 342–346
  SNMP lab source code, 332
Python SNMP libraries, 290, 291
Python SSH library, 53
Python stand-alone applications, 649
Python's telnetlib, 119
Python's threading module, 591
python telnet_lab2_check.py command, 78
Python Telnet scripts, 58
Python tools, 649
python upgrade_c8000v_bt4.py, 666
python upgrade_c8000v_bt8.py, 683
python-V command, 253
Python virtualenv
  Ansible (see Ansible)

pyATS, 350
Python Web Framework, 353

Q

quicksnmp_v3_get_bulk_auto.py, 346

R

RAA key fingerprint, 133
Random access memory (RAM), 564
re module, 394
Reachable IP addresses, 524
reachable_ips, 520
read_info7.py, 552
Red Hat's Ansible tools, 57
Regular Expression, 540, 567, 568
Regular Expression pattern ^end, 78
Reloading device, 630–645
reload_yes_no1.py, 616, 688
reload_yes_no2.py, 615, 642
reload_yes_no2.py script, 627, 628
Resource-saving tool, 409
REST API SMS message tool, 447, 448
Return Material Authorization (RMA), 239
Router failure simulation, P1 network recovery
  GNS3, 239
  GNS3 topology, 239, 240
  new router powered-on, 241
  OOB, 239
  R2, 238
  R2's GigabitEthernet 0/0 interface, 242
  rCisco CML router replacement, 240
  router replacement, failed IP service router, 239
  routers boot up, 238
  startup-config command, 242

INDEX

Router failure simulation, P1 network recovery (*cont.*)
    tcl shell script, 243
    TFTP server, 241, 242
Router IOS upgrade lab preparation
    Cisco Catalyst 8000v, 493
    Cisco IOS XE software, 492
    IOS/IOSvL3 images, 492
    virtualized Cisco routers and switches, 492
Router's flash memory, 578
Routing protocols, 647
Routing tables, 570–575
rpm-q cronie command, 262
RSA key fingerprint, 138
R2, 238
run_change_boot_var(), 694
running-config backup, 570–575

S

Sandboxed environment, 407
SCP debugging, 589
SCP protocol, 677
Script, 179
Secure Copy Protocol (SCP) file transfer, 556
Secure Shell (SSH)
    Cisco devices, 127, 129
    cloud-based solution physical switches, 122
    Linux Server, 124–127
    Python networking, 123
    remote enterprise device management, 122
Security evolution, 491
Self-paced learning, 411
Sendmail Docker lab
    account registering, 412
    components, 412
    containerization concept, 410
    containers, 411, 412
    develop, ship, and run anywhere, 410
    docker-version command, 413
    enable command, 413
    functionality, 412
    hello-world command, 414
    installation, 412
    operates, 412
    PAAS software, 410
    package, 413
    self-contained units, 410
    self-paced learning, 411
    sudo commands, 414
    test-driving docker, 415–423
    topology, 411
    Ubuntu or Fedora Server, 411
Sendmail email notification
    connectivity, 441
    Docker environment, 437
    enterprise network, 436
    IT enterprise ticketing, 436
    ITSM systems, 436
    Link flap, 437
    monitoring and supplementary email scripts, 443
    monitoring_logs.txt, 446
    monitor_sw1.py, 440, 441
    new directory, 438
    Port 22, 444, 445
    reattach it to Docker instance, 438
    sent folder, 446
    smtplib Python module file, 442
    SNMP monitoring, 437
    socket-checking script, 439
    Spam folder, 445
    SW1's console, 443

system monitoring, 436
24/7 service desk team, 436
Sendmail Python lab using Docker
   Less Secure Apps, 425
   MFA, 424–436
   pre-installed, 424
   Python's smtplib library, 424
   SMTP traffic, 424
   Yahoo, 435, 436
Server-client model, 282
Serverless applications, 57
Serverless platforms, 57
Service-oriented architecture, 491
S/FTP server, 118
show commands, 79, 573, 574
show cdp neighbor command, 389
show clock, 556, 629
show dir output, 616
show ip ospf neighbor command, 69
show ip ssh command, 127
show version command, 382
show vlan command, 87
Simple Gateway Monitoring Protocol (SGMP), 276
Simple Network Management Protocol (SNMP)
   agent, 282, 291
   application protocol, TCP/IP protocol stack, 284, 285
   cloud-based monitoring systems, 275
   critical network and systems infrastructure, 275
   device monitoring, 281
   enterprise networks, 275
   features, versions, 284
   history, 276, 277
   message, 288
   message types, 285
   MIB, 286
   OID, 286
   Polling *vs.* Trap (Event Reporting) method, 283
   Python, 274, 275
   Python applications, network automation, 273
   read-only community string, 287
   read-write community string, 287
   SMI, 286
   standard operation, 281, 282
   synchronous *vs.* asynchronous communication, SNMP Traps, 287, 288
   Syslog and API comparison, 279–281
   system monitoring types, 277–279
   tools and python integrations, 288–290
   trap community string, 287
   versions, 283
Single Telnet interface, 53
Site Reliability Engineer (SRE), 58
smtplib module, 436
smtplib library, 424
SNMP agents, Cisco devices
   clse condition, 316
   network-testing capabilities, install fping, 314
   precheck_tool.py, nano text editor, 315
   R1, 322
   R2 configuration, 313
   r3, 322
   r3, SNMP user information, 324
   script, 319
   show clock command, test_ssh3.py to write memory, 324
   show snmp user command, 313
   snmp_config.py, 317
   user and group configurations, 321

INDEX

snmpd service, 294
snmpget command, 326, 328
SNMP lab source code
    cast() function, 334
    construct_object_types function, 333
    fetch function, 333
    get(), 332
    SNMPv3, 334
    SNMPv3 Python code (*see* SNMPv3 Python code)
SNMP manager (server), 287
SNMP monitoring, 437
SNMP Trap methods
    asynchronous mode, 288
    synchronous mode, 287
SNMPv2, 283
SNMPv2c, 296, 332
SNMPv3
    CML network devices, 297
    Python script, 297
    SNMP agents, Cisco devices (*see* SNMP agents, Cisco devices)
    SNMP authentication, 296
    SNMPwalk, Linux Server (*see* SNMPwalk, Linux Server)
    technological challenges, 297
SNMPv3 Python code
    pysnmp library, 335
    pysnmp version 4.4.12, 340
    quicksnmp.py, 335
    quicksnmp_v3.py, 335, 341
    SNMPUser1, 337
    SNMP user data, 342
SNMPv3 queries
    getCmd function, 290
    MIB, 291
    OID, 291
    pysnmp, 290

SNMP Libraries, 290, 291
SNMP Server, Fedora
    configuration, 293–296
    installation, 293–296
    SNMPwalk, Linux Server, 292
SNMPwalk, 273, 292, 293
snmpwalk command, 295, 296, 325, 328
snmpwalk script, 458
SNMPwalk, Linux Server
    ifAdminStatus.1, 327
    ifOperStatus, 328
    MIB information, 326
    OID, 328
    OID information, vIOS, 329
    rmon OID, 329
    snmpget command, 326, 328
    snmpget command queries, 329
    snmpwalk command, 325, 326
    sysName.0 or sysName, 327
socket.AF_INET, 204
Socket application, 631, 632
socket.close(), 204
Socket validation, 519–537
Software-defined environments, 409
Software-driven networking devices, 123
SolarWinds, 436
SolarWinds Network Performance
    Monitor, 290, 292
"Somebody else's computer" concept, 123
Source code, 651
SRE engineers, 122
ssh command, 203
SSH 1.99, 311
SSH 2.0, 311
SSH config file, 134
SSH connections, 527
SSH console, 520
sshd_config file, 124

INDEX

SSH key exchange methods, 123
~/.ssh/known_hosts file, 139
ssh-l user_name IP_address
    command, 157
SSH netmiko Lab 1
    IOS (.bin) files, 194
    Delete_me file, 200
    Delete_me_2.bin file, 200
    devices, 191
    direct console interaction, 192
    disk formatting, 192
    fping/ping R2, 198
    input statements and getpass
        functions, 189
    IOS upgrade, 192
    JSON format, 188
    mkdir flash/directory_name, 193
    netmiko_delete_me.py, 195, 199
    Python interpreter, 189
    R2 directory, 202
    structured information, 188
    Tcl shell, 193
    Tcl shell commands, 191
SSH netmiko Lab 2
    code validation, 212
    devices, 204
    nano netmiko_disable_telnet.py,
        207, 209
    netmiko_disable_telnet.py, 204
    netmiko_disable_telnet_save.py, 212
    port-scanner tool, 203
    port-scanning task, 206
    port-scanning tool, 213
    running-config, 211
    scanning network devices, 203
    scan_open_ports.py, 206, 210
    socket.socket(family, type), 204, 205
    vty lines, 203

SSH paramiko Lab 1
    AAA RADIUS server, 129
    checkAuth_2.py, 143–145
    Cisco IOS commands, 131, 132
    CML to CML SSH key exchange, 158
    devices, 130
    Python dir() module, 152
    json library, 148, 149
    module's packages, 152
    Python scripts, 145, 146
    SSH authentication
        application, 141–143
    SSHClient and AutoAddPolicy
        modules, 150
    ssh user_name@IP_address
        command, 132–136
    troubleshooting, SSH key
        exchange, 136–140
SSH paramiko Lab 2
    ch3_ssh directory, 163
    devices, 162
    NTP service, 168, 169
    NTP services, 163
    NTP synchronization, 171–173
    ssh_ntp_lab.py script, 170
    SSH scripting, 163–166
    TCP port 22, 166, 167
    tools, 162
SSH paramiko Lab 3
    devices, 176
    end-to-end task duration, 179
    TFTP services, f38s1, 178
    fping, 178
    getpass module, 181
    interactive paramiko SSH Script, 175
    print statement, 181
    python interactive_backup.py
        command, 184, 185

INDEX

SSH paramiko Lab 3 (*cont.*)
    python timer_tool.py, 180
    running-config backups, 182
    /var/lib/tftpboot/ directory, 184
SSH PuTTY session, 609
ssh-Q commands, 125
ssh-Q key commands, 126
ssh-Q Query Options, 126
ssh server kex algorithm, 160
ssh username@host_ip_address commands, 361
ssh-v command, 125
Stop/reload the routers
    boot system, 611
    ch10_tools_dev2 directory, 612
    PEP 8 style guide, 614
    reload command, 620
    reload_yes_no1.py, 616–619
    reload_yes_no2.py, 615
    reload_yes_no2.py script, 627, 628
    sequential/linear application, 622
    show commands, 620, 621
    threading module, 622–626
    time manual measure, 628–630
    yes_or_no function, 612, 613
Storage area network (SAN), 476
Structured Management Information (SMI), 286
Subscription-based CML, 3
sudo commands, 252, 414
sudo crontab-e, 252
sudo crontab-u user_name-l command, 256
sudo /home/jdoe/.ssh/config command, 361
sudo service redis-server status, 726
Systemd (systemctl command), 256
System monitoring
    service, 277
    types, 277–279

T

TACACS server, 536, 581
tclsh commands, 51
Tcl shell commands, 51
Telnet, 527, 537
Telnet Lab 1
    interactive Telnet session to Cisco devices
        ch2_telnet, 64
        Cisco Telnet sessions, 63
        debugging message, 67
        debug telnet command, 66
        Devices in Use topologies, 61
        nano add_lo_ospf1.py, 65
        objectives, 61
        output, 69, 70
        powered on devices, 60
        python3 command, 62, 63
        python add_lo_ospf1.py command, 66
        R2 configurations, 61
        script creations, 61
        show IP interface brief command, 67, 68
        logging in, u22s1 server, 62
        undebug all command, 71
Telnet Lab 2
    SW1 configuration, Python Telnet template
        add_lo_ospf1.py, 72
        debug telnet command, 75
        devices in use, 72
        GigabitEthernet ports, 72
        IP address, 73
        network connectivity, 75

INDEX

objectives, 73
python add_vlans_single.py
   command, 76
show flash Telnet script, 77–79
SW1's console window, 77
SW1's switch ports, VLANs, 77
u22s1 Linux server, 72
SSH, u22s1 server, 73
Telnet Lab 3
   random VLAN configurations, for
      loop, 79
   add_vlans_for_loop.py, 80
   add_vlans_single.py file, 80
   objectives, 80
   python3 command, 83
   random VLAN configurations, for
      loop, 80
   script update, 81
   show vlan command, 84
   SW1, 80
   telnet_labs directory, 81
   u22s1, 85
Telnet Lab 4
   random VLANs configuration,
      while loop
      add_vlans_for_loop.py file, 86
      add_vlans_while_loop.py, 86
      altering script, 87
      cut-and-paste method, 89
      devices, 86
      objectives, 86
      Server is responsive message, 89
      sh_vlan1.py, 91
      sh_vlan2.py, 93
      telnet_labs directory, 87
      u22s1, 86
      VLAN execution, while loop script, 90
Telnet Lab 5

100 VLANs configuration, ~ in range
   loop method
   add_100_vlans.py script, 97, 99
   devices, 96
   objectives, 96
   reliable network connectivity, 98
   reverse_100_vlans.py, 101
   sh_vlan2.py, 101
   sh_vlan2.py script, 100
   SW2 switches, 96
   whitespace and script's code
      blocks, 97
Telnet Lab 6
   privilege 3 user, IP address
      add_junioradmin.py, 104, 108, 109
      check_junioradmin.py, 109–111
      devices powered-on, 102
      fping script, 106
      ip_addresses.txt file, 105
      junior network administrator
         privilege, 102
      objectives, 103
      PATH variable, 107
      ping script, 105, 106
      PuTTY, Windows host PC, 111, 112
      Python 3, 107
      running devices, 103
      show commands, 103
      u22s1 Python server, 103
Telnet Lab 7
   backed-up running-config files
      devices, 113
      fping command, 115
      ls-lh 20* command, 116
      modifications, 114
      objectives, 113
      save_run.py file, 117, 118
      take_backup.py script, 115

787

INDEX

Telnet Lab 7 (*cont.*)
    take_backups.py, 113
    timestamps, 116
    u22s1's local directory, 112
    write memory /copy running-config
        startup-config commands, 117
Telnet labs, 123
telnetlib library, 53
telnetlib Python script, 132
Test-driving Docker
    commands, 415, 416
    container, 418
    container ID/names, 420
    containers, 419
    Docker Hub ID, 416
    Docker images, 417
    docker pull, 417
    docker rmi image_name:version, 423
    docker start instance_name, 421
    docker system prune, 422
    image version, 419
    Linux server's directory, 420
    Login Succeeded, 416, 417
    Python modules, 419
    Python version, 419
    testfile999.txt, 419
    Ubuntu Server, 415
testfile999.txt, 419
TFTP/FTP server, 556
TFTP/FTP/SFTP/SCP protocol, 544
TFTP server, 175
Threading method, 627
Threading module, 577
tool5_fsize_cisco, 565
Traditional network engineers, 649
Trap method, 283
Troubleshoot SSH communication issues,
    Fedora and Cisco devices

crypto key, 304
install paramiko and netmiko
    libraries, 300
IP addresses, 310
paramiko library, 301
Python as network tool, 298
R2, 299
ssh jdoe@IP_address command, 299
SSH protocol versions/security
    features evolution, 312
SSH settings, key exchanges, 303
SW2, 306
test_ssh1.py, 308
test_ssh1.py script, 302
Twilio
    account creation, 449
    account registration, 449, 450
    account SID and token, 452
    API account, 470
    cloud communications
        platform, 448
    CPU utilization overloading lab
        topology, 456
    credentials.py, 453
    Docker container, 454
    Docker instance, 453
    get phone number, 451
    pip3 command, 453
    product survey to complete account
        setup, 450
    pynetauto_ubuntu20twilio Docker
        container, 458
    Request section, 452
    send SMS messages, 447, 448
    SMS message received, 455
    SMS notifications, 405
    SMS scripts, 459
    SMS sent from trial account, 451

web service APIs, 448
Type-2 hypervisors, 175

U

u22s1, 184, 519, 525
u22s1 Python server, 517
u22s1 server, 587
u22s1 virtual server, 251
u22s1 IOS upload, 679
Ubuntu, 415
Ubuntu 22.04 LTS Linux servers, 56
Ubuntu/Fedora Server, 411
Ubuntu LTS task scheduler (crontab)
    create directory, 252
    cron.log file, 255, 257
    crontab-e command, 252, 254, 256
    crontab for user_name, 256
    crontab-l command, 255
    "hello" statement, 252
    nano text editor, 254
    as non-root Linux user, 252
    pwd command, 253
    service or systemd (systemctl)
        commands, 256
    sudo crontab-u user_name-l
        command, 256
    u22s1 virtual server, 251
    which python3 command, 253
Ubuntu Server, 412, 413, 723
ufw command, 176
undebug_all.py file, 75
Unreachable IP addresses, 524
unreachable_ips, 520
upload_ios1.py file, 581, 585
upload_ios1.py Python file, 581
upload_ios1.py script, 590
User input collector, 536–543

V

/var/lib/tftpboot directory, 185
venv Python module, 357
virtualenv
    Ansible, 352–355
    create isolated environments, 355
    isolation, 354
    pyATS (*see* pyATS (Genie))
    Python ecosystem, 354
    usage, 355
virtualenv installation, Ansible
    activate virtual environment, 358
    ad hoc command, 366
    ansible-playbook execution
        command, 372
    ansible_venv, 357
    deactivate command, 358
    deactivate Python virtual
        environment, 374
    first_playbook.yml, 367
    install packages and development
        tools, 357
    install Python module, 365
    inventory file creation, 369
    IP address, ansible-playbook
        command, 368
    known_hosts, 361
    ls ansible_venv/bin/ansible*, 360
    mkdir, 357
    network module documents, 366
    pip3 command, 364
    pip3 install commands, 358
    Python setup, 374
    reactivate environment, 358
    run Python–V or python3–V, 358
    show_command.yml file, 369
    show_command.yml file, update hosts
        to switches, 371

INDEX

virtualenv installation, Ansible (*cont.*)
    ssh username@host_ip_address commands, 361
    sudo /home/jdoe/.ssh/config command, 361
    SSH key, u22s1 server, 359
    update and upgrade the Ubuntu packages, 356
    venv Python module, 357
    YAML playbook, 370
Virtual environment, 406
Virtualized Cisco routers and switches, 492
Virtual machines (VMs), 407
Virtual router, 237
VMware console, 604
VMware ESXi7.0/8.0 virtual machine, 493
VMware Workstation, 473
VMware Workstation 17 Pro, 494
VMware Workstation snapshot, 701
vRouter, 237

W

which python3 command, 253
Windows 10, 56
Windows Notepad++, 59
Windows Subsystem for Linux (WSL), 56
Windows Task Scheduler, 230, 250, 251
WinMD5.exe, 556
WinSCP, 397, 545, 557, 580, 644
write memory method, 117

X

.xlsx file, 545

Y

Yahoo, Spam folder, 435
YAML Ain't Markup Language (YAML), 351
Yellowdog Updater, Modified (YUM), 266

Z

Zabbix, 288, 292
Zero Trust security model
    CVEs, 492
    IP services, 491
    network and security engineers perspective, 491
    proactive OS patch management, 491

GPSR Compliance

The European Union's (EU) General Product Safety Regulation (GPSR) is a set of rules that requires consumer products to be safe and our obligations to ensure this.

If you have any concerns about our products, you can contact us on

ProductSafety@springernature.com

In case Publisher is established outside the EU, the EU authorized representative is:

Springer Nature Customer Service Center GmbH
Europaplatz 3
69115 Heidelberg, Germany

www.ingramcontent.com/pod-product-compliance
Lightning Source LLC
LaVergne TN
LVHW082021260326
834688LV00062B/1013